Essentials of Econophysics Modelling

Essentials of Econophysics Modelling

František Slanina

Institute of Physics, Academy of Sciences of the Czech Republic, Prague

OXFORD

UNIVERSITY PRESS

OXFORD
UNIVERSITY PRESS

Great Clarendon Street, Oxford, OX2 6DP,
United Kingdom

Oxford University Press is a department of the University of Oxford.
It furthers the University's objective of excellence in research, scholarship,
and education by publishing worldwide. Oxford is a registered trade mark of
Oxford University Press in the UK and in certain other countries

First Edition published in 2014

Impression: 1

Published in the United States of America by Oxford University Press
198 Madison Avenue, New York, NY 10016, United States of America

British Library Cataloguing in Publication Data

Data available

Library of Congress Control Number: 2013942055

ISBN 978–0–19–929968–3

Printed and bound by
CPI Group (UK) Ltd, Croydon, CR0 4YY

Rzeczywistość, jak zawsze, jest bardziej skomplikowana,
ale i mniej złośliwa.[1]

Stanisław Lem

---

[1] As usual, the reality is more complex, but less mischievous.

Contents

List of Boxes

List of Figures

Introduction

The science of complexity

Throughout history there have always been scientists who sought unification of all scientific knowledge. They dreamt about having a small, easily remembered set of principles that would provide us with all we need to understand any phenomenon and to give us all the answers we want. No doubt this reductionist point of view is aesthetically pleasing, but every day we witness the failure of such an effort. As an ideologically neutral example let us mention the long-term unpredictability of weather. The prophets of reductionism are not discouraged, though. They especially like to believe that social phenomena can be described as completely as physical ones if only we knew the right set of laws and were smart enough to do the calculations involved.

This belief is actually quite an old one. Auguste Comte [1, 2] was its first prophet, in the early 19th century. Comte coined the term 'social physics' as an explicit reference to the success of the Newtonian mechanics. Though Comte himself abandoned the term social physics, it was called to life again by his successors, most notably Adolphe Quételet [3], and it has survived in various disguises up to the present time.

Naive and unjustified claims to handle scientifically complex systems like society were especially popular in totalitarian systems of various kinds. The most acute analysis of the arrogance of this pseudo-science and the harms it induces when applied to society was set forth by Friedrich von Hayek [4]. He argued that complex systems, and human society in particular, cannot be described as an assembly of simple elementary parts. There is always something irreducible that emerges when these parts are put together. Moreover, complex systems are characterised by a high level of self-organisation, which is actually the most prominent feature of human society. This precludes all means of directing the society from a single centre using pure and infallible deduction based on a first principle theory. In Hayek's line of thought the so-called 'hard science' becomes in fact a pseudo-science when applied outside of its domain. It promises something it can never fulfil. On the other hand, science cured of its hubris can prove to be very useful. This is the moment where the science of complexity comes in [5].

The science of complexity pays attention to exactly the same features of complex systems which are stressed by Hayek. The first one is emergence: an ensemble of interacting units exhibits new features that cannot be immediately deduced from the properties of its constituents. Moreover, properties of the elements may be to a large extent (but not fully, of course!) irrelevant for the emergent phenomenon, as if the microscopic and macroscopic levels of description were quasi-independent.

Another feature of complex systems is self-organisation. The processes within a complex system often lead to the creation of structures that are surprisingly stable under external perturbations. The system does not need to be governed from outside.

Nowadays, considerable knowledge has been accumulated about self-organisation on the levels of physical, chemical, and biological as well as social phenomena. If we know what question to ask and if we are aware of the limitations of the answers, the study of complexity can be extremely useful. This book intends to contribute to the science of complexity in the specific field of economics.

Econophysics

Society is complex, and in order to succeed everybody has to make a good deal of rational decisions. The area where complexity and rationality meet is economic activity, a kind of process which is of concern for every inhabitant of our planet. That is why scientists investigating complex systems soon focused their attention on economics, and this is the area where the discipline we now call econophysics was born. We do not intend to write a review on its history, but let us mention a few important milestones.

At the beginning of the 19th century, at about the same time that the concept of social physics appeared, a country nobleman in southern Bohemia contemplated applying the principles of classical mechanics to economics. His name was Georg Graf von Buquoy, the Count of Nové Hrady. His knowledge of Newtonian mechanics was very profound, and he contributed to several fields of science [6, 7]. His bulky treatise *Die Theorie der Nationalwirthschaft nach einem neuen Plane und nach mehrern eigenen Ansichten dargestellt* [8] appeared in 1815 and, with a moderate amount of mathematics, presented something which now would be a chapter in a textbook on economics. He explicitly refers to mechanics, and such reasoning was revolutionary in his time.

It was not until the second half of the 19th century when a similar attempt was undertaken by Vilfredo Pareto. An Italian engineer who spent long years of his career in Switzerland, Pareto turned to economics equipped with a good knowledge of classical Newtonian mechanics and a good deal of enthusiasm to describe social movements in the same way that physics describes the motion of planets. Indeed, Pareto himself liked to compare Kepler's laws to new economic laws that were yet to be discovered. The Pareto law of wealth distribution is now a standard piece of economic knowledge. However, the reduction of economic laws to something akin to Newton's laws of mechanics remained a dream for some people, a Fata Morgana for others.

Physics, however, changed completely following the discovery of the theory of relativity and quantum mechanics at the beginning of the 20th century. The question naturally emerged: if the old Newtonian physics was found inadequate in explaining complex systems like society and economics, does the same apply to the new physics of Einstein and Schrödinger? Or does the new physics open new ways to grasp complexity scientifically? Ettore Majorana, a genius of theoretical physics, tried to respond to these questions in his famous last paper [9, 10].

One of the decisive breakthroughs which influenced the whole course of interactions between physics and economics was the introduction of the concept of fractals [11] by Benoît B. Mandelbrot. But many other developments went on in parallel. There is an often forgotten event that played a decisive role in the transfer of the ideas and language of physics into other branches of human knowledge. A conference was scheduled to take place in Moscow from 1 to 5 July 1974. Scientists both from the West and the USSR were invited to discuss the implications of physics in other fields, including social sciences and humanities. The organising committee included people like Kenneth Arrow, a Nobel laureate in economics, and Hans Bethe, a Nobel laureate in physics. However, the communist leaders found the subject of the meeting incompatible with the ruling ideology. The conference was banned, most of the Russian participants were arrested, and a majority of them eventually left the USSR, going mainly to Israel. But many drafts scheduled for the conference talks were successfully smuggled from the USSR to the West and eventually were published in a proceedings volume [12]. A tiny portion of it appeared in Ref. [13].

After the non-event in Moscow, the attempts to merge physics with other disciplines began to flourish. In the 1970s and 1980s, Hermann Haken developed a very general concept of synergetics [14, 15]. About the same time, the science of complexity became increasingly popular, with the Santa Fe Institute playing the key role in this research. It was the place where the theory of complex systems began to be systematically applied to economics. Another physics Nobel laureate, P. W. Anderson [16], began to be deeply involved in promoting this line of research. Finally, in 1991 the journal *Physica A* published what is now considered to be the first paper that can truly be attributed to the newborn field of econophysics [17].

What is so special about econophysics? In what respect is it different from other approaches trying to mix natural sciences and economics? The answer is not very easy. Econophysics is certainly not a simple mixture of disciplines, and it definitely is not just economics done by people who have obtained a PhD in theoretical physics. Neither it is correct to label econophysics as an interdisciplinary science living in a shadow area between the kingdoms of Economics and Physics. We prefer to call it transdisciplinary. It looks like a little difference, but it is not. Econophysics does not aim at obtaining a place between these disciplines but rather tries to find principles that are as true in one discipline as in the other. It is a quest for common principles, common tools, and common consequences found both in physics and economics, and it is by no means evident that such common principles exist. But let us just say that one might try and see. Suppose we find the common principles; then we can check whether they can be useful. We hope that the reader, after having gone through this book, finds that they are useful indeed.

Another feature which is characteristic for econophysic endeavours is the way it applies physics to economics: physics is used systematically and proceeds from thorough empirical analysis and model building, using advanced mathematical techniques for real-world applications. It is not like taking a snippet of a physical theory and re-interpreting it in the language of economics. Econophysics tries to be comprehensive and coherent.

Which part of econophysics does this book belong to?

Econophysics is an internally structured field. The first branch relies fully on empirical data and tries to apply sophisticated methods to their analysis. Physics is an empirical rather than speculative science, and it has developed powerful techniques and know-how that prove very fruitful when applied to economic data like time series of prices, etc. The first chapter of this book is devoted to an overview of basic empirical findings the econophysicists collected. However, empirics is not the main subject of this book. We present here only what we consider absolutely necessary for understanding the facts and motivations contained in the rest of the book.

The second branch of econophysics concentrates on modelling, and this is the proper theme of this book. Modelling uses the information provided by empirical econophysics and tries to explain the collected facts by means of more or less sophisticated models. Modelling tries to make a copy of reality, and, as in every process of copying, something is always lost. The ability not to lose essential details is what distinguishes a good model from a bad one. However, it can rarely be known a priori whether the model in question keeps the essential and throws away the unessential, or the other way round. Therefore, modelling is always an interactive process. The econophysics models resemble ad hoc solutions, and the coat they bear is patchy in many places. We do not consider it a disadvantage. Instead, the lack of systematics is very often a sign of vitality. Even within the modelling branch of econophysics there is an enormous amount of material. Models abound, and the quantity of achievements is very large. It means that we had to make very careful selections which ones to include. In doing that, we were guided by a certain vision of what the econophysics modelling should look like.

But let us go back to the structure of econophysics. There is a large area we have deliberately left out. It may be called the practical fruits of econophysics, as it deals with problems such as the evaluation of options, investment strategies, predictions of crashes, and similar things. There are two good reasons for not including it in this book. First, and most importantly, the author feels there are good books on the subject already; see for example these classical titles [18, 19]. Second, this book is devoted to ideas, to discovering why things are happening and not to provide recipes for becoming rich. The author does not feel qualified in the latter.

The structure of the book

We expect readers of this book to have a basic knowledge of either economics or physics. We strove to make the book accessible and interesting to both communities. Both disciplines are mathematically based, so we also suppose the reader has no fear of mathematical formulae. At some places, however, we felt that certain concepts and techniques deserve some explanation. We put these explanations into boxes, in order that an advanced reader could easily skip them while on the other hand undergraduate students could use the book without waiting until they learned these concepts later in their university curriculum. We also hope the boxes may help to bridge the remaining cultural gap between economics- and physics-based readers.

To stress that the field is open and to encourage the reader to think independently, we placed a Problems section at the end of each chapter (with the exception of the first one). Some of the problems are fairly trivial to solve, but some are more like little research projects. Each chapter is closed with a section named What remains. It contains a short annotation of the material which was not included, mainly due to the lack of space, and relevant references. We tried to include just the most useful references, not a complete bibliography of the subject.

The book consists of eight chapters. In the first one we present fundamental empirical facts necessary to understand the motivation of the models. It starts with a brief historical introduction and then covers the findings on price fluctuations, order books, and non-trivial correlations among prices. The cornerstone of mainstream economics, the efficient market hypothesis, is called into question. The second chapter shows how price fluctuations and other phenomena can be modelled using stochastic processes. From random walks we go to GARCH and cascade processes and finish with models with stochastic volatility. The third chapter opens the matter of agent-based models, showing the wide variety of their variants. The fourth chapter is slightly different. Agents are nearly forgotten. We focus on modelling the order book, and the methods more closely resemble modelling a snowfall than rational people. We call that approach zero-intelligence agent modelling. In the fifth chapter the intelligence reappears. We deal with a prominent model of econophysics, the minority game. Both simulation results and solutions using the advanced replica method are explained in depth. In the sixth chapter we explain how the network structures appear in economics, how they are modelled, and what their relevance is. In the seventh chapter we show how economic activity organises society into groups of rich and poor people, what the distribution of wealth is, and what dynamics this distribution has. The eighth chapter deals with other aspects of social structures that emerge in human society. What is the origin of cooperation? Should it appear at all? How is consensus reached, and how much do people trust each other?

Where to read more

Some material contained in this book has already been published by the author elsewhere. Besides the journal articles, which will be cited at appropriate places, this category includes the article *Social Processes, Physical Models of*, contained in the *Encyclopedia of Complexity and Systems Science* [20], and the review article in the edited volume *Order, disorder, and criticality, vol. 3* [21].

A good number of sources exist that cover various aspects of econophysics, including several good books. Let us present a short list of existing econophysical literature which the reader is encouraged to consult. The original sources will be cited at proper places in corresponding chapters. Here we mention only general literature.

Among the basic econophysics books let us name (in alphabetical order of the first author) the following ones: Bouchaud and Potters [18, 19]; Johnson, Jefferies, and Hui [22]; Levy, Levy, and Solomon [23]; Mantegna and Stanley [24]; Roehner [25–27]; Roehner and Syme [28]; Schulz [29]; Sinha, Chatterjee, Chakraborti, and Chakrabarti [30]; Sornette [31]; and Voit [32]. There are two special books on minority game: the first by Challet, Marsili and Zhang [33], and the other by Coolen [34]. Also, the above mentioned book [22] mostly deals with minority game. Complexity of social structures is thoroughly investigated in the monograph on active Brownian agents by Schweitzer [35]. Econophysical applications are also discussed in the book on thermodynamics by Šesták [36] and in the book on new applications of statistical physics by Wille [37].

More for historical than factual interest, the reader may wish to review the already-mentioned proceedings volume of the Moscow conference [12] and the book on economics written from the point of view of a theoretical physicist, by M. F. M. Osborne [38].

There are many proceedings from conferences on econophysics. The one certainly worth mentioning originated from the very first conference with the word econophysics in its name. It was held in 1997 in Budapest and the proceedings never appeared in print but are available on the Internet [39]. There are also several volumes from the Nikkei econophysics symposia [40–42] and several others from the conferences held in Kolkata [43–47]. The Santa Fe Institute produced one of the first edited volumes of econophysical papers [16], which was followed by at least two others [48, 49]. Among other edited volumes let us mention Refs. [50–57].

Besides these books, a lot of information is contained in numerous special journal issues devoted to econophysics and related areas [58–69]. Many review articles have appeared where the results are presented in a compact form [70–74]. Numerous papers pose general questions about the proper place of econophysics among other fields [75–84]. Finally, there are also popular articles, in which the general public can get an idea of what econophysics is all about [85–92].

The message

This book emerged from several years of my activities related to various fields of econophysics. Originally, I planned to write a much shorter book, containing only the really essential issues in the modelling branch of econophysics. As time went on, the material grew and the book swelled until it reached its current size. I would never have succeeded without the initial and lasting inspirations from Yi-Cheng Zhang, who introduced me to the jungle of econophysics in the late 1990s. And I would not have started writing a single letter of this book without the kind and friendly encouragement from Sorin Solomon. I am extremely grateful to both of them. I wish to thank numerous friends and colleagues whom I have had the chance to speak to and work with during the time I have been involved in the econophysics business. I am also indebted to Jan Klíma, who helped me to improve the style. The original results contained in this book were supported by the MŠMT of the Czech Republic, grant no. OC09078.

It was not my intention to help to raise an amount of gold clinking in readers' pockets. But I will be happy if a single person after reading this book feels enriched by the understanding of the world around us. A. M. D. G.

Prague, June 2012   *František Slanina*

1

Empirical econophysics and stylised facts

'Daß alle unsere Erkenntnis mit der Erfahrung anfange, daran ist gar kein Zweifel.' (That all our knowledge begins with experience there can be no doubt.) This is the famous sentence opening Immanuel Kant's monumental treatise, *The Critique of Pure Reason*. If the highly abstract and subtle philosopher values experience above all his reasoning, what else should the scientists of today, who are supposed to be practical and base all their knowledge on empirical facts, do? Yet there is a constant temptation to build speculative theories rather than keep in contact with reality. There are numerous examples showing that theorising is good. Indeed, the theory of relativity was highly speculative at first, only to become a daily business of many researchers now. But there are nearly as many examples of how the exaggerated disposition for rigorous mathematical apparatus void of any empirical basis was harmful, at a minimum by guiding the enthusiastic young people into a dead end.

You will see that this book is not free of such temptations either. To keep balanced as far as we can, we shall expose at least some of the most important empirical findings, which will provide motivation for the models discussed in later chapters. We encourage the reader to frequently compare the outcomes of the models with the data shown here, in order to develop her own opinion on the trustworthiness of the abstract theories we shall elaborate in the rest of the book.

1.1 Setting the stage

1.1.1 A few words on history

People have been engaged in economics since the Stone Age. Tribes were trading in flint, arrowheads, and even gold. It is certain that the rate at which these articles were exchanged varied in space and time, and these variations soon became driving forces for profound changes in human society. Since the Neolithic Revolution we live in an environment of incessant trade. The emergence of money, i.e. a certain good accepted by anybody although having little use in itself, marked another revolution. Since then, exchange rates could be expressed in the universal language of a currency (or currencies).

First records

People probably always kept records on prices and exchange rates, but it was not until the 17th century that they realised how important it was to do it systematically.

The first modern stock markets emerged in Amsterdam and London [93, 94]. London became the hub of the information network communicating share prices for joint-stock companies, exchange rates, and various other information relevant for people busy with investing capital. The international links led chiefly to Amsterdam, but also to Paris and other commercial centres. After some ephemeral periodicals containing price lists, the first long-lasting source of reliable information on exchange was 'The Course of the Exchange and other things', published in London by 'John Castaing, Broker, at his Office at Jonathan's Coffee-house' as stated at the bottom of the page [93]. It appeared twice a week from March 1697 and contained daily stock and commodity prices, as well as exchange rates and other financial information. Figure 1.1 shows a facsimile of the issue dated 4 January 1698, as well as graphs of the historical share prices of the Bank of England and the South Sea Company.

This newspaper became a widely accepted standard and continued under the heirs of Castaing for more than a century. John Castaing himself was a French Huguenot who arrived in England in the 1680s and worked as a broker at the Royal Exchange during the 1690s. At that time the coffee houses were the usual places where stock brokers met with their clients and where most transactions were negotiated and settled. What is now the London Stock Exchange evolved gradually and painfully from these unorganised coffee-house businesses. It is an early example of the self-organising potential of the free market [95], and as such it is cited in textbooks, although nowadays hardly any person is willing to wait for decades or a century until a confused chaos self-organises into a shiny, frictionless economic machine.

In any case, the broad dispersion of information was a crucial ingredient in the self-organisation of the early modern international stock market. The transfer of economic news relied on the already existing information networks consisting of many newspapers, which mushroomed in England in large numbers after the de facto freedom of press came into effect in 1695. The rapid dispersion of exchange information not only became indispensable for stockbrokers and stockjobbers, the professionals in share trading, but from the outset it provided fertile ground for various kinds of fraud. Dissemination of false information, bribery of journalists, and other artificial manipulations of exchange courses were common, as testified by many contemporary writers, such as Daniel Defoe [96] (quoted in [97]).

First bubbles

The manias, fuelled by either true or false news of fantastic gains, resulted inevitably in short-lived bubbles, devastating the major emerging financial markets in the early 18th century [98, 99]. The first in the series occurred in Paris in 1719–1720. In the decades that preceded, John Law, a Scottish financier hired by the Regent of France, was in charge to cure the French state finances, burdened by enormous war debts. As a part of his scheme, he reorganised the former Mississippi Company and transformed it in several steps into the Compagnie des Indes, which was granted a monopoly for virtually all overseas trade. Irrational expectations of enormous profits that the Compagnie would offer quickly pushed the prices of its shares to the skies. Not only that—all economy suddenly seemed to flourish; but the dream lasted only a few months before everything crashed.

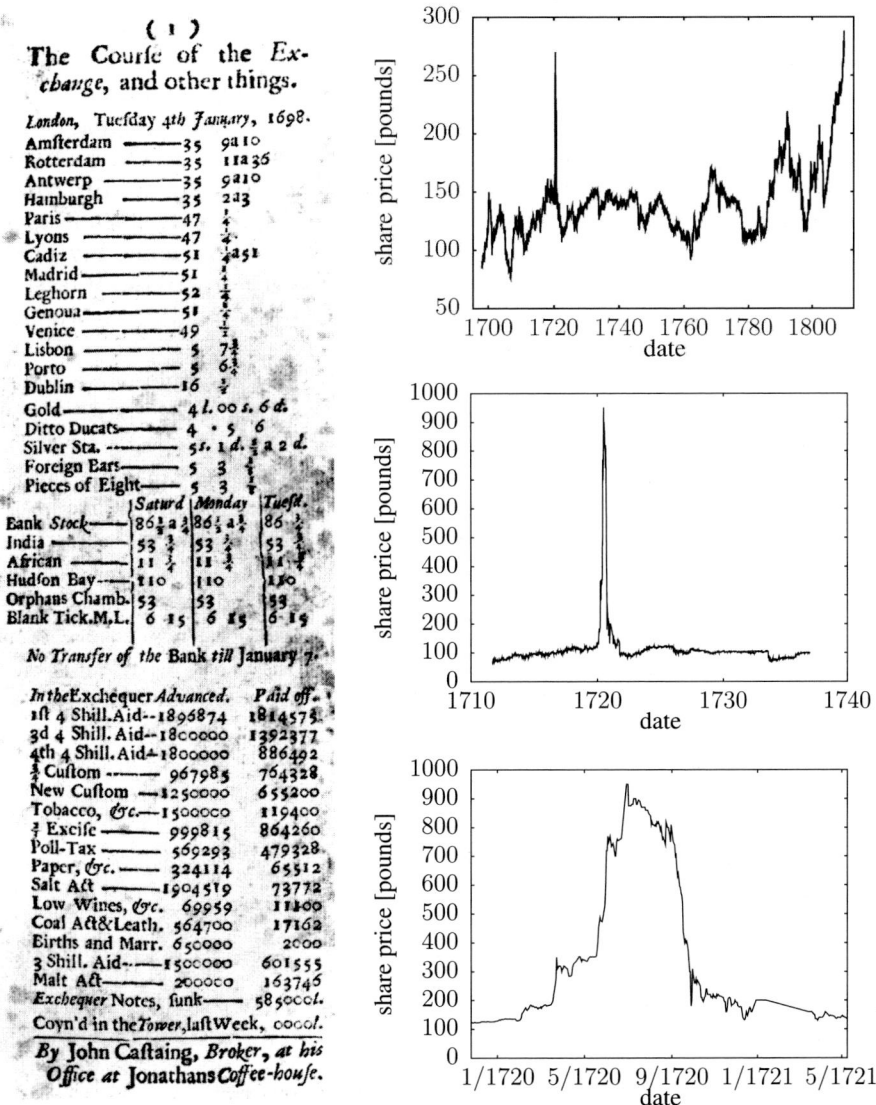

Fig. 1.1 In the left panel, a facsimile of Castaing's Course of the Exchange from 4 January 1698 (Julian calendar). Reprinted with the permission of The Lordprice Collection. In the right column, three graphs of stock prices, based on the data compiled from the Course of the Exchange by L. Neal, accompanying the book [93] and accessible at www.le.ac.uk/hi/bon/ESFDB/NEAL/neal.html. In the top panel, the price of a share of the Bank of England from 1698 to 1809. Note the sudden blip in 1720, related to the South Sea Bubble. In the middle panel, the price of the South Sea Company shares, from 1711 to 1734. In the lower panel, the detail of the South Sea Company prices during the bubble period in 1720.

Law's idea was based on the hope that high share prices could be sustained perpetually by issuing a large amount of paper currency. Alas, in the absence of proper regulation it resulted in a disaster, which was then only made worse by government directives. Immediately after the collapse of Law's scheme, the attention of the speculators was drawn to London, where the shares of the South Sea Company started to rise in an unprecedented manner. In complete analogy with Paris, the South Sea Company was intended to liberate the State Treasury from the burden of war debts. Within several months the shares of the South Sea Company rose nearly ten-fold and then plummeted back to nearly the starting value, leaving crowds ruined and angry, some individuals fleeing safely abroad and a few arrested. In Fig. 1.1 we show the course of the South Sea Bubble, as documented in the share prices published in the Course of the Exchange. Ironically, just before the bubble started, Defoe had published his pamphlet [96] severely condemning the 'knavish practices' of stockjobbers as 'treason'. All warnings, including a few courageous voices from the Parliament speaking counter to the overwhelming current, were of no effect.

The bubbles in France and England were also followed by some repercussions in Amsterdam and Hamburg before the business settled down to its usual pace. It is important to realise that the spread of information, either true or false, on the state of the stock market, on current prices, ship arrivals or shipwrecks, was essential in the emergence of bubbles. Equally important was the ease of travel between the three principal centres, Paris, London and Amsterdam. The bubbles were all too visible signs that the markets had become both information-driven and truly international. The information not only describes the events but at the same time causes them to appear.

The moral is that...

This was the first (or at least the first well-documented) occasion for humanity to taste the boisterous elements of information economy. Since then, we live in incessant fear of a crash [100]. In an information-driven market, catastrophic events occur much too often, more than anybody would like, more than any person would predict using only common sense. For a scientist, it is a challenge to express the fact of events that are too large and happen too frequently in a mathematical language. It is highly probable that the disasters would come regardless of whatever effort we make to tame them. Science may help us to allocate them a proper place in our understanding of the outside world.

1.1.2   About efficient markets

Efficient market hypothesis

Every person engaged in the analysis of prices, studying the question how and why they evolve, must be confronted with the cornerstone of financial theory, which is the efficient market hypothesis (EMH) [101]. Briefly stated, it claims that all information you might try to use to make a profit from price movements has already been incorporated into the price. This has the serious consequence that the future price cannot be predicted from the observation of past prices alone. Moreover, EMH in its strongest formulation means that you would not be able to predict the price even with all other public sources of information, not just the past price record. Partisans of EMH call that condition 'no free lunch' and are happy with that.

Roots

In brief, EMH implies that price makes random and unpredictable movements up and down, much like the path of a drunkard. This idea is not new at all. Already in 1900 the French mathematician Louis Bachelier, in his thesis entitled *Théorie de la spéculation* [102], introduced the idea that the prices at the Bourse follow a random walk and from this was able to make some exact predictions. The comparison with real stock price data was rather scarce but considered satisfactory.

Yet Bachelier was not the first. Already in 1863, Jules Regnault in his book *Calcul des chances et philosophie de la bourse* [103, 104] assumed that prices rise or fall with equal probabilities of 1/2, thus reacting to random items of information. He deduced that the mean deviation of price during a certain interval of time is proportional to the square root of that time. It is probably the first correct formulation of the law of the random walk, although it was more of an empirical observation than a mathematical theorem. Unfortunately, the mathematical argument put forward by Regnault was misleading.

Regnault was, in turn, profoundly influenced by the Belgian Adolphe Quételet and his concept of social physics [3]. Quételet's aim was to establish quantitative laws governing all human behaviour at the level of individuals as well as societies, nations, and states. The methods for achieving this goal were based on accumulating large amounts of empirical data followed by the statistical analysis of that data, which was as sophisticated as the science of that time permitted. What is now empirical econophysics and what we try to cover at least partially in this chapter would fit excellently into the programme of social physics put forward by Quételet more than a century and half ago.

Let us make one more dive deeper into history before we stop. In fact, the notion of social physics was first used by Auguste Comte [1], in a much wider sense than Quételet's. Comte's original idea corresponds much better to the modern field of sociophysics, to be discussed at length in Chapter 8. Why Comte later abandoned the term 'social physics' and started to use 'sociology' instead is a question unrelated to the theme of this book. However, it might be useful for a physicist to be aware of this terminological shift and of the fact that the respectable discipline of sociology is somehow rooted also in physics.

Random walk or not?

EMH was explicitly formulated by Eugene F. Fama in the early 1960s [105]. Since then it has many times been criticised, and its formulation was just as many times corrected. It has had an undeniable positive influence in two ways. First, it stressed that the principal impossibility of obtaining systematic profit from stock market fluctuations is a scientific statement, not mere babble. It was shown empirically a long time ago that stock-market forecasting companies are right as often as they are wrong [106]. With EMH, the price signal itself became the main object of study. Indeed, it was soon established that statistical regularities indicating the presence of remaining information are very weak and can be explained as consequences of external hindrances [101]. Hence the conclusion that the market becomes more efficient as fewer regulations are imposed on it.

The second benefit from EMH is the increased interest in the precise statistical analysis of price fluctuations, which can go into the most minute details. The focus of this chapter is to show some of the results. The consequence was drawn immediately from the very first formulation of EMH [105] that stock prices must follow a random walk. This means a price change during time δt is independent of all price changes in the past. If it is true that all information is incorporated into the price immediately, without any delay, this should hold for any time δt, however short it may be.

There is, moreover, a mathematical theorem [107] stating that a continuous time sequence with increments which are independent of each other at arbitrarily short time intervals must be a continuous-time limit of a random walk, i.e. a Brownian motion, combined perhaps with a Poisson process, if we allow instantaneous finite jumps. Therefore, the price changes Δz over a fixed time interval δt must be normally distributed, that is,

$$P_{\text{price change}}(\delta t; \Delta z) = \frac{1}{\sqrt{2\pi \, \sigma \, \delta t}} \exp\left(-\frac{(\Delta z)^2}{2\sigma \, \delta t} \right) \qquad (1.1)$$

where σ is a parameter to be found empirically. If EMH holds exactly, nothing else can be expected.

This is a falsifiable statement and thus a sound basis for a scientific theory, if it passes empirical testing. Indeed, the time series of prices was found to be unpredictable in the sense that the price increments are uncorrelated [101]. On the other hand, the fundamental property of the random walk, which is the normal distribution of the deviation from its initial value, was refuted. A refined, much more rigorous formulation of the mathematical consequences of EMH was found on the basis of the martingale property, by P. A. Samuelson [108]. We shall resist the temptation to dwell on martingales and other beauties of mathematical finance [109]. There are many excellent books on this subject, e.g., the classical one by R. C. Merton [110], and we wish to move swiftly toward distant and less orthodox fields, specific to the discipline of econophysics.

The milestones we cannot pass by, however, are the empirical and theoretical works on the distribution of price changes. We already noted that the normal distribution must be abandoned. What are we left with?

Perhaps the most fundamental deviation was announced by M. F. M. Osborne [111], a former theoretical physicist. He observed that it is not the price but the logarithm of it which is normally distributed—at least approximately. Moreover, there are also two purely theoretical arguments in favour of such a result. First, the normal distribution leaves a non-zero probability for negative prices, which is contrary to any common-sense intuition. We do not have this problem with the logarithm of price. Second, it was found in many areas that human perception of external signals is scaled according to a logarithmic law. For example, we perceive a fixed difference in the luminosity of two light signals if their actual intensities have fixed ratios, or

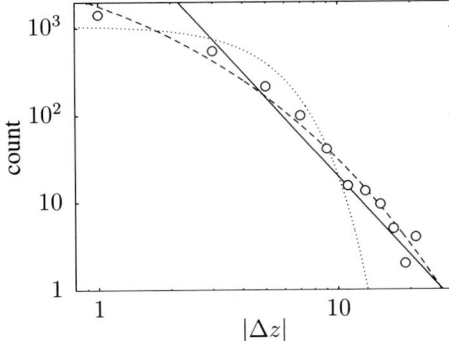

Fig. 1.2 Weekly price change distribution for wheat traded at Chicago. On the horizontal axis, absolute value of the price change in cents per bushel; on the vertical axis, number of such events in the time series starting in 1883 and ending in 1934. The data were extracted from Ref. [112], Table 1. The solid line is the power-law fit $\propto |\Delta z|^{-3}$, the dashed line is the stretched-exponential fit $\propto e^{-(|\Delta z|/a)^\gamma}$ with $a = 0.13$ and $\gamma = 0.43$, and the dotted line is the fit to the normal distribution (1.1).

equivalently, the logarithms of the intensities have fixed differences. The generalisation of this observation states that it is the logarithm of the physical quantity that is subjectively felt. This empirical rule is called the Weber-Fechner law, and Osborne claims that it also holds for stock prices, so that it is the logarithm of the price that is naturally observed by the participants in the stock market. Then, EMH holds for the logarithms of price, and it is the logarithm which is normally distributed. The distribution of price changes is log-normal.

Unfortunately, more detailed studies showed that neither normal nor log-normal distribution of price changes is compatible with the data. Too many large changes were observed consistently. As an early example of such a result we plot in Fig. 1.2 the distribution of weekly price changes, taken from Ref. [112]. We can see there that the normal distribution is indeed a very bad approximation of the data. Much better agreement is achieved with the fit to the stretched-exponential distribution, which has the form

$$P_{\text{price change}}(\Delta z) \propto \exp\left(-\left(|\Delta z|/a\right)^\gamma\right) \tag{1.2}$$

where a and γ are suitable parameters [113]. Surely, many different distributions can be tried also. But whatever is the form of the best fit, the tails are much fatter than predicted by the normal distribution. The emergence of fat tails in price-change distribution is the basic non-trivial fact to be explained in any theory of stock-market behaviour.

The stretched-exponential function has two free parameters. There is another possible fit, nearly as good as the former one, which has the advantage that it requires only one free parameter. It is the fit by a power law

$$P_{\text{price change}}(\Delta z) \propto |\Delta z|^{-1-\alpha} \tag{1.3}$$

where α is the only parameter characterising the distribution. This fit is especially appropriate at the tails, i.e. for large price changes, as can also be seen from Fig. 1.2, where we used $\alpha = 2$.

This finding deserves our full attention. A great deal of the research in econophysics revolves around a power-law dependence for this or that quantity [114, 115].

Mandelbrot and fractals

Power-law distribution is indeed very far from the normal one. If EMH implies normality, how could the empirical data be consistent with it? Is there any way to reconcile EMH and power laws? Here comes the groundbreaking idea of B. B. Mandelbrot. In his paper [116] published in 1963 he analysed in detail the fluctuations in the price of cotton. He compared three distributions of price changes. First, daily changes in the period 1900–1905; second, daily changes in the period 1944–1958; and third, monthly changes, 1880–1940. In all three cases he found that the tail of the cumulative distribution of price changes is well fitted to a power law, as in (1.3), with $\alpha \simeq 2$. So far, he repeats and stresses older results, including those of Ref. [112], which we have seen in Fig. 1.2. But Mandelbrot goes much farther than his predecessors.

As a brilliant mathematician, Mandelbrot knew that the theorem we mentioned above holds only under strict conditions. The continuous-time process with independent increments results in normal distribution only provided that the average of the square of the increments, the variation, is finite. How could it be that the variation is infinite? Does it happen only under truly pathological circumstances? From a mathematical point of view, such a situation occurs when the distribution of increments has a power-law tail like (1.3), with $\alpha \leq 2$. For a long time people were not familiar with real-life examples of such situations. Mandelbrot showed that fluctuations in the price of cotton fall within that category. And later he devoted much of his life to the development of the concept of fractals, geometrical objects with possibly fractional dimensionality (hence the name) [11], which are the best pedagogical examples of situations that lead to power-law distributions. It grasped the essence of power-law distributions, scattered by that time in various disciplines. Note, for example, the works of Pareto in economics [117], Zipf in linguistics [118], Korčák in geography [119], and more. It became common wisdom to look for fractals everywhere a power-law distribution pops up. This popular knowledge is not always justified, but the widely acknowledged ubiquity of fractals (everybody knows rivers, rocks, coastlines, etc. ad infinitum) makes it much easier for us to accept the fact of power-law distributions than it was before Mandelbrot.

The mathematical theorem which apparently leads inevitably to the normal distribution must be modified if the distribution of increments has power-law tails. In this case, the distribution of price changes (or changes in the logarithm of price, according to Osborne) must follow one of the Lévy-stable distributions. There is no compact formula for them, except for several special cases (see Box 7.1 for more information). One of them is the Cauchy distribution (also called the Lorentz distribution by physicists)

$$P(\Delta z) = \frac{1}{\pi} \frac{a}{a^2 + (\Delta z)^2}. \tag{1.4}$$

The tail of the Cauchy distribution decreases as a power $\sim (\Delta z)^{-2}$, which corresponds to $\alpha = 1$ in (1.3). All Lévy distributions are characterised by power-law tails with $\alpha < 2$.

We shall discuss them further in Sec. 7.1.3. For now, it is enough to know that the Lévy-stable distributions for price changes are consistent with the requirements of independent increments at arbitrarily short times δt and with the power-law tail of such increments. The random process with such properties is called the Lévy walk, as opposed to the usual random walk. What is important is that the mathematical content of the EMH is saved, at least for the moment. On the other hand, important implications for investment strategies were noted immediately [120].

More on this topic can be found in the review [121] by Mandelbrot himself. Unfortunately, reality is not very kind to us. As we shall soon see, the empirical tails of the distribution of price increments do follow a power law, but with too large an exponent, $\alpha \simeq 3$, outside the range of Lévy-stable distributions. Things seem to be even more complicated. It is difficult to hold EMH any longer. What does remain from Mandelbrot's analysis is the idea that the graph of price fluctuations, seen as a geometric object, is a fractal.

Critique

It is not only the empirical finding, but also sound reasoning that undermines EMH. Sure, EMH is extremely intellectually appealing. It is easy to formulate. It is easy to understand (at least it seems so). It is very intuitive. It seems to have far-reaching consequences. It looks like a gem of human thought. Yet it cannot withstand even most naive questions. The first: if it is true that nobody can earn money systematically on an efficient market, where does the wealth of the biggest speculators come from? The second: if nobody can gain in the long run, why do people keep buying and selling stock all the time? According to EMH, no systematic profit can be expected from any trade, so why did trading not completely stop a long time ago? The third: how should one understand the claim that 'information is immediately incorporated into the price'? Does 'immediately' mean within milliseconds or femtoseconds, or even quicker? And what is the mechanism of the 'incorporation'? Is the price adjusted without any trade? If not, does it mean that all agents on the stock market act in the same manner within the same instant (within femtoseconds?), so that all of them collectively readjust the price and none of them has any advantage over the others? If yes, it means that either all of them buy or all of them sell. But then, where is the other partner of the trade?

We could continue in this way for quite a long time, suggesting the impossibility of the EMH in its classical form. The proponents of EMH have invented many improvements and weakened the statement of EMH so that it does not look that absurd. (For example, it is argued that in fact the markets are not efficient, and we do not even have the means to check whether or not they really are, but for all practical purposes they look like they are efficient; and that is all we need.) Still, there are serious criticisms which suggest that it is much wiser to abandon EMH as an intellectual dead end, instead of repairing the wreck again and again. Most importantly, practical implications follow from that stance [122–124].

It is not our intention to go into detail here, but we must mention the fundamental argument by Grossman and Stiglitz, which goes as follows [125]. If the information is to be incorporated into the price, one should first have that information. But gathering and analysing information requires some resources and takes some time. The cost of information processing may be so large that nobody can afford to incorporate the information into the price. Therefore, informationally efficient markets are not attainable. In fact, the importance of the economics of information was stressed long before by Stigler [126], and the consequences of imperfect and asymmetric information were investigated by Akerlof [127].

This idea has been amply discussed, and from the many ramifications which have appeared since, (see e.g. Ref. [128]) let us mention only the theory of marginally efficient markets, developed by Yi-Cheng Zhang [129], for it is deeply rooted in genuinely physical reasoning. This theory is a beautiful analogy of the third law of thermodynamics, stating that it is impossible to reach absolute zero of temperature. Because heat engines are the more efficient the closer we can cool them to absolute zero, the third law implies the impossibility of constructing an engine with 100 per cent efficiency. Analogously, exploiting some pieces of information in the stock market costs something. Exploiting more information costs more. Our resources are finite, so we must stop at certain point, leaving some information out of the price determination. Exploiting all information would require infinite information processing capacity, and we would spend an infinite amount of money for that. Hence, some margin of inefficiency must always remain. Some people have asked if that implies the vanishing of Adam Smith's 'invisible hand' of the market [130]. We leave such speculations to the reader's self-study and proceed further to empirical facts.

1.1.3   Stylised facts

Empirical econophysics engages the analysis of large amounts of various data characterising economic activity. Certain universal features exist independent of the details of the trading, of the place on Earth and, within reasonable historical horizon, independent of time. Those features constitute the so-called stylised facts. Each model aimed at explaining and predicting economic phenomena must be matched with these stylised facts. If the model fails in such examination, it must be discarded, or at least it must not be trusted too much. We shall see many times in the course of this book that this is the unfortunate fate of numerous, otherwise beautiful, models.

The most immediate characteristics are the prices of various commodities, goods, rights, etc. We shall start our overview of basic empirical findings using the price statistics in Sec. 1.2. If we go deeper into the workings of the market, we ask not only about the price, but also about how much of each commodity was sold at given moment, i.e. the volume of the transactions. But this is still only the tip of the iceberg. To fully describe the trading, at least in the stock and commodity markets, we would need the history of each and every order to buy and sell, including the identity of the agent who issued the order. Some of the information is inaccessible, and some will be sketched in Sec. 1.3. Finally, in Sec. 1.4 we shall show how the correlations between various prices can serve to extract structural information on the market, for example the classification of companies into business sectors.

1.2   Price fluctuations

1.2.1   Notation

Let us first introduce some notation we shall use throughout this chapter. We shall analyse time sequences for various quantities. Suppose we know a time-dependent quantity A_t at times $t \in \mathcal{T} = \{T_1, T_2, \ldots, T_M\}$, where $T_i < T_{i+1}$. The number $M = |\mathcal{T}|$ is the length of the data set. We denote the time average

$$\langle A_t \rangle = \frac{1}{M} \sum_{t \in \mathcal{T}} A_t. \tag{1.5}$$

Sometimes, the quantity A_t will be a product of two other quantities, say, $A_t = B_t\, C_t$. Then, the average $\langle B_t\, C_t \rangle$ can be considered as a scalar product of the two time sequences B_t and C_t, viewed as M-component vectors. Indeed, it is easy to verify that the properties of the scalar product are satisfied, most notably the triangle inequality. It will play an important role later.

We shall frequently look at cumulative distributions of various quantities. For A_t it would be defined as

$$P_A^>(a) = \langle \theta(A_t - a) \rangle \tag{1.6}$$

where $\theta(x)$ is the Heaviside function, equal to 1 for $x > 0$ and 0 otherwise. Less frequently, we shall use the distribution function, defined as

$$P_A(a) = \langle \delta(A_t - a) \rangle \tag{1.7}$$

where $\delta(x)$ is here the Kronecker delta, equal to 1 if $x = 0$ and 0 otherwise. Obviously, in practice the distribution function is plotted as a histogram, rather than an ensemble of discrete ticks of height $1/M$ as the function (1.7) would suggest. However, we shall see later a plot (Fig. 1.12), in which these ticks do appear and coexist with a histogram.

It will frequently happen that A_t depends implicitly on a parameter, say, w. In that case, we shall include it into the notation for the distribution and cumulative distribution as $P_A(w; a)$ and $P_A^>(w; a)$, respectively.

To complete the definitions, we also introduce the conditional time average, including two time sequences A_t and B_t. We write

$$\langle A_t | B_t = b \rangle = \frac{\langle A_t\, \delta(B_t - b) \rangle}{\langle \delta(B_t - b) \rangle} \tag{1.8}$$

where $\delta(x)$ is again the Kronecker delta. Further, more specialised notation will be introduced below where needed.

1.2.2   Basic analysis

Scaling and power-law tail in return distribution

Thus, we suppose that the empirical data set we have at our disposal is a finite sequence of prices Z_t, where $t \in \mathcal{T} = \{T_1, T_2, \ldots, T_M\}$. The instants T_i may or may not be equidistant, depending on how the data were collected. In the most precise studies, the times T_i correspond to the moments when individual transactions took place.

Then, of course, the intertrade times are not all equal. They themselves form a time sequence

$$\Delta T_t = T_i - T_{i-1} \qquad (1.9)$$

where $t = T_i$. In more coarse-grained studies, the prices are recorded regularly at times that are multiples of an elementary time unit.

Following the widely accepted idea of Osborne, explained earlier, we consider the logarithm of the price $Y_t = \ln Z_t$ as the natural variable for further statistical analysis. The first quantity derived from the price logarithm is the return. We can define either the return at fixed time lag δt, which is

$$X_t = Y_t - Y_{t-\delta t} \qquad (1.10)$$

or the returns realised at individual trades

$$X_t = Y_t - Y_{t'} \qquad (1.11)$$

where $t = T_i$ and $t' = T_{i-1}$. In the former case, the time lag δt is an implicit parameter of the time sequence of returns X_t.

In both cases, the first question we ask is: what is the distribution of returns? In the early 1990s, Rosario N. Mantegna pioneered the field of econophysics in his investigation of the distribution of returns in the Milan stock exchange index [17]. He found quite good agreement with a Lévy-stable distribution with the tail exponent $\alpha = 1.16$. However, a closer look at the data showed that at the ultimate tails were not as fat as the Lévy distribution would have predicted. Following this study, Mantegna and H. Eugene Stanley showed that the distribution of returns exhibits scaling, i.e. it is invariant under rescaling of time and price [131]. In physics, scaling is the most typical phenomenon accompanying phase transitions and critical behaviour. Scaling is the daily business for many branches of complexity science. When a physicist comes across a situation where scaling plays a role, she immediately smells something extremely appealing. That is why physicists got excited once they learnt that scaling appears in economics. So, what is it all about?

Scaling means that the distribution of returns at various time lags δt, which is a function of two variables, x and δt, can be written using a function of one variable only, as

$$P_X(\delta t; x) = (\delta t)^{-H} g\left(x\,(\delta t)^{-H}\right) \qquad (1.12)$$

where the parameter H is the Hurst exponent and $g(u)$ is the common scaling function. In Ref. [131] scaling was confirmed in the range of times from $\delta t = 1$ min to $\delta t = 1000$ min. This is long enough to be assured of the existence of the scaling phenomenon. One should note, however, that for very long times, on the order of years, scaling cannot hold any longer, and a crossover time scale, at which scaling breaks down, must exist.

For the success of the analysis of scaling it is vital to establish the Hurst exponent reliably. The easiest way to measure it is through the dependence of the probability of zero return at time δt, or, in other words, the probability that the price comes back to its initial value after that time. From the function (1.12) we can see that it decreases according to a power law

$$P_X(\delta t; 0) = g(0)\,(\delta t)^{-H}. \tag{1.13}$$

In Ref. [131], the Hurst exponent was estimated as $H = 0.712 \pm 0.025$ using this method. In later analyses the value of the Hurst exponent varied from $H \simeq 0.5$ to $H \simeq 0.7$, with $H \simeq 0.6$ being the most frequent estimate. It seems certain that H is larger than the value $1/2$, corresponding to the random walk, Brownian motion, and similar 'simple' random processes. Irreducible complexity of price fluctuations is looming here. Sometimes, it is speculated that the fractional value $H = 2/3$ is the proper universal value. However, this hypothesis is far from being an established fact. We shall see later how the Hurst exponent can be measured more precisely.

The value $H > 1/2$ already indicates that the fluctuations of price are far from trivial. But the Hurst exponent still bears only a little portion of the information on the price movements. We can learn more from the form of the scaling function $g(u)$. As already noted, the original hypothesis was that the scaling function is a Lévy-stable distribution [17, 116]. If that were true, the tail exponent α of the Lévy distribution and the Hurst exponent would be connected by the simple scaling relation

$$H = \frac{1}{\alpha}. \tag{1.14}$$

For the value of H reported in Ref. [131] it means that the tail exponent would have the value $\alpha \simeq 1.4$.

However, in reality the fit to a Lévy-stable distribution with this exponent is reasonably good only in the central part of the distribution, but the tails are thinner. The Lévy distribution overestimates the probability of very large price changes. This is a significant discrepancy, because it is just the tail of the distribution which is most important for the estimation of the risks induced by the price fluctuations.

There is no consensus on the form of the scaling function, but the best data seem to indicate rather unequivocally that the tail of the scaling function is a power law, $g(u) \sim u^{-1-\alpha}$, for $u \to \infty$, with exponent $\alpha \simeq 3$, outside the Lévy range [132–134]. We show in Fig. 1.3 the cumulative distribution of returns for several values of δt, indicating both the scaling phenomenon and the power-law tail of the distribution. In this figure, the scaling is achieved by dividing the returns by the average quadratic volatility $\sigma_X(\delta t)$, defined as

$$\sigma_X^2(\delta t) = \langle (X_t - \langle X_t \rangle)^2 \rangle. \tag{1.15}$$

The dependence on δt on the right hand side is hidden as an implicit parameter of the returns X_t. The quadratic volatility increases as a power, $\sigma_X(\delta t) \sim (\delta t)^H$, where H is again the Hurst exponent introduced in (1.12). The value reported in Ref. [133] is $H \simeq 0.67$.

It would be nice if the tail exponent α was universal, i.e. identical for all stocks and commodities traded on the stock market. Unfortunately, it does not seem so. Also in Fig. 1.3 we show the histogram of values of α found by a systematic study of a large set of stocks. The histogram has a maximum of around $\alpha = 3$, so this value is

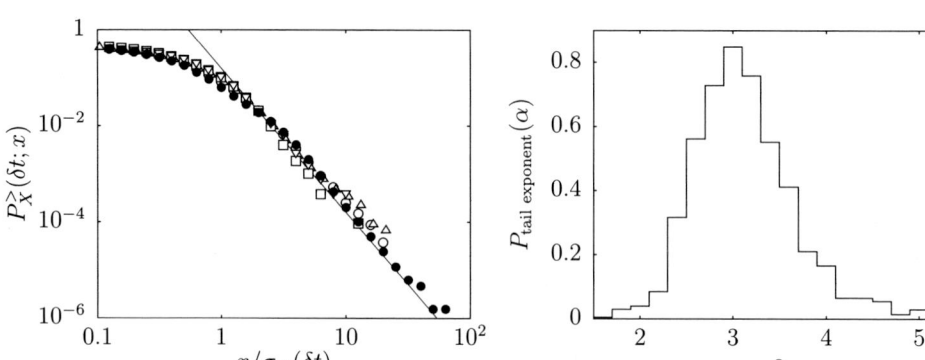

Fig. 1.3 Left panel: Cumulative distribution of returns for the S&P index, at time intervals $\delta t = 1$ min ($\bullet$), $\delta t = 16$ min ($\bigcirc$), $\delta t = 32$ min ($\triangle$), $\delta t = 128$ min ($\square$), and $\delta t = 512$ min ($\triangledown$). The straight line is a power law with exponent $\alpha = 3$. Data were extracted from Ref. [133]. Right panel: Histogram of fitted tail exponents α of the cumulative return distributions (at $\delta t = 5$ min) for 1000 companies traded in the American stock exchanges. The data were extracted from Ref. [134].

considered as the most typical value of the tail exponent. However, the peak is rather broad. Values ranging from $\alpha = 2.5$ to $\alpha = 4$ are found in a significant number of cases.

In any case, the scaling relation (1.14) is violated. This means that Mandelbrot's idea of prices following a Lévy walkis wrong. The weak point is the assumption of independence of price increments (i.e. returns) at short time intervals. On the contrary, we shall soon see that the returns are strongly and non-trivially dependent even at quite large time distances.

So, we can see that the price signal is indeed rather complex. The power-law tail in the return distribution is the simplest of all the complex features. All sensible models of stock-market fluctuations must reproduce, at least qualitatively, these power-law tails.

Indeed, since the beginning of the empirical econophysics studies [17, 131, 135–139], the amount of data supporting the basic results outlined here is enormous, confirming their generic validity within the span of centuries [140] all over the entire globe [141–166], comparing places as diverse as Hong Kong [144], Warsaw [157], and Tehran [147]. The precise form of the tail in the return distribution was a subject of debate for some time, and suggestions ranged from stretched-exponential [167] to truncated Lévy distributions [168–175], to exponential [176, 177].

Properties of volatility

If stock-market fluctuations are not well described by either a classical random walk or a Lévy walk,we should ask what the crucial ingredients are which make them different. One of them certainly is the fact that the amplitude of the fluctuations of prices is not stationary but depends on time, while both random and Lévy walks can be characterised by the time-independent typical size of fluctuations.

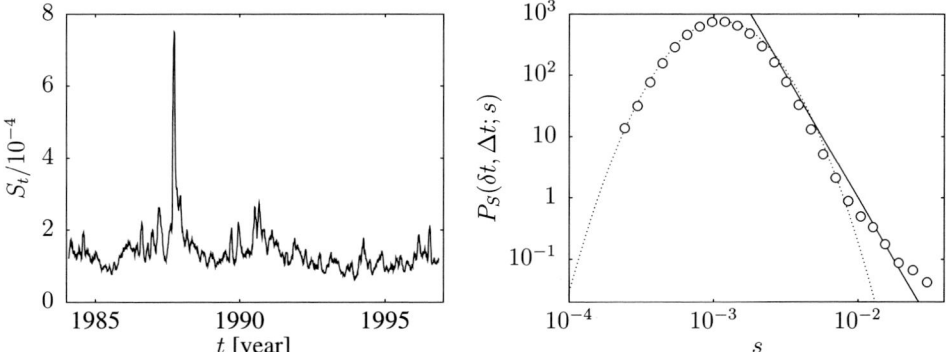

Fig. 1.4 In the left panel, time sequence of the volatility for S&P 500 index, at $\delta t = 30$ min, $\Delta t = 8190$ min (i.e. 1 month). Note the sharp peak indicating the crash in October 1987. In the right panel is the distribution of volatility for the S&P 500 index, calculated at $\delta t = 30$ min, $\Delta t = 120$ min. Dotted line is the log-normal distribution, solid line is the power law $P_S \sim s^{-1-\alpha}$ with $\alpha = 3$. All the data were extracted from Ref. [178].

For measuring this scale, or size, there is no single universally applicable quantity. We already encountered the quadratic volatility $\sigma_X(\delta t)$, defined in Eq. (1.15), which measures the average amplitude of the fluctuations. To see how the fluctuations swell or diminish in time, we can use the sequence of volatilities computed over the fixed time window $\Delta t = n\delta t$. It is defined as (see Ref. [178])

$$S_t = \frac{1}{n} \sum_{k=0}^{n-1} |X_{t-k\delta t}|. \tag{1.16}$$

Note that it depends on two implicit parameters: the time lag δt at which the returns are computed and the length of the time window Δt. Clearly, other measures are also possible; for example, instead of the absolute value of return, as in the sequence (1.16), we can take its square, or in general any power of its absolute value. Then each power provides somewhat different information.

In analogy with the sequence of returns, the sequence (1.16) can also be statistically analysed [178–181] in the same manner. The most immediate property is the distribution of volatilities. We show an example of the volatility sequence as well as its distribution in Fig. 1.4. We can clearly see that the volatility is not stationary. There are long periods, several months or even more, of high volatility, followed by other long intervals of low volatility. In probabilistic language, volatilities at different times are positively correlated. Qualitatively, this phenomenon is called volatility clustering. It may be expressed quantitatively in the slow decay of the correlation function, as will be shown below.

The distribution of the volatility is essentially log-normal, as shown in Fig. 1.4. However, the right tail deviates markedly from log-normality and usually is well fitted to a power law. For the data shown here, which originate from Ref. [178], we have the same value of the tail exponent $\alpha = 3$ as found for the distribution of returns, although the fit is not excellent at all. However, the evidence for the power-law tail was strengthened by other studies [19, 182], and now it is considered very probable. It is also very likely that it is not a mere coincidence that the tail exponents in return and volatility distributions are close to each other. Indeed, the excess probability of periods with high returns can serve as a simple (and cheap) explanation of the anomalously large probability of high returns. This idea is formally elaborated in the stochastic volatility models, treated in Sec. 2.4.

Further non-trivial features of the time sequence of volatility were discovered in the study of time intervals between successive returns of volatility to the same level [183–185]. There is a subtle point in this study. On average, volatility follows a specific intraday pattern. It is higher close to the opening and closing times of the stock market, while in the middle of the day, the volatility is lower. Such a 'U' shape must inevitably be taken into account when we ask when the volatility reaches a prescribed level. In the analysis, we first calculate the intraday volatility pattern by averaging over many trading days and then divide the true volatility by this pattern. The time sequence of scaled volatility $\overline{S}_t$ obtained in this way is further analysed. We find the instants $\overline{T}_i$ at which it reaches the fixed threshold $\overline{s}_{\mathrm{thr}}$ and obtain the waiting times $\overline{\Delta T}_t = \overline{T}_i - \overline{T}_{i-1}$, where $t = T_i$ in this formula. If the fluctuations in volatility were random, the distribution of the waiting times would be Poissonian $P_{\overline{\Delta T}}(\overline{\Delta t}) \sim \exp(-a\,\overline{\Delta t})$. However, the data show something different. The distribution is much closer to the stretched-exponential form $P_{\overline{\Delta T}}(\overline{\Delta t}) \sim \exp(-a\,(\overline{\Delta t})^\gamma)$, with $\gamma \simeq 0.3$ [184]. This is a sign of strong temporal correlations in the sequence of volatilities.

Volatility defined as in the sequence (1.16) is a temporal average of absolute returns over a certain time window. Instead of the time average we can also consider an ensemble average over a set of N_{sto} different stocks [186–191]. Each stock, numbered by the index a, has its own time sequence of returns $X_t^{(a)}$. To measure the differences between returns of different stocks, we define the quantity R_t called variety [186] as

$$R_t^2 = \frac{1}{N_{\mathrm{sto}}} \sum_{a=1}^{N_{\mathrm{sto}}} \left(X_t^{(a)} - \frac{1}{N_{\mathrm{sto}}} \sum_{a=1}^{N_{\mathrm{sto}}} X_t^{(a)} \right)^2. \tag{1.17}$$

We can again study the distribution of variety, and generically we find that it has features similar to the distribution of volatility. The central part of the distribution, around the maximum, is close to log-normal, while for large values a power-law tail develops. However, compared to the volatility, the tail exponent is much larger. For example in Ref. [186] the value is about $\alpha \simeq 6$.

1.2.3   More advanced analysis

Anomalous diffusion

There are also other scenarios of how the movement of price can become different from ordinary Brownian motion. One of them is related to the physical phenomenon of anomalous diffusion [194]. To know more about what this term means in physics, see Box 2.8. The basic difference of the anomalous diffusion from the ordinary one is that, for the former, the waiting times between subsequent steps or jumps have a non-trivial distribution. The input data are, first, the sequence of times of individual trades T_i,

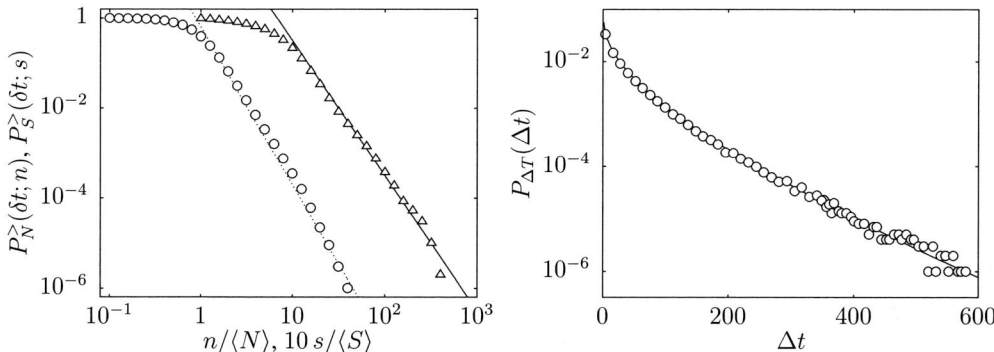

Fig. 1.5 Left panel: Cumulative distributions of the number of trades (○) and quadratic volatility (△) during the time $\delta t = 39$ min, obtained by averaging over approx. 200 individual stocks of U. S. companies traded in 1994 and 1995. The dotted line is the power law $\propto n^{-3.5}$, the solid line is the power law $\propto s^{-3}$. For better visibility, the distribution of quadratic volatility was shifted by factor 10. The data were extracted from Ref. [192]. Right panel: Distribution of intertrade times, in seconds, for Boeing traded from January 1993 to December 1996. The line is the Weibull distribution with $\tau = 27$ and $\eta = 0.73$. The data were extracted from Ref. [193].

from which we get the sequence of waiting times defined in Eq. (1.9); and second, the sequence of prices established in these trades, which yields the sequence of returns (1.11). The properties of the waiting times can be observed in two complementary quantities either directly in the waiting-time distribution $P_{\Delta T}(\Delta t)$, or in the number of trades which took place within a time interval of fixed length. The latter quantity counts the number of such trade times T_i which fall within the interval $[t - \delta t, t]$, i.e. formally

$$N_t = \sum_i \theta(T_i - t + \delta t)\,\theta(t - T_i). \tag{1.18}$$

We show the distributions for both ΔT_t and N_t in Fig. 1.5. The distribution of the number of trades exhibits a power-law tail with exponent $\alpha \simeq 3.5$. This means that there is an anomalously large probability of finding a time interval within which a very large number of trades occur. An attentive look at the distribution of the waiting times instead reveals power-law dependence for short times. The tail of the waiting-time distribution is not very important, and its form is better fitted on a stretched exponential. Together with the power law at short times, this information leads to the hypothesis that the waiting times are described by the Weibull distribution

$$P_{\Delta T}(\Delta t) = \frac{\eta}{\tau}\left(\frac{\Delta t}{\tau}\right)^{\eta - 1} \exp\left[\left(-\frac{\Delta t}{\tau}\right)^{\eta}\right]. \tag{1.19}$$

The data plotted in Fig. 1.5 are consistent with the value of the exponent $\eta = 0.73$.

We may also look at how much the price fluctuates in the interval $[t - \delta t, t]$. To calculate a measure of these fluctuations, it is natural to work with returns of individual trades, defined in Eq. (1.11). However, the data we show in Fig. 1.5 are based on

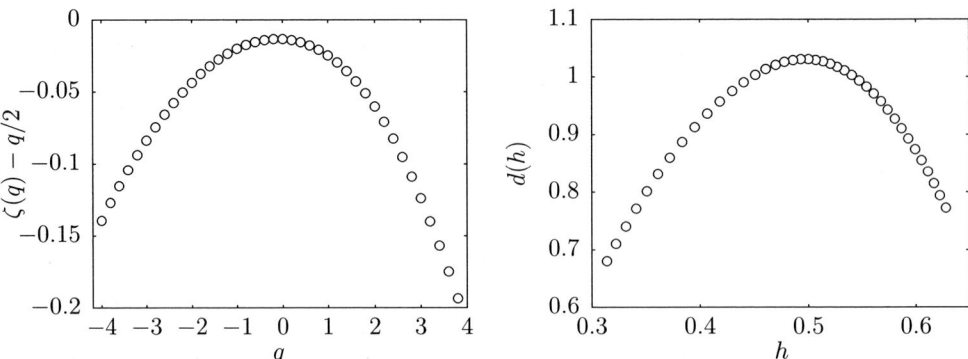

Fig. 1.6 In the left panel, the exponents for the moments obtained by multifractal detrended fluctuation analysis are shown, for approx. 2500 stocks. The length of the time series was from 10 to 30 years and the length of the segments $\Delta t = 10$ to 100. To make the nonlinearity more visible, we subtracted the linear dependence $q/2$. In the right panel, the multifractal spectrum calculated by the Legendre transform from the data shown in left panel. All the data were extracted from Ref. [209].

Ref. [192], where the price changes in individual trades are used instead, defined as $\Delta Z_t = Z_t - Z_{t'}$, with $t = T_i$ and $t' = T_{i-1}$. The fluctuations are then measured by the time-dependent quadratic volatility

$$S_t = \left[\frac{1}{N_t}\sum_i \theta(T_i - t + \delta t)\,\theta(t - T_i)\big(\Delta Z_{T_i}\big)^2\right]^{1/2}. \tag{1.20}$$

Its cumulative distribution is shown in Fig. 1.5. It has again a power-law tail, with exponent $\alpha \simeq 3$.

Thus, we prove that the fluctuations of prices are described by an anomalous diffusion, rather than an ordinary one [195]. However, we have here several quantities with non-trivial behaviour, and it would be very interesting to see if we could combine them into a quantity which could perhaps be artificial, yet would exhibit ordinary fluctuations following a normal (Gaussian) distribution. It was found [192] that such a combination is

$$G_t = \frac{1}{S_t\sqrt{N_t}}\sum_i \theta(T_i - t + \delta t)\,\theta(t - T_i)\Delta Z_{T_i} \tag{1.21}$$

i.e. the price change during the interval $[t-\delta t, t]$ rescaled so that the effects of a variable number of trades and the variable amplitude of the fluctuations are eliminated.

Finally, let us mention that the distribution of waiting times between individual trades is an extremely important quantity, and its distribution [193, 196–208] serves as an input for stochastic models based on the idea of the continuous-time random walk, to be explained in Sec. 2.1.3.

Multifractality

We have already seen two measures for the size of price fluctuations: the volatility (1.16) and its quadratic version. It is not a priori clear whether the two bear the same information or not. Moreover, we could use an arbitrary power of the return in the definition, not just the first or the second. This way we could obtain an entire ensemble of volatilities

$$S_t^{(q)} = \frac{1}{n} \sum_{k=0}^{n-1} |X_{t-k\delta t}|^q. \qquad (1.22)$$

The parameter δt is the time interval at which the return is calculated, and $\Delta t = n\delta t$ is the time window over which the fluctuations are measured. The exponent q can be an arbitrary real number. Calculating (1.22) we can compare the properties of volatilities for different q. A non-trivial dependence on q would provide us with further information on the complexity of price fluctuations.

We shall look here at another smart way to perform such an analysis. It relies on the method called multifractal detrended fluctuation analysis (DFA) [209–216]. If the series of price logarithms Y_t extends from time t_1 to t_2, we first divide the interval $[t_1, t_2]$ into segments of equal length Δt. In the ith segment, $\mathcal{T}_i = [t_1+(i-1)\Delta t, t_1+i\Delta t]$, we first find the linear trend $y_t(i) = a_i + b_i\,t$ by the least square fit to Y_t. Note that the segments can each contain a different number of points N_i. Then, we calculate the mean square fluctuation in each segment

$$m_2(i; \Delta t) = \frac{1}{N_i} \sum_{t \in \mathcal{T}_i} \left(Y_t - y_t(i) \right)^2 \qquad (1.23)$$

and the qth moment of the detrended fluctuation, defined as

$$m_q(\Delta t) = \frac{\Delta t}{t_2 - t_1} \sum_i \left(m_2(i; \Delta t) \right)^{q/2}. \qquad (1.24)$$

The number q is not necessarily a positive integer. On the contrary, it can be any real number, positive or even negative. The most important information is the dependence of the moments on the length of the time segments. It is supposed that it grows as a power with an exponent which depends on q, i.e.

$$m_q(\Delta t) \sim (\Delta t)^{\zeta(q)}. \qquad (1.25)$$

If the time sequence of price logarithms Y_t was a simple fractal, the function $\zeta(q)$ would be linear, or $\zeta(q) = Hq$. In that case, H would be the Hurst exponent of the time series. However, it has been established that $\zeta(q)$ is nonlinear, indicating that the time sequence has a more complicated structure. It is a multifractal, characterised by not just one, but more or less a wide spread of exponents. The fluctuations are given by a mixture of many components, each of them having a different Hurst exponent. To stress the difference of the Hurst exponent H computed from the entire time series from those computed from the components, we denote the latter by lowercase h. In order to measure how much each of the values of h is represented in the mixture, we introduce

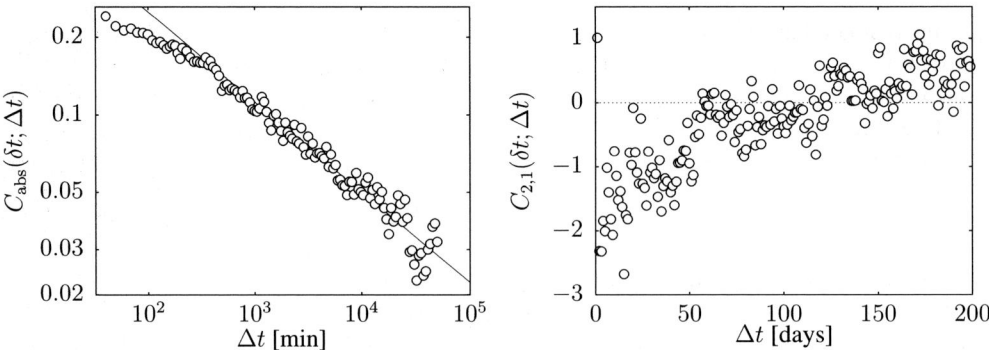

Fig. 1.7 In the left panel, autocorrelation of absolute values of return are shown for the S&P 500 index in the period 1984-1996 at $\delta t = 1$ min. The straight line is the power-law decay $\propto (\Delta t)^{-0.35}$. The data were extracted from Ref. [133]. In the right panel, the leverage effect is illustrated for returns calculated at $\delta t = 1$ day. The correlation function was averaged over an ensemble of 500 European stocks, traded in the period 1990-2000. The data were extracted from Ref. [251].

a quantity $d(h)$ called a multifractal spectrum (for a more precise explanation see Sec. 2.3). It is obtained from $\zeta(q)$ by the Legendre transform

$$d(h) = \min_q \big(qh + 1 - \zeta(q) \big). \tag{1.26}$$

The figure '1' appearing in this formula refers to the one-dimensionality of time.

We can see the behaviour of the moments and the multifractal spectrum in Fig. 1.6. The nonlinearity of the function $\zeta(q)$ is clearly visible. After subtracting the linear trend, the dependence of $\zeta(q) - q/2$ is a concave function with a clear maximum. The multifractal spectrum $d(q)$ obtained by numerical implementation of the Legendre transform (1.26) also has a maximum, at a value of h very close to $1/2$, the Hurst exponent of the ordinary random walk. Speaking very vaguely, the larger value of $d(h)$ means that this value of h is present in the fluctuations with a greater weight. The maximum of $d(h)$ at $h \simeq 1/2$ implies that the trivial random walk's behaviour is still dominant, but from behind its shoulders complex fluctuations peep out, with a wide range of exponents h for which $d(h) > 0$. The minimum of such values of h corresponds to the slope of the function $\zeta(q)$ at the lowest values of q. In principle we should calculate this slope in the limit $q \to -\infty$, but this is impossible in practice. Analogically, the maximum observed value of h corresponds to the slope of $\zeta(q)$ at the highest measured values of q.

Multifractality is one of the most subtle features of the price signals and difficult to detect reliably [217, 218], but it has been studied extensively [145–147, 212, 219–247]. To reproduce it in a model, we shall resort to analogies with aerodynamic turbulence [248–250]. The physics of wind will serve as an inspiration for multifractal cascade models of price fluctuations, investigated in Sec. 2.3.

Nonlinear correlations

In one respect the EMH is absolutely true. The direction of the price change in the nearest future is practically independent of all the past price changes. This is the reason why systematic gain on the stock market is impossible.

The words 'practically independent' can be given very sound mathematical meaning in terms of the autocorrelation function of returns

$$C_{1,1}(\delta t; \Delta t) = \langle X_t \, X_{t+\Delta t} \rangle. \tag{1.27}$$

The dependence on the time lag δt at which the returns are computed is implicit on the right-hand side. We call these correlations linear as both X_t and $X_{t+\Delta t}$ enter linearly into the average (1.27). Within the level of statistical noise, the function $C_{1,1}(\delta t; \Delta t)$ is zero, except for short time distances $\Delta t \lesssim 20$ min [133]. The length of such a time interval is determined by transaction costs. Indeed, a large transaction fee discourages frequent trading, and returns may remain correlated during times within which few trades, if any, occur. The length of the interval of non-negligible correlations can be considered as a practical measure of the market's efficiency. It is interesting to note that the residual correlations during that short interval of inefficiency have a negative sign [19]. The returns are short-time anticorrelated, meaning that after each casual price fluctuation the market forces tend to quickly restore the price to its original level.

So far it has seemed that there is no problem with the validity of EMH. However, EMH does not restrict its claims to the mere absence of correlations. It requires independence of future returns over the course of history. Empirical evidence shows that such pretended independence is very far from reality, but this is not revealed in the linear correlations. To see it, we must calculate nonlinear correlations. The simplest one is the autocorrelation function of absolute values of returns

$$C_{\mathrm{abs}}(\delta t; \Delta t) = \langle |X_t \, X_{t+\Delta t}| \rangle. \tag{1.28}$$

It has been known for quite a long time [252] that it decays very slowly, the typical time scale being a few days. This effect is sometimes called volatility clustering, evoking the optical impression already visible in the time sequence of volatility in Fig. 1.4 that periods of high volatility tend to be clustered together, leaving other long periods for low-volatility stages.

In the data shown in Fig. 1.7 we observe that not only is the typical scale of the decay of correlations long, but the form of the decay is also a slow function. It is a power law with a very small exponent

$$C_{\mathrm{abs}}(\delta t; \Delta t) \sim (\Delta t)^{-\tau} \tag{1.29}$$

with $\tau \simeq 0.35$. The precise value of τ varies from stock to stock, but always remains small, $\tau \lesssim 1/2$. The smallness even makes the precise fit difficult, and as an alternative it has been suggested that the decay is not power-law, but logarithmic [253–255]. Whatever the correct functional form of the slow decay is, it means that the memory contained in the price signal is preserved for a very long time. However, this memory cannot be used for gaining profit. EMH is true up to the point that all the profitable information is very quickly (although not 'immediately', since that would be physical nonsense) incorporated into the price signal. But there is a huge amount of non-profitable information remaining. That makes the price signal complex, definitely different from any form of random or Lévy walk, and in this respect EMH completely misses the point.

An example of even more subtle correlations in the price sequence is the leverage effect [251, 256–260]. In words, it is a negative correlation between volatility and past returns. It is measured by the correlation function

$$C_{2,1}(\delta t; \Delta t) = \frac{\langle (X_{t+\Delta t})^2 X_t \rangle}{\langle (X_t)^2 \rangle^2}. \tag{1.30}$$

For $\Delta t < 0$ the correlation function is zero within the scope of statistical noise but becomes negative for $\Delta t > 0$ and decays to zero as a rather slow exponential, within a typical scale of months. Thus, we can write

$$C_{2,1}(\delta t; \Delta t) \simeq -A\,\theta(\Delta t)\,e^{-\Delta t/a}. \tag{1.31}$$

This means that price drops bring about an increase in volatility, but the causality does not go in the other direction. Increased volatility in the past does not have any influence on price movement in the future. We can see a typical example of the correlations in Fig. 1.7, where the decay time is well fitted to $a \simeq 40$ days [251]. Interestingly, the amplitude of the correlations is often very close to $A \simeq 2$, at least for the price sequences of individual stocks. The same correlation function computed for financial indices, such as the S&P 500, shows much higher A, for reasons that are not completely clear [256].

1.3 Order-book statistics

1.3.1 Order placement

The prices announced at a stock market are only a tiny tip of the iceberg. Many things must happen prior to the trade which fixes the price. There are many potential buyers who stood idle because the offered price seemed too high for them. On the opposite end of the market we can see as many people waiting in hope that they will sell later at a more favourable price.

Whatever the structure of the stock market is, the participants always issue orders, either buy or sell, which are to be executed under various conditions. These conditions usually fix the limit price, i.e. the minimum price at which shares can be sold or the maximum at which they can be bought. The issuer can also specify how long the order can last, if not executed yet, and can also cancel the order directly. Therefore, not all orders are eventually satisfied. The orders with fixed prices are called limit orders. Besides those, there are also market orders that require buying or selling at any price available in the market. While the limit orders last for some time and therefore constitute the memory of the market, the market orders are executed immediately.

There are many stock markets around the globe, and they differ in their ways of organising the trade. Some of them rely on a market maker who collects the orders and

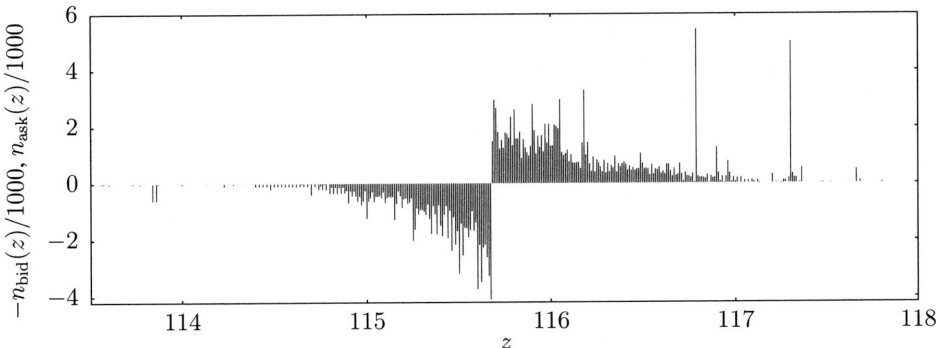

Fig. 1.8 A snapshot of the order book for FGBL (futures contracts on Euro Bund, a bond issued by the Federal Republic of Germany), traded at Eurex (European financial derivatives exchange) on 23 March 2007 at 15:21. For each price z we show the cumulative size of all orders $n_{bid,ask}(z)$ at this price. For better visibility, the size of buy orders is plotted with a negative sign. For this plot, the data freely accessible at www.eurexchange.com were used.

after a certain interval of time, she/he makes the best effort to settle the accumulated package of buy and sell orders against each other. As a result, it is the market maker who decides the price. Some other markets are fully automatic and each order is executed instantly, as soon as an order of the opposite type is issued which can match the price limits.

Snapshot

In any of the cases, the order book is a list of orders to sell (ask) and buy (bid) with a specified price. The list is dynamic and is updated at a rate of seconds. Obviously, all bids have lower prices than all asks. The distance between the lowest ask and highest bid is called spread, and its fluctuations are nearly as important as the fluctuations of the price.

The price at which an order is placed is not completely arbitrary but is usually a multiple of a fixed discrete unit. Therefore, there may be several, or even many, orders placed at the same price. The orders also differ in size, measured in the number of shares to be sold or bought. Thus, at each moment we can draw a diagram showing total size of the orders waiting at each price. An example of such a snapshot of the order book is shown in Fig. 1.8. We can see that the instantaneous configuration of orders is very uneven, with high peaks at some prices. Generally, however, the number of orders is highest close to the actual price and decreases when we go farther from it.

Regularities in order placement

The placement of the orders is a matter of the investors' strategy. An order placed very close to the current price is likely to be executed rather soon, while an order far away will probably wait quite a long time before it is satisfied, but the expected profit is higher. The balance between tiny but frequent short-time gains versus rare

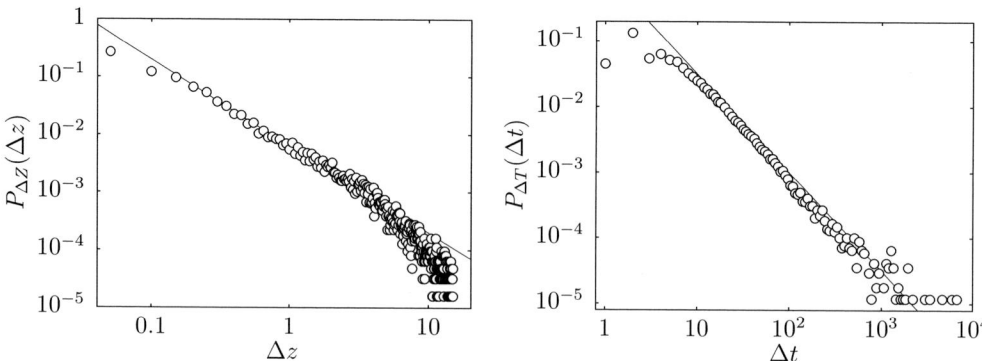

Fig. 1.9 Left panel: Distribution of orders placed at distance Δz from the current best price (in Euros), for France Telecom traded at Paris Bourse in 2001. The line is the power law $\propto (\Delta z)^{-1.5}$. The data were extracted from Ref. [261]. Right panel: Distribution of the lifetimes of orders (in seconds), from placement to execution. The line is the power law $\propto (\Delta t)^{-1.5}$. The data were extracted from Ref. [262].

but large ones makes the order placement a complex optimisation problem [263]. Here we discuss the empirically observed pattern which results from it.

First, it has been established that the size of the orders V is distributed according to a power law $P_V(v) \sim v^{-1-\gamma}$ with exponent $\gamma \simeq 1$ [264]. This holds both for limit and market orders. Furthermore, we can observe how far from the current price the limit orders are placed. The distribution of such distance ΔZ is also governed by a power law $P_{\Delta Z}(\Delta z) \sim (\Delta z)^{-1-\mu}$, as shown in Fig. 1.9. The value of the exponent μ is controversial [261, 263, 265, 266], ranging from $\mu \simeq 0.5$ [261] to $\mu \simeq 1.4$ [266].

So far, we have investigated the 'spatial' characteristics of the order-placement process. One of the important temporal characteristics is the lifetime of the orders, i.e. how long a specific order lasts within the order book. Here we must distinguish two ways in which the existence of the order can come to an end. Either the order is executed by matching a complementary order, or the order is removed from the book without being satisfied. In the former case a trade follows, accompanied by a newly fixed price. Therefore, the occurrence of such an event is detectable in the time series of prices. The latter case results in no visible effect, but it affects future price movements through an indirect and subtle mechanism. The removed orders, either cancelled by the brokers who placed them or erased automatically after their pre-defined expiry time has elapsed, leave an empty space which may result in enhanced price movement in the future.

The lifetime ΔT of both executed and removed orders is power-law distributed, $P_{\Delta T}(\Delta t) \sim (\Delta t)^{-1-\eta}$ [262, 267–270]. The value of the exponent differs in these two cases, being $\eta \simeq 1.1$ for removed or $\eta \simeq 0.5$ for executed orders, as also shown in Fig. 1.9. A non-trivial spatio-temporal pattern in the placement and removal of orders was revealed in the dependence of the probability of an order being cancelled on the distance Δz from the current price. It was found [265] that the cancel rate decays with Δz, and except for the closest vicinity of the price, the decay roughly follows the power

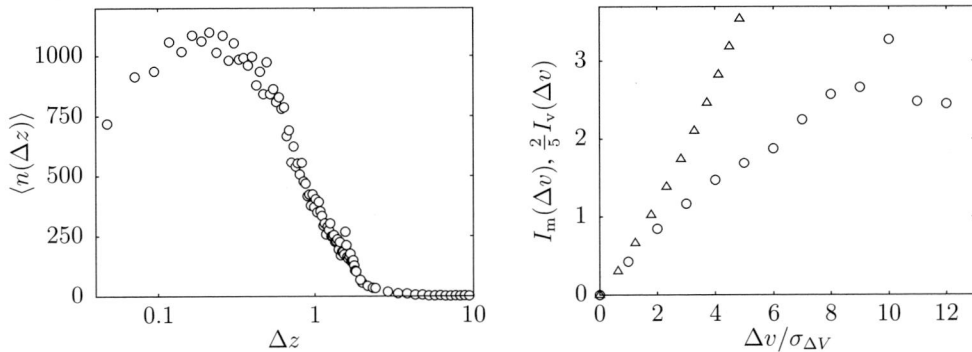

Fig. 1.10 The left panel shows the average order-book profile, i.e. the average cumulative volume of all orders placed at distance Δz (in Euros) from the current best price, for France Telecom at Paris Bourse on 2001. The data were extracted from Ref. [261]. In the right panel, the averaged virtual price impact ($\triangle$) and market price impact ($\bigcirc$) are shown for $\delta t = 5$ min. The data were averaged over 20 frequently traded stocks in NASDAQ from the year 2002. Note that the virtual impact was rescaled by the factor 2/5 for better comparison with the market impact. The data were extracted from Ref. [271].

law $\simeq (\Delta z)^{-1.2}$. This finding is very important for any attempt to model the order book, as it implies that close to the current price the orders are added and removed extremely quickly. Such rapid rearrangement of the order book is like a complex dance of gear trains inside a clock, from which we perceive only the number indicated by the clock hand, which is the price.

1.3.2 Price impact

Order-book profile

In the first approximation we can forget about the fluctuations and try to understand the behaviour of the order book from its averaged properties. The key quantity is the averaged profile of the order book, which is found by superposing the instantaneous profiles, like that of Fig. 1.8, in such a way that the current price is fixed at the origin of the coordinates. Each profile is shifted to the right or left according to the actual price. We must note that this procedure is not as innocent as it might look. An order which is, say, placed at a moderate distance from the price and waits there for some time appears in several superposed profiles. In each of these profiles it is located at a different position, because the profiles are shifted by different amounts. Therefore, one single order contributes to the average profile at several different places.

In Fig. 1.10 we show the dependence of time-averaged total size of orders $\langle n(\Delta z) \rangle$ as a function of the distance Δz from the actual price. The general impression, drawn from the snapshot in Fig. 1.8, namely, that the density of orders decreases with Δz, is nearly correct. However, in the region very close to the price the profile is suppressed, so that a maximum develops at a finite distance from the price. In the example shown in Fig. 1.10 the maximum lies around 0.3 EUR. It is also possible to look at the form of the profile's tail. The results seem to be compatible with a power-law decay $\langle n(\Delta z)\rangle \sim (\Delta z)^{-2}$ [261, 265, 272].

Virtual vs. market price impact

At first glance we might think the statistics of price changes can be deduced directly from the average profile and distribution of order sizes. To this end, the function called virtual price impact is defined. Intuitively, it tells us how far we must go on the price axis in order to remove a specific volume of orders from the average order-book profile. We shall now disclose the point of this story. The actual market impact, i.e. the measured change in price induced by placing a given volume of orders, is very much different from the virtual impact predicted from the average profile.

Let us observe the problem more closely. As the data we show here are based on the work [271], in which the authors use a logarithm of price $Y = \ln Z$ as an independent variable instead of the price itself, we now switch the notation: in the following we shall express the average profile as a function of Δy, the distance of log-prices.

We denote V_t the volume of all market orders which were issued from a time t_0 in the distant past up to the time t. Because we shall always deal only with differences of the process V_t, the precise choice of t_0 is irrelevant. The study of the market impact is nothing else than studying the mutual correlation between the processes Y_t and V_t.

We measure the fluctuations in the flow of orders within the time interval of length δt by the quantity

$$\sigma_{\Delta V} = \langle |V_{t+\delta t} - V_t - \langle V_{t+\delta t} - V_t\rangle|\rangle. \tag{1.32}$$

For convenience, we shall measure the price impact in the units of the standard deviation of log-price fluctuations, defined as

$$\sigma_{\Delta Y} = \sqrt{\left\langle \left(Y_{t+\delta t} - Y_t - \langle Y_{t+\delta t} - Y_t\rangle\right)^2\right\rangle}. \tag{1.33}$$

This facilitates the comparison of price impacts for different stocks, each with a different level of trading activity.

If $\langle n(y)\rangle$ is the average order-book profile as a function of log-price, the expected log-price change Δy after arrival of market orders of volume Δv is calculated by inverting the integration, from the equation

$$\int_0^{\Delta y} \langle n(y')\rangle \, dy' = \Delta v. \tag{1.34}$$

Then, the virtual price impact is

$$I_v(\Delta v) = \frac{\Delta y}{\sigma_{\Delta Y}}. \tag{1.35}$$

This is to be compared with the actual market price impact, defined as the conditional average of log-price shift

$$I_m(\Delta v) = \frac{\langle Y_{t+\delta t} - Y_t | V_{t+\delta t} - V_t = \Delta v\rangle}{\sigma_{\Delta Y}} \tag{1.36}$$

on the condition that the volume of the orders supplied was Δv.

In Fig. 1.10 we compare the empirical data for virtual and market impacts [271]. We see not only that the market impact is much smaller than the virtual impact, but that the curvature of the two functions also has the opposite signs. While the virtual impact is convex, which follows immediately from the decaying tail in the average book profile, the actual market impact is concave. This discrepancy is the sign of a complex relationship between price movements and order placement.

On a basic level of understanding, we would like to know if the shape of the actual market impact follows a universal law, as the tail of the return distribution seems to do. As we already said, the function is concave [273–277]. Zhang conjectured [129] that the market impact should follow a square-root form, i.e.

$$I_{\mathrm{m}}(\Delta v) \sim (\Delta v)^{\nu} \qquad (1.37)$$

with $\nu = 1/2$. This was indeed found consistent with the data in several studies [278, 279] and a theory of power-law tails in return distribution was built on it [278–282], but the story does not seem so simple. First, other functional forms were tried, often with better success, like the logarithmic function [265] or a power with exponent $\nu < 1/2$ [283–285]. Moreover, the market impact does not seem universal, as it varies from market to market and even from stock to stock [283], despite the attempts to find a single master curve for market impact [286–288]. On the other hand, a convincing theory supported by empirical evidence was set up [289], showing that the impact function has a square-root shape.

Second, a detailed study showed that, while the market impact grows with the volume of the orders, the return distribution conditioned to a fixed volume is nearly independent of the volume. This suggests that the form of the return distribution is determined by the fluctuations in liquidity, i.e. fluctuations in the instantaneous order-book profile rather than the averaged market impact [290, 291]. A fairly convincing argument is the comparison of the distribution of returns generated by market orders and the distribution of first gaps, i.e. distances between the highest and second highest ask (or lowest and second lowest bid). These two distributions are indeed nearly identical [290]. This finding later inspired modelling order books as sequences of gaps between limit orders, as shown in Chap. 4.

1.3.3 Long memory in the order flow

Another puzzling aspect of the complexity of order-book dynamics is implied by long-time correlations in the signs of the orders, which lead to immediate trades (in our terminology market orders) [292–294]. In these studies time is usually measured in number of trades. If we define $\epsilon_n = \pm 1$ for a buy (+ sign) or sell (− sign) market order corresponding to trade number n, we can calculate the correlation function $C_{\mathrm{sig}}(\Delta n) = \langle \epsilon_n \epsilon_{n+\Delta n} \rangle - \langle \epsilon_n \rangle \langle \epsilon_{n+\Delta n} \rangle$. Alternatively, we can investigate order imbalance within a fixed interval of time. Let $n \in (t, t+\delta t)$ have the meaning that the market order number n was issued within the time interval from t to $t + \delta t$. Then we can define the normalised order imbalance $E_t = (\sum_{n \in (t, t+\delta t)} \epsilon_n)/(\sum_{n \in (t, t+\delta t)} 1)$ and its correlation function $C_{\mathrm{sig}}(\delta t; \Delta t) = \langle E_t E_{t+\Delta t} \rangle$. Both of the correlation functions consistently show a long-term power-law behaviour

$$C_{\text{sig}}(\Delta n) \sim (\Delta n)^{-\theta}$$
$$C_{\text{sig}}(\delta t; \Delta t) \sim (\Delta t)^{-\theta'}. \tag{1.38}$$

Both exponents have very low values $\theta, \theta' < 1$ indicating extremely long memory in the sequence of market orders. The precise values seems to vary among different stocks, for example $\theta = 0.4$ for Vodafone [293], $\theta = 0.53$ for Shell [294], or $\theta = 0.2$ for France Telecom [292]. Also θ and θ' are not quite the same for the same stock, for example $\theta' = 0.3$ for Vodafone again [293]. There is a consensus that the source of the long memory lies in the presence of hidden orders, which are split into many smaller orders to be revealed and submitted into the order book [294, 295]. Under the assumption of power-law distribution of the sizes of the hidden orders, we get power-law correlations as in (1.38).

Superficially, we are tempted to trivialise it. One power law more or less, who cares among the plethora of others? But this time it is not the case. If the sell orders are followed by sell again, and buys by buys, why does the price keep fluctuating unpredictably? There must be a delicate mechanism which balances the correlated market order flow with subtle correlations in liquidity. Some information is certainly contained in the fluctuations of the bid-ask spread and its response to sudden perturbations [296–300]. However, the mechanisms are more complex and still not fully understood. We prefer to stop here and suggest the reader to refer to the original sources for the rest [285, 292, 294, 301–332].

1.4 Correlation matrices and economic sectors

1.4.1 Hierarchies among stocks

Correlations

We have seen enough evidence of the very complicated behaviour of price sequences. Yet another level of complexity emerges when we observe the prices of several stocks (or commodities or financial indices) simultaneously. There are lots of things one may be curious about. How much do the price changes differ between different stocks? Are there groups of stocks whose prices follow each other faithfully? Is it possible to trace some information transfer through prices from one stock to the other?

The most fundamental notion enabling at least a partial answer to these questions is the cross-correlation matrix. Let us have N_{sto} stocks and record their price sequences $Z_t^{(a)}$, for $a = 1, 2, \ldots, N_{\text{sto}}$. The returns realised at time interval δt and denoted $X_t^{(a)}$ are computed as usual. First, we define the reduced returns

$$\widetilde{X}_t^{(a)} = \frac{X_t^{(a)} - \langle X_t^{(a)} \rangle}{\sqrt{\langle (X_t^{(a)} - \langle X_t^{(a)} \rangle)^2 \rangle}}. \tag{1.39}$$

Then, the elements of the cross-correlation matrix are defined as averages

$$M_{ab} = \langle \widetilde{X}_t^{(a)} \widetilde{X}_t^{(b)} \rangle \tag{1.40}$$

which can be interpreted as scalar products of the time sequences $\widetilde{X}_t^{(a)}$ and $\widetilde{X}_t^{(b)}$. The information contained in the cross-correlation matrix M can be extracted using various methods. We shall show three of them in the following paragraphs.

Distance

The correlations tell us which stocks are closer, in some sense, to each other. We can arrange the stocks into groups which evolve more or less homogeneously. Or, based on the sequence of returns, we can draw a hierarchy of stocks, much like biologists draw genealogical trees of species, based on the sequence of base pairs in their genetic code. This methodology was used many times for analysing dependencies among stocks, as well as indices and interest rates [333–362].

Since the elements of M are scalar products, they should satisfy an appropriate triangle inequality, and therefore it must be possible to define a distance between stocks. The set of stocks forms a metric space.

The most convenient way to define the distance is by the formula

$$d(a, b) = \frac{1}{2}(1 - M_{ab}). \tag{1.41}$$

Then, for each pair of stocks a and b, we have $d(a, b) \in [0, 1]$.

Minimum spanning tree

The distance is but an instrument for finding a structure which visualises the complex dependencies between stocks [333]. The structure in question is the minimum spanning tree. What is it? We can make a graph out of the set of stocks by first assigning a point to each stock and then joining pairs of such points with lines. Depending on how many lines (called edges) we draw and which points (called vertices) we connected by edges, we construct different graphs. We shall read much more about graphs and related things in Chap. 6. Here we need only a few basic notions.

A tree is such a graph that does not contain cycles, i.e. we cannot find a sequence of edges which starts in a vertex and returns back again to the same vertex. A spanning tree is a tree in which any two vertices are connected by a path along the edges of the tree. Every vertex is connected by an edge to another vertex. None is left isolated. Any spanning tree on N_{sto} stocks has exactly $N_{\text{sto}} - 1$ edges.

There are still many possible ways to make a spanning tree with a given set of stocks. If the edges are not all equivalent but each bears some weight or length, we can say which spanning tree is better than the other by summing the weights, or lengths, of all the edges involved. For this purpose we assign the distance $d(a, b)$, defined above, to the edge connecting stock a with stock b. The minimum spanning tree is a spanning tree that has the least possible sum of distances between all the pairs of stocks connected by an edge.

The construction of the minimum spanning tree is easy. We start with an empty graph, i.e. there are N_{sto} vertices and no edges. Then we find the minimum distance $d(a, b)$ between all possible pairs of vertices and add an edge between a and b. Then we find the second smaller, third smaller, etc. distances and add edges one by one. In

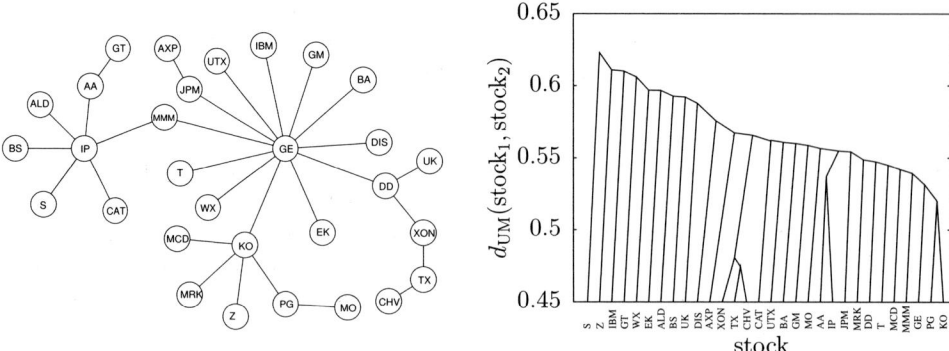

Fig. 1.11 In the left panel, we show the minimum spanning tree constructed from the correlation matrix of the 30 stocks which are the components of the Dow Jones Industrial Average index. In the right panel, the hierarchical structure of stocks generated by the ultrametric distance d_{UM} associated with the minimum spanning tree. The acronyms of the stocks are: AA–Alcoa, ALD–Allied Signal, AXP–American Express Co, BA–Boeing Co, BS–Bethlehem Steel, CAT–Caterpillar, CHV–Chevron, DD–Du Pont, DIS–Walt Disney, EK–Eastman Kodak, GE–General Electric, GM–General Motors, GT–Goodyear Tire, IBM–IBM Corp., IP–International Paper, JPM–Morgan JP, KO–Coca Cola, MCD–Mc-Donalds, MMM–Minnesota Mining, MO–Philip Morris, MRK–Merck & Co, PG–Procter & Gamble, S–Sears Roebuck, T–AT&T, TX–Texaco, UK–Union Carbide, UTX–United Tech, WX–Westinghouse, XON–Exxon, and Z–Woolworth. In both panels, we reproduce the data shown in Ref. [333].

so doing, we must avoid the situation in which adding an edge would produce a cycle in the graph, which would spoil the tree structure. Therefore, we skip this 'bad' edge and proceed to the next smallest distance. When $N_{sto} - 1$ edges are added, we know that the tree is spanning, and we stop here.

The structure of the minimum spanning tree shows dependencies between different stocks. Two stocks whose prices follow very similar courses in time are placed next to each other in the tree. An example of a minimum spanning tree constructed from the correlations of 30 important stocks is shown in Fig. 1.11. We observe a big hub in the middle, which is the General Electric (GE) stock. Many other stocks are joined directly to it, which means that GE can be considered as the centre of a cluster of stocks which follow each other on the market. There are also several minor hubs, like Coca Cola, International Paper, and Du Pont. These can be readily interpreted as centres of secondary clusters. Therefore, the minimum spanning tree reveals the structure of emergent business sectors, not necessarily related to a priori classification of businesses according to their fields of activity. The a posteriori classification according to the minimum spanning tree is also much more relevant for the investors, as it tells which stocks are tied together and which ones are more independent. From the point of view of diversification of the portfolio, it is useful to invest simultaneously in stocks far apart on the tree, while investing in neighbours on the tree makes little sense.

Ultrametricity

There is also another way to visualise the same data. The function $d(a, b)$ is not the only way of defining a distance on the set of stocks. If we find the minimum spanning tree, we can subsequently define a new distance $d_{\mathrm{UM}}(a, b)$ using the following procedure. For a pair of vertices a and b we find a path on the minimum spanning tree connecting a with b, i.e. the sequence of edges on the tree along which we can walk from a to b (and back, of course). Each edge along this path joining, say, c with d, bears the distance $d(c, d)$. We can find the maximum of all the distances along the path from a to b and call this number $d_{\mathrm{UM}}(a, b)$.

This function again has all the properties we require from a distance. Moreover, the triangle inequality is replaced by a stronger statement that, for any triple a, b, and c, the following inequality holds:

$$d_{\mathrm{UM}}(a, b) \leq \max(d_{\mathrm{UM}}(a, c), d_{\mathrm{UM}}(c, b)). \tag{1.42}$$

A set of points endowed with a distance satisfying such inequality is called ultrametric space [363]. Every ultrametric space is also a metric space, but the opposite does not generally hold. The most important property of every ultrametric space is that its points can be organised hierarchically. It is possible to draw a tree which visualises such hierarchy, so that the points of the space—in our case the stocks—are end points of the tree. The branches of the tree are drawn in such a way that the ultrametric distance $d_{\mathrm{UM}}(a, b)$ between stocks a and b is just the length from the end points to the point inside the tree where the branch ending at a departs from the branch ending at b. We show in Fig. 1.11 a tree constructed in this way, from the same data as the depicted minimum spanning tree. We can see, for example, a closely tied cluster of Exxon-Texaco-Chevron. On the other hand, the stock of Sears Roebuck seems to be related only loosely to the rest of the other stocks.

The analysis of the dependencies among stocks can be made even more detailed. Instead of correlations between returns at the same moment, as in (1.40), we can define the two-time correlation function $M_{ab}(\Delta t) = \langle \widetilde{X}^{(a)}_{t+\Delta t} \widetilde{X}^{(b)}_{t} \rangle$. If we find that this function has its maximum at a positive value of Δt, we can conclude that the movements of the prices of stock a are delayed with respect to the fluctuations in the stock b. There is a directed influence from b to a. This way we can draw a directed graph of mutual influences, showing a complex cascade of dependencies in the stock market [364–366].

1.4.2 Spectrum and localisation

Density of eigenvalues

A completely different method of identifying important clusters of stocks is based on spectral analysis of matrix M [367–400]. It is inspired by quantum mechanics, where the set of eigenvalues of the Hamiltonian determines all allowed values of energy. On the one hand, from the analysis of spectra of heavy atomic nuclei, it evolved into the abstract theory of random matrices [401] and on the other hand, the study of electronic spectra of organic compounds [402] gave rise to the spectral graph theory [403]. A brief explanation can be found in Box 1.1.

Eigenvalue problem and inverse participation ratio   Box 1.1

Let A_{ij} be elements of an $N \times N$ matrix. If we find a number λ and a column vector with elements e_j such that $\sum_j A_{ij} e_j = \lambda e_i$, then λ is an eigenvalue of the matrix A, and e is the corresponding eigenvector. It is chosen so that $\sum_i e_i^2 = 1$. The set of eigenvalues is the spectrum of the matrix A. In quantum mechanics, the system (e.g. an atom or an electron in a metal) is characterised by a matrix called Hamiltonian, and the eigenvalues of this matrix are the only allowed energy levels of the system. We shall often deal with matrices whose elements A_{ij} are random variables. Then its eigenvalues $\Lambda_1, \Lambda_2, \ldots, \Lambda_N$ are also random. For any finite N, the density of eigenvalues is the ensemble of δ-functions $P_\Lambda(\lambda) = \frac{1}{N} \sum_{i=1}^{N} \delta(\lambda - \Lambda_i)$, which can, however, approach a continuous function in the limit $N \to \infty$. We also define the integrated density of eigenvalues $P_\Lambda^>(\lambda) = \frac{1}{N} \sum_{i=1}^{N} \theta(\Lambda_i - \lambda)$.

The example of an electron in a solid material reveals the importance of the quantity called inverse participation ratio. If the eigenvalue λ corresponds to the eigenvector e, then inverse participation ratio is defined as

$$q^{-1}(\lambda) = \sum_i e_i^4.$$

An electron which can move freely in the entire volume of the solid is said to be in a delocalised state. If we take its eigenvector and calculate the inverse participation ratio, it decreases with increasing size of the system as $q^{-1} \sim N^{-1}$. On the contrary, if the electron remains trapped within a limited area, its state is called localised. Inverse participation ratio does not tend toward zero but has a finite limit for $N \to \infty$.

The matrix elements M_{ab} are scalar products (1.40) of fluctuating empirical data. Thus, they are to a large extent random numbers. The matrices composed of random numbers, the so-called random matrices, are the subject of a special discipline, the random-matrix theory. It is relatively difficult, but good books are available, which cover the essential results, e.g. Ref. [401]. For a large, real, $N \times N$ symmetric matrix with zero diagonal, whose off-diagonal elements are independent and identically Gaussian distributed random variables with zero mean, the density of eigenvalues approaches, for $N \to \infty$, a semi-circular form $P_\Lambda(\lambda) = \sqrt{\sigma^2 - \lambda^2}/(\pi \sigma^2)$. Results are also known for other distributions and other kinds of matrices, but we shall not mention them here, except for a rather special type which is relevant for the correlation matrices we investigate.

Marčenko-Pastur and beyond

The point is that the elements M_{ab} are not all mutually independent, because they are constructed from N time sequences of returns. As a model for the correlation matrix, we introduce a matrix $A_{ij} = \sum_{k=1}^{P} D_{ik} D_{jk}$, where the elements of the $N \times P$ rectangular matrix D are already all independent Gaussian variables with variance $1/N$. We may think of k as time and consider D_{ik} and D_{jk} as time series for stocks i and j. It is possible to calculate the density of eigenvalues of A in the limit $N \to \infty$, $P \to \infty$, with $r = P/N$ kept fixed. The result is the so-called Marčenko-Pastur density [404]

$$P_{\Lambda \; \mathrm{MP}}(\lambda) = \frac{1}{2\pi r \, \lambda} \sqrt{4r \, \lambda - (\lambda + r - 1)^2} \qquad (1.43)$$

characterised by a unique parameter r. This is the first benchmark against which the spectra of empirical correlation matrices are scrutinised. An important feature is that the Marčenko-Pastur density is non-zero only within the interval $\lambda_- < \lambda < \lambda_+$, where

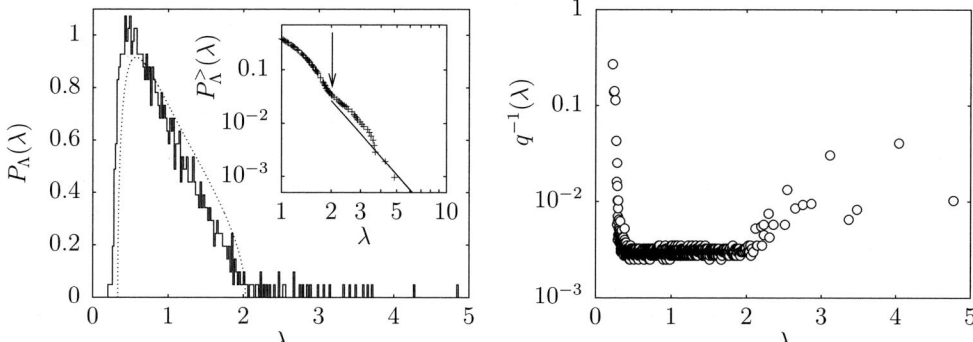

Fig. 1.12 In the left panel, the histogram of eigenvalues is shown for the correlation matrix for 1000 large US companies traded in 1994–1995. The time series of returns were calculated using time distance $\delta t = 30$ min. The dotted line is the Marčenko-Pastur eigenvalue density with $r = 0.18$. In the inset, cumulative distribution of eigenvalues. The line is the power law $\propto \lambda^{-3.5}$, and the arrow indicates the upper edge of the Marčenko-Pastur density. In the right panel, the inverse participation ratio is shown for the same matrix. Note that the localised states lie outside the upper and lower edges of the Marčenko-Pastur density. The data in both panels were extracted from Ref. [367].

the endpoints are $\lambda_\pm = (1 \pm \sqrt{r})^2$. For an improvement over the Marčenko-Pastur density, see Ref. [405].

We show in Fig. 1.12 the density of eigenvalues for a correlation matrix of 1000 stocks, compared with a fit by Marčenko-Pastur density (1.43). We can clearly see that in the empirical data there are many eigenvalues beyond the Marčenko-Pastur bounds. The most important difference occurs at the upper end of the spectrum, as demonstrated by the integrated density in the inset of Fig. 1.12. While the Marčenko-Pastur density ends sharply at $\lambda = \lambda_+$, the empirical tail extends beyond this edge and decays approximately as a power-law $P_\Lambda^>(\lambda) \sim \lambda^{-3.5}$. This crucial difference is partially explained by the fat tails in the return distribution [406–411]. Indeed, fat tails in the distribution of random elements D_{ik} may be translated into a power-law tail modifying the Marčenko-Pastur form (1.43). The results of [409, 411] indicate that, if the distribution of returns is characterised by a power-law tail with exponent α, the density of eigenvalues should decay as $P_\Lambda(\lambda) \sim \lambda^{-\eta-1}$, where $\eta = \alpha/2$. Unfortunately, this is at odds with the empirical findings that $\alpha \simeq 3$ and $\eta \simeq 3.5$. It seems that other mechanisms are also at play, influencing the exponent of the power-law tail.

Inverse participation ratio

An even more important difference from Marčenko-Pastur behaviour is found in the localisation properties of the eigenvectors, as measured by the inverse participation ratio, denoted $q^{-1}(\lambda)$. We can see in Fig. 1.12 that the inverse participation ratio is very low inside the Marčenko-Pastur edges, but outside these bounds inverse participation ratio is much greater. This is an indication of localised states, both in the lower and the upper tail of the spectrum. These states were used to identify business sectors

[374, 375, 381, 382, 386]. The heuristic argument behind this method follows. Imagine that the stocks are points in an abstract space, and there is a walker in that space, jumping randomly from one point to another. We require that the probability of the jump from stock a to stock b is proportional to the correlation M_{ab} between these two stocks. We can ask what the probability is that, after some time, the walker is found at this or that point. The time evolution of this probability is described by the diffusion matrix $\overline{M}_{ab}$ which only slightly differs from the correlation matrix M_{ab}. The difference lies in the diagonal elements. While the correlation matrix has $M_{aa} = 1$, for the diffusion matrix we require conservation of probability, which results in the condition $\overline{M}_{aa} = -\sum_{b(\neq a)} \overline{M}_{ba}$.

If we neglect for the moment the difference between M and $\overline{M}$, we can (vaguely) argue that those eigenvectors of M which are localised, i.e. only have a few non-negligible elements, correspond to such states in which the walker remains for a prolonged time within a certain small set of points. These are just the points at which the eigenvector has larger elements. The set of the stocks represented by these points is then interpreted as a cluster, or business sector.

Surely such argumentation is far from being exact. However, when used in practice, it does reveal clusters which typically belong to the same type of business [375]. This gives strong empirical support to the method for finding the clusters through localised eigenvectors. Let us also note that this method was successfully used for finding clusters in social networks [412].

1.4.3   Clustering in correlation networks

Which clusters?

The groups of stocks highlighted by localisation of eigenvectors are by definition small compared to the whole set of stocks. Very often, in fact most often, we ask a different question. What are the clusters of strongly tied stocks into which the ensemble can be meaningfully divided? What are the modules which make up the stock market?

In this formulation of the problem, the clusters, or modules, are expected to cover a significant portion of the stocks. Thus, we are not looking for small, compact groups as we were when using the localisation method. Here we want to see large sectors. Typically, we expect to find a few large and important clusters, accompanied by many small ones with a low level of significance.

Threshold algorithm

The problem of finding functional modules and structural clusters is extensively elaborated in the theory of random networks [413]. Not all of them are suitable for finding clusters in the 'network' of correlations among stocks. It is not a proper network, as all stocks are interconnected with all others; only the weights and signs of the connections vary. But using a procedure of suppressing weak ties and fixing the strong links we can obtain a network, or graph, which grasps some important structures. If that procedure is smart, it visualises the clusters immediately. Perhaps the easiest and still very efficient way to obtain such a graph is the filling method. We start with an ensemble of N_{sto} vertices, each representing one stock, and 'pour' into it links between stocks until the number of edges L reaches the prescribed fraction p of all possible edges, so $L = p N_{\text{sto}}(N_{\text{sto}} - 1)/2$. The order in which the edges are added is given by the absolute value of the correlations M_{ab} between stocks, i.e. the largest $|M_{ab}|$ is added first, the second largest next, etc. Therefore, the construction of the graph is in a sense similar to the way we obtained the minimum spanning tree, but now we do not care about the appearance of loops, and the number of edges to be added is a free parameter.

We show in Fig. 1.13 such a graph, replotted from the data given in Ref. [386]. Different symbols for the vertices correspond to different business sectors, according to the Forbes classification. Generally, the result fulfils our expectations. There are several relatively large clusters and many small ones, mostly composed of an isolated stock. Within each of the big clusters, we find stocks from the same sector, with only a few exceptions. This confirms that the clusters bear sensible information.

The sensitive point of this clustering procedure is determination of the most suitable value of p. Sometimes, instead of prescribing the target concentration of edges p, we fix the threshold value of the correlation M_{thr} and connect by edges all pairs of stocks with a correlation stronger than the threshold $|M_{ab}| > M_{\text{thr}}$ [414]. The two approaches are of course equivalent. Both suffer from the uncertainty of a threshold variable, either p or M_{thr}. Reliable information may come only from comparison of the resulting cluster structure for a series of different thresholds [386]. Such a comparison, supported by independent information, for example on the business sector classification, can help us establish the proper value of the threshold.

Potts-model algorithm

Among other possible methods for finding clusters, let us briefly mention only the procedure based on the Potts model [415, 416]. This method is routinely used for analysing large sets of data in medicine [417, 418].

In this scheme, each stock should belong to one of q groups. There are many ways to distribute the stocks among groups. We define a function, which may be called energy or cost, for each of these distributions. Distributions with low energy (or cost) should be favoured. To calculate the energy, we go through all pairs of stocks and add $-M_{ab}$ if the stocks belong to the same group. Otherwise, the contribution is zero.

This done, we make the usual calculations of statistical physics with the energy function defined above, which can be formally written as

$$E\big(\sigma(1), \sigma(2), \ldots, \sigma(N_{\text{sto}})\big) = -\sum_{a<b} M_{ab}\delta_{\sigma(a),\,\sigma(b)} \tag{1.44}$$

where $\sigma(a) \in \{1, 2, \ldots, q\}$ denotes the group to which the stock a belongs, and $\delta_{\sigma,\,\sigma'}$ is the usual Kronecker delta. Minimisation of energy prefers that all stocks belong to the same group. This is the low-temperature ordered state of the system. At high temperatures this state will be destroyed by thermal fluctuations. If the correlations among stocks are such that the stocks can be split into clusters, we can find a third state at an intermediate temperature in which the system is ordered within clusters, i.e. all stocks belong to the same group but different clusters belong to different groups. To find the clusters we must decide whether the intermediate regime exists at all; and if it

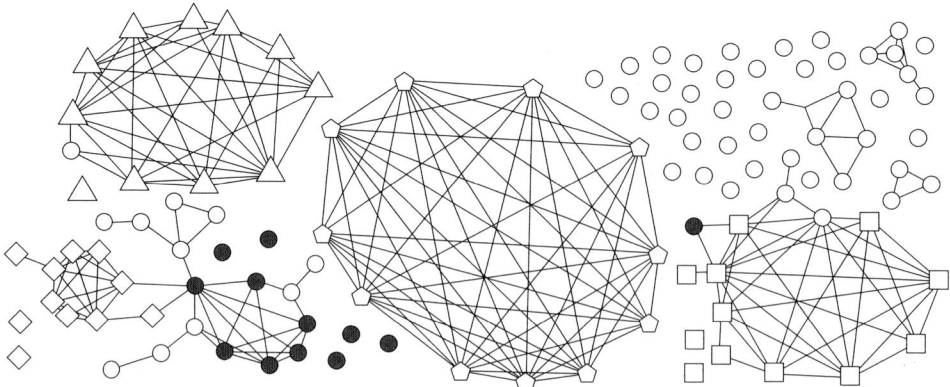

Fig. 1.13 Network of companies constructed from the correlation matrix of approx. 100 stocks traded on the New York Stock Exchange in the period 1997–2000. The links were added in the order of their strength until the fraction of links reached the value $p = 0.03$. Different symbols denote different business sectors according to the Forbes classification (www.forbes.com), namely: basic materials (□), health care (◇), energy (○), electric utilities (△), financial and conglomerates (●), and other sectors ○. The network was plotted from the data shown in Ref. [386].

does, what interval of temperatures it spans. To this end, we calculate the fluctuation of the energy; more precisely, how the average fluctuations depend on temperature. The temperature dependence typically has at least one maximum. Each maximum marks the temperature at which the state of the system changes. It is not a proper phase transition, because the system is finite, but it resembles a phase transition to some extent. The desirable situation occurs when there are two maxima. The interval between these maxima represents just the intermediate regime we are looking for. Now, the only remaining task is to look at the typical distribution of stocks among groups in the intermediate regime, and the clusters are found.

1.5 What remains

We shall stop here. A comprehensive review of the empirical basis for econophysics would be much too much [74, 114, 371, 372, 419–506]. Let us only mention a few more or less randomly chosen themes we were not able to cover in depth but want to bring to the reader's attention.

In the study of autocorrelations of price signal and correlations between different prices, many subtle phenomena are revealed [507, 508]. One of them is the Epps effect [509], which is expressed as a decrease of correlations between two different stocks when the sampling frequency increases. Other interesting phenomena were found in the investigation of the time evolution of correlation matrices [510–512]

The prices of stock and commodities are not the only economic signals studied. Others include interest rates [513], real estate prices [514, 515], duration of calm-time periods [516], duration of recessions and periods of prosperity [517–524], and business cycles [525]. The inverse statistics of the price signal investigates the time needed to reach a prescribed increase or decrease of price [526–531]. The behaviour of prices after a stock-market crash is also investigated [532], and a similarity to the aftershocks following an earthquake (governed by the Omori law) was observed [533]. In a more peculiar area, the laws governing the prices of precious stones [534] were empirically observed.

The empirical studies often apply more sophisticated methods than those mentioned in this chapter, for example the n-Zipf analysis [213, 214, 535], detrended cross-correlation analysis [536], or mapping on the non-extensive statistical physics [537, 538].

A very important area, although not yet satisfactorily explored, is the phenomenology of bubbles and crashes. The generic pattern of booms followed by crashes has been watched throughout history [539–558]. To mention some more specific studies, postage-stamp bubbles [559] and responses to 11 September 2001 [560] were investigated. In Ref. [561] the generic patterns characterising the course of economic recession were identified. To this line of research, we should also add the broad literature on the log-periodic precursors of stock-market crashes [215, 562–574]. This is intimately related to the finding that the truly extreme events on the market are 'outliers', i.e. they are governed by different mechanisms than the ordinary, however large, fluctuations [552, 566].

Finally, let us stress that our list is by no means representative, and many other results would certainly deserve the same attention. However, we must hasten now to the core of this book, which covers building and investigating models for the above-described empirical phenomena.

2
Stochastic models

Random factors play a significant role in any aspect of economic activity. We cannot eliminate the influence of weather; people fall ill at unpredictable moments; our decisions are overbalanced by quantum fluctuations in the synapses joining neurons in our heads. All of this makes the processes at work indeterministic, and any model of what is happening in the economy and in human society in general must essentially be a stochastic process. The econophysics modelling is therefore much different from calculating electrical conductivity of carbon nanotubes or solving the stability of a 100-storey building. These are essentially deterministic situations, although some casual factors may be taken into account as corrections.

We may also say that these problems are 'merely' complicated, while economics is complex. The two words have a common stem, but denote different things. Sending a human crew to Mars and bringing it back again is an extremely complicated task, but it is not complex. It may need a lot of work, a huge amount of money, and ranks of dedicated people. But still, in principle, all that is to be done relies on well-established knowledge and experience. On the other hand, predicting precisely what happens when one damages a certain protein within the cell is a complex problem. Previous knowledge can give us some hints, but almost by definition every protein has a distinct function and different interactions; so we must investigate all consequences of its malfunction each time anew. And if we want to generalise the complex interactions of all proteins throughout the immense variety of living organisms, we are deeply bogged down in the science of complexity.

Econophysics also contrasts with, for example, modelling of water flowing along a ship's sides, or a flame streaming from a gas burner. Seemingly similar in terms of poor predictability, the latter situations follow the paradigm of deterministic chaos rather than that of stochastic evolution. Certainly, chaotic systems can be considered as complex in a broader sense. For some time economics was thought to be chaotic as well, all its complexity stemming from deterministic, but hardly predictable, rules. Nowadays, this attitude has become rather rare, as people recognise the prevalence of true chance over deterministic chaos in economics and social phenomena.

Therefore, the proper language for econophysics modelling is that of probability and stochastic processes. The more random they are, the more faithful are the models born in the theoretician's brain. But it is useful to make a distinction between two kinds of these, which we can call 'bare' and 'involved'. To see what we have in mind consider, e.g. fluctuations in the price of silver. We can suppose that all information on the external influences are contained in the past record of prices, and therefore changes in the future are determined by the past, plus stochastic noise. The noise may appear

Random variable $\boxed{\text{Box 2.1}}$

can be intuitively understood as a mathematical object X assuming random values. To give a random variable a sensible meaning, we must be able to calculate the probability that X falls within a given interval. All information on X resides in its probability distribution function, which is just given by the probability
$$F_X(x) = \text{Prob}\{X \le x\}.$$
Obviously, the distribution function is non-decreasing, and its values lie within the interval $[0, 1]$. Discrete random variables have distribution functions composed of a countable number of jumps. On the contrary, if the distribution function can be expressed as an integral,
$$F_X(x) = \int_{-\infty}^{x} P_X(x')\mathrm{d}x',$$
we call X a continuous random variable and the function $P_X(x)$ its probability density. Mostly we work with several random variables simultaneously. For random variables X and Y we define their joint probability distribution function
$$F_{X,Y}(x, y) = \text{Prob}\{X \le x, Y \le y\}.$$
The key concept of the theory of probability is independence. We call the random variables X and Y independent, if their joint probability distribution can be factorised
$$F_{X,Y}(x, y) = F_X(x)\,F_Y(y).$$
If both of them are continuous, their joint probability density can also be factorised in the same manner: $P_{X,Y}(x, y) = P_X(x)\,P_Y(y)$.

in various ways, but essentially it is constructed from a sequence of independent and statistically identical random events. The complexity stems from the possible coupling between the random variables describing the events and the past price. For example, the price may follow a random walk, but the length of the steps may vary. It can grow if the position fluctuated much in the past, but remain small if the past fluctuations were low. This way, we introduce a kind of positive feedback in the price fluctuations, leading to non-trivial consequences. This type of model deserves the name 'bare' stochastic process, as we do not care much about the source of the randomness or how and why the coupling with past values of the process comes into play. This chapter will be entirely devoted to these 'bare' stochastic processes.

On the contrary, the 'involved' processes focus precisely on those mechanisms taken for granted in the 'bare' processes. We try to build a model mimicking the true sources of the noise and emulating the real mechanisms leading to feedback loops. Necessarily, we make assumptions about hidden variables and processes, fluctuations in the silver price being only the tiny tip of an iceberg. Generically, these models make some hypotheses on the nature and behaviour of economic agents, whether they are customers, brokers, housekeepers, or multinational companies. These 'involved' processes will be investigated in later parts of this book, starting with Chap. 3.

2.1 Random walk and beyond

The simplest stochastic processes do not look back. Whatever the past was, the random influences do not care. The most important example of these memoryless processes is the random walk. At each time, the quantity in question increases or decreases by a random amount, the differences at all times being statistically independent. No wonder that this simple dynamics was tried first as a model for price fluctuations.

Stochastic process Box 2.2

Generally, a stochastic process is a collection of random variables $\{X_t | t \in T\}$ parameterised by time t, which may be continuous, $T = [0, \infty)$, or discrete, $T = \{0, 1, 2, \ldots\}$. Full information on the stochastic process is contained in the joint probability distribution of all variables X_t. This is so complicated an object that we rarely compute it in practice. Mostly we ask about the time dependence of the moments $\langle X_t^k \rangle \equiv \sum_x x^k \operatorname{Prob}\{X_t = x\}$, and correlation functions $\langle X_t X_{t'} \rangle \equiv \sum_{xx'} x\, x' \operatorname{Prob}\{X_t = x \wedge X_{t'} = x'\}$. Often we need the probability distribution at one specific time $P_t(x) \equiv \operatorname{Prob}\{X_t = x\}$ and its limit when the time goes to infinity. If this limit exists, it provides the stationary one-time distribution for the process in question.

2.1.1 Bachelier's programme

Probabilistic approach

If anybody merits the fame of the founding father of stochastic modelling in economics, it is certainly Louis Bachelier. In his thesis *Théorie de la spéculation* [102], published in 1900, he notices that the changes in the prices quoted at the Bourse occur very often without any apparent relation to external events. This prevents the dynamics of the Bourse from becoming an exact science. On the other hand, the theory of probability makes it possible to establish the probabilistic laws which must be obeyed by variations of the prices. Knowing these laws, we can reliably predict how often various events happen within a prolonged period of time.

To be sure, Bachelier stresses that there are two sources of the random movements of the Bourse. The first one cannot be determined a priori without knowledge of the actual state of the market, while the second has its origin in specific facts which are themselves unpredictable, but once they occur, their influence on the prices can be more or less reliably established. It is the latter randomness the speculators rely upon, while the former is the subject of mathematical study. Therefore, we can see from the very beginning that the mathematical theory Bachelier tries to develop cannot make the people playing with their fortune on the Bourse rich, just as the classical theory of probability cannot help gamblers win in a casino, although it can say a lot of things about the outcomes of roulette. Mathematics can provide understanding, thus giving you a comparative advantage, but not an unearned gain.

Towards Gaussian distribution

At each time the price of a commodity traded on the Bourse is a random variable. Altogether, the sequence of prices at all limes t makes up a stochastic process Z_t. (If obscure, see Boxes 2.1 and 2.2, which explain random variables and stochastic processes, respectively). The difference of the price from its starting value is again a stochastic process $X_t = Z_t - Z_0$; and Bachelier asks, what should the probability density be for the price difference at time t? It can be written as

$$P_{X,t}(x) = \frac{\mathrm{d}}{\mathrm{d}x} \operatorname{Prob}\{X_t \le x\}. \tag{2.1}$$

The first and most essential assumption is that the price increments in subsequent time intervals are independent, i.e. $X_{t',t} = Z_{t'} - Z_t$ and $X_{t'',t'} = Z_{t''} - Z_{t'}$ are independent random variables. Furthermore, Bachelier assumes that the Bourse is stationary, so that the probability distributions do not depend on the absolute beginning of time, but only on time differences. So, the variables $X_{t',t}$ and $X_{t'',t'}$ have the same distributions as variables $X_{t'-t}$ and $X_{t''-t'}$, respectively.

Evidently, the price change from time 0 to t is the sum of two independent random variables $X_t = X_{t,t'} + X_{t'}$ for any intermediate time t', and to find the probability density for it we need to calculate the convolution

$$P_{X,t}(x) = \int P_{X,t-t'}(x - x') \, P_{X,t'}(x') \, \mathrm{d}x'. \tag{2.2}$$

We assume that the probability density has the same general form for all time differences. This statement is somewhat vague but we can give it a precise meaning, requiring that the function does not change when we simultaneously rescale time and price. More precisely, we desire that there exists a function of one variable $G(u)$ so that

$$P_{X,t}(x) = t^{-1/\mu} \, G(x \, t^{-1/\mu}) \tag{2.3}$$

for all times, with a suitably chosen exponent μ. The scaling assumption (2.3), although not stated with full emphasis, was one of the main ingredients of Bachelier's work, and, as we have seen in Chap. 1, it was not until the 1990s that its validity as well as its limitations were put on a solid empirical basis.

Let us turn now to slightly more technical questions. In fact, it is much more convenient to work with the characteristic function of the variable X_t, instead of the probability density, as the convolution in (2.2) becomes a simple multiplication. In terms of the function $G(u)$, or rather its Fourier transform $\widetilde{G}(w)$, we can write

$$\widetilde{G}(t^{1/\mu}w) = \widetilde{G}((t - t')^{1/\mu}w)\widetilde{G}(t'^{1/\mu}w), \tag{2.4}$$

and the only remaining task is to find a function satisfying this functional relation for any choice of times t and t'.

It needs only a little guesswork to find a solution. It is easy to see that the normal distribution

$$G(u) = \frac{1}{\sqrt{2\pi}\,\sigma} \, \mathrm{e}^{-\frac{u^2}{2\sigma^2}} \tag{2.5}$$

satisfies Eq. (2.4) with $\mu = 2$. This is the result announced by Bachelier: the distribution of price changes is Gaussian.

Lévy-stable distributions

However, it is clear that other solutions of the functional equation (2.4) can also be found. To this end we convert the multiplicative formula (2.4) into an additive one taking the logarithm $g(v) = \ln \widetilde{G}(v^{1/\mu})$, so that

$$g((t + t') \, w^\mu) = g(t \, w^\mu) + g(t' \, w^\mu). \tag{2.6}$$

Differentiating first with respect to t and then with respect to t', we find that the second derivative of $g(v)$ is zero for any v. Moreover, taking the limit $w \to 0$ we obtain

Central limit theorem   Box 2.3

states that the probability distribution of a sum $S_n = \sum_{i=1}^{n} X_i$ of many independent and identically distributed random variables X_i approaches the normal distribution, with the only condition that the distribution of the individual variables X_i is not 'too wild'. More precisely, if the average $\langle X_i \rangle = \mu$ and the second cumulant $\langle (X_i)^2 \rangle - \langle X_i \rangle^2 = \sigma^2$ are finite, then

$$\lim_{n \to \infty} \text{Prob}\{S_n - n\mu \le \sqrt{n}\,\sigma\,x\} = \int_{-\infty}^{x} e^{-\frac{1}{2}y^2} \frac{dy}{\sqrt{2\pi}}.$$

$g(0) = 0$. Therefore, the general solution of (2.6) is the linear function $g(v) = -av$ with a constant a. However, we must be careful when extending the result to negative values of v, because the power v^μ may not exist. The problem can be cured by always taking the modulus of v, at the price that the function $\widetilde{G}(w)$ may not be analytic at $w = 0$. Having this in mind, we conclude that the most general form of the Fourier transform of the distribution function is the stretched exponential

$$\widetilde{G}(w) = e^{-a|w|^\mu}. \tag{2.7}$$

For $\mu = 2$ we recover the previous result, but any value of $\mu < 2$ should work equally well. For example, for $\mu = 1$ we get a simple exponential $\widetilde{G}(w) = e^{-a|w|}$ which is the Fourier transform of the well-known Cauchy distribution

$$G(u) = \frac{a}{\pi} \frac{1}{u^2 + a^2}, \tag{2.8}$$

also called the Lorentz distribution by physicists. However, this is only the simplest-ever example of the so-called Lévy-stable distributions, which all share the common feature that they keep their form upon adding the underlying random variables, i.e. if X_1 and X_2 are independent random variables with the same Lévy distribution, then the sum $X_1 + X_2$ again has the same Lévy distribution. For $\mu < 2$ the Lévy distributions have another very important property. They decay as a power for large values of the argument

$$G(u) \simeq A_\pm |u|^{-1-\mu} \quad \text{for} \quad u \to \pm\infty \tag{2.9}$$

where $A_\pm$ are constants. For the general value of the parameter μ, the Lévy distribution cannot be expressed using only elementary functions, but it is always possible to write it as an inverse Fourier transform of a stretched-exponential function, as we have seen above. For further information on Lévy distributions see also Box 7.1.

Random walk

Thus, the ensemble of all Lévy distributions, together with the Gaussian (or normal), constitute all possible solutions of the problem posed by Bachelier, to find a distribution of price changes which preserves its form as time proceeds onward. We may ask why only the normal distribution was accepted as the correct result. Most probably it was just the simplest possibility. Moreover, as Bachelier notes, the same result can be obtained by another, rather intuitive procedure. One of the principles Bachelier postulates about the movements of the stock market states that the true price, i.e.

price minus the long-time trend, rises or decreases with exactly the same probability. If the changes within a very short time interval Δt can assume only two values $\pm \Delta x$, the price performs a symmetric random walk with step size Δx. If there are n steps total in such a walk, the probability that k of them will go up, so that the walk moves the distance $X_{n\Delta t} = (2k - n)\Delta x$ from the initial position, is given by the binomial distribution

$$\text{Prob}\{X_{n\Delta t} = (2k - n)\Delta x\} = \frac{1}{2^n}\binom{n}{k}. \tag{2.10}$$

If we observe the price movements on a 'mesoscopic' time scale, where individual tiny price jumps are invisible but the relevant price changes are relatively slow, we can make the limit of continuous time, $\Delta t \to 0$, $n \to \infty$, with $t = n\Delta t$ and $(\Delta x)^2/\Delta t = 2\sigma^2$ kept constant. Then, the binomial coefficient turns into a Gaussian

$$\frac{1}{2^n}\binom{n}{k} \simeq \sqrt{\frac{2}{\pi n}}\, e^{-\frac{(2k-n)^2}{2n}}, \tag{2.11}$$

and the binomial distribution (2.10) for discrete variable $X_{n\Delta t}$ converges to the normal one, Eq. (2.5), for its continuous counterpart $U = X_t/\sqrt{t}$. This is one of the basic results of elementary probability theory. It also exemplifies the well-known and deep piece of knowledge called the central limit theorem (see Box 2.3). The elementary changes in price may not have the same size. Indeed, we know nearly nothing about the minute influences pushing the stock market a little up or down. The point is that we can well do without that knowledge. Whatever are the elementary movements which compose the price change at longer time scales, they result in the ubiquitous normal distribution, except for some 'pathological' cases, neglected for decades. Nowadays, we know that just these pathologies constitute the essential (and most interesting) features of price fluctuations.

Empirical hints

Bachelier gives numerous examples of how the theory can predict empirical findings, if we know just the parameters of the normal distribution of price changes, i.e. the average growth and the volatility. The simplest one is perhaps calculation of the interval of price changes such that the normal distribution predicts equal probabilities $1/2$ of finding the price inside and outside the interval. What is interesting in Bachelier's numbers is the systematic deviation from the prediction. For the time interval of one month, he finds that the price change lies outside in $\frac{27}{60} = 0.45$ of the cases recorded, instead of 0.5, and for the interval of one day this fraction is even lower and equals 0.4387. If Bachelier had used the Lévy-stable distributions instead of the Gaussian, the agreement would have been better, and he would have gained immortal fame as the discoverer of fat tails in price fluctuations!

2.1.2 Geometric random walk

There is one thing in Bachelier's theory which is serious enough not to be cast aside. If the price moved as a random walk, there would be no guarantee that it would remain positive. Neither does use of the Lévy distributions cure the flaw. Although negative

---

Stochastic differential equations | Box 2.4

They look similar to normal (deterministic) differential equations but there are some crucial differences. Any given stochastic differential equation defines a stochastic process X_t, or a family of processes, distinguished by initial conditions prescribed for the solution of the stochastic differential equation. The time is assumed to be continuous, $t \in [0, \infty)$, and the solution is considered to be a limit of processes obtained by discretisation of the time $t = i\Delta t$, $i = 0, 1, \ldots$. (We shall always suppose $\Delta t > 0$.) The essential ingredient of all non-trivial stochastic differential equations are the independent increments ΔW_t, representing the sources of external noise. They are random variables, independent as long as they correspond to different times t. We typically assume that their mean value is zero $\langle \Delta W_t \rangle = 0$. An example of a discretized equation is
$$\Delta X_t \equiv X_{t+\Delta t} - X_t = a(t)\Delta t + b(t)\Delta W_t$$
with given functions $a(t)$ and $b(t)$. The problem with taking the limit $\Delta t \to 0$ in this equation is that, contrary to normal differential equations, we do not possess any sensible recipe for defining the 'derivative' $\lim_{\Delta t \to 0} \Delta W_t / \Delta t$. This fact has two main consequences. The first one is rather formal. The standard notation of stochastic differential equation corresponding to the $\Delta t \to 0$ limit of the above discretised process is
$$\mathrm{d}X_t = a(t)\mathrm{d}t + b(t)\mathrm{d}W_t,$$
which may seem peculiar at first sight but it is actually both rigorous and useful in practice.

The second consequence is deeper. The quantity ΔW_t does not behave as an ordinary differential in the limit $\Delta t \to 0$. Instead, its square is exactly $[\mathrm{d}W_t]^2 = D\,\mathrm{d}t$ where D, the diffusion constant, is the only parameter characterising the noise $\mathrm{d}W_t$. (We neglect the further complication that the constant D may in fact depend on t.) This looks really odd. But as soon as we get used to such ridiculous behaviour from the independent increments $\mathrm{d}W_t$, we have essentially all the machinery of stochastic differential equations in our hands. For more details we refer the reader to the books [575, 576].

---

prices did occur in various obscure privatisation campaigns in ex-communist countries in the East, in a normal economy it is sheer nonsense. Random walk could perhaps describe a short-time price evolution, but we must look for more adequate modelling on a wider time horizon.

Logarithm of price

Let us now remember M. F. M. Osborne, a physicist who worked on the superfluidity of thin films of liquid helium. In 1959 he suggested [111] that it is the logarithm of the price of a commodity which performs a random walk, not the price itself. He supported the idea with two complementary arguments. From an empirical point of view, he showed that the distribution of changes in log prices is symmetric around its maximum and that its shape is nearly Gaussian. From an abstract stance, Osborne claimed the applicability of the Weber-Fechner law, which states that equal ratios of physical stimuli, like sound frequency or light intensity, correspond to equal intervals of subjective sensation. If it is commonly accepted in physiological optics and acoustics, why not use the same principle when we describe the subjective impact of price changes?

If Z_t is the price at time t, Osborne's assumption implies that it is the difference of price logarithms $X_t = \ln Z_t - \ln Z_0$ that is normally distributed, which in turn means, by trivial substitution, that the price itself has log-normal distribution

Fokker-Planck equation | Box 2.5

is a partial differential equation governing time evolution of the one-time probability density $P_t(x) = \frac{d}{dx}\text{Prob}\{X_t \leq x\} = \langle\delta(X_t - x)\rangle$ for the process generated by a stochastic differential equation. A more precise notation for this function would be $P_{X,t}(x)$, indicating that the probability density in question corresponds to the random variable X. We shall in most cases omit the index X, unless it is necessary to avoid ambiguity. Let us show the derivation of the Fokker-Planck equation for the equation $dX(t) = a\,dt + dW(t)$, where the noise W_t has the diffusion constant D. For any twice-differentiable function $\phi(x)$ we have, in Itô convention, $d\phi(X_t) = \phi'(X_t)\,dX_t + \frac{1}{2}\phi''(X_t)\,[dX_t]^2 = \phi'(X_t)\,dW_t + \left[a\,\phi'(X_t) + \frac{1}{2}D\,\phi''(X_t)\right]dt$. So, $\partial_t\langle\phi(X_t)\rangle = \int \phi(x)\,\partial_t P_t(x)\,dx = \int \left(a\phi'(x) + \frac{1}{2}D\,\phi''(x)\right)P_t(x)\,dx = \int \phi(x)\left(-a\,\partial_x + \frac{1}{2}D(\partial_x)^2\right)P_t(x)\,dx$. As this identity should hold for any function $\phi(x)$, we obtain
$$\partial_t P_t(x) = -a\,\partial_x P_t(x) + \frac{1}{2}D\,(\partial_x)^2 P_t(x),$$
which is the Fokker-Planck equation for the above stochastic differential equation.

$$P_Z(z) = \frac{1}{\sqrt{2\pi}\,\sigma z}\exp\left(-\frac{(\ln z - \ln z_0)^2}{2\sigma^2}\right). \tag{2.12}$$

Nowadays, knowing all about fat tails and other subtle stylised facts, this result may seem irrelevant, but historically it had an enormous impact. The beginnings of almost all mathematical finance can be traced to Osborne's result, expressed formally as Eq. (2.12).

Stochastic differential equations

In the decades following the work of Osborne, continuous time finance modelling grew enormously [110]. The underlying mathematical apparatus relies on the very powerful machinery of stochastic differential equations (see Box 2.4).

Let us first investigate the simpler case advocated by Bachelier, which is the random walk. In the continuous time setup it becomes the Brownian motion. Denote Z_t the price at time t. The underlying stochastic differential equation is perhaps the simplest ever possible

$$dZ_t = \sigma\,dW_t, \tag{2.13}$$

where we supposed that the diffusion constant for the noise dW_t is $D = 1$ and we introduced instead the parameter σ, quantifying the amplitude of the price fluctuations. Then, obviously, $\langle dZ_t\rangle = 0$ and $\langle[dZ_t]^2\rangle = \sigma^2 dt$. We can write the corresponding Fokker-Planck equation for the distribution of the price at time t (see Box 2.5 if unclear)

$$\frac{\partial}{\partial t}P_t(z) = \frac{\sigma^2}{2}\frac{\partial^2}{\partial z^2}P_t(z). \tag{2.14}$$

A trained physicist immediately recognises the diffusion equation. No wonder, indeed, for Brownian motion is the microscopic mechanism of the macroscopic effect we call diffusion. Solving Eq. (2.14) requires specification of initial conditions, which depend on the particular question we pose to ourselves. If we want to know the probability that the price changes by a certain amount from time 0 to time t, the appropriate initial condition is $P_0(z) = \delta(z - z_0)$, where z_0 is the assumed price at time 0. We immediately get the solution, which is again the well-known normal (or Gaussian) distribution

$$P_t(z) = \frac{1}{\sqrt{2\pi t}\,\sigma} \exp\left(-\frac{(z-z_0)^2}{2\sigma^2\,t}\right). \tag{2.15}$$

This is Bachelier's result, as we expected.

The noise in Eq. (2.13) appears as an additive term. In each time interval it adds something to the random variable in question, in our case to the price. The price process suggested by Osborne is different. The change in price is proportional to the price itself, and the stochastic differential equation describing such situation is

$$\mathrm{d}Z_t = \sigma Z_t\,\mathrm{d}W_t. \tag{2.16}$$

We call this random influence multiplicative noise, as opposed to additive noise in Eq. (2.13).

Now we cannot avoid choosing between the Itô and Stratonovich conventions (see Box 2.6). It is generally accepted that stochastic differential equations describing economic processes should use the former convention. This leads to the following Fokker-Planck equation

$$\frac{\partial}{\partial t}P_t(z) = \frac{\sigma^2}{2}\frac{\partial^2}{\partial z^2}\left(z^2\,P_t(z)\right). \tag{2.17}$$

Before solving it, we shall try to extract some information directly from Eq. (2.16). Quite trivially, the average price $\langle Z \rangle$ does not change in time due to the Itô convention we adopted. Indeed, $\mathrm{d}\langle Z \rangle = \sigma\langle Z_t\,\mathrm{d}W_t \rangle = \langle Z_t \rangle\langle \mathrm{d}W_t \rangle = 0$. On the other hand, the logarithm of price does change, as

$$\mathrm{d}\langle \ln Z \rangle = \left\langle \frac{1}{Z}\mathrm{d}Z - \frac{1}{2Z^2}(\mathrm{d}Z)^2 \right\rangle = \sigma\langle \mathrm{d}W \rangle - \frac{\sigma^2}{2}\mathrm{d}t = -\frac{\sigma^2}{2}\mathrm{d}t. \tag{2.18}$$

This implies a (downward) trend in the logarithm of price. Such trend modifies the normal distribution we expect for the logarithm of price. In fact, the trend can be tuned by hand, introducing an additional term $\mu Z\mathrm{d}t$ on the right-hand side of Eq. (2.16). However, such modification does not bring anything essentially new and we omit it for the sake of simplicity. The explicit solution of the Fokker-Planck equation (2.17) is then

$$P_t(z) = \frac{1}{\sqrt{2\pi t}\,\sigma z} \exp\left(-\frac{\left(\ln z + \frac{\sigma^2}{2}t\right)^2}{2\sigma^2 t}\right). \tag{2.19}$$

We recovered the distribution postulated by Osborne, justifying to a large extent the assumption that the process underlying price fluctuations can be adequately modelled by multiplicative stochastic differential equations.

The conclusion we have just reached is the point of departure for a large segment of classical mathematical finance. The gem amongst its results is the Black-Scholes equation and the whole option pricing theory built around it. However, this topic is so well and amply covered in finance books [110] that we feel quite comfortable skipping it here.

Itô vs Stratonovich convention | Box 2.6

Imagine a simple stochastic differential equation $dX_t = X_t\,dW_t$, with noise dW_t characterised by the diffusion constant D. The problem we encounter here is that it is not clear in the right-hand side of the equation whether X_t and dW_t should be considered independent or not. To understand what is happening here we must return to the discretised equation and write the difference $\Delta W_t \equiv W_{t+\Delta t} - W_t$ in place of dW_t. Then if the factor X_t is taken exactly at time t even before taking the limit $\Delta t \to 0$, then it is independent of dW_t and, e.g. $\langle X_t\,dW_t \rangle = \langle X_t \rangle \langle dW_t \rangle = 0$. This is the Itô convention. However, there are other possibilities. If we take $X_{t+\Delta t/2}$ instead of X_t, we might naively expect the same result for $\Delta t \to 0$. This is wrong, though. Expanding in the Taylor series, we have $X_{t+\Delta t/2} \simeq X_t + \frac{1}{2}dX_t\,\Delta t/dt$, so $\langle X_t\,dW_t \rangle \equiv \lim_{\Delta t \to 0}\langle X_{t+\Delta t/2}\,\Delta W_t \rangle = \langle X_t \rangle \langle dW_t \rangle + \frac{1}{2}\langle X_t\,[dW_t]^2 \rangle = \frac{1}{2}D\langle X_t \rangle\,dt$. This is the Stratonovich convention. Jumping back and forth between the two conventions is a large part of the job in solving stochastic differential equations. Which convention is appropriate for a particular situation cannot be decided a priori by mathematics alone but needs some physical or practical insight into the problem in question.

2.1.3 Continuous-time random walk

Time is granular

There is an obvious reason why the continuous-time finance formalism presented above is principally flawed. The use of stochastic differential equations relies on the assumption that the source of the noise emits independent increments for arbitrarily small time intervals Δt. However, in reality the independence cannot persist taking the limit $\Delta t \to 0$, and a certain time scale Δt_c, on which the actions of economic agents are correlated, comes into play. To be sure, when we observe the stock market at times much longer than Δt_c, the 'granularity' of time has a negligible effect, and stochastic differential equations do the financial experts a pretty good service. However, in recent decades the focus of the analysts moved to high-frequency and tick-by-tick data, where the time is measured in seconds. On this scale of resolution the assumption of independent increments breaks down, and stochastic differential equations are of little use.

One might try to overcome this obstacle by using a discrete time variable, postulating the correlation time Δt_c as the length of a single time step. We shall see later how this idea works in the so-called GARCH models. On the other hand, physical time is indeed continuous, and a formalism would be welcome that would reconcile the continuity of time with the non-negligible correlation in increments. We shall show here how it can be implemented within the formalism of the continuous-time random walk [197, 200, 201, 577–583].

The walker is supposed to perform instantaneous jumps of lengths X_i, at moments T_i, $i = 1, 2, \ldots$. The waiting times $\Delta T_i = T_i - T_{i-1}$ are independent and identically distributed random variables, and so are the jump lengths. However, we admit there is a dependence between a waiting time and the jump occurring immediately afterwards. All input information on the process is then contained in the joint probability density $P_{X\Delta T}(x, t)$ for X_i and ΔT_i. We have seen in Sec. 1.2.3 the empirical data on waiting time distributions. In most reported cases the distribution seems to decay considerably faster than a power law.

Fourier and Laplace transforms | Box 2.7

In this book, we use the following notation
$$\widehat{f}(z) = \int_0^\infty \mathrm{e}^{-zt}\, f(t)\, \mathrm{d}t$$
for the Laplace transform of a function $f(t)$, and
$$\widetilde{f}(p) = \int_{-\infty}^\infty \mathrm{e}^{\mathrm{i}px}\, f(x)\, \mathrm{d}x$$
for the Fourier transform of a function $f(x)$. We also use the notation
$$\widetilde{f}(p) = \sum_{x\in\Lambda} \mathrm{e}^{\mathrm{i}px}\, f(x)$$
for the Fourier transform on d-dimensional hypercubic lattice $\Lambda = \{(L-1)/2, \ldots$
$\ldots, -1, 0, 1, \ldots, (L-1)/2\}^d$. Its inverse is, for $L \to \infty$,
$$f(x) = \tfrac{1}{(2\pi)^d} \int_{-\pi}^\pi \mathrm{e}^{-\mathrm{i}px}\, \widetilde{f}(p)\, \mathrm{d}^d p.$$
Among the elementary properties of the Laplace transform we shall need the following
$$\text{if}\quad g(t) = \mathrm{e}^{-at} f(t), \quad \text{then}\quad \widehat{g}(z) = \widehat{f}(z+a)$$
$$\lim_{t\to\infty} f(t) = \lim_{z\to 0+} z\widehat{f}(z)$$
$$\widehat{f'}(z) = z\widehat{f}(z) - f(0^+).$$
The behaviour of $f(t)$ for $t \to \infty$ is related to the position and type of the singularity of $\widehat{f}(z)$ closest to the point $z = 0$. Roughly speaking $\widehat{f}(z) \sim (z+z_0)^\alpha$ means that $f(t) \sim t^{-1-\alpha} \mathrm{e}^{-z_0 t}$. Some examples are shown in the following table.

$f(t)$	1	t	$t^{-1/2}$	$(t+1)^{-3/2}$
$\widehat{f}(z)$	z^{-1}	z^{-2}	$(\pi/z)^{1/2}$	$2\left(1 - \mathrm{e}^z \operatorname{erfc}(\sqrt{z})\sqrt{\pi z}\right)$

We also need the Laplace transform of various Bessel functions (assume $\nu \in \mathbb{N}$).

$f(t)$	$I_0(t)$	$I_\nu(t)$	$I_0^2(t)$
$\widehat{f}(z)$	$1/\sqrt{z^2-1}$	$(z - \sqrt{z^2-1})^\nu/\sqrt{z^2-1}$	$2K(2/z)/(\pi z)$

The number of jumps N_t in the time interval $[0, t]$ is also a random number. The distance travelled meanwhile is then

$$Y_t = \sum_{i=1}^{N_t} X_i \tag{2.20}$$

and we would like to interpret the stochastic process Y_t as change in the logarithm of price from time 0 to t. It is easy to write a recurrence equation for its probability density $P_{Y,t}(y)$. Two cases must be distinguished. Either $N_t = 0$, i.e. the price never changed, or $N_t > 0$ and we can relate the distribution to the state of one jump earlier. These two cases combined make the following equation

$$P_{Y,t}(y) = \delta(y)r(t) + \int_0^t \int_{-\infty}^\infty P_{X\Delta T}(x, t')\, P_{Y,t-t'}(y - x)\, \mathrm{d}x\, \mathrm{d}t' \tag{2.21}$$

where $r(t) = \int_t^\infty \int_{-\infty}^\infty P_{X\Delta T}(x, t')\, \mathrm{d}x\, \mathrm{d}t'$ relates to the probability of no-jump. A formal solution of this equation is possible using the Laplace transform in time domain and the Fourier transform in 'space', i.e. along the log-price axis (see Box 2.7 for notation). We get

$$\widehat{\widetilde{P}}_{Y,z}(p) = \widehat{r}(z)\, \frac{1}{1 - \widehat{\widetilde{P}}_{X\Delta T}(p, z)}. \tag{2.22}$$

Now, all that remains is to (numerically) invert both the Fourier and Laplace transforms.

A more explicit expression is found in the decoupled case: suppose that X_i and ΔT_i are independent, so $P_{X\Delta T}(x,t) = P_X(x)P_{\Delta T}(t)$. Then, we can separately calculate the probability $P_{N,t}(n)$ that the price makes n jumps from time 0 to t and the probability density $P_Y(y|n)$ of the sum of lengths over the n jumps, for n fixed. Combining the two quantities we get

$$P_{Y,t}(y) = \sum_{n=0}^{\infty} P_Y(y|n)\, P_{N,t}(n). \tag{2.23}$$

The ingredients can be calculated by the n-fold convolution

$$
\begin{aligned}
P_Y(y|0) &= \delta(y) \\
P_Y(y|n) &= \int_{-\infty}^{\infty} P_Y(y-x|n-1)P_X(x)\mathrm{d}x \\
P_{N,t}(0) &= r(t) \\
P_{N,t}(n) &= \int_0^t P_{N,t-t'}(n-1)P_{\Delta T}(t')\mathrm{d}t'.
\end{aligned}
\tag{2.24}
$$

As a typical and solvable example, let us investigate the case of normally distributed jump lengths and exponentially distributed waiting times

$$
\begin{aligned}
P_{\Delta T}(t) &= \rho\,\mathrm{e}^{-\rho t} \\
P_X(x) &= (2\pi\sigma^2)^{-1/2}\mathrm{e}^{-x^2/(2\sigma^2)}.
\end{aligned}
\tag{2.25}
$$

The multiple convolutions can be obtained explicitly and the expression (2.23) becomes

$$P_{Y,t}(y) = \mathrm{e}^{-\rho t}\delta(y) + \sum_{n=1}^{\infty} \mathrm{e}^{-\rho t}\frac{(\rho t)^n}{n!}\,(2\pi n\sigma^2)^{-1/2}\mathrm{e}^{-y^2/(2n\sigma^2)}. \tag{2.26}$$

We are mainly interested in the tails of the distribution. For large y and t the sum can be approximated by an integral, namely

$$P_{Y,t}(y) \simeq \int_1^{\infty} \mathrm{e}^{-\rho t}\frac{(\rho t)^n}{n!}\,(2\pi n\sigma^2)^{-1/2}\mathrm{e}^{-y^2/(2n\sigma^2)}\mathrm{d}n = \mathrm{e}^{-\rho t}\int_1^{\infty}\mathrm{e}^{-\psi(n)}\,\mathrm{d}n \tag{2.27}$$

where $\psi(n) \simeq n\big(\ln\frac{n}{\rho t}-1\big)+\frac{1}{2}\ln n+\frac{y^2}{2n\sigma^2}$. The integral can be approximately computed by the saddle-point method (see Box 2.9). We find that the exponent has a unique minimum, $\psi'(n^*) = 0$ for $n^* \simeq \rho t$, and at the end of the calculations we recover the normal distribution for the change of price logarithm

$$P_{Y,t}(y) \simeq \frac{1}{\sqrt{2\pi\sigma^2\rho t}}\,\mathrm{e}^{-\frac{y^2}{2\sigma^2\rho t}}\ ,\ t\to\infty. \tag{2.28}$$

The continuous-time random walk, at least for the simple choice of waiting-time distribution (2.25), has the same long-time behaviour as both the usual random walk and the Brownian motion. We can easily guess why it is so. The exponential distribution (2.25) exhibits a characteristic time scale $t_s = \rho^{-1}$, and if we observe the

---

Anomalous diffusion | Box 2.8

In genuinely physical systems it can occur due to a lack of characteristic scale in the inter-event times. A textbook example is the movement of a particle trapped in the regular lattice of eddies in a liquid. Convective cells formed in a not-too-thick layer of oil in a saucepan heated from below may serve for demonstration. If we drop a little ground pepper into the oil we can observe that the small pepper particles remain for a long time within the same convective cell before they jump to the next, only to be imprisoned there for another length of time. The overall movement of such particles is therefore anomalously slow, and the mean distance x from the original to the final position does not scale with time according to $x \sim t^{1/2}$ as in ordinary diffusion but obeys a different law $x \sim t^\tau$ with $\tau < 1/2$. For a thorough review of the phenomenon, see Ref. [194].

---

system at much longer times, $t \gg t_s$, the difference between random walk in discrete time, continuous-time random walk, and Brownian motion becomes irrelevant. The same conclusion would hold for any distribution of waiting times, on the condition a single typical finite time scale t_s can be identified. Trivially it applies if all waiting times are equal, and a slightly less trivial example is provided by the distribution $P_{\Delta T}(t) = \rho^3 \sqrt{2/\pi}\, t^2 e^{-(\rho t)^2/2}$, which again has the characteristic scale $t_s = \rho^{-1}$. Mathematically it is reflected by the fact that in such cases the distribution $P_{N,t}(n)$ resulting from the many-times-repeated convolution develops a sharp peak around the value $n^* \simeq t/t_c$ for large enough times t; hence the applicability of the saddle-point method in the integral (2.27).

Anomalous diffusion

However, as soon as the waiting times do not possess a characteristic scale, the situation becomes quite different. Instead of ordinary diffusion of price we encounter anomalous diffusion (see Box 2.8 and Ref. [194]). Such a situation occurs if the waiting time distribution has a power-law tail with exponent $1 + \mu < 2$, resulting in a divergent average.

In technical language, the difference consists in the fact that the multiple convolutions in (2.24) approach a Lévy distribution with a power-law tail, rather than a narrow peak. The use of the saddle-point method is no longer justified. Nevertheless, we can guess at least some features of the continuous-time random walk in this case. Indeed, the distribution $P_{N,t}(n)$ is essentially the n-fold convolution of the distribution of waiting times $P_{\Delta T}(t)$, and if $P_{\Delta T}(t)$ is a Lévy distribution with parameter $\mu < 1$, making convolutions with itself yields back the Lévy distribution with the same parameter μ. Therefore, it behaves like $P_{N,t}(n) \simeq n^{-1/\mu} P_{\Delta T}(t n^{-1/\mu})$.

Now we must distinguish the regimes of large and small n. In so doing we keep in mind that we are interested in long-time behaviour, so the argument $t n^{-1/\mu}$ is large for a fixed and not too big n. Here, the power-law tail of the Lévy distribution prevails, and we have $P_{N,t}(n) \simeq A_+ n^{-1/\mu}(t n^{-1/\mu})^{-1-\mu} = A_+ t^{-1-\mu} n$. On the other hand, the Lévy distribution $P_{\Delta T}(t)$ approaches a constant (let us denote it B) for $t \to 0$, so for a large enough n we have $P_{N,t}(n) \simeq B n^{-1/\mu}$. We can estimate the position n^* of the maximum of the distribution $P_{N,t}(n)$ by equating the just-established asymptotic regimes

$$B(n^*)^{-1/\mu} \simeq A_+ t^{-1-\mu} n^*. \tag{2.29}$$

Saddle-point method | Box 2.9

is an approximation used for calculation of integrals like
$$I = \int e^{-Nf(x)} dx$$
for large N. The trick consists in expanding the function $f(x)$ around its minimum, for this is just the region which dominates the integral. We proceed by first solving the equation $f'(x^*) = 0$ for the location of the minimum (with condition $f''(x^*) > 0$), then neglecting the terms of order $(x - x^*)^3$ and higher, and performing the remaining Gaussian integration. The result is
$$I = \int e^{-N\left[f(x^*) + \frac{1}{2}f''(x^*)(x-x^*)^2 + \cdots\right]} dx \simeq e^{-Nf(x^*)} \sqrt{\frac{2\pi}{N f''(x^*)}} \simeq e^{-Nf(x^*)}.$$
The last approximate equality is based on the observation that, for large N, the exponential factor varies much faster than the algebraic factor $N^{1/2}$. Indeed, the saddle-point method works better the larger N is. What is 'large N' in a particular case mostly depends on the value of the second derivative $f''(x^*)$, but rigorous analysis of this question may become a surprisingly difficult problem.

Hence the most probable number of jumps performed in time t is

$$n^* \simeq \left(\frac{B}{A_+}\right)^{\frac{\mu}{1+\mu}} t^{\mu}. \tag{2.30}$$

Since the length of the jumps is normally distributed, the variance of the price change is scaled with the number of jumps as $\langle Y_t^2 \rangle \simeq \sigma^2 n$. Assuming that the typical number of jumps coincides with the most probable value n^* given by (2.30), we find that the price change is scaled anomalously with time

$$\langle Y_t^2 \rangle \sim t^{\mu}. \tag{2.31}$$

For example, if the tail of the waiting time distribution decays as $P_{\Delta T}(t) \sim t^{-3/2}$, we have $\mu = 1/2$, and the typical price change perceived after time t behaves like $t^{1/4}$, so the price diffusion is slower compared to the ordinary random walk. This observation holds for continuous-time random walk in general. If the waiting times are anomalously long, which is mathematically expressed by their power-law distribution, the price movements become subdiffusive, i.e. anomalously slower than ordinary diffusion. We have $\langle Y_t^2 \rangle \sim t^{2H}$ with $H = \mu/2$ for $\mu < 1$ and $H = 1/2$ for $\mu \geq 1$. On the other hand, in reality we observe rather the opposite. The stock-market prices are superdiffusive, and $H \simeq 0.6$ in most measurements, as we have seen in Sec. 1.2.2. To account for this phenomenon, we must keep trying to get something else.

2.2 GARCH and related processes

2.2.1 Markovian or not? Bad question

In all models described so far we have assumed that the change in price in a certain interval of time $(t, t + \Delta t)$ does not influence the later price change, from time $t + \Delta t$ to $t + 2\Delta t$; or, at least, the influence is 'weak'. In mathematical parlance, we assumed that the price fluctuations are described by a Markov process, which is memoryless, or at least the memory is finite and of fixed length. The consecutive price differences are independent random variables, and the underlying stochastic process has independent increments.

Markov process   Box 2.10

is a special type of stochastic process. Roughly speaking, a stochastic process has the Markov property if it does not remember its past. More precisely, the multi-time conditional probabilities can be simplified to two-time ones:
$$\mathrm{Prob}\{X_t \leq x | X_{t_1} \leq x_1 \wedge X_{t_2} \leq x_2 \wedge \ldots \wedge X_{t_n} \leq x_n\} =$$
$$= \mathrm{Prob}\{X_t \leq x | X_{t_1} \leq x_1\}$$
for any $n \in \mathbb{N}$ and any ordered collection of times $t > t_1 > t_2 > \ldots > t_n$.
A Markov process for which X_t takes values in a finite or countable set is called a Markov chain. All information on a Markov process is in principle contained in the initial condition X_0 and the transition probabilities (for discrete time) or rates (for continuous time). The transition probabilities from state x_1 to x_2 are simply
$$W(x_1, x_2) = \mathrm{Prob}\{X_{t+1} = x_2 | X_t = x_1\}.$$
The transition rates $w(x_1, x_2)$ are defined for $x_1 \neq x_2$ by
$$\mathrm{Prob}\{X_{t+\tau} = x_2 | X_t = x_1\} = w(x_1, x_2)\,\tau + o(\tau)$$
as $\tau \to 0^+$. In general, the transition probabilities/rates may depend on the time t, but we shall not come across a model with that feature in this book.
The one-time probability for the process to be in the state x, which is $P_t(x) \equiv \mathrm{Prob}\{X_t = x\}$, satisfies a master equation, which for discrete time has the form
$$P_{t+1}(x) = \sum_y W(y, x)\, P_t(y)$$
while for continuous time it is
$$\frac{\mathrm{d}}{\mathrm{d}t} P_t(x) = \sum_{y(\neq x)} \left[w(y, x)\, P_t(y) - w(x, y)\, P_t(x) \right].$$

Besides practical simplicity there are good principal reasons to build the theory of stock-market fluctuations on Markov processes, or processes with independent increments. And we can equally well find principal reasons against that choice. So, what are the pros and cons? First, as we remarked in Chap. 1, when we look at the correlations between price changes at different times, we find values below the inevitable statistical noise. We can consider it proved that the price increments are uncorrelated. As a mental shortcut we may deduce that the increments are also independent. Alas! As every student of elementary probability remembers, examples of random variables which are uncorrelated but not independent can be constructed (quite easily, after all), and the stock-market fluctuations are a real-life incarnation of such a phenomenon. Moving swiftly from an apparent pro to a marked con, we must notice that not only do the price increments exhibit mutual dependence, but this dependence is very protracted. The correlation function of absolute returns decays very slowly, showing the effect we usually call volatility clustering. It is not crucial now whether the decay follows a power law with a small exponent, or is logarithmic, or assumes yet another form. In any case, the decay is much slower than exponential, which means that not only is there a non-negligible memory effect, but this memory is very long, maybe potentially infinite. The idea of modelling the stock prices directly by a Markov process, however ingeniously designed, must be abandoned. On the other hand, however complex the intrinsic dynamics of the economy may be, the driving forces of the stochastic behaviour, as opposed to deterministic chaos, if there is any, are random external 'shocks', usually very small, sometimes large, but much less mutually correlated than the resulting movements of the price. It is quite reasonable to suppose that the external influences are represented by independent random variables, but the response of the stock market to these influences is far from straightforward and produces strong dependence in the prices even when there was no statistical dependence in the input. We may formalise the idea by saying that the stochastic model of the stock market is a kind of a thermal processor. At the input, there is a random, completely featureless noise with normal distribution; and at the output, we obtain a signal with many non-trivial properties, including fat tails, scaling, volatility clustering, or even multifractality.

To be sure, one may say that the external influences themselves can have behaviour as complex as the price fluctuations. But even if it were so, the argument does not bring us any closer to answering the question 'why?' Indeed, if the price only mechanically follows the complex dynamics of the outside world, then we should embark on modelling the outside world. Most probably, if we remain within the realm of stochastic processes, the thermal processor will again be our No. 1 choice. Simplicity in, complexity out: that is the goal.

2.2.2 Heteroskedasticity

The phenomenon of volatility clustering implies that the variance of price increments is not constant but changes in time. This feature is called heteroskedasticity (sometimes spelled heteroscedasticity), as opposed to homoskedasticity, characterised by constant finite variance of the increments. One may be tempted to simply choose the variance in each step anew, from some well-suited distribution. But the long memory present in the economic signals disqualifies such a simple idea. The variance is indeed a random variable, but should depend strongly on the past development. Imagine yourself in the skin of a nervous investor (investors are always nervous). You observe large price movements in yesterday's listings. Stress is mounting: does it mark the beginning of a bubble or is it the first sign of a crash? You do not know. Surely you will be even more vigilant and most probably increase your trading activity. This way you will contribute to the rising turmoil in the market. The volatility will grow. The mechanism we have just described motivates the study of the so-called Autoregressive Conditional Heteroskedasticity (ARCH) and Generalised ARCH (GARCH) processes [172, 422, 532, 584–589].

ARCH process

In 1982 Robert F. Engle introduced a stochastic process commonly referred to by its acronym ARCH [584]. The process runs in discrete time t and in fact consists of two coupled processes, the price increment X_t of an asset at time t and its instantaneous variance S_t. The past prices influence the present variance. To be specific, the new variance is the sum of an a priori contribution with a linear combination of squares of the past price changes. If we go q steps back (but not more), the process belongs to the category denoted ARCH(q). Formally, we write

$$X_t = \sqrt{S_t}\, W_t$$
$$S_t = \alpha_0 + \sum_{l=1}^{q} \alpha_l X_{t-l}^2 \tag{2.32}$$

where W_t are independent and equally distributed random variables with zero mean, $\langle W_t \rangle = 0$, and unit variance $\langle W_t^2 \rangle = 1$. A normal (i.e. Gaussian) distribution is often chosen for them, but other distributions may serve equally well. If we recall our idea of the thermal processor, the trivial process W_t can be considered the input and the process X_t describing the price movements the output. The volatility S_t appears as an auxiliary variable, or a hidden internal layer, providing us with the desired complexity. Sophisticated designs may include many such hidden layers, but for our purpose one is enough.

Of these processes, the simplest one is ARCH(1). It exposes all typical behaviours of the whole ARCH family, and we shall now show its essential properties. The second equation of (2.32) is simplified to

$$S_t = \alpha_0 + \alpha_1 X_{t-1}^2. \tag{2.33}$$

The first quantity we are interested in is the volatility. We must distinguish between the conditional volatility, calculated for a given S_t

$$\langle X_t^2 | S_t \rangle = S_t \tag{2.34}$$

and unconditional volatility, calculated from the former by averaging over S_t. This is easy to perform and a straightforward calculation leads to a linear equation for this quantity

$$\langle X_t^2 \rangle = \langle S_t \rangle = \alpha_0 + \alpha_1 \langle S_{t-1} \rangle. \tag{2.35}$$

If $\alpha_1 < 1$, the stationary state exists and the volatility is

$$\sigma^2 \equiv \lim_{t \to \infty} \langle X_t^2 \rangle = \frac{\alpha_0}{1 - \alpha_1}. \tag{2.36}$$

Higher moments are slightly more involved. Again, we start with the conditional average

$$\langle X_t^{2k} | S_t \rangle = c_k S_t^k. \tag{2.37}$$

The set of constants $c_k = \langle W_t^{2k} \rangle$, $k = 1, 2, \ldots$ depends on the distribution of the variable W_t. If it is Gaussian, we have $c_k = 1 \cdot 3 \cdot 5 \cdot \ldots \cdot (2k - 1)$, while for $W_t = \pm 1$ with equal probabilities, we obtain $c_k = 1$ for all k. In any case, $c_1 = 1$. Following the steps which lead to the expression for volatility, we obtain the set of equations for averages in stationary state $a_m \equiv \lim_{t \to \infty} \langle S_t^m \rangle$. In fact, it is rather a sequence of recursion formulae

$$a_m = \sum_{l=0}^{m} \binom{m}{l} \alpha_0^{m-l} \alpha_1^l c_l a_l \tag{2.38}$$

which can be solved iteratively. The first variable $a_1 = \sigma^2$ was already calculated in (2.36). The next two are

$$
\begin{aligned}
a_2 &= \alpha_0^2 \frac{1 + \alpha_1}{(1 - \alpha_1)(1 - \alpha_1^2 c_2)} \\
a_3 &= \alpha_0^3 \frac{1 + 2\alpha_1 + 2\alpha_1^2 c_2 + \alpha_1^3 c_2}{(1 - \alpha_1)(1 - \alpha_1^2 c_2)(1 - \alpha_1^3 c_3)},
\end{aligned}
\tag{2.39}
$$

and we could, in principle, calculate as many moments as we wish. All of them are rational functions of the parameter α_1. The most important piece of information is contained in the denominators, showing that the $2k$-th moment a_k has poles at values $\alpha_1 = 1$ and $\alpha_1 = c_l^{-1/l}$, for $l = 2, 3, \ldots, k$. To get some idea about the position of the poles, assume that W_t is normally distributed. For a large enough l we approximate the factorial by the Stirling formula and obtain $c_l^{-1/l} \simeq e/(2\,l)$. Thus, the smallest pole in the $2k$-th moment behaves like $\sim k^{-1}$, and we can find an arbitrarily small pole if we go to a high enough moment.

What can be inferred about the return distribution averaged over all times

$$P_X^{>}(x) \equiv \lim_{T \to \infty} \frac{1}{T} \sum_{t=0}^{T} \text{Prob}\{|X_t| > x\} \tag{2.40}$$

from what we know about the moments a_k? Unfortunately, the precise form is not yet accessible, but we can still say something about the tail of the distribution.

Suppose we slowly increase the value of the parameter α_1 and observe the behaviour of a fixed moment a_k. The simplest example is the volatility $\sigma^2 = a_1$. As long as α_1 remains smaller than 1, volatility stays finite. Therefore, the distribution of returns is certainly not described by a Lévy distribution. Those who would hastily conclude that the distribution lacks a power-law tail would be wrong, though. Higher moments testify to the error. The poles indicate that $2k$-th moment is finite only for $\alpha_1 < c_k^{-1/k}$ and there will always be a moment which diverges, together with all higher ones. The source of the diverging moments is in fact the power-law tail of the return distribution, which is always there. For special values $\alpha_1 = c_k^{-1/k}$ we can conclude that

$$P_X^{>}(x) \sim x^{-\gamma} \tag{2.41}$$

with $\gamma = 2k$. Although the analytic form of the return distribution is not known exactly, it is extremely easy to check the power-law tails in a numerical implementation of the ARCH process. In Fig. 2.1 we can see an example of the outcome of such a simulation. In this case the variable W_t was allowed to assume one of the four values $\pm g$, $\pm 2g$, with equal probability $\frac{1}{4}$. The parameter g is adjusted to $g = \sqrt{2/5}$ in order that W_t has unit variance. Hence we calculate the constants c_k and obtain the following expression which relates the exponent γ in the distribution (2.41) to the value of the parameter α_1 of the ARCH(1) process

$$\alpha_1 = \frac{5}{2}\left(\frac{2}{1 + 2^\gamma}\right)^{\frac{2}{\gamma}}. \tag{2.42}$$

The calculation of moments guarantees validity of the formula (2.42) only for an even integer γ, but there seems to be no hindrance for analytical continuation to all real values. Indeed, the inset in Fig. 2.1 confirms that the power-law tails seen in computer realisations of the ARCH(1) process do obey Eq. (2.42).

Besides the return distribution, we are also interested in the volatility autocorrelation. It can be calculated, provided that the fourth moment is finite. The key ingredient is the conditional probability density for the variable X_t if the value X_{t-1} of one step earlier is prescribed. That is

$$P_{X_t|X_{t-1}}(x|x') = \frac{1}{\sqrt{\alpha_0 + \alpha_1\,x'^{\,2}}}\, P_W\left(\frac{x}{\sqrt{\alpha_0 + \alpha_1\,x'^{\,2}}}\right) \tag{2.43}$$

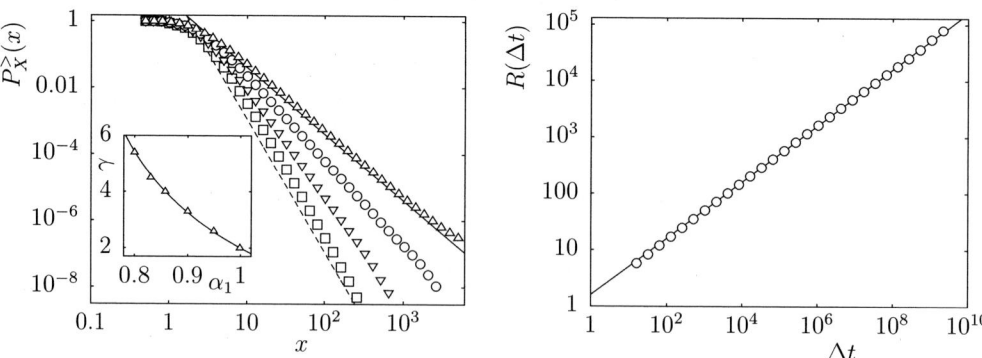

Fig. 2.1 Results of the simulation of ARCH(1) process. In the left panel, return distribution for the values of the parameter $\alpha_1 = 5/\sqrt{34}$ ($\square$), 0.9 ($\triangledown$), 0.95 ($\bigcirc$), and 0.999 ($\triangle$). The lines are the power laws $\propto x^{-2}$ (solid) and $\propto x^{-4}$ (dashed). The inset shows the values of the tail exponent γ obtained from the simulations. The line is the analytical prediction (2.42). In the right panel we show the Hurst plot for the price fluctuations. The line is the power $\propto (\Delta t)^{1/2}$

where $P_W(z)$ is the (time-independent) probability density for the variable W_t. Because the ARCH(1) process is Markovian in the variable X_t, the conditional probability density (2.43) bears all necessary information.

To see the trick, let us first calculate the volatility autocorrelation at time separation $\Delta t = 1$. In a stationary state, using (2.36), we get

$$\langle X_t^2 X_{t-1}^2 \rangle = \int P_{X_t|X_{t-1}}(x|x')\, x^2\, x'^2\, P_{X_{t-1}}(x')\, \mathrm{d}x\, \mathrm{d}x'$$
$$= \alpha_0 \langle X_{t-1}^2 \rangle + \alpha_1 \langle X_{t-1}^4 \rangle \tag{2.44}$$
$$= (1-\alpha_1)\langle X_t^2 \rangle \langle X_{t-1}^2 \rangle + \alpha_1 \langle X_{t-1}^4 \rangle.$$

Hence

$$\langle X_t^2 X_{t-1}^2 \rangle_c \equiv \langle X_t^2 X_{t-1}^2 \rangle - \langle X_t^2 \rangle \langle X_{t-1}^2 \rangle = \left(\langle X_{t-1}^4 \rangle - \langle X_t^2 \rangle \langle X_{t-1}^2 \rangle \right) \alpha_1, \tag{2.45}$$

and we can easily generalise the calculation to any Δt. We apply the chain of conditional probabilities from time $t - \Delta t$ to t and obtain

$$\langle X_t^2 X_{t-\Delta t}^2 \rangle = \alpha_0 \left(1 + \alpha_1 + \ldots + \alpha_1^{\Delta t - 1} \right) \langle X_{t-\Delta t}^2 \rangle + \alpha_1^{\Delta t} \langle X_{t-\Delta t}^4 \rangle; \tag{2.46}$$

and finally we get

$$\langle X_t^2 X_{t-\Delta t}^2 \rangle_c = \frac{(c_2 - 1)\,\alpha_0^2}{(1-\alpha_1)^2(1 - c_2\,\alpha_1^2)}\,\alpha_1^{\Delta t}. \tag{2.47}$$

The autocorrelation of square returns decays exponentially, and the same should also be expected from absolute returns. The characteristic time is determined solely by the parameter α_1, and is $t_c \simeq (\ln \alpha_1)^{-1}$. This is a serious discrepancy with the empirical data. We might expect that the agreement becomes somehow better when the parameter α_1 approaches 1 and the characteristic decay time blows up. But before that, the fourth moment of the distribution of returns diverges for $\alpha_1 \to c_2^{-1/2}$ and so does the amplitude of the square returns autocorrelation. Within the ARCH(1) process, we are unable to consistently model the power-law decay of autocorrelations.

Let us now turn to the question of how the typical size of price fluctuations depends on the time scale at which we make the observations. To this end, we combine the process X_t indicating the returns, into a process describing the logarithm of price

$$Y_t = \sum_{t'=0}^{t} X_{t'}. \tag{2.48}$$

The scale of fluctuations of the signal Y_t at time distances Δt is described by the Hurst plot, i. e. the dependence of the quantity $R(\Delta t)$ on the time difference Δt (if unclear, see Box 4.3 for definition). Unfortunately, it is not accessible analytically, but numerical simulations of the ARCH process provide quite unambiguous results. We can see an example in Fig. 2.1, indicating that the Hurst exponent is exactly $H = 1/2$, the same as in the random walk and below the empirical value. Once again, we can testify how difficult it is to set up a process with realistic fluctuation properties.

GARCH process

We have seen that all important properties of the ARCH(1) process are determined by the single parameter α_1. The other one, α_0, only plays a very marginal role. The general ARCH(q) process is more plastic, but not enough. The memory is q steps long, so the set of q subsequent returns is subject to a Markov process. There is a way to extend the memory effectively to infinity (although only apparently, as we shall soon see). Instead of coupling the volatility with past returns, we may also couple it with past volatilities. This way we arrive at the generalised ARCH, or GARCH, process, introduced by Bollerslev [585]. The simplest variant, the GARCH(1,1) process, replaces the recursion (2.32), (2.33) by the pair

$$\begin{aligned} X_t &= \sqrt{S_t}\, W_t \\ S_t &= \alpha_0 + \alpha_1 X_{t-1}^2 + \beta_1 S_{t-1}. \end{aligned} \tag{2.49}$$

The generalisation to GARCH(p,q) process is obvious; one should go q steps into the past returns X_{t-l}, $l = 1, \ldots, q$ and p steps into the past volatilities S_{t-l}, $l = 1, \ldots, p$. As we did with the ARCH process, here we also limit our analysis to the simplest case defined by (2.49). Repeating the already familiar steps, we obtain for the moments in the stationary state, $\lim_{t\to\infty}\langle X_t^{2m}\rangle = c_m\, a_m$, that

$$a_1 = \frac{\alpha_0}{1 - \alpha_1 - \beta_1}$$

$$a_2 = \alpha_0^2 \frac{1 + \alpha_1 + \beta_1}{\left(1 - \alpha_1 - \beta_1\right)\left(1 - c_2\,\alpha_1^2 - 2\alpha_1\,\beta_1 - \beta_1^2\right)} \tag{2.50}$$

$$\vdots$$

We can see that the moments diverge for specific combinations of parameters α_1 and β_1. We find the same situation as in the ARCH process. The return distribution exhibits generic power-law tails, fatter or thinner, depending on the adjustment of the free parameters.

And similar to what we did in the case of ARCH process, we compute the volatility autocorrelation function. A little thought reveals that GARCH(1,1) is a Markov process for the pair of stochastic variables (X_t, S_t). Therefore, in the calculation of the autocorrelation function we can proceed through a chain of conditional probability densities

$$
P_{X_t, S_t | X_{t-1}, S_{t-1}}(x, s | x', s') = \frac{1}{\sqrt{s}} P_W\left(\frac{x}{\sqrt{s}}\right) P_{S_t | X_{t-1}, S_{t-1}}(s | x', s')
$$
$$
P_{S_t | X_{t-1}, S_{t-1}}(s | x', s') = \delta\left(\alpha_0 + \alpha_1 x'^2 + \beta_1 s' - s\right).
$$
(2.51)

The correlation between times t and $t - \Delta t$ is then

$$
\langle X_t^2 X_{t-\Delta t}^2 \rangle =
$$
$$
\int w_0^2 s_0 \, w_{\Delta t}^2 \, s_{\Delta t} \left[\prod_{t'=0}^{\Delta t - 1} P_W(w_{t'}) P_{S_t | X_{t-1}, S_{t-1}}\left(s_{t'} | w_{t'+1}\sqrt{s_{t'+1}}, s_{t'+1}\right) \right]
$$
$$
\times P_W(w_{\Delta t}) P_{S_{t-\Delta t}}(s_{\Delta t}) \prod_{t'=0}^{\Delta t} \mathrm{d}w_{t'} \, \mathrm{d}s_{t'},
$$
(2.52)

and in the stationary state we find

$$
\langle X_t^2 X_{t-\Delta t}^2 \rangle_c = \left[\frac{c_2 \alpha_1 + \beta_1}{\alpha_1 + \beta_1} a_2 - a_1^2 \right] (\alpha_1 + \beta_1)^{\Delta t}.
$$
(2.53)

The decay of volatility autocorrelations is again exponential, as it was in the ARCH process. We gained some more degrees of freedom, but the principal disagreement with the empirical data is not lifted.

The reason why the correlations remain exponentially dampened is that both ARCH and GARCH processes are essentially Markovian. The consequences are even deeper. In all main characteristics, ARCH and GARCH processes behave in a qualitatively very similar way. To obtain something fundamentally new, it is necessary to generalise the ARCH process in a much more drastic way than the GARCH process does. Indeed, many more variants were tried [252, 590–592]. We shall not go into details and instead limit ourselves to citation of the fractionally integrated GARCH (FIGARCH) process [593], which successfully reproduces the power laws in volatility autocorrelation. It is reasonable to expect that constructing smart enough generalisations of the ARCH process, we may finally reproduce most of the known stylised facts about price fluctuations. However, two lessons should simultaneously fuel our optimism and damp it down. First, various stylised facts seem to be independent. Power-law return distribution with a specific exponent seems to be totally unrelated to the value of the Hurst exponent. The same holds also for the volatility autocorrelation. Second, there seems to be no obstacle for constructing phenomenological stochastic models with many, if not all, features imitating reality. But still, having these models at hand, do we indeed understand more about the processes running not in a computer but on a trading floor? Perhaps here we touch the limits of stochastic modelling of the stock market. If we want to do better, we must inevitably resort to models we called 'involved' at the beginning of this chapter, i.e. the models including the details of the microscopic mechanisms of price formation.

Nevertheless, if we want to simply use a reasonably working phenomenological model of stock market fluctuations, for example when we are asked to estimate the investment risks, we might be faced with in the next month, the GARCH type models can be an excellent choice.

2.3 Cascades

Looking for explanations of various stylised facts, we often end up exhausted. Not only do we feel tired, but sometimes we run out of ideas. For example, we have seen how difficult it is to reproduce volatility clustering in a model without very special features which are difficult to justify. Long memory in the volatility signal, quantified by power-law (or maybe logarithmic) decay of autocorrelations cannot, for principal reasons, be found in any model with the Markov process at its core. Completely different ideology is indispensable. It will come now under the name of cascade processes. We shall proceed by a long detour through multifractals, and the slow decay of autocorrelations will come at the end as a bonus.

2.3.1 Motivation

Multifractality in price signals

Yet another empirical feature of fluctuations of economic indicators is multifractality [594]. Often it is ignored as something, well, real and, yes, interesting, but certainly marginal. There are also some arguments that the multifractality seen in price signals may be a spurious by-product of the analysis of incomplete data [217]. However, here we adopt the view that it is a real phenomenon, measured through the moments of the price increments. Let Z_t be the price at time t. We have seen in Sec. 1.2.3 the empirical data [209, 211, 212] indicating that the returns $X_t(\delta t) = \ln Z_t - \ln Z_{t-\delta t}$ realised during the time δt behave as

$$\langle |X_t(\delta t)|^q \rangle \simeq b_q \, (\delta t)^{\zeta(q)}, \quad \delta t \to \infty. \tag{2.54}$$

Note that $X_t(\delta t)$ is a stochastic process in time t, while δt is a fixed parameter. The average runs over the complete realisation of the process and all times t; b_q are parameters independent of the time lag δt. Of course, if the probability density for the quantity $X_t(\delta t)$ has a power-law tail with exponent $\alpha + 1$, the coefficients b_q are finite only for $q < \alpha$. Therefore, $\zeta(q)$ is also defined only for $q < \alpha$. Ordinary fractal behaviour with Hurst exponent H corresponds to $\zeta(q) = qH$. Any nonlinearity in the function $\zeta(q)$ is an unmistakable signal of multifractality. And this is just what is observed empirically.

Let us have a slightly deeper look at it. For a single realisation of the process Z_t, i.e. for one given time series of prices, we try to classify the time instants t according to how violently the price fluctuates around the time t. Let $\Omega(h)$ be the set of all such times t where the dependence of the size of the fluctuations on the time lag δt is characterised by the exponent h, i.e. $|X_t(\delta t)| \sim (\delta t)^h$. These sets corresponding to different h may be fractals. Denote $d(h)$ the fractal dimension of the set $\Omega(h)$. The function $d(h)$ is called a multifractal spectrum. How can we relate it to the empirically measured exponents $\zeta(q)$?

Imagine that the investigated time range is divided into N intervals of length l, centred at times t_i, $i = 1, 2, \ldots, N$. We want to know the extent to which the prices fluctuate in the vicinity of these times. To this end, we fix one of the times, say, t_j, increase the length l' of the interval length around this time, and observe the size of the price change over that interval. If it is scaled as $|X_{t_j}(l')| \sim (l')^{h_j}$ when l' increases, then h_j is the local exponent corresponding to the time t_j. As the next step, let us identify the ensemble of the sets of times characterised by the same exponent, $\Omega_l(h) = \{t_j | h_j = h\}$. We want to know how the sizes of these sets depend on how refined the division of the whole time interval is, i.e. how they change with l. In so doing, we tacitly assume that the limit of the sets $\Omega_l(h)$ exists in some sense with h fixed and $l \to 0$. If $|\Omega_l(h)| \sim l^{-d(h)}$, when $l \to 0$, then the limiting set $\Omega(h) = \lim_{l \to 0} \Omega_l(h)$ has the fractal dimension $d(h)$. Of course, if we sum the sizes of all the sets, we should get $\sum_h |\Omega_l(h)| = N \propto l^{-1}$.

The moments introduced in (2.54) can be estimated as

$$
\begin{aligned}
\langle |X_t(l)|^q \rangle &\simeq \frac{1}{N} \sum_{i=1}^{N} |X_{t_i}(l)|^q = \frac{1}{N} \sum_h \left[\sum_{t \in \Omega_l(h)} |X_t(l)|^q \right] \\
&\sim \frac{1}{N} \sum_h |\Omega_l(h)| \, l^{qh} \sim \sum_h l^{qh+1-d(h)}.
\end{aligned}
\tag{2.55}
$$

If the discretisation is fine-grained enough, the last sum in (2.55) can be replaced by an integral over h, and this is, in turn, easily computed by the saddle-point method, giving

$$
\langle |X_t(l)|^q \rangle \sim l^{\zeta(q)}
\tag{2.56}
$$

where

$$
\zeta(q) = \min_h \left(qh + 1 - d(h) \right).
\tag{2.57}
$$

The minimum is sought over a range of h for which the set $\Omega(h)$ is non-empty, or $d(h) > 0$. For example, for an ordinary fractal, put $d(h) = D \leq 1$ for $h = H$ and $d(h) = 0$ otherwise. From (2.57) we obtain the linear dependence $\zeta(q) = 1 - D + qH$ as expected.

The equation (2.57) relating the moment exponents $\zeta(q)$ with the multifractal spectrum $d(h)$ is called the Legendre transform. The reader may perhaps remember that from a mathematical point of view it is the same Legendre transform as that which occurs in rational thermodynamics to relate various thermodynamic potentials to each other, or in quantum field theory where it connects the generating function of the connected correlation functions (the free energy) and the generating function of the vertex functions. It is also important to notice that the inverse of the Legendre

> **Turbulence** Box 2.11
>
> is the mechanism of energy dissipation in fluids [595, 596]. The kinetic energy of macroscopic movement of the fluid on large scales (e.g. the Gulf Stream) is converted into energy of eddies on smaller and smaller scales, until it is transformed into heat at the smallest scale, determined by the viscosity. Mathematically, understanding turbulence amounts finding stable solutions of the Navier-Stokes equations for the velocity vector $\mathbf{v}(x,t)$ as a function of space and time, $\dot{\mathbf{v}} + \mathbf{v} \cdot \nabla\mathbf{v} = -\nabla p + \nu\Delta\mathbf{v}$, where p is the pressure and ν the viscosity. (The difficulty of the problem may be quantified by the $\$10^6$ prize offered by the Clay institute for solving the Navier-Stokes equations.) The first significant insight into the complex structure of eddies within the several orders of magnitude separating the largest scale, on which energy is supplied, from the viscous scale, on which the energy is dissipated, was provided by Kolmogorov in 1941 [597] (the so-called K41 theory). In essence, it is based on the idea of scale invariance. On the assumption that at every intermediate scale no energy is dissipated but everything that flows in from a larger scale flows homogeneously out to a lower scale, the law for typical velocity difference on the length scale l is $\delta v \sim l^{1/3}$. The most important statement of the K41 theory says that the energy contained in velocity fluctuations with wavevector k is scaled as $E(k) \sim k^{-5/3}$ [596].

transform can be written in exactly the same form, except that the role of the functions $\zeta(q)$ and $d(h)$ is interchanged

$$d(h) = \min_q \Big(qh + 1 - \zeta(q) \Big). \tag{2.58}$$

As a real-life example, recall the work [209], where moments are computed from empirical data in the range $-1 \leq q \leq 4$ and the multifractal spectrum is non-zero at least in the interval $h \in (0.3, 0.6)$ for stock prices and $h \in (0.25, 0.9)$ for commodity prices. Therefore, the multifractality is so marked a feature that it should not be disregarded in modelling. Let us see now what physical systems may provide an inspiration.

Turbulence

Incidentally, multifractality is extensively studied and relatively well understood in the field of fully developed turbulence [595, 596] (see Box 2.11). The word 'turbulence' is used frequently in economics on a rather superficial level, denoting up-and-down price movements larger than usual. Perhaps the discomfort felt by business-class travellers when the aircraft enters the area of developed atmospheric turbulence is the source of the metaphoric use of the term. Indeed, turbulence is characterised by very large spatio-temporal changes in the velocity field $v(x,t)$ (hence the jumps of the plane), measured, e.g. by anemometry. At a fixed time, we can analyse the moments of the velocity differences, which are supposed to be scaled with distance as [602, 603]

$$\langle |v(x) - v(x-l)|^q \rangle \sim l^{\zeta(q)} \tag{2.59}$$

in full analogy with the moments (2.54) measured in the price signal. This observation led very early to attempts to base the models for price fluctuations on the theory of turbulence [604, 605]. The rather superficial analogies met immediate deep criticism [248, 606, 607]. Since then, the matter has been fairly clarified, and useful concepts,

---

Turbulence: beta- and random beta-models | Box 2.12

In reality the dissipation in turbulent flow is not homogeneous in space, as assumed in the K41 theory, but takes place on a complicated set of singularities with a fractal dimension lower than 3. To account for this fact, the beta-model was introduced by Frisch et al [598] and further exposed from a multifractal perspective by Frisch and Parisi [599]. It consists in a simple regular cascade of scales. On the n-th level of the cascade, the largest scale l_0 is divided into boxes of sizes $l_0/2^n$. Boxes containing singularities are called active. Going from level n to $n+1$, every active box is divided into 2^3 smaller boxes, the fraction β of which is now active. By iteration, a classical fractal emerges, with a definite fractal dimension of $D < 3$, on which the singularities are located. This scheme was later developed into the random beta-model [600, 601], which also reproduces in a schematic way the multifractality of the set of singularities. The point is that, allowing β random from a fixed distribution, we obtain a whole family of singularities which can be classified according to their 'strengths'. In both K41 and the beta-model the velocity is scaled as $\delta v \sim l^{1/3}$ everywhere in the singular set. However, we can investigate sets with singularities of a more general type, $\delta v \sim l^h$. More precisely, we require that on such a set $\lim_{l\to 0} |\delta v|/l^h \neq 0$. Its fractal dimension is denoted $d(h)$. In K41 theory we have $d(h) = 3\,\theta(h-1/3)$, while in the beta-model $d(h) = D\,\theta(h-(D-2)/3)$. (Here, $\theta(h)$ is the Heaviside function.) The function $d(h)$ characterises the multifractal spectrum. In practice it is extracted from the moments $\langle (\delta v)^q \rangle \sim l^{\zeta(q)}$. Their relation is provided by the Legendre transform $\zeta(q) = \min_h \left(hq + 3 - d(h) \right)$. The term '3' here comes from the fact that the space is three-dimensional.

---

first used in turbulence, have been adapted to adequately reflect the different reality of stock markets [249, 253]. We especially have in mind the random-beta model of fully developed isotropic turbulence [598, 600] (see Box 2.12), which can be seen as a three-dimensional variant of the Mandelbrot multifractal cascade described below.

2.3.2 Realisation

In multifractal models of price fluctuations, the starting point is the ordinary Brownian motion, or random walk. The new ingredient added is the deformation of time. Time is not homogeneous, but sometimes it is stretched, sometimes compressed. In reality it corresponds to the varying overall frequency of trading. At each trade, the return distribution is the same (e.g. Gaussian with fixed variance), but the number of trades in the given time interval δt may vary substantially. Intervals with a high frequency of trades mark high-volatility periods and vice versa. This way the complexity of the price signal is traced back to the complex pattern of human activity, with steady 'business as usual' periods punctuated by bursts of fever not unlike the gold rush.

Replacing the physical time t by the deformed time $\Theta(t)$, the logarithm of price follows the compound stochastic process $Y_{\Theta(t)}$, where Y_t is the ordinary Brownian motion. Given a particular realisation $\theta(t)$ of the process $\Theta(t)$ the distribution of log-price changes from time t_1 to t_2, denoted $X_{t_1}(t_2 - t_1) = Y_{\Theta(t_2)} - Y_{\Theta(t_1)}$, is

$$P_{t_1,t_2}(x|\theta) = \frac{1}{\sqrt{2\pi(\theta(t_2) - \theta(t_1))}\,\sigma} \exp\left(-\frac{x^2}{2\sigma^2\,(\theta(t_2) - \theta(t_1))} \right). \tag{2.60}$$

Thus, the only non-trivial point consists in a smart choice of the time deformation $\Theta(t)$. It is also a stochastic process, and various models of the multifractal price fluctuations differ by the method of constructing the process $\Theta(t)$.

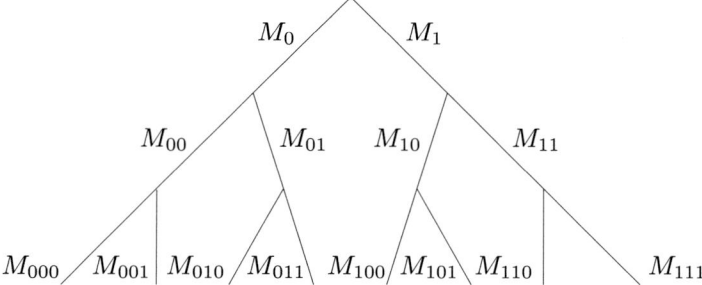

Fig. 2.2 Scheme of the Mandelbrot binary cascade at level $k = 3$. Each endpoint of the tree represents one of the $2^k = 8$ intervals. Its measure is computed by multiplying all M's we meet when we go from the root of the tree to the endpoint.

Regular Mandelbrot cascade

The easiest procedure carrying multifractal properties is the binary cascade introduced by Mandelbrot, Calvet, and Fisher [608–610]. It is directly related to the random beta-model used in the theory of fully developed turbulence.

Let us observe the price fluctuations on the time interval $[0, T]$. For simplicity we set the units of time so that $T = 1$. The deformed time will be anchored at the starting and ending point, $\Theta(0) = 0$, and $\Theta(1) = 1$ for all realisations. We may alternatively interpret the time deformation in terms of a measure. Indeed, the measure of the interval $[t_1, t_2]$ can be defined as $\mu([t_1, t_2]) = \Theta(t_2) - \Theta(t_1)$. Reverting the procedure, we start with constructing the (random) measure on the interval $[0, 1]$ and then deduce the random process $\Theta(t)$ uniquely determined by it. The measure is constructed iteratively. Initially, we set the measure of the entire interval to 1, so $\mu([0, 1]) = 1$. Next, we divide the interval into two equal parts and attribute measures M_0 and M_1 to the left and right half, respectively. In order to preserve the measure of the whole, the measure of the parts must sum to 1, so $M_0 + M_1 = 1$. Apart from this constraint, M_0 and M_1 are random non-negative numbers.

We proceed further by doing the same with the halves what we did with the whole interval. The interval $[0, 1]$ is now divided into four quarters with measures $M_0 \, M_{00}$, $M_0 \, M_{01}$, $M_1 \, M_{10}$, and $M_1 \, M_{11}$, respectively. This process can then be repeated as long as we please. We illustrate it using a binary tree in Fig. 2.2. Generally, at the k-th level, we express the measure of the interval $[t, t + 2^{-k}]$ using the binary notation for the first point $t = \sum_{i=1}^{k} a_i \, 2^{-i}$, $a_i \in \{0, 1\}$, as $\mu([t, t + 2^{-k}]) = M_{a_1} M_{a_1 a_2} \cdots M_{a_1 a_2 \ldots a_k}$. The measure-conservation condition is $M_{a_1 a_2 \ldots a_{k-1} 0} + M_{a_1 a_2 \ldots a_{k-1} 1} = 1$, for any k. Except for that, all coefficients $M_{a_1 a_2 \ldots}$ are assumed to be independent and identically distributed. In the following, we shall, as the simplest choice, use the bimodal distribution

$$\text{Prob}\{M = m\} = \text{Prob}\{M = 1 - m\} = \frac{1}{2} \tag{2.61}$$

with $m \in (0, \frac{1}{2})$ the only parameter of the model. We show in Fig. 2.3 an example of the random measure after $k = 15$ iterations for $m = 0.4$.

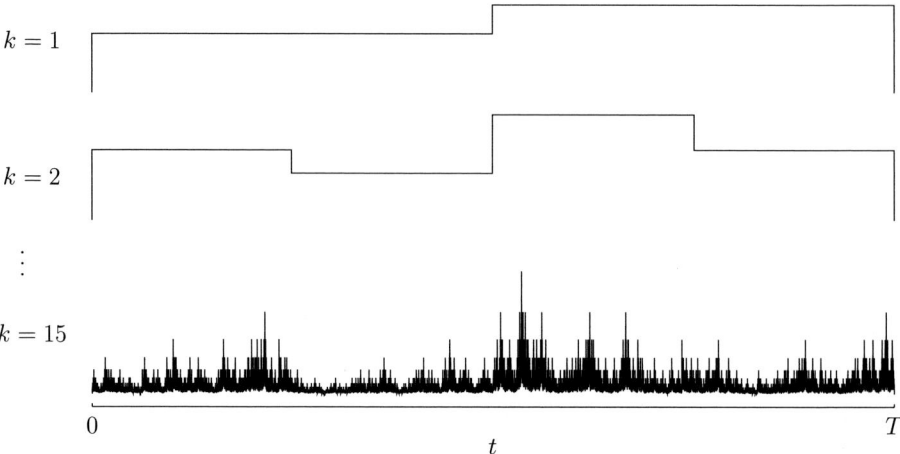

Fig. 2.3 An example of a random measure realisation on the interval $[0, T]$. It is a result of a random binary cascade with $k = 15$ levels and the parameter $m = 0.4$. In the top two rows, there is the measure after the first and second step of the iteration, and in the bottom, the final measure.

The construction can be generalised in several obvious ways. Binary divisions of the intervals give rise to a binary cascade. Equally well we can construct n-ary cascades by iteratively dividing the intervals into more than two parts. At this point it is perhaps opportune to note that in the random beta-model of turbulence, the number of parts at each branching of the cascade must be at least three to get non-trivial results. Furthermore, the number of divisions may itself be a random number. The probability distribution for M's can assume various forms. For example choosing log-normal distribution has the advantage that all the products of the Ms again have the log-normal distribution, only with a different average and width. All of these modifications bring nothing but computational complications without conceptual novelty and we shall not discuss them any more.

Let us now see how multifractality arises in the cascade. We start assuming the normal distribution (2.60) for the difference of the logarithm of price. The deformation of time $\Theta(t)$ is given by the random measure μ on the interval $[0, 1]$. For a given random measure μ we compute the moments of the change in log-price from time t to $t + \delta t$, with $\delta t = 2^{-k}$ and t equal to an integer multiple of 2^{-k}. Then we average over the realisations of the measure and over all starting times t. The fact that Ms at different levels are all independent and equally distributed facilitates the calculation. The result is

$$\langle |X(\delta t)|^q \rangle = c_q \, \sigma^q \, (\delta t)^{\zeta(q)} \tag{2.62}$$

where $c_q = \frac{\Gamma(q+1)}{2^{q/2} \, \Gamma(1+q/2)}$ originates from the moments of the normal distribution with unit variance. The most important part is the q-dependence of the exponent, which is generally

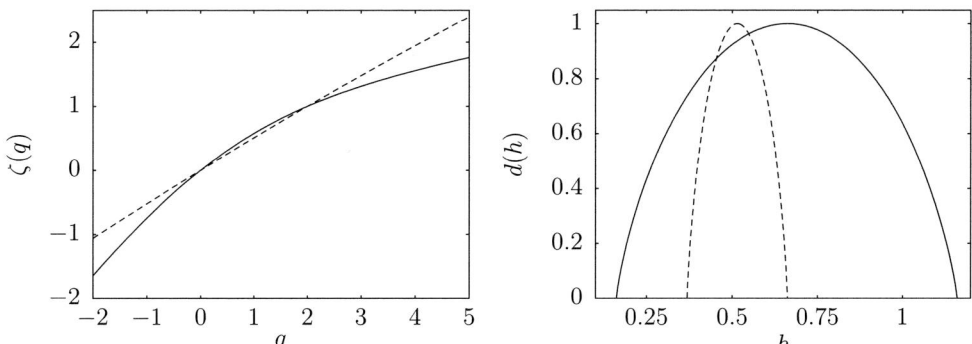

Fig. 2.4 Multifractality in the Mandelbrot binary cascade. In the left panel, the moment exponents are shown; in the right panel, the multifractal spectrum. The distribution of factors was bimodal with parameter $m = 0.2$ (solid lines) and $m = 0.4$ (dashed lines).

$$\zeta(q) = -\frac{\ln\langle M^{q/2}\rangle}{\ln 2} \; ; \qquad (2.63)$$

and in the simplest case of the bimodal distribution (2.61) we have

$$\zeta(q) = 1 - \frac{\ln\left(m^{q/2} + (1-m)^{q/2}\right)}{\ln 2}. \qquad (2.64)$$

It is easy to invert the Legendre transform and calculate the multifractal spectrum using (2.58). We leave it to the reader as Problem 3. In Fig. 2.4 we show an example of both the exponents according to (2.64) and the corresponding multifractal spectrum for two values of m. Clearly, the multifractality is the more pronounced the more m differs from the value $m = 0.5$, corresponding to the ordinary Brownian motion.

Further we shall look at the autocorrelation function of returns. The returns themselves will be calculated at the time distance δt, and we shall observe correlations of two returns separated by time Δt. Therefore, two other scales enter the calculation besides the global time scale T. However, at the end of the calculation we shall assume that $\delta t \le \Delta t \ll T$, so that only the fraction $\Delta t/\delta t$ remains relevant. For simplicity we assume that both time scales are powers of two, say $\delta t = 2^{-k}$ and $\Delta t = 2^{s-k}$, with $0 \le s < k$. We define the average autocorrelation

$$C_q(\Delta t) = \frac{1}{2^k - 2^s} \sum_{i=0}^{2^k-2^s-1} \langle |X_{i\delta t}(\delta t)\, X_{i\delta t + \Delta t}(\delta t)|^q\rangle, \qquad (2.65)$$

keeping in mind that eventually we shall take the limit $k \to \infty$.

The value of the average under the summation sign depends on times $t = i\delta t$ and $t + \Delta t$. The tree structure of the cascade implies that the value depends only on the number of levels one should climb up the tree if we want to get from time instant $t = i\,2^{-k}$ to $t + \Delta t = j\,2^{-k}$. This number is called ultrametric distance $u(i,j)$ [363]. For example, in our binary cascade $u(0,2) = u(0,3) = u(1,3) = 2$, $u(0,4) = u(3,4) =$

$u(2, 7) = 3$, and so on, as can be seen in Fig. 2.2. The sum over the time t reduces to a sum over all possible ultrametric distances, with factors indicating how many times a given ultrametric distance occurs in (2.65). The correlation function can be then expressed as

$$C_q(\Delta t) = \frac{1}{2^k - 2^s} \sum_{m=1}^{k} a_s(m) \, K_q(m). \tag{2.66}$$

The number of time pairs t, $t + \Delta t$ at distance m is

$$a_s(m) = \begin{cases} 0 & \text{for } 1 \leq m \leq s \\ 2^{k-m+s} & \text{for } s < m \leq k \end{cases} \tag{2.67}$$

and the correlation at exactly this distance,

$$K_q(m) \equiv \langle |X_0(2^{-k}) \, X_{(2^m-1)2^{-k}}(2^{-k})|^q \rangle, \tag{2.68}$$

can be calculated after noticing that, among the factors M hidden in the correlation function, which are all independent at levels $> m$, we have once M and once $1 - M$ at level m, and we have in total $m - 1$ mutually independent factors M^2 at levels $< m$. Finally, we get

$$K_q(m) = c_q^2 \, \sigma^{2q} \, \frac{\langle M^{q/2}(1 - M)^{q/2} \rangle}{\langle M^{q/2} \rangle^2} \langle M^q \rangle^k \left(\frac{\langle M^{q/2} \rangle^2}{\langle M^q \rangle} \right)^m. \tag{2.69}$$

The exponential dependence on m already indicates that the autocorrelation function may decay as a power of the time difference Δt. To confirm this intuition we must perform the sum in Eq. (2.66). Finally, after taking the limit $k \to \infty$, we conclude that the autocorrelation function relative to same-time correlation $C_q(0)$ decays as a power when time is measured in units of δt, namely

$$C_q(\Delta t) = C_q(0) \frac{c_q^2}{c_{2q}} \frac{\langle M^{q/2}(1 - M)^{q/2} \rangle}{2\langle M^q \rangle - \langle M^{q/2} \rangle^2} \left[\frac{\Delta t}{\delta t} \right]^{-\ln\left(\langle M^q \rangle / \langle M^{q/2} \rangle^2\right)/\ln 2}. \tag{2.70}$$

This is the promised feature of the multifractal cascade model. By construction, the cascade comprises correlations at all time scales. The long-time memory manifested in the volatility clustering as in Eq. (2.70) follows automatically.

Are there fat tails?

Another feature we would like to check in the cascade model is the presence of power-law tails in the return distribution. A clear indication would be the divergence of certain moments of the distribution. However, as shown in Eq. (2.62), when we keep δt fixed, all moments remain finite as long as $M \geq 0$ and $\text{Prob}\{M > 0\} > 0$. This looks bad, but a small modification of the cascade improves things. We shall relax the condition of strict conservation of measure, requiring only that it is conserved on average. At each branching of the tree (see again Fig. 2.2) the factors going to the right and to the left are now independent. Therefore, all factors M within the cascade are independent and equally distributed random variables with average $\langle M \rangle = \frac{1}{2}$. The result is that the measure of an interval is modified to $\mu\big([t, t+2^{-s}]\big) = \Omega_{a_1 a_2 \ldots a_s} M_{a_1} M_{a_1 a_2} \cdots M_{a_1 a_2 \ldots a_s}$ and the moments of the distribution of returns realised during the time $\delta t = 2^{-s}$ are now

$$\langle |X_t(\delta t)|^q \rangle = c_q \, \sigma^q \langle \Omega_{a_1 a_2 \ldots a_s}^{q/2} \rangle \langle M^{q/2} \rangle^s. \qquad (2.71)$$

The divergence of the moments of the return distribution can originate solely from the divergence of the moments of the additional factor $\Omega_{a_1 a_2 \ldots a_s}$. Let us look at it now.

Suppose that the tree has in total k levels. The distribution of the factor $\Omega_{a_1 a_2 \ldots a_s}$ depends on how many levels there are below the s-th, i.e. on the number $k - s$. All factors on the same level are independent random variables. Denote by $\Omega_{(k-s)}$ the generic factor on this level. We can compute its arbitrary moment iteratively. Clearly, at the lowest level the factor is non-random, $\Omega_{(0)} = 1$. Proceeding one level upwards we have $\Omega_{(1)} = M_0 + M_1$ and the moments are $\langle \Omega_{(1)}^{q/2} \rangle = \sum_{j=0}^{q/2} \binom{q/2}{j} \langle M^j \rangle \langle M^{q/2-j} \rangle$. Generally, when climbing the tree we obtain the following sequence of formulae

$$\langle \Omega_{(m+1)}^{q/2} \rangle = \sum_{j=0}^{q/2} \binom{q/2}{j} \langle M^j \rangle \langle M^{q/2-j} \rangle \langle \Omega_{(m)}^{j} \rangle \langle \Omega_{(m)}^{q/2-j} \rangle. \qquad (2.72)$$

Note that $\langle \Omega_{(m)} \rangle = 1$ for any m. This is the mathematical expression of the measure conservation on average.

We are interested in the limit $k \to \infty$ with $\delta t = 2^{-s}$ kept fixed. The variables $\Omega_{(m)}$ approach to a limit $\Omega = \lim_{m \to \infty} \Omega_{(m)}$ and its moments can be inferred recursively from (2.72) for any even q. We get

$$\langle \Omega \rangle = 1$$
$$\langle \Omega^2 \rangle = \frac{1}{2} \frac{1}{1 - 2\langle M^2 \rangle}$$
$$\vdots \qquad\qquad\qquad\qquad (2.73)$$
$$\langle \Omega^{q/2} \rangle = \frac{1}{1 - 2\langle M^{q/2} \rangle} \sum_{j=1}^{q/2-1} \binom{q/2}{j} \langle M^j \rangle \langle M^{q/2-j} \rangle \langle \Omega^j \rangle \langle \Omega^{q/2-j} \rangle;$$

and therefore the q-th moment of the return distribution diverges if $q > 2$ and $\langle M^{q/2} \rangle = \frac{1}{2}$. From (2.63) we can see that this is equivalent to the condition $\zeta(q) = 1$. It never happens for the bimodal distribution (2.61), but it is possible if we assume a log-normal distribution for M. It is characterised by two parameters, say $\langle M \rangle$ and $\langle M^2 \rangle$. As $\langle M \rangle = \frac{1}{2}$, all properties of the cascade can be expressed in terms of $\langle M^2 \rangle$. For example, we find that the diverging moment determined by $\zeta(q_c) = 1$ is $q_c = \frac{4 \ln 2}{\ln(4\langle M^2 \rangle)}$. The most probable empirical value $q_c \simeq 3$ implies $\langle M^2 \rangle \simeq 0.63$. Furthermore, the exponent in the power-law decay of the autocorrelation function $C_2(\Delta t) \propto (\Delta t)^{-\tau}$ is, according to (2.70), $\tau = 4/q_c \simeq 1.33$. Alas! This value is much too large compared to empirical results. Moreover, the value of the multifractal exponent $\zeta(3) \simeq 1$ is also refuted by the actual price data [209]. We must conclude that the cascade process alone cannot serve as an explanation of the fat tails in the return distribution. Multifractality, volatility clustering, and power-law tails do not have one common cause, or at least the cause does not reside in the hierarchical cascade-like structure of the time in economic processes [611]. A cheap way out is to simply suppose that the normal distribution of the a priori price changes (2.60) is replaced by a Lévy distribution. This would work fine, but instead of getting the power-law tails in result, we would enter them by hand. Certainly, this makes the model much less attractive from a theoretical point of view, although it may become rather useful in practice.

Lux and Markov-switching models

The most important disadvantage of the regular cascade model we have presented so far is the limitation of the time to the fixed interval $[0, T]$. What happens beyond the point T? Of course, we could stack many identical cascades one after the other, but it is rather strange to suppose that at certain times T, $2T$, $3T$, ... everything is forgotten and the price evolution starts anew. To cure this flaw and continue the price evolution within the cascade model to an indefinite future, two very similar models were suggested.

Again, the definition of the stochastic process is reduced to the construction of the random time stretching $\Theta(t)$. It is influenced by events on several time scales, $1, 2, \ldots, k$, where level 1 is the slowest and k the fastest. On level j we sow random breakpoints at times $t_{js} < t_{js+1}$, $s = 0, 1, 2, \ldots$, where at level j the breakpoints are on average more distant than on level $j + 1$. We use the convention $t_{j0} = 0$ for all levels j. We assign a random number M_{js} to the interval between breakpoints $[t_{js}, t_{js+1}]$. The independent and identically distributed variables M_{js} play a role similar to the Ms in the Mandelbrot cascade, but now we assume that their average is $\langle M_{js} \rangle = 1$.

In the Markov-switching multifractal model [223, 612–621] of Calvet and Fisher the breakpoints are chosen at each level j independently according to the Poisson point process with intensity λ_j. This means that the average number of breakpoints in the interval of length δt is $\lambda_j \delta t$, and the probability that there are exactly n breakpoints in this interval is given by the Poisson distribution

$$P_{\text{break}\, j}(\delta t; n) = \mathrm{e}^{-\lambda_j\, \delta t}\, \frac{(\lambda_j\, \delta t)^n}{n!}. \tag{2.74}$$

The second of the models we present here was introduced by Thomas Lux [622]. (Attention! This stochastic Lux model must not be confused with the agent-based Lux-Marchesi model we shall discuss in Sec. 3.2.3.) The breakpoints are determined in a slightly more complicated manner. First, we randomly scatter the breakpoints according to the Poisson distribution (2.74). Then, for each breakpoint at time t_{js} at level j we add other breakpoints at the same time on all lower levels $j' > j$. This feature makes the Lux model more similar to the Mandelbrot cascade than the Markov-switching model. On the other hand, we lose the advantage of having the levels mutually independent.

The stretching process $\Theta(t)$ is again defined through the measure $\mu([t, t']) = \Theta(t') - \Theta(t)$. Suppose now that none of the breakpoints at any level fall inside the interval $[t, t']$, although we allow that the endpoints t and t' may coincide with some breakpoints. Any general interval on the time axis can be constructed as a union of non-overlapping (except common endpoints) intervals with such property, so for definition of $\Theta(t)$ it is sufficient to establish the measure of such special intervals. Thus, at each level we find the breakpoints closest to the interval, $t_{jr} \leq t < t' \leq t_{jr+1}$ and relate the measure to the product of the corresponding M-factors

$$\mu([t,t']) = (t'-t) \prod_{j=1}^{k} M_{jr}. \qquad (2.75)$$

In addition to the measure (2.75) we introduce partial measures, comprising only levels from j to k. Again, for an interval $[t,t']$ which does not contain any breakpoints, we define $\mu_j([t,t']) = \prod_{j'=j}^{k} M_{j'r}$, and we extend the definition to a general interval in such a way that the measure μ satisfies the additivity property. Of course, $\mu([t,t']) = (t'-t)\mu_1([t,t'])$. The partial measures depend on each other recursively. Suppose that the interval $[t,t']$ contains l breakpoints $t_{j1}, t_{j2}, \ldots, t_{jl}$ on the j-th level and denote $t_{j0} = t$ and $t_{jl+1} = t'$. Let us also introduce the relative positions of the breakpoints $u_{jr} = (t_{jr}-t)/(t'-t)$. Then, the j-th partial measure of this interval is a combination of the $j+1$-th measures of the intervals between the adjacent breakpoints

$$\mu_j([t,t']) = \sum_{r=0}^{l} (u_{jr+1} - u_{jr}) M_{jr} \, \mu_{j+1}([t_{jr}, t_{jr+1}]) \qquad (2.76)$$

where for convenience we defined $\mu_{k+1}([t,t']) = 1$ for any t and t'.

In the following, we shall use the notation for the moments of the partial measures $\overline{\mu_j^p}(t'-t) = \langle (\mu_j([t,t']))^p \rangle$. Then, the moments of the return distribution are

$$\langle |X(\delta t)|^q \rangle = c_q \, \sigma^q \, (\delta t)^{q/2} \, \overline{\mu_1^{q/2}}(\delta t) \qquad (2.77)$$

and the deviation from the ordinary Brownian motion behaviour is concentrated in the non-trivial dependence of $\overline{\mu_1^{q/2}}(\delta t)$ on the time difference δt. If it is a constant, the Brownian regime is recovered.

The exact calculation of it is rather tricky, as we must take into account that within the time interval of length δt there can be an arbitrary number of breakpoints $t_{j1}, t_{j2}, \ldots, t_{jl}$ at any level j. In some cases the situation is simplified. For example if $q = 2$, we start with $\overline{\mu_{k+1}}(\delta t) = 1$, and using (2.76) we show that $\overline{\mu_j}(\delta t) = 1$ at any level j, because $\langle u_{j1} M_{j0} + (u_{j2} - u_{j1}) M_{j1} + \ldots (1 - u_{jl}) M_{jl} \rangle = 1$. This implies $\zeta(2) = 1$, which is exactly the same exponent as for an ordinary Brownian motion. Other values of q can be treated in the limit of a large number of levels, $k \to \infty$ and accepting a certain assumption about the waiting times between breakpoints. We shall show how it works in the case of the Lux model [229, 623].

We suppose that the frequency of the breakpoints increases in a geometric sequence, $\lambda_j = \lambda_0 Q^j$. The Lux model with quotient $Q = 2$ can then be considered a random counterpart of the Mandelbrot binary cascade. If we also suppose that the number of levels is very high, $k \to \infty$, or at least $k \gg j$, the partial measures possess an important discrete scaling symmetry. Indeed, when we multiply all times by the factor Q, it is the same as advancing one level down. The measure $\mu_j[Qt, Qt']$ must have the same statistical properties as $\mu_{j+1}[t, t']$. It is possible to show by induction that the moments of these measures obey

$$\overline{\mu_j^p}(Q\delta t) = \overline{\mu_{j+1}^p}(\delta t) \tag{2.78}$$

for any p, as long as the limit $k \to \infty$ exists.

The q-th moment of the return distribution is proportional to the function $\overline{\mu_1^{q/2}}(\delta t)$. Let us restrict the calculation to the intervals of the lengths increasing by powers of Q, $\delta t = t_0 Q^m$ and look at the moments of an arbitrary partial measure. The interval $[t, t + \delta t]$ may contain a variable number n of the breakpoints at level j. Thus

$$\overline{\mu_j^{q/2}}(\delta t) = \sum_n P_{\text{break } j}(\delta t; n) \sum_{p_0\, p_1 \ldots p_n} \binom{q/2}{p_0\, p_1 \ldots p_n}$$

$$\times \prod_{r=0}^n \langle M^{p_r} \rangle \Big\langle (u_{jr+1} - u_{jr})^{p_r} \overline{\mu_{j+1}^{p_r}} \big((u_{jr+1} - u_{jr})\delta t\big) \Big\rangle_{\text{over } u}. \tag{2.79}$$

The average $\langle \ldots \rangle_{\text{over } u}$ on the right-hand side is taken over random positions of the points u_{jr} within the interval $[0, 1]$. The situation is considerably simplified if the probability of having $n > 0$ breakpoints at level j is negligible, $P_{\text{break } j}(\delta t; 0) \simeq 1$. This happens if $\delta t \ll 1/\lambda_j$. In that case the sum in (2.79) contains only one term, and we have a very simple expression of the measure at level j in terms of the measure at level $j + 1$ in the form $\overline{\mu_j^{q/2}}(\delta t) \simeq \langle M^{q/2} \rangle \overline{\mu_{j+1}^{q/2}}(\delta t)$. We can use it for $j = 1$ together with the discrete scale invariance (2.78) and the relation (2.77) to relate the moments of the return distribution at time interval δt with those at time interval $Q\, \delta t$:

$$\frac{\langle |X(\delta t)|^q \rangle}{\langle |X(Q\, \delta t)|^q \rangle} = Q^{-q/2} \langle M^{q/2} \rangle. \tag{2.80}$$

Assuming that the moments of the return distribution behave like $\langle |X(\delta t)|^q \rangle \propto (\delta t)^{\zeta(q)}$ we conclude that the exponents are

$$\zeta(q) = \frac{q}{2} - \frac{\ln \langle M^{q/2} \rangle}{\ln Q}. \tag{2.81}$$

This should hold for times not exceeding the characteristic time of the slowest level of the cascade, i.e. $\delta t \ll 1/\lambda_1$. For longer times, the random-walk behaviour is restored. This is in fact compatible with the empirical observation that the multifractality holds for a fairly extended, but still limited time span. Thus, the frequency of the slowest level can be directly read off from the empirical data. As for the frequencies of the subsequent levels, it may prove difficult, but not impossible, to fit them on the data [613, 615].

Note that for $Q = 2$ this is exactly the same dependence (2.63) as in the case of the Mandelbrot cascade, if we only take into account the difference in the notation ($\langle M \rangle = 1$ here, $\langle M \rangle = 1/2$ there). This coincidence stems from the particular choice of the breakpoint frequencies λ_j as a geometric sequence.

Now we turn to volatility clustering. We shall show the calculation within the Markov-switching model. As for the Lux model, the procedure is only slightly more difficult and the reader may try her fortune herself or look at Refs. [622,623].

We shall relate the returns realised during time intervals of length δt, separated by time distance Δt. The autocorrelation function is

$$
\begin{aligned}
C_q(\Delta t) &= \langle |X_t(\delta t)\, X_{t+\Delta t}(\delta t)|^q \rangle \\
&= c_q^2\, \sigma^{2q}\, (\delta t)^q \left\langle \Big(\mu_1\big([t, t+\delta t]\big)\, \mu_1\big([t+\Delta t, t+\Delta t+\delta t]\big) \Big)^{q/2} \right\rangle.
\end{aligned}
\tag{2.82}
$$

We suppose that the scale δt is sufficiently small so that we have many levels with no breakpoints within intervals of length δt. In other words, we can identify the level j^* so that $P_{\text{break}\,j}(\delta t; 0) \simeq 1$, that is $\delta t \ll 1/\lambda_j$, for all $j < j^*$. We can estimate such level as $j^* \simeq -\ln(\lambda_0\, \delta t)/\ln Q$. An important point is that j^* is independent of Δt. So, we can afford some level of ambiguity in the determination of what the proper level j^* is if we just investigate the dependence of the correlation function on the time distance Δt without bothering about the dependence on δt.

What we do need to suppose is that Δt is large enough to have some breakpoints at level j^* within the time Δt, so $\Delta t \gg 1/\lambda_{j^*}$. But this is safely satisfied if $\Delta t \gg \delta t$. With all of this in mind, we can write

$$
C_q(\Delta t) \simeq c_q^2\, \sigma^{2q}\, (\delta t)^q \prod_{j=1}^{j^*-1} \left\langle \big(M_{j0}\, M_{j1}\big)^{q/2} \right\rangle \left(\overline{\mu_{j^*}^{q/2}}(\delta t) \right)^2
\tag{2.83}
$$

where we used the independence of M's at different levels, so the average is done at each level separately. In this formula, each M_{j0} comes from the measure on the interval $[t, t+\delta t]$, while M_{j1} originates from the interval $[t+\Delta t, t+\Delta t+\delta t]$. If there is a breakpoint in between, these two are different, $M_{j0} \neq M_{j1}$; otherwise they are equal. The latter case occurs with the probability $P_{\text{break}\,j}(\Delta t; 0) = \mathrm{e}^{-\Delta t\, \lambda_j}$. Thus,

$$
\left\langle \big(M_{j0}\, M_{j1}\big)^{q/2} \right\rangle = \langle M^{q/2}\rangle^2 + \mathrm{e}^{-\Delta t\, \lambda_j} \big(\langle M^q\rangle - \langle M^{q/2}\rangle^2 \big),
\tag{2.84}
$$

and the correlation function is proportional to the product

$$
C_q(\Delta t) \propto \prod_{j=1}^{j^*-1} \frac{1 + m_q\, \mathrm{e}^{-\Delta t\, \lambda_j}}{1 + m_q}
\tag{2.85}
$$

where $m_q = \langle M^q\rangle / \langle M^{q/2}\rangle^2 - 1$. To compute it, we should note that for large Δt the factor $\mathrm{e}^{-\Delta t\, \lambda_j}$ jumps swiftly around the level $j = j^+$ from a value very close to 1 for $j < j^+$ to a very small value for $j > j^+$. The transition level is about $j^+ \simeq -\ln(\lambda_0\, \Delta t)/\ln Q$. Therefore, the product is well approximated by $(1+m_q)^{j^+-j^*}$. This consideration works only if $j^+ \geq 0$, which imposes the upper bound $\Delta t \lesssim 1/\lambda_0$ on the time difference. Now we can express j^* and j^+ in terms of δt and Δt, respectively, and obtain power-law decay of the autocorrelations

$$
C_q(\Delta t) \simeq C_q(0)\, \frac{c_q^2 \left(\overline{\mu_{j^*}^{q/2}}(\delta t) \right)^2}{c_{2q}\, \overline{\mu_{j^*}^q}(\delta t)} \left[\frac{\Delta t}{\delta t} \right]^{-\ln\left(\langle M^q\rangle / \langle M^{q/2}\rangle^2 \right)/\ln Q},
\tag{2.86}
$$

which should hold for time differences in the range $\delta t \ll \Delta t \ll 1/\lambda_0$. For times larger than the characteristic time of the slowest level, i.e. for $\Delta t \gg 1/\lambda_0$, the correlation function relaxes exponentially to its asymptotic value.

Comparing this result with (2.70) we can see that for $Q = 2$ the exponent coincides with the regular Mandelbrot binary cascade. This is again due to the special choice of the frequencies λ_j. Other choices may naturally lead to a variety of different laws for the decrease of the autocorrelation function. This makes the Markov-switching model a much more flexible tool when we try to mimic real price data using a cascade model. To see how the parameters λ_j can be inferred from empirical observations, look at Refs. [613, 615].

2.4 Stochastic volatility models

In the preceding sections we elaborated in various ways on the concept that volatility is not constant but varies in time in a random manner. In Sec. 2.2.2 we a priori discarded the idea that volatility at each time may be an independent random variable drawn from a prescribed distribution. However, if we think it over, we should admit that the idea was not so bad, if only we could abandon the requirement of independence. We can rather prescribe a stochastic process governing the time evolution of the volatility. The important thing is that the stochastic process for the volatility affects the process describing the price, but not the other way round. The properties of the volatility may then be studied independently; for example its probability distribution can be found; then, the properties of price fluctuations are computed on top of the known properties of the volatility. In the design of the model, we can be led by the empirically known distribution of volatility, which fluctuates much less wildly than the returns [19, 178, 180, 182].

This is the basic setup of the family of processes called stochastic volatility models. Although it would be possible to formulate them in discrete time, as it is done for the GARCH model, almost all implementations rely on stochastic differential equations. We shall adopt this approach here as well. In all that follows, the Itô convention is assumed.

2.4.1 Systematics

Vasicek model

In 1977 Oldřich Vašíček, a Czech-born mathematical economist, introduced a diffusion process describing the evolution of interest rates, R_t [624]. As a simple solvable example he used an Ornstein-Uhlenbeck process in the form

$$dR_t = a\,(r - R_t)\,dt + g\,dW_t \tag{2.87}$$

where r, a, and g are the parameters of the model. The solution is indeed simple, and in the stationary state the interest rate R has Gaussian distribution with average r and width governed by the quantity $g^2/(2a)$. The approach to equilibrium, as well as the decay of correlations, follows an exponential law with rate a (be patient, we shall show the calculations in a while).

> **Ornstein-Uhlenbeck process** | Box 2.13
>
> After the groundbreaking paper of A. Einstein on Brownian motion [628] there was a great surge of interest in the movement of small colloidal particles suspended in water, in an attempt to make the analysis neat and clear. In one of such efforts, L. S. Ornstein [629] together with G. E. Uhlenbeck [630] wrote a stochastic differential equation properly describing the velocity of the colloidal particle. The viscous environment damps the velocity to zero, but incessant impacts of surrounding water molecules sum up to a random force which drives the colloidal particle out from the rest. This idea is embodied into the stochastic differential equation
> $$\mathrm{d}V_t = -a\,V_t\,\mathrm{d}t + g\,\mathrm{d}W$$
> fully—and relatively easily—solvable using the Fokker-Planck equation. The two main results are, first, that velocity is normally distributed in the stationary state and, second, that the average velocity exponentially decays from its initial value to zero, i.e. if $\langle V_0 \rangle = v_0 \neq 0$, then $\langle V_t \rangle = v_0\,\mathrm{e}^{-a\,t}$.

If you look at Eq. (2.87) more closely, you will quickly find a bit of nonsense, which may seem of minor relevance, but it is there. The Gaussian distribution extends from minus to plus infinity, whatever the mean and variance may be. But a negative interest rate hardly makes sense in modern economics. Cox, Ingersoll, and Ross [625] suggested a remedy, consisting of suppressing the fluctuations when the interest rate approaches zero. Therefore, it is dynamically prevented from entering the range of negative values. The modified stochastic differential equation describing the Cox-Ingersoll-Ross model is

$$\mathrm{d}R_t = a\,(r - R_t)\,\mathrm{d}t + g\,\sqrt{R_t}\,\mathrm{d}W_t. \tag{2.88}$$

In fact, the same problem was investigated, on the level of the Fokker-Planck equation, much earlier by Feller [626], so sometimes you may find it called the Feller process.

To add more plasticity, Hull and White [627] formulated a very general process

$$\mathrm{d}R_t = \big(b(t) + a(t)\,(r - R_t)\big)\,\mathrm{d}t + g(t)\,R_t^{\beta}\,\mathrm{d}W_t \tag{2.89}$$

with $\beta = 0$ or $\beta = 1/2$. The functions $a(t)$, $b(t)$, and $g(t)$ are determined using additional assumptions in combination with fitting to empirical data. We shall not employ such a sophisticated model and concentrate on equations with constant coefficients, exemplified by Eqs. (2.87) and (2.88).

Stein-Stein model

It is tempting to reinterpret the Vasicek model as describing the evolution of price volatility, instead of the interest rate, and people did pursue this track [631, 632]. A problem remains here: how to treat the negative values of the quantity for which an equation like (2.87) is written. In the model of Stein and Stein [633–635] a natural solution is adopted, saying that the volatility is the square of the process described by Eq. (2.87), not the process itself. Thus we arrive at a set of two coupled stochastic equations, together describing the evolution of the logarithm of price Y_t. They are

$$\begin{aligned} \mathrm{d}Y_t &= S_t\,\mathrm{d}W_{1t} \\ \mathrm{d}S_t &= a\,(\sigma - S_t)\,\mathrm{d}t + g\,\mathrm{d}W_{2t} \end{aligned} \tag{2.90}$$

where the two sources of noise W_{1t} and W_{2t} are mutually independent and $(\mathrm{d}W_{1t})^2 = (\mathrm{d}W_{1t})^2 = \mathrm{d}t$. The momentary volatility in the Stein-Stein model is S_t^2, and it is obviously positive. That is good. A less welcome thing comes out only after we solve the equations and find that the stationary probability density for the volatility diverges at small values, in stark contrast with the empirical findings. Nevertheless, the Stein-Stein model can serve as a useful starting point and a paradigm for all other stochastic volatility models. We shall analyse it in more depth later.

Heston model

In a similar manner as the Vasicek model serves as a basis for the Stein-Stein stochastic volatility model, we can start with the Cox-Ingersoll-Ross version (2.88) and adapt it to price evolution. This way we get the Heston model [636], which became very popular [176, 634, 637–640]. Since there is no need to bypass negative volatilities, we can identify the volatility with just S_t, instead of S_t^2 as Stein and Stein did. This way we avoid the nuisance with the diverging probability density at small volatilities. The set of equations for the Heston model is then

$$\begin{aligned} \mathrm{d}Y_t &= \sqrt{S_t}\,\mathrm{d}W_{1t} \\ \mathrm{d}S_t &= a\,(\sigma - S_t)\,\mathrm{d}t + g\,\sqrt{S_t}\,\mathrm{d}W_{2t}. \end{aligned} \tag{2.91}$$

Very generally, we can study a whole family of stochastic volatility models, parameterised by the exponents α, β, and γ and described by the equations

$$\begin{aligned} \mathrm{d}Y_t &= S_t^\gamma\,\mathrm{d}W_{1t} \\ \mathrm{d}S_t &= a\,(\sigma - S_t)\,S_t^\alpha\,\mathrm{d}t + g\,S_t^\beta\,\mathrm{d}W_{2t}. \end{aligned} \tag{2.92}$$

Besides the two special sets of exponents mentioned so far, several other combinations are of special interest. We make their overview in the following table.

α	β	γ	name or acronym of the model
0	0	1	Stein-Stein
0	0	1/2	'Ornstein-Uhlenbeck'
0	1/2	1/2	Heston
0	1	1/2	'GARCH'
1	1	1/2	'geometric Ornstein-Uhlenbeck'
0	3/2	1/2	3/2-model

(The names in quotation marks ' ' are somewhat abuses of notation, because properly speaking they are already in use in a more or less different sense.) In the next two sections we shall show the properties of most, if not all, of the items in the above table. As a historical remark, let us note that probably the first who investigated the Stein-Stein model was L. O. Scott [631], and the pioneers of the 'Ornstein-Uhlenbeck' model were Hull and White [632].

2.4.2   Price fluctuation moments

There are essentially two ways to look at the process defined by (2.92). We can either compute the moments, or the joint probability density for the processes Y_t and S_t.

The first method leads to solving the set of coupled ordinary differential equations, the second requires solution of a partial differential equation. We shall see that for the long-time behaviour of the stochastic volatility model the former method is more appropriate, while for short times we can proceed more conveniently using the latter way.

One-time quantities

The general moment is defined as $\mu_{m,n}(t) = \langle Y_t^m S_t^n \rangle$, and its time evolution according to the process (2.92) is contained in the equation

$$\frac{\mathrm{d}}{\mathrm{d}t}\mu_{m,n}(t) = na\left(\sigma\,\mu_{m,n+\alpha-1}(t) - \mu_{m,n+\alpha}(t)\right)$$
$$+ \frac{1}{2}m(m-1)\,\mu_{m-2,n+2\gamma}(t) + \frac{g^2}{2}n(n-1)\,\mu_{m,n-2+2\beta}(t). \tag{2.93}$$

There is no closed-form solution in a general case, but we can investigate each model separately. We shall essentially follow the order in which the models are listed in the table above.

Let us start with the Stein-Stein model and proceed from small values of m and n to higher ones. For $m = 0$ the equation is

$$\dot\mu_{0,n} + na\,\mu_{0,n} = na\sigma\,\mu_{0,n-1} + \frac{g^2}{2}n(n-1)\,\mu_{0,n-2}, \tag{2.94}$$

and we can see that knowing $\mu_{0,n-1}$ and $\mu_{0,n-2}$ we obtain $\mu_{0,n}$ by trivial integration. The starting point $\mu_{0,0}(t) = \langle 1 \rangle = 1$ is obvious. The next two moments are

$$\mu_{0,1}(t) = \sigma + \left(\mu_{0,1}(0) - \sigma\right)\mathrm{e}^{-at}$$
$$\mu_{0,2}(t) = \sigma^2 + \frac{g^2}{2a} + \left[\mu_{0,2}(0) - \left(\sigma^2 + \frac{g^2}{2a}\right)\right]\mathrm{e}^{-2at} \tag{2.95}$$
$$+ 2\sigma\left(\mu_{0,1}(0) - \sigma\right)\left(\mathrm{e}^{-at} - \mathrm{e}^{-2at}\right),$$

and we can continue as far as we please. The generic feature of the time evolution of moments, seen in (2.95), is an exponential relaxation to a stationary value. Setting $\dot\mu_{0,n} = 0$ in (2.94) we obtain a chain of equations for the stationary values of the moments $\mu_{0,n}(\infty)$. The first two of them, $\mu_{0,1}(\infty) = \sigma$, $\mu_{0,2}(\infty) = \sigma^2 + g^2/(2a)$ can be seen in the result (2.95); the next two are

$$\mu_{0,3}(\infty) = \sigma^3 + 3\sigma\frac{g^2}{2a}$$
$$\mu_{0,4}(\infty) = \sigma^4 + 6\sigma^2\frac{g^2}{2a} + 3\left(\frac{g^2}{2a}\right)^2; \tag{2.96}$$

and again, we can continue iteratively.

The moments with $m > 0$ are just a bit more complicated. We need to only investigate even m, because all moments are zero for odd values of m. Integrating Eq. (2.93) we can get the moments for higher and higher m systematically. The most interesting part of the result is the behaviour of the moments for large time values. For example, we find, using (2.95), that

$$\mu_{2,0}(t) = \mu_{2,0}(0) + \int_0^t \mu_{0,2}(\tau)\,\mathrm{d}\tau \simeq \left(\sigma^2 + \frac{g^2}{2a}\right)t\,, \quad \text{as} \quad t \to \infty. \tag{2.97}$$

Generally, it is possible to show that $\left(\frac{\mathrm{d}}{\mathrm{d}t}\right)^l \mu_{2l,n}(t)$ has a finite limit when $t \to \infty$, and after some easy algebra we arrive at a formula summarising the long-time behaviour of all moments:

$$\lim_{t\to\infty} t^{-l}\,\mu_{2l,n}(t) = \frac{(2l)!}{2^l\,l!}\left(\sigma^2 + \frac{g^2}{2a}\right)^l \mu_{0,n}(\infty). \tag{2.98}$$

We recognise very well the factor $(2l)!/(2^l\,l!) = 1 \cdot 3 \cdot 5 \ldots (2l-1)$. It is the $2l$-th moment of the normal distribution with unit variance, and therefore the result (2.98) indicates that for long time intervals the returns have Gaussian distribution with variance $\left(\sigma^2 + g^2/(2a)\right)t$.

Analogous steps ought to be taken when we want to find the moments in the case of the 'Ornstein-Uhlenbeck' model. Note that the moments including only the process S_t, i.e. those with $m = 0$, are identical to the Stein-Stein model. Therefore, the formulae (2.95) and (2.96) remain in force. The long-time asymptotic behaviour of the remaining moments is obtained in the same way as in Eq. (2.98). This time, the result is

$$\lim_{t\to\infty} t^{-l}\,\mu_{2l,n}(t) = \frac{(2l)!}{2^l\,l!}\,\sigma^l\,\mu_{0,n}(\infty) \tag{2.99}$$

and again suggests a Gaussian distribution of the returns. One must be very careful, though. The trouble lies in the already mentioned problem of negative volatilities, which cannot be excluded, as long as the volatility is described by the pure Ornstein-Uhlenbeck process. The increment of the price logarithm Y_t is proportional to the square root of the volatility, hence it can acquire a non-zero imaginary part. This is certainly something beyond real-world economy. In the calculation of the moments the imaginary part is averaged out to zero, so the problem is concealed.

We can avoid it entirely if we impose a reflecting boundary at $S_t = 0$. If the volatility process S_t comes close to zero it never drops below it, but bounces back to positive values. We shall see in the next section that in this case we can easily find the stationary distribution of volatility and also the short-time properties of the log-price. However, the moments are not subordinated to simple linear equations like (2.93). What, then, is the information contained in the result (2.99)? The error we make when neglecting the negative volatilities is small if $2a\,\sigma^2 \gg g^2$. In that case we can consider (2.99) a good approximation to the 'Ornstein-Uhlenbeck' model with a reflecting boundary.

The next in line is the Heston model. The moments evolve according to the set of equations

$$\dot{\mu}_{m,n} + na\,\mu_{m,n} = na\left(\sigma + \frac{g^2}{2a}(n-1)\right)\mu_{m,n-1} + \frac{1}{2}m(m-1)\,\mu_{m-2,n+1}; \tag{2.100}$$

and solving them for the lowest non-trivial moments we get

$$\mu_{0,1}(t) = \sigma + \left(\mu_{0,1}(0) - \sigma\right) e^{-at}$$

$$\mu_{2,0}(t) = \mu_{2,0}(0) + \sigma t + \frac{1}{a}\left(\mu_{0,1}(0) - \sigma\right)\left(1 - e^{-at}\right). \tag{2.101}$$

The long-time behaviour is again obtained assuming that the derivatives $\left(\frac{d}{dt}\right)^l \mu_{2l,n}(t)$ have a finite limit for $t \to \infty$ and checking a posteriori that this assumption does not contain any contradiction. We can do even better than we could for the Stein-Stein and 'Ornstein-Uhlenbeck' models. The limits of the moments are expressed by the closed formula

$$\lim_{t \to \infty} t^{-l} \mu_{2l,n}(t) = \frac{(2l)!}{2^l \, l!} \sigma^l \left(\frac{g^2}{2a}\right)^n \frac{\Gamma\left(\frac{2a\sigma}{g^2} + n\right)}{\Gamma\left(\frac{2a\sigma}{g^2}\right)}, \tag{2.102}$$

showing again without any doubt that the long-time distribution of returns is Gaussian.

At this moment the reader surely has a question on the tip of her tongue: Is it a generic feature of all stochastic volatility models that, if we wait long enough, the return distribution becomes Gaussian? Well, the answer is no! The same calculations as shown above, when done for the 'GARCH' stochastic volatility model, reveal a power-law tail of the return distribution through exponential divergence of certain moments when $t \to \infty$. To leave the reader some fun, we suggest that she performs all of the algebra herself (Problem 6).

Autocorrelations

On the level of moments, it is much more difficult to calculate the autocorrelation function of log-price increments $X_t(\delta t) = Y_t - Y_{t-\delta t}$. The expression

$$C_q(\Delta t) = \langle |X_t(\delta t)\, X_{t+\Delta t}(\delta t)|^q \rangle \tag{2.103}$$

depends on four different time instants, and, even if we take time homogeneity into account, the number of independent time variables is three. The calculation thus involves solving a partial differential equation in three variables. This is no less difficult than solving directly the Fokker-Planck equation, which in principle contains complete information, including the ingredients needed for determination of the general autocorrelation function (2.103).

Instead of pursuing this painful path, we simplify the task and suppose that we are interested only in autocorrelation of returns realised over infinitesimally small times, i.e. $\delta t \to 0$. But for short times we can approximately write $|Y_t - Y_{t-\delta t}| \propto |S_t|^\gamma$. Thus, we scratch the definition (2.103) and below we use

$$C_q(\Delta t) = \langle |S_t\, S_{t+\Delta t}|^{\gamma q} \rangle \tag{2.104}$$

instead. Continuing the series of definitions, we also introduce more general correlations, which will be useful in the course of the computation, $C_{r,s}(\Delta t) = \langle S_t^r\, S_{t+\Delta t}^s \rangle$. Of course, $C_q(\Delta t) = C_{q\gamma,q\gamma}(\Delta t)$ if $q\gamma$ is an even integer. (For the Heston model it also holds for $q\gamma$ odd, because $S_t \geq 0$.)

It is easy to see that the set of functions $C_{r,s}(\Delta t)$ obeys the same differential equations as the moments $\mu_{0,s}$. We have solved several of them already. The difference from the present case consists only in the initial conditions. Because we assume time homogeneity, i.e. independence of the autocorrelation function on the time t, we should have $C_{r,s}(0) = \mu_{0,r+s}(\infty)$. We have collected many results for these asymptotic values of the moments, so we can readily combine them into results for the function $C_{r,s}(\Delta t)$. For example, for the Stein-Stein model, as well as for the 'Ornstein-Uhlenbeck' model, we get

$$
C_{1,1}(\Delta t) = \sigma^2 + \frac{g^2}{2a}\,\mathrm{e}^{-a\,\Delta t}
$$

$$
C_{2,2}(\Delta t) = \left(\sigma^2 + \frac{g^2}{2a}\right)^2 + 4\sigma^2 \frac{g^2}{2a}\,\mathrm{e}^{-a\,\Delta t} + 2\left(\frac{g^2}{2a}\right)^2 \mathrm{e}^{-2a\,\Delta t}.
$$

(2.105)

A common feature of the autocorrelation functions is exponential relaxation, just as it was observed in the ARCH and GARCH models. Volatility clustering is present, but significantly weaker than what is observed in actual data. This was to be expected, as it is the direct consequence of the Markov property characteristic for all finite sets of stochastic differential equations.

2.4.3   Volatility and return distributions

The long-time properties of the stochastic volatility models studied so far seem somewhat boring. The return distribution is mostly Gaussian, with the exception of the 'GARCH' model, where power-law tails develop. At short times, however, more diversity emerges. In this section we shall look at short-time return distribution. An ingredient needed for these investigations will be the stationary distribution for the volatility process S_t. Recall that the instantaneous volatility influences the price process X_t, but is not affected by it, so we can find the distribution for S_t separately.

Fokker-Planck equation

We would be happy if we could obtain full information on the coupled pair of processes (2.92) for price and volatility. In principle, the straightforward way is to solve the Fokker-Planck equation for the joint probability density for both quantities $P_{YS,t}(y,s) = \langle \delta(Y_t - y)\delta(S_t - s)\rangle$. In our case, we obtain

$$
\begin{aligned}
\frac{\partial}{\partial t} P_{YS,t}(y,s) = {}& a\frac{\partial}{\partial s}\left[(s-\sigma)s^\alpha\, P_{YS,t}(y,s)\right] \\
& + \frac{1}{2}\frac{\partial^2}{\partial y^2}\left[s^{2\gamma}\, P_{YS,t}(y,s)\right] \\
& + \frac{g^2}{2}\frac{\partial^2}{\partial s^2}\left[s^{2\beta}\, P_{YS,t}(y,s)\right].
\end{aligned}
$$

(2.106)

Because the volatility process S_t is not influenced by the price process Y_t, we can easily write the equation for the distribution of S_t alone. It can be viewed as the marginal probability density derived from $P_{YS,t}$ as $P_{S,t}(s) = \int P_{YS,t}(y,s)\,\mathrm{d}y$. The equation in question is obtained from (2.106) by omitting the term containing derivatives with respect to y. (It vanishes upon integration over y.) Thus, the equation is

$$\frac{\partial}{\partial t} P_{S,t}(s) = a \frac{\partial}{\partial s} \left[(s - \sigma) s^\alpha \, P_{S,t}(s) \right]$$
$$+ \frac{g^2}{2} \frac{\partial^2}{\partial s^2} \left[s^{2\beta} \, P_{S,t}(s) \right]. \tag{2.107}$$

While the solution of (2.106) cannot be given in a closed form, except for a few special values of the parameters α, β, and γ, it is vastly easier to solve the equation (2.107), especially if we are interested only in the stationary state, where $\partial P_{S,t}/\partial t = 0$. In the next paragraph we shall see what can be inferred from this partial information.

Short-time limit

The study of autocorrelations in the preceding section showed that the instantaneous volatility changes with typical time $\simeq 1/a$. We shall investigate the price change during a much shorter time interval $\Delta t \ll 1/a$, so that the volatility process S_t stays approximately constant. Of course, the magnitude of the price fluctuations in the interval $(t_1, t_1 + \Delta t)$ may differ for different starting times t_1, according to the value of the volatility process at that time S_{t_1}. Therefore, the method of computing the short-time return distribution consists of three steps.

First, we calculate the stationary distribution $P_S(s)$ of the variable S_t by solving the Fokker-Planck equation (2.107). Second, from the first equation of (2.92), we calculate the distribution of log-price change $X_t = Y_t - Y_{t-\Delta t}$ during time interval of length Δt, with S_t replaced by a constant s. This step is trivial, since the result is just the normal distribution

$$P_X(x|s) = \frac{1}{|s^\gamma| \sqrt{2\pi \, \Delta t}} \exp\left(-\frac{x^2}{2 \, s^{2\gamma} \, \Delta t} \right). \tag{2.108}$$

Finally, we average over the distribution of S_t,

$$P_X(x) = \int P_X(x|s) P_S(s) \, \mathrm{d}s. \tag{2.109}$$

This kind of approach is sometimes related [641] to the Born-Oppenheimer approximation used in quantum chemistry.

We shall show the calculation explicitly for some of the combinations of the parameters α, β, γ.

As a first case, let us examine again the Stein-Stein model, $\alpha = \beta = 0$, $\gamma = 1$. We cannot avoid solving the Fokker-Planck equation, but this is the easier step. We shall show it here once in more detail. The reader should be able to easily repeat this procedure for any other values of α and β. In a stationary state the left-hand side of (2.107) is zero. The right-hand side can be written as the derivative of a function. Hence, that function must be a constant c. This reduces the partial differential equation into an ordinary differential equation of the first order

$$\frac{g^2}{2} \frac{\mathrm{d}}{\mathrm{d}s} P_S(s) + a \, (s - \sigma) \, P_S(s) = c. \tag{2.110}$$

We naturally suppose that $P_S(s)$ tends toward zero for both $s \to \infty$ and $s \to -\infty$. The only value of c compatible with both of these requirements is $c = 0$. Then, the solution of (2.110) is trivial:

$$P_S(s) = \sqrt{\frac{a}{\pi g^2}} \, e^{-\frac{a}{g^2}(s-\sigma)^2}, \tag{2.111}$$

and proves the already-mentioned fact that the Ornstein-Uhlenbeck process in a stationary state has a Gaussian distribution. Substituting it into (2.109) we get the short-time return distribution in the form

$$P_X(x) = \sqrt{\frac{a}{2\pi^2 \, g^2 \, \Delta t}} \int \exp\left(-\frac{x^2}{2 \, s^2 \, \Delta t} - \frac{a}{g^2}(s-\sigma)^2\right) \frac{1}{|s|} \, ds. \tag{2.112}$$

There is no simple closed formula for this integral, except for $\sigma = 0$. This case may become relevant if we are interested in the distribution of very large returns, when x is typically much larger than σ. Indeed, we shall compute here the shape of the tails of the distribution $P_X(x)$, with the hope of seeing something that resembles the fat tails known from empirical studies.

Before doing that, we cannot help noticing that the opposite limit of very small x reveals a pathology of the Stein-Stein model. The factor $|s|^{-1}$ makes the integral (2.112) divergent for $x = 0$. For x non-zero but small, the factor $\exp(-x^2/(2 \, s^2 \, \Delta t))$ ensures convergence, but implies that the return distribution develops a logarithmic singularity $P_X(x) \sim \ln(|x|/\sqrt{\Delta t})$ as $x \to 0$. Surely in reality the minuscule price changes are not infinitely more probable than the large ones, as the Stein-Stein model tries to tell us. That is one of the reasons why the Stein-Stein model, despite its mathematical appeal, fell into disfavour among the people modelling the price fluctuations.

We shall ignore this problem when investigating the distribution of large returns. This is what we are going to do now. The formula (2.112) is considerably simplified if we introduce new parameters

$$\xi = \sqrt{\frac{2a \, x^2}{g^2 \, \Delta t}}$$

$$\eta = \sigma \sqrt[4]{\frac{2a \, \Delta t}{g^2 \, x^2}} \tag{2.113}$$

and change the integration variable appropriately. The return distribution is then

$$P_X(x) = \sqrt{\frac{a}{2\pi^2 \, g^2 \, \Delta t}} \, e^{-\frac{1}{2}\xi \eta^2} \sum_{\rho=\pm 1} \int_{-\infty}^{\infty} e^{-\frac{1}{2}\xi \, \psi_\rho(u,\eta)} \, du \tag{2.114}$$

where

$$\psi_\rho(u,\eta) = e^{-2u} + e^{2u} - 2\rho \, \eta \, e^u. \tag{2.115}$$

Note that the single integral in Eq. (2.112) is transformed into the sum of two integrals in Eq. (2.115), distinguished by the sign $\rho = \pm 1$, due to the presence of the absolute value $|s|$ of the integration variable in the formula (2.112).

A large return implies a large value of the parameter ξ. Therefore, we can calculate the integrals in (2.115) by the saddle-point method. At the same time, the other parameter η becomes small, and we can safely expand all quantities in powers of η and retain only the lowest-order terms. This is first used when we calculate the position of the saddle point u^* by solving the equation $\partial\psi_\rho(u^*,\eta)/\partial u = 0$. We can easily find that $e^{u^*} = 1 + \eta/4 + O(\eta^2)$ and hence $\psi_\rho(u^*,\eta) = 2 - 2\rho\eta + O(\eta^2)$ and $\partial^2\psi_\rho(u^*,\eta)/\partial u^2 = 8 - 2\rho\eta + O(\eta^2)$. Putting everything together and neglecting the terms of the order $O(\eta^2)$ we obtain the following asymptotic formula for large returns:

$$P_X(x) \simeq \sqrt{\frac{a}{\pi\, g^2\, \Delta t}}\ e^{-\xi(1+\eta^2/2)}\, \xi^{-1/2}\, \cosh\xi\eta\ ,\ \xi \to \infty. \tag{2.116}$$

If the formula (2.116) is still not very transparent to the reader, note that it implies the following type of large-return distribution:

$$P_X(x) \sim \frac{1}{\sqrt{|x|}}\, e^{-|x|} \tag{2.117}$$

where we omitted all the dependence on various parameters of the model. The fundamental conclusion is that the Stein-Stein stochastic volatility model is characterised by exponentially decaying tails of the return distribution. Therefore, no true fat tails are present, but on the other hand the tails are considerably fatter than that of the Gaussian distribution predicted by a simple random walk with fixed volatility. This may seem a considerable improvement and indeed it may be, and was used to predict investment risks with higher precision than the bare diffusion models do [631, 632, 634, 636, 642, 643].

Encouraged by this progress, let us look at other species of stochastic volatility models. Perhaps they will work even better. At a minimum, we want to get rid of the pathological behaviour of the Stein-Stein model at $|x| \to 0$. Thus, let us look at the 'Ornstein-Uhlenbeck' model. Formally, it differs from the Stein-Stein model by setting $\gamma = 1/2$ instead of $\gamma = 1$. In words, the volatility is assumed equal to random variable S_t instead of S_t^2. There is a practical problem here; namely, S_t can take positive as well as negative values in the Ornstein-Uhlenbeck process. To avoid negative volatilities, we can impose a reflecting boundary at the point $S_t = 0$. It was difficult to implement the boundary when we investigated the process using the moments, so we decided to sweep the problem under the carpet. Now we approach the process through the Fokker-Planck equation, and it turns out that the introduction of the reflecting boundary is trivial. We just solve the Fokker-Planck equation without the boundary and then cut out the part corresponding to negative volatilities. This is possible because of the mirror symmetry of the stationary distribution for S_t. In so doing, we must not forget that the normalisation of the distribution is now changed. In fact, the Fokker-Planck equation for the volatility was already solved a few paragraphs above, since it is identical to the Stein-Stein case. Therefore, the solution will be a slight modification of the distribution (2.111), namely

$$P_S(s) = \sqrt{\frac{a}{\pi g^2}}\ \frac{2\,\theta(s)}{\operatorname{erfc}(-\sigma\sqrt{a/g^2})}\ e^{-\frac{a}{g^2}(s-\sigma)^2} \tag{2.118}$$

where $\theta(s)$ is the Heaviside function, and the short-time return distribution, in analogy with (2.112), is

$$P_X(x) = \frac{\mathcal{N}}{\sqrt{\Delta t}} \int_0^\infty \exp\left(-\frac{x^2}{2\,s\,\Delta t} - \frac{a}{g^2}(s-\sigma)^2\right) \frac{ds}{\sqrt{s}}. \qquad (2.119)$$

We absorbed all the necessary but bothersome normalisation into the constant $\mathcal{N}$. Note that for $\Delta x \to 0$ the integral converges, so the peculiarity of the Stein-Stein model is indeed cured. This will be true for all models with $\gamma = 1/2$.

Now we are ready to proceed in the same way we did with the Stein-Stein model. We introduce the following new variables

$$\xi = \left(\frac{2a}{g^2}\right)^{\frac{1}{3}} \left(\frac{x^2}{\Delta t}\right)^{\frac{2}{3}}$$

$$\eta = \sigma \sqrt[3]{\frac{2a\,\Delta t}{g^2\,x^2}} \qquad (2.120)$$

then we make an appropriate substitution in the integration variable and perform the saddle-point approximation, together with the expansion in powers of η. The result is

$$P_X(x) \simeq \frac{\mathcal{N}}{\sqrt[3]{\Delta t\,|x|}}$$

$$\times \exp\left[-\frac{3\xi}{2^{5/3}}\left(1 - \frac{2^{4/3}}{3}\eta + \frac{2^{5/3}}{9}\eta^2\right)\right], \quad |x| \to \infty \qquad (2.121)$$

where we again gathered all normalisation into a (new) constant $\mathcal{N}$. Schematically, we can say that the tail of the return distribution behaves like a stretched exponential

$$P_X(x) \sim \frac{1}{|x|^{1/3}}\,e^{-|x|^{4/3}}. \qquad (2.122)$$

The same procedure can now be repeated for the Heston model, for 'GARCH' and as many other variants of the stochastic volatility models one might like. For the Heston model, the tail of the return distribution is exponential

$$P_X(x) \sim \frac{1}{|x|^{1-2a\,\sigma/g^2}}\,e^{-|x|} \qquad (2.123)$$

while the 'GARCH' features the power-law tail

$$P_X(x) \sim \frac{1}{|x|^{1+4a/g^2}}. \qquad (2.124)$$

In fact, compared to the Stein-Stein or 'Ornstein-Uhlenbeck' models, the Heston and 'GARCH' models are even easier to deal with, as the integrals analogous to (2.112) and (2.119) can be computed explicitly. We encourage the reader to look up the formulae, e.g. in Ref. [644].

To summarise the results we have so far obtained for the stochastic volatility models, we can say that they are so plastic that it may be fairly easy to find agreement with the empirical return distribution. We can also easily perform an independent check of the obtained volatility distribution against the empirical data, [19, 178, 180, 182, 488]. On the other hand, we must constantly keep in mind that the real data are much more rich, and especially that the long-time correlations are missing in any of the stochastic volatility models.

2.5 What remains

The view of price fluctuations as a Brownian motion must resound in the physicist's mind, and various associations pop up. First, we already mentioned that the Fokker-Planck equation for such a process is just the heat conduction equation. Or, it might be considered to be like the Schrödinger equation in imaginary time. Suddenly we are in the realm of the quantum!

When we want to deal with quantum mechanics on an advanced level, we resort to path-integral formalism. Although it is fully equivalent to the standard treatment via the Schrödinger equation, in practice it brings some advantages. The way initiated by Bachelier finds natural continuation in the use of path integrals to the study of the stock market.

Perhaps the first pioneer in this direction was J. W. Dash, a former particle physicist. In a series of three technical reports [645–647] he adapted the idea of path integrals to the language of finance and showed how it may be useful in one of the central issues, pricing of financial derivatives. Several followers picked up the baton in the late 1990s [648–651], and the methods based on path integrals have kept being developed through the present day [652–663]. There are also at least three books describing the technique [664–666].

Without denying the achievements of the 'continuous-time finance', i.e. modelling based on stochastic differential equations, the approach using continuous-time random walk is superior by at least one grade. The basic-level continuous-time random walk described in Sec. 2.1.3 was developed in many directions. For example, there is an analogy of the Fokker-Planck equation where the usual derivatives are replaced by fractional ones. This seemingly formal algebraic acrobatics finds natural interpretations, describing certain continuous-time random walks; and, conversely, fractional derivatives seem to be a natural manipulating tool for continuous-time random walk. The advantage is that non-trivial power-law dependences immediately follow [197, 578].

The continuous-time random walks can be coupled and mixed in various ways [667, 668]. One should also not forget that so far the waiting times were prescribed by hand. For a more reliable results it is necessary to make a more fundamental model which would produce these waiting times and would be based on a simple mechanism, as in Refs. [669–671].

Another important ramification consists of applying many time scales in continuous-time random walk. Several independent sequences of jump times are produced, each sequence having its own characteristic waiting times and characteristic lengths of jumps. The sequences are then combined to make up the price-movement process. Its properties crucially depend on the distribution of time scales and on the relation between the characteristic waiting time and the characteristic jump length. It was shown [672] that such a model can provide a superdiffusive ($H > 1/2$) price movement. The relation of this multiple-scale model to a Markov-switching multifractal cascade immediately comes into mind.

There are also direct ways to obtain the power-law distributions found empirically in return distributions. Unfortunately, these paths to success usually lack a deeper understanding of the mechanisms behind the scene. One of the direct approaches is the use of multiplicative-additive random processes, already shown by Kesten [673] to exhibit power-law tails [674, 675]. The application of the idea to finance and economics is rather straightforward, and there are many works developing it to a considerable level of complexity [170, 173, 676–692]. Another such approach is based on the non-extensive statistical physics introduced by Tsallis [693], which is appropriate for non-equilibrium systems and is also characterised by power-law distributions [694] replacing the common exponential Boltzmann factors appearing in ordinary equilibrium statistical mechanics. Its relation to economics is investigated, e.g. in [695–698].

There have also been attempts to relate the non-extensive statistical physics to GARCH-like stochastic processes [699]. The GARCH process and relatives derived from it would deserve an entire book. For example, the simplest GARCH we presented here does not show power-law autocorrelations of absolute returns. But it is possible to have this feature if we postulate that the coefficients in the GARCH recurrence formula decay themselves as a power law [700]. From the opposite direction, people tried to 'derive' the ARCH or GARCH processes from more fundamental microscopic models [701]. Let us also mention the possibility of introducing multiple time scales here [702], just as it was done in continuous-time random walk, mentioned above. Again, the tempting possibility arises of uniting GARCH-based models with multifractal cascades. Of practical importance is the problem of fitting the GARCH parameters to the empirical data [703]. Comparison with the S&P500 Index can be found, e.g. in Ref. [422], and the work [532] compares the predictions of GARCH models with the observed behaviour of prices after a crash. In Ref. [704] the relation to the best linear forecast problem is investigated.

The physics of multifractal cascades is especially interesting and extends beyond econophysics modelling (see, e.g. Ref. [705]). It also poses non-trivial mathematical problems of intrinsic value, like the proper definition of a multifractal random walk and the use of wavelets in a more precise formulation of the cascades [253–255, 706–708]. On the other hand, it is not clear how to justify the cascade starting from a microscopic description of the market. An investigation of generic multifractality can be found, e.g. in Ref. [709].

With the section on stochastic volatility models we returned back to the 'continuous finance' methodology, albeit on a more sophisticated level. We do not have enough space to show all developments. Moreover, an encyclopedia of solved stochastic volatility models would very soon become boring. The interested reader may look at Ref. [488] for a systematic comparison of wide range of stochastic volatility models. An interesting comparison of volatility autocorrelations in stochastic volatility and ARCH-like models can be found in [710]. From the plethora of variants [711–719], let us mention explicitly only the works on the exponential Ornstein-Uhlenbeck process [488, 631, 643, 720–722] and the explanation of the leverage effect, using two correlated sources of noise for the price and the volatility processes [256, 258–260, 723]. A related approach also provides the basis for the 'string' theory of the forward interest rate curve [724, 725]. Finally, there is also a book on option pricing with stochastic volatility models [726].

Problems

1. The key ingredient of continuous-time random walks is the waiting time distribution. One should always make some hypothesis about how these waiting times occur. If we must wait for a single random event which happens with average frequency λ, the distribution is exponential with characteristic time λ^{-1}. We can imagine a slightly more involved scenario. Two events must happen one after the other before the price changes. The first event has a larger frequency λ_1, and the other is slower, with the frequency $\lambda_2 < \lambda_1$. The distribution of waiting times is then $P_{\Delta T}(t) = \frac{\lambda_1 \lambda_2}{\lambda_1 - \lambda_2} \left(e^{-\lambda_2 t} - e^{-\lambda_1 t} \right)$. Investigate this case.

2. The Mandelbrot binary cascade is assumed symmetric, i.e. M and $1 - M$ have the same distribution. Show that it implies that $\langle M \rangle = \frac{1}{2}$, and hence $\zeta(2) = 1$. Conclude that observing only the second moment of the returns, it is impossible to distinguish this multifractal signal from an ordinary random walk. Generalise the argument for multinomial cascades.

3. Calculate the multifractal spectrum produced in the binary cascade with exponents given by (2.64). Show that $\max_h d(h) = 1$.

4. Show that in the cascade model, the autocorrelation function of the logarithms of volatility decays logarithmically with the time difference [707]. Note: this fact was observed empirically, see, e.g. Refs. [253–255].

5. What is the decay of absolute return autocorrelations in the Markov-switching multifractal model if we suppose that the breakpoint frequencies λ_j increase linearly with j? What if they increase quadratically?

6. Calculate the moments $\mu_{2l,n}$ for the 'GARCH' stochastic volatility model and show that they diverge exponentially when $t \to \infty$ for large enough $l + n$. Conclude that the return probability distribution has a power-law tail. What is the dependence of the tail exponent on parameters a and g?

7. The short-time limit for the return distribution in a stochastic volatility process can in principle be found for an arbitrary combination of parameters α, β, and γ. Complications arise from the fact that the argument of the exponential in the integral analogous to (2.112) and (2.119) contains several different powers of s, and the saddle point cannot be found explicitly. However, there are certain combinations of α, β, and γ for which s occurs only in two different non-zero powers, the saddle point is easily found, and the tail of the return distribution can be calculated exactly. What are these combinations? Show that the Heston and 'GARCH' models belong to these special cases.

3

Basic agent models

3.1 Aggregation models

In the classical view of an efficient market there are two sources of price fluctuations. First, there are randomly arriving items of news, indicating that the price should rise or go down. This is the component of price movements which one may call deterministic. Indeed, knowing the mechanism of the rational price adjustment, provided the news is known, there is no ambiguity as to the value of the new price. Second, the individual differences between the economic agents and imperfection of their actions contribute to a stochastic element.

However, both of these contributions have in common that they result, after a sufficiently long time, in normal distribution of price changes, due to the central limit theorem. But how does it go with the observed fat tails of the return distribution? It is evident that some of the assumptions should be replaced by more sophisticated ones. The problem is how to justify the new principles, and how to see that they are more sensible than the classical efficient market hypothesis.

The first attempt to fix this weak point is the idea that the external news influencing price changes itself has a distribution with power-law tails, and this distribution is automatically transferred into the distribution of returns. This is the path Mandelbrot followed in his suggestion [727] that the returns follow the Lévy-stable distribution. We can equally well suppose that it is not the external influence but the uncertainty about the agent's reaction that is Lévy-distributed. However, it is no less troublesome to explain to an unbiased curious audience why we consider it a success to be explaining fat tails by other fat tails.

Perhaps it is wiser to add one more level of complexity into our system and suppose that the agents do not act independently. Instead, they can form aggregates of various sizes, and the agents within the same aggregate perform an identical action. Formally, we can say that the assumption of independent random variables, which leads to normal distribution according to the central limit theorem, is lifted. Thus, we may expect a much wider variety of distributions, and, if we are lucky, the power-law tails of return distributions will follow.

The ways the agents can aggregate with each other are multiple. We shall present several possibilities in the following sections.

Self-organised criticality Box 3.1

is a concept introduced in 1987 as an attempt to explain the so-called $1/f$ noise, which was observed in many different systems but whose origin was unclear [728]. Self-organised criticality occurs in open dynamical systems with infinitesimally small external driving and is generally connected with the existence of one or more absorbing states, i.e. states which remain unchanged in the model dynamics. In the kinetic Ising model in 1 dimension the absorbing states are the two configurations with all spins identical (either all $+1$ or all -1). Analogically, in the voter model it is the homogeneous state. In the sandpile model (see Box 3.2) it is any of the stable configurations in which no site exceeds the threshold height. If we kick off the system by an infinitesimal perturbation (in the 1D Ising model by flipping a single spin, in the sandpile model by adding one grain on a randomly chosen site, etc.), we may observe what is going on before the system gets stuck in an absorbing state again, or, as we call it, during an avalanche. We can measure the duration of the avalanche t, its size s, defined as the number of elementary changes in the course of the avalanche (such as flipping a single spin, toppling on a single site), etc. The system is self-organised critical if the distribution of avalanche sizes (and durations) has a power-law tail, $P(s) \sim s^{-\tau}$, $s \to \infty$. For example in the 1D kinetic Ising model, or the sandpile model above its upper critical dimension, the exponent is $\tau = 3/2$.
Many real systems were supposed to be self-organised critical in the above-mentioned sense, and models exhibiting power-law distributions of avalanche sizes were developed for their description. Examples include earthquakes [729, 730], biological evolution [731], dislocation dynamics in solids [732], internal mechanical collapse [733], traffic jams [734, 735], friction [736], and many more.

3.1.1 Economic sandpile

Every child knows sand very well. If wet, it so easily obeys our desire of form that children play with it indefinitely. If dry, it is much less exciting, but still one can heap up large piles of sand, observing how the grains added at the top roll down the slope.

The dynamic of the sandpile is highly nonlinear. Adding one grain usually does not make any change. Occasionally, however, the grain can trigger an avalanche involving a large number of grains sliding down. The point is that there is a threshold slope, and if exceeded locally, the configuration of the grains is no longer stable and prompts reorganisation of the surface of the sandpile.

What if we identify the economic system with the sandpile and avalanches with bursts of activity resulting in the change of price? The distribution of avalanche sizes would then be reflected in the distribution of price changes. The ensemble of grains moving side by side can be compared to the concerted action of several, or many agents forming dynamical aggregates which act together as large super-individuals and then break up again, only to wait until another economic avalanche emerges.

Self-organised criticality

This idea was first put forward in 1993 by Per Bak and his coworkers [737, 738] as a ramification of the earlier work [728] of Bak et al introducing an incredibly fruitful concept of self-organised criticality (see Box 3.1). In the theory of self-organised criticality, the power-law distributions emerge spontaneously without need of parameter tuning, and there was a time when econophysicists shouted with joy when hearing this good news. The sandpile model was the first and became the most typical abstract modelling scheme in self-organised criticality.

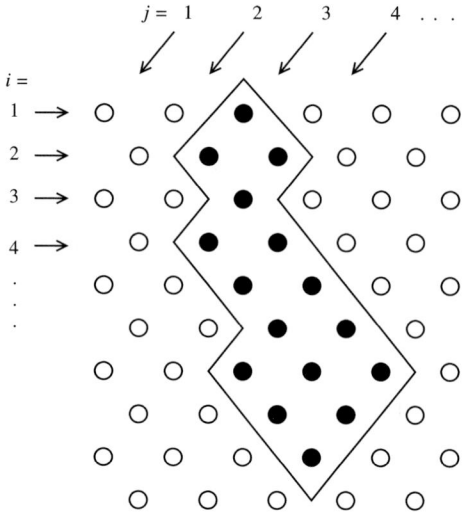

Fig. 3.1 Sketch of the economic sandpile model. The activity proceeds from the top to the bottom. The example of an avalanche starting at the point $(1, 3)$ is shown by filled circles. The lines delimiting the area covered by the avalanche can be regarded as traces of two random walkers annihilating at the end. Note that the 'columns' numbered by index j are in fact tilted by 45 degrees.

There is its simplified version, also called the directed sandpile model [739], and we shall now formulate it in the language of economics. For details on the original sandpile model, we refer the reader to Box 3.2.

Imagine that the function of the economic agents consists in sending items of goods to their customers and ordering goods from their suppliers. The agents sit on the nodes of an oriented production network where the edges are oriented from the customers to the suppliers. For simplicity we shall assume that the network has the form of a regular square lattice rotated by 45 degrees, as depicted in Fig. 3.1. So, the rows in the production networks are the diagonals in a square lattice.

The topmost row represents the final consumers. One layer below we have the agents who send their products to final consumers and order the material from their suppliers on the next row below. The crucial assumption making the model non-trivial is the nonlinear response of the agents to the orders they receive. Denote z_{ij} the number of unsatisfied orders the agent in row i and column j has in her agenda. She remains idle until the heap of orders on her desk reaches a threshold value $z_c = 2$. If that happens, she wakes up and solves the situation by sending one order to each of her 2 suppliers. After that, she considers 2 orders in her business resolved and her state variable diminishes by 2. Such event is called toppling and is formally described by the following update rules for the numbers of waiting orders

---

Sandpile model Box 3.2

of Per Bak, Chao Tang, and Kurt Wiesenfeld was the first example of a self-organised critical system [728, 740]. It is an infinitesimally slowly driven cellular automaton with open boundaries. It is defined on a d-dimensional hypercubic lattice ($d \geq 2$) generated by elementary translation vectors e_α, $\alpha = 1, \ldots, d$. The state of each lattice point x is determined by the time-dependent height variable $z_x(t)$. The lattice is supposed to be large but finite: if x_α are the coordinates of the point x, we work within the cube $0 \leq x_\alpha < L$. Outside the cube we fix $z_x(t) = 0$ for all times, which implements the free boundary conditions. There is a threshold z_c determining whether the site is active ($z_x \geq z_c$) or idle. These dynamics proceed in a sequence of avalanches, according to the following two rules.

1. External driving. If $\forall x : z_x(t) < z_c$ holds, the system is in a static (absorbing) state. If there had been an avalanche active in the last step $t - 1$, it is terminated now. Then, we choose a lattice point x_0 at random and drop a 'grain' there:
$$z_{x_0}(t + 1) = z_{x_0}(t) + 1.$$
If the threshold is reached at the selected site, $z_{x_0}(t + 1) \geq z_c$, a new avalanche starts, and we proceed to rule 2. Otherwise we repeat rule 1.

2. Cellular automaton updating. If $\exists x : z_x(t) \geq z_c$, we are in an avalanche. We update the state of all sites:
$$z_x(t + 1) = z_x(t) - 2d\,\theta(z_x(t) - z_c) + \sum_{\alpha=1}^{d} \sum_{\sigma \in \{+1,-1\}} \theta(z_{x+\sigma e_\alpha}(t) - z_c)$$
with the notation $\theta(z) = 1$ if $z \geq 0$ and $\theta(z) = 0$ if $z < 0$. The second term on the right-hand side accounts for redistribution of $2d$ grains from the site x to its $2d$ neighbours, if the threshold z_c is reached or exceeded. Such an event is called toppling. The last term corresponds to the grains received from all toppling neighbours.

The number of topplings during one avalanche is called the size s of the avalanche. Numerically it was found that the distribution of avalanche sizes satisfies a power law, $P(s) \sim s^{-\tau}$ with exponent $\tau = 1.27 \pm 0.01$ in $d = 2$ [741]. The mean-field approximation, based on the theory of branching processes, gives the value $\tau = 3/2$ [742–748].

$$z_{ij} \to z_{ij} - 2$$
$$z_{i+1\,j} \to z_{i+1\,j} + 1 \tag{3.1}$$
$$z_{i+1\,j+1} \to z_{i+1\,j+1} + 1$$

Now, the suppliers on the next row must do their job, but they react in the same nonlinear way, considering it worthless to switch on before they have enough, i.e. at least $z_c = 2$, items of work to do. If they reach the value 2, they become active, send the orders to their suppliers, and the turn is now on the next layer of agents. An important point is that the production chain is open and finite, however large it might be, so the agents in the last layer, who produce the most basic raw materials, do not have anybody to send their orders to. When they become active, they diminish their state variable by 2 as everybody else, but without passing the order further. We can imagine that they dig into the earth and find oil. This guarantees that the activity will eventually stop, at most at the lowest layer. Only after all agents become inactive (we call that configuration an absorbing state), a new order is sent to a randomly chosen agent in the topmost layer.

Avalanches and price changes

The picture which follows is characterised by avalanches of activity propagating down the production chain. We can view the avalanches as longer or shorter jumps from one absorbing state to another. The activity wave travels in a specific direction and never returns within a single avalanche. That is why the model is also called the directed sandpile model. As usual in the self-organised critical models, the distribution of avalanche sizes obeys a power law. Before proceeding with details on the power-law distribution, we must say a few more words on the economic interpretation of the model.

Each of the avalanches can be interpreted as a temporary coalition of agents responding to a single external stimulus. Such a response will influence the prices of various goods and the price impact will be larger as the avalanche becomes larger. For simplicity we can assume that the relative price difference, or return, is proportional to the avalanche size s, defined as the number of agents which were active during the avalanche. To avoid confusion, we should note that we distinguish between two separate time scales. From the viewpoint of the sandpile dynamics, some topplings occur in each time step until the absorbing state is reached; and the avalanche stops after some time t_a, which is the duration of the avalanche. On the other hand, from the perspective of somebody who observes the price changes, the avalanches occur instantaneously and in each time the size of the avalanche $s(t)$ (which might also be zero) is determined. We might say that the time is measured in the number of grains dropped on the sandpile from the outside, because each avalanche is triggered by dropping a single grain.

This way we can prescribe how much the logarithm of price $Y_t = \ln Z_t$ should change due to the avalanche. As for the sign $\epsilon_t = \pm 1$ of the price difference, we assume it is random and uncorrelated from time to time, $\langle \epsilon_t \epsilon_{t'} \rangle = \delta_{tt'}$. Hence the return is assumed to be

$$X_t = Y_t - Y_{t-1} = b\,\epsilon_t \frac{s(t)}{N} \tag{3.2}$$

where $s(t)$ is the size of the avalanche that occurred at time t, N is the total number of agents, and the parameter b quantifies the strength of the price impact. With such simple relationship, the distribution of returns will be a copy of the distribution of avalanche sizes. We may begin to feel satisfaction: the autocorrelation of returns is zero by definition, and the distribution of returns obeys the power-law characteristic for self-organised critical systems. This agrees perfectly with the empirical data, but this agreement is only qualitative. We would like to be more specific and know, for example, the value of the exponent in the power law.

Exact solution

Fortunately, the model is exactly solvable based on a straightforward consideration. First, we should note that the avalanche always covers a compact set of lattice points, free of any voids. To formalise this observation, we state that for each avalanche there are two sequences of column indices $l_1, l_2, l_3, \ldots, l_t$, and $r_1, r_2, \ldots, r_t$, such that all sites (i, j) with $1 \le i \le t$ and $l_i \le j \le r_i$ toppled exactly once and all other sites never

> **Discrete Laplace transform** | Box 3.3
>
> of a function $f(s)$ of an integer variable $s \geq 0$ is defined as
> $$\hat{f}(z) = \sum_{s=0}^{\infty} z^s f(s) \qquad (*)$$
> for $z \in \mathbb{C}$. If $f(s)$ is a probability distribution, then the normalisation $\sum_s f(s) = 1$ ensures $\hat{f}(1) = 1$. The behaviour of $f(s)$ for $s \to \infty$ corresponds to the behaviour of $\hat{f}(z)$ at $z \to 1$. For example, the power-law tail $f(s) \sim s^{-\gamma}$, $s \to \infty$ is translated to the singularity $1 - \hat{f}(z) \sim (1-z)^{\gamma-1}$, $z \to 1^-$, as can be hinted by approximating the discrete sum $(*)$ by an integral. The exponentially truncated power-law tail $f(s) \sim s^{-\gamma} \exp(-hs)$ occurs when the singularity is shifted to the point $z = e^h$, i.e. $\hat{f}(e^h) - \hat{f}(z) \sim (e^h - z)^{\gamma-1}$. In many cases investigated in this book the singularity is a square root, $\gamma - 1 = 1/2$, which corresponds to the tail with exponent $\gamma = 3/2$. In probability theory, the discrete Laplace transform is usually called a characteristic function.

toppled during the avalanche. These sequences are paths delimiting the avalanche from the left and right sides, as we also show schematically in Fig. 3.1.

To prove that, we proceed by induction. Indeed, the avalanche starts on a single point $(1, j_0)$ in the first layer, so $l_1 = r_1 = j_0$. Supposing that the statement is true for row i, we prove that it also holds for $i+1$ by constructing the appropriate l_{i+1}, r_{i+1}. We know that all sites in columns $l_i \leq j \leq r_i$ toppled. Therefore all columns $l_i + 1 \leq j' \leq r_i$ in the row $i + 1$ received 2 grains from the row i and therefore also toppled. Four cases are possible now, depending on the height variables in columns l_i and $r_i + 1$. If $z_{i+1\,l_i} = z_{i+1\,r_i+1} = 0$, then $l_{i+1} = l_i + 1$, and $r_{i+1} = r_i$; if $z_{i+1\,l_i} = z_{i+1\,r_i+1} = 1$, then $l_{i+1} = l_i$, and $r_{i+1} = r_i + 1$. In the latter case the avalanche widens on both sides, while in the former it shrinks from both the left and the right. We are sure the reader gets the idea and already knows the result in the remaining two cases. This completes the proof, but we are not yet ready to determine the statistic of avalanches. To this end, we must know more about the sequences determining the two edges of the avalanche. Clearly, both of them can be considered as walks on a square lattice, with steps of length 0 or 1. Both of the walks start at the same point $(1, j_0)$. It is evident from the proof above that they must also end at the same point (or at the bottom of the lattice; we ignore this finite-size effect for now, since it occurs very rarely in a large system), and the avalanche is therefore delimited by the paths of two walkers who departed from one point and annihilated each other when they first met. The next crucial piece of the whole picture is the fact that they are in fact random walkers, because all four possibilities of continuing with the left and right edges of the avalanche to the next row are equally probable. This follows from the finding that all inactive configurations $z_{ij} \in \{0, 1\}$ of the economic sandpile occur with the same probability values. We shall not prove this property here and refer to the original literature [739] presenting an exact solution of the directed sandpile model.

Random walker returning to the origin

Calculating properties of the avalanches equals to investigation of paths traced by pairs of random walkers starting at the same point at time 0 and annihilating each other at later time t. When we observe the second walker relative to the position of the first walker, the situation is equivalent to calculating the distribution of times the walker spends up to her first return to the point of departure. Now, our aim will be to prove that the distribution of first-return times, i.e. as well as the duration of the avalanches, obeys a power law

$$P_{\text{dur}}(t) \sim t^{-3/2}, \ t \to \infty. \tag{3.3}$$

From this result it follows that the distribution of avalanche sizes is

$$P_{\text{siz}}(s) \sim s^{-\tau}, \ s \to \infty$$
$$\tau = \frac{4}{3}. \tag{3.4}$$

Indeed, if the walker spent time t before returning back, her typical departure from the origin was $\sim t^{1/2}$, and the area covered by the avalanche is therefore $s \sim t^{3/2}$. Substitution in the result (3.3) immediately gives (3.4).

We describe the position of the walker at time t by random integer-valued variable $N_t \in \mathbb{Z}$. Our single walker representing the original pair of walkers can make a step to the left or right with equal probabilities $1/4$, or remain at the same point with probability $1/2$. This is exactly the same as if we let a walker move only left or right, but observe her position every second step.

So, we investigate the first return times of a simple random walk with probabilities $\text{Prob}\{N_{t+1} - N_t = \pm 1\} = 1/2$. Initial condition is $N_0 = 0$, and the time of the first return to origin is also a random variable, denoted T. So, $N_T = N_0 = 0$ and $N_t \neq 0$ for $0 < t < T$. Denoting $P(n; t) \equiv \text{Prob}\{N_t = n, N_0 = 0, N_{t'} \neq 0 \text{ for } 0 < t' \leq t\}$ the probability that the walker who started at the origin arrived to point n at time t but never visited the origin again, we can write the following recursive relation

$$P(n; t) = \delta_{t0}\delta_{n0} + \frac{1}{2}(1 - \delta_{t0})(1 - \delta_{n0})\left(P(n - 1; t - 1) + P(n + 1; t - 1)\right). \tag{3.5}$$

The Kronecker deltas ensure the obvious initial and boundary conditions $P(0; t) = 0$ for $t > 0$ and $P(0; 0) = 1$. Knowing $P(n; t)$ we can immediately deduce the distribution of first return times to origin, as

$$P_{\text{f.r.}}(t) = \frac{1}{2}\left(P(1; t - 1) + P(-1; t - 1)\right). \tag{3.6}$$

It is easier to work with the discrete Laplace transform (see Box 3.3). After a bit of algebra on equations (3.5) and (3.6) we get the following simple expression

$$\hat{P}_{\text{f.r.}}(z) = 1 - \sqrt{1 - z^2}. \tag{3.7}$$

The time dependence can be recovered by expanding it into a power series in z. Recalling that the time of the avalanche proceeds twice as fast compared to the walker described by the relation (3.5), we deduce the exact distribution of avalanche durations

$$P_{\text{dur}}(t) = \frac{(2t)!}{2^{2t+1}\, t!\, (t + 1)!}. \tag{3.8}$$

The Stirling formula (see Box 3.5 if obscure) for large t yields the asymptotic behaviour (3.3). In fact, the asymptotic power behaviour with exponent $3/2$ follows immediately from the square-root singularity of $\hat{P}_{\text{f.r.}}(z)$ at $z \to 1^-$.

One may ask how the result would change if we put the economic agents on a different supply network. The obvious generalisation is to use a d-dimensional hypercubic lattice instead of the planar square network investigated so far. It turns out that the directed sandpile model is exactly solvable in any dimension [739], but for all dimensions $d \geq 3$ we get the same exponent in the distribution of avalanche durations, $P_{\mathrm{dur}}(t) \sim t^{-2}$ and avalanche sizes $P_{\mathrm{siz}}(s) \sim s^{-3/2}$.

Results and objections

The list of questions posed by empirical econophysics always start with the problem of fat tails in the return distributions. The self-organised critical model of economic activity seems to provide a transparent explanation of the power-law tails in terms of avalanches of various sizes spreading through the economy. The economic sandpile model exhibits the return distribution according to the law

$$P_X(x) \sim x^{-1-\alpha}, \text{ for } x \to \infty \tag{3.9}$$

with the return exponent α related to the avalanche size exponent τ by

$$\alpha = \tau - 1 = \frac{1}{3}. \tag{3.10}$$

A smart enough researcher would come up with many more variants of the theme exemplified by us with the directed sandpile model. All of them would result in the same power-law behaviour (3.9), only differing in the value of the exponent τ. For example, as already mentioned, if we only increase the dimensionality of the supply network in the economic sandpile model to 3 or more, the exponent changes to $\tau = 3/2$. It is therefore vital to discriminate between the candidate models by comparing the return distribution quantitatively with the real data.

Unfortunately, it must be stated that the sandpile model does not pass this simple test. Indeed, as we have seen in Sec. 1.2.2, the exponents found in reality are much higher, with values around $\alpha \simeq 3$. Even worse, it was found that all self-organised critical models are generically characterised by avalanche exponents $\tau \leq 3/2$, the maximum value $3/2$ being reached in the mean-field approximation, or above the upper critical dimension (if the latter exists). The conclusion is that despite the qualitative success, the self-organised criticality cannot serve as a direct explanation of the observed price fluctuations. The exponent is too far from the data. Even if we used a different prescription for the price impact of avalanches, i.e. we postulate a nonlinear function, e.g. $X \propto \sqrt{s}$ instead of the linear relation (3.2), the situation would be only slightly improved.

On the other hand, it does not mean that self-organised criticality should be judged irrelevant for describing economic phenomena. Actually the framework of self-organised criticality helps to understand some of the models of social organisation, as will be shown in Chap. 8. It is also possible that other power-law dependencies found in economics, such as distribution of transaction volumes, can be traced to avalanche phenomena pertinent to self-organised criticality. Finally, let us mention one open question connected to our economic sandpile model. It is expected that all types of supply networks result in exponent τ not exceeding the mean-field value $3/2$. But

Percolation | Box 3.4

is a phenomenon occurring in various physical systems. As an example of site percolation we can consider a container full of closely packed balls, some of them being metallic, some of them insulating. Voltage is applied to the opposite sides of the container, and we ask whether the particular configuration of the balls is conducting or not. An affirmative answer means that we have here a cluster of mutually touching metallic balls extending from one side of the sample to the opposite side. The probability P that a given site belongs to such a spanning cluster depends on the concentration $p \in [0, 1]$ of the metallic balls. In the thermodynamic limit there exists a critical concentration p_c called the percolation threshold, such that $P = 0$ for $p < p_c$, $P > 0$ for $p > p_c$, and the function $P(p)$ is continuous at $p = p_c$. So, the percolation threshold marks a second-order phase transition with the density as a control parameter.

Completely analogous is the dual phenomenon of bond percolation. Here, we may think of molecules sitting in the nodes of a regular lattice (but any graph can be used as well). Each pair of neighbours on the lattice is connected by a bond with probability p. The mechanical properties of the system change dramatically at a critical concentration p_c of bonds, when a cluster occurs extending from one side of the sample to the other side. In this case the system becomes stiff. Otherwise it is a viscous liquid.

The critical behaviour at the percolation threshold, $p \to p_c$, is described by a host of critical exponents. For example the number of clusters of size s is scaled as $n_s \sim s^{-\tau} e^{-bs}$, where $b \sim |p - p_c|^{1/\sigma}$; the fraction of nodes contained in the largest cluster behaves as $P \sim (p - p_c)^\beta$; the mean cluster size as $S \sim |p - p_c|^{-\gamma}$, etc. Scaling relations hold for the values of the exponents: $\beta = (\tau - 2)/\sigma$, $\gamma = (3 - \tau)/\sigma$. Both bond and site percolation are exactly solvable in one dimension and on the Bethe lattice. In the latter case $\tau = 5/2$ and $\sigma = 1/2$, and these values are exact in dimensions $d \geq 6$. For regular lattices in dimensions $d \geq 2$ the renormalisation group methods are applied to calculate the critical exponents. Many results are obtained only thanks to numerical simulations. In $d = 2$ the hypothetically exact values are $\tau = 187/91$ and $\sigma = 36/91$. A detailed account of percolation can be found, e.g. in the textbook [751] or the review article [752].

what will happen if we allow the topology of the network to evolve in response to the activity of the agents? A motivated reader can take inspiration, for example, from the work [749].

3.1.2 Cont-Bouchaud model

In the sandpile model the agents are grouped dynamically, each time anew. Sometimes, however, we can find long-lasting coalitions, which act together as large meta-agents. Alternatively, we can imagine herds of agents imitating each other and forming large clusters bound together by strong pairwise ties. To mimic this herding effect, Rama Cont and Jean-Philippe Bouchaud introduced a model [750] closely related to the effect known in physics as bond percolation, while mathematicians speak about Erdős-Rényi random graphs.

Random graphs and bond percolation

When cooked gelatin cools down and forms a gel, when cement paste thickens so that the builders can remove the formwork, or latex vulcanises due to sulphur being added, we are always witnesses to the phenomenon of percolation (see Box 3.4). The units making up our system bind gradually one to the other, and when the number of bonds exceeds a certain critical threshold, the fluid assembly of particles turns into a solid, though spongiform, body. If we let the process continue, the concentration of bonds increases further and the material hardens more and more, resulting in quality rubber or in a first-class concrete.

A similar mechanism can also be at work when investors on the stock market are herding under the influence of a rumour or a collective panic. We can imagine that bonds are added between agents, and once bonded, agents always perform identical actions. All agents bound together form a single herd, or cluster. Generally, many clusters of various sizes are formed this way. To asses the influence of the herds on the movements of the stock market we must know something about the sizes of the clusters.

First of all we must determine which bonds between agents are a priori allowed. We assume there is a virtual social network connecting the agents, and it is only on the links of that network where the bonds can be placed. Then, as the clusters are determined by the configuration of bonds, we speak about bond percolation on that network. For example, one can consider a very unrealistic, but theoretically appealing case of agents placed on the nodes of a regular network, be it a linear chain, a square lattice or a hypercube. Or, one can give up the complications of the social structure and assume the network is a complete graph, which means that everybody is connected to everybody else, and the bonds can be established between any pair of agents. The latter case corresponds to bond percolation on the complete graph, and that is the case we shall develop below.

So, let us have N nodes, and each pair of nodes be connected by a bond with probability $p \in [0, 1]$. The average number of bonds is therefore $\overline{N_B} = p\,N\,(N-1)/2$. It is also evident that the average degree (number of attached bonds) for any of the nodes is $\bar{k} = p\,(N-1)$.

In fact, the structure created by bonds distributed randomly among the pairs of nodes was investigated in late 1950s by the Hungarian mathematicians Pál Erdős and Alfréd Rényi. The concept they introduced was called a random graph [753]. We shall return to various ramifications of the theory of random graphs in Chap. 6. For now, we mention only a few results relevant to our percolation problem.

Indeed, we asked about the distribution of cluster sizes. Certainly it will depend on the parameter p. The first thing we should clarify is the size N_1 of the largest cluster. We assume that the number of agents N is large, thus we need to know what happens with the cluster sizes in the limit $N \to \infty$. We also suppose that p depends on N as $p = c/N$, with a certain constant c independent of N. The largest cluster can comprise a finite fraction of nodes, $N_1/N \to g(c) > 0$. In this case it is called a giant component. Or, its size can grow only logarithmically with the number of nodes, $N_1 = O(\log N)$. The former case happens for $c > 1$, the latter for $c < 0$. The value $c = 1$ marks the percolation threshold, characterised by the emergence of a giant cluster which extends over a macroscopic part of the system. In fact, the relative size of the giant component $g(c)$ is the order parameter of the percolation transition.

It is not realistic to assume that a very large portion of the market participants herd together and buy or sell unanimously. If that happens, the market crash seems inevitable, which in turn shakes all opinions, and coalitions have to be rebuilt from scratch. This implies that we can exclude the situation above the percolation threshold,

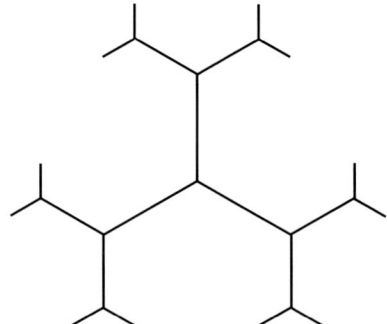

Fig. 3.2 An example of a finite Bethe lattice with node degree $k = 3$. Note that there are 12 nodes on the surface and only 10 nodes in the bulk of the lattice. It is a common feature of all Bethe lattices that the surface contributes by a finite fraction of the nodes even when the overall size goes to infinity.

$c > 1$, from our considerations. However, it is not too silly to suppose that the traders hazard approaching the percolation transition as much as possible. So, we want to calculate the distribution of cluster sizes below but close to the percolation threshold.

Loops and Bethe lattices

To find exact information on cluster sizes is not an easy task. The treatment is substantially simplified if we acknowledge that the percolation clusters are essentially trees. Indeed, a tree is characterised by the absence of closed loops, and it was shown that the length of typical cycles grows with the number of the nodes as $\ln N$ (see Sec. 6.1.2). So, we shall not be much mistaken if we ignore the loops entirely.

The cleanest way to avoid cycles in our clusters is to study the percolation problem on a regular tree structure, that is, the Bethe lattice. An example of the Bethe lattice is depicted in Fig. 3.2. We suppose that all nodes, except those at the extremities of the graph, have a fixed degree k. Let us now compute the probability that a randomly chosen node i belongs to a cluster of size s, i.e. containing s nodes connected by $s - 1$ bonds. The central node i can be connected to $l \leq k$ of its neighbours. If we now cut off the node i, the graph splits into k disconnected branches and so does the percolation cluster. Now we have l sub-clusters, each of them headed by one of the neighbours of the central node i. The sub-clusters are independent of each other. Knowing the probabilities $P_{\mathrm{sub}}(s_{1j})$ that the sub-clusters headed by the neighbour j of site i contain s_{1j} nodes, we combine the probability that the entire cluster contains s nodes

$$P(s) = \sum_{l=0}^{k} \binom{k}{l} p^l (1-p)^{k-l}$$

$$\times \sum_{s_{11}} \sum_{s_{12}} \cdots \sum_{s_{1l}} \prod_{j=1}^{l} P_{\mathrm{sub}}(s_{1j}) \, \delta \left(1 + \sum_{j=1}^{l} s_{1j} - s \right).$$

(3.11)

Using the discrete Laplace transform we obtain the more compact expression

$$\hat{P}(z) = z\left(1 - p + p\,\hat{P}_{\text{sub}}(z)\right)^{k}. \tag{3.12}$$

In a similar way we proceed one step further and remove the heading node from each of the sub-clusters. They in turn split into disjoint sub-subclusters, and if we suppose our system is large enough, the size distribution for sub-clusters coincides with the distribution of sub-subclusters. We therefore get a closed equation for $P_{\text{sub}}(s)$, which in Laplace transform reads

$$\hat{P}_{\text{sub}}(z) = z\left(1 - p + p\,\hat{P}_{\text{sub}}(z)\right)^{k-1}. \tag{3.13}$$

Obviously $\hat{P}_{\text{sub}}(1) = 1$. The tail of the cluster size distribution corresponds to the behaviour of $\hat{P}(z)$ for $z \to 1$, and close to the point $z = 1$ we can expand Eq. (3.13) in powers of $\hat{P}_{\text{sub}}(z) - 1$. This leads to the approximate quadratic equation

$$z - 1 + (zp(k-1) - 1)(\hat{P}_{\text{sub}}(z) - 1) + zp^2\frac{(k-1)(k-2)}{2}(\hat{P}_{\text{sub}}(z) - 1)^2$$
$$+ O((\hat{P}_{\text{sub}}(z) - 1)^3) = 0, \tag{3.14}$$

implying a square-root singularity in $\hat{P}_{\text{sub}}(z)$ located at a certain point z_0. For the percolation transition the singularity occurs at $z_0 = 1$; hence the percolation threshold is

$$p_c = \frac{1}{k-1}. \tag{3.15}$$

Below the transition we find $z_0 - 1 \simeq (p - p_c)^2/\left(2\,p_c(1 - p_c)\right)$. Using the expression (3.12) the location and the square-root type of the singularity in $\hat{P}_{\text{sub}}(z)$ is directly inherited also in $\hat{P}(z)$. So, the probability that a given node is contained within a cluster of size s follows an exponentially truncated power law with exponent $3/2$. Note that we get the same value of the exponent as when we investigated the first returns of the random walker to the origin. In both cases the value $3/2$ can be traced to the square-root singularity of the discrete Laplace transform, which is a typical feature of various exactly solvable models.

We are now close to the end of our calculations, but we still need to make a few additional steps before we can write the cluster size distribution for our original percolation problem on a complete graph. First, the degree of nodes in our Bethe lattice is k, but in the original graph it was $N - 1$; so we must substitute $k = N - 1$ in our formulae. Next, in the thermodynamic limit the size of the graph goes to infinity, and so does also the degree k. To keep relevant quantities finite, we must rescale the probability of establishing a bond $p = c/N$, as suggested before. Then, in the limit $N \to \infty$ the percolation threshold occurs at $c = 1$ (and we can start breathing again, as we are compatible with the Erdős-Rényi rigorous result). The probability that a given site happens to fall within a cluster of size s is proportional to s; therefore we can conclude that the distribution of cluster sizes slightly below the percolation threshold is

$$P_{\text{clu}}(s) \sim s^{-\frac{5}{2}} \exp\left(-\frac{1}{2}(1 - c)^2\,s\right), \quad s \to \infty. \tag{3.16}$$

Herds in a stock market

Now we know how large the herds of market sheep are, but we still keep watching with curiosity, to see what the animals are going to do. To teach them something simple enough, we shall prescribe the following rule of conduct.

Suppose there are N_c clusters of sizes s_l, $l = 1, \ldots, N_c$. In each step of trading every cluster of agents decides either to buy a unit of stock with probability a, or sell a unit of stock with the same probability a, or stay inactive with probability $1 - 2a$. Denoting $\sigma_l(t) \in \{-1, 0, +1\}$ the action of the cluster l at time t (with obvious meaning of the values $0, \pm 1$), we assume that the change in the logarithm of price due to the excess demand or supply is given by the same linear impact function

$$X_t = Y_t - Y_{t-1} = b \sum_{l=1}^{N_c} \sigma_l(t)\, s_l, \tag{3.17}$$

as was postulated in the economic sandpile model.

If the overall trading activity is very small, $a \to 0$, the return (3.17) is dominated by the action of a single cluster. The distribution of returns therefore truly reproduces that of the cluster sizes (3.16). The most important finding is that for the intermediate returns, i.e. large enough for the validity of the asymptotic formula (3.16) but not too large, so that the exponential cut-off is not yet effective, the return distribution follows a power law

$$P_X(x) \sim x^{-1-\alpha} \tag{3.18}$$

with a specific value of the return exponent

$$\alpha = \frac{3}{2}. \tag{3.19}$$

As we can see, we are now much closer to the empirical value of the return exponent than in the economic sandpile model, but still not quite at the desired number. In fact, the linear price impact (3.17) is not very realistic, and the square root dependence, based on both theoretical speculation [129] and empirical evidence [279, 289], seems to be a better hypothesis. Supposing still that only one cluster is active at the moment and that the return is scaled as $X \propto \sqrt{s}$ with the size s of the active cluster, we get the return exponent

$$\alpha = 3, \tag{3.20}$$

in perfect accordance with the empirical data we saw in Chap. 1.

It seems we have a good reason to be happy with that result. After all, we were able to quantitatively reproduce the power-law return distribution without excessive ad hoc assumptions, and certainly the resulting exponent was not arbitrarily tuned by hand. However, we cannot be satisfied, for multiple reasons. First, it is not fully clear why the agents must keep the game close to the percolation threshold. The argument that the higher volatility provided by such attitude can bring them more opportunities for profit is rather shaky and remains on verbal level only. Second and perhaps the most important flaw is the absence of long-time correlations in volatility. Indeed, in the Cont-Bouchaud model not only are the returns uncorrelated from time to time, but so are the absolute returns, because in each step the choice of active cluster(s) is made independently of the actions taken one step before. The model does not have any memory mechanism which would lead to volatility clustering. Finally, the herds in real markets, if they form, are dynamical objects. The model can be considered realistic only provided we introduce a mechanism of cluster formation and dissolution, based on some plausible assumptions about the behaviour of the agents. Let us now show how some of the above-mentioned difficulties can be alleviated.

Introducing fluidity

The power-law distributions are not always signs of being in a critical state. It is quite sufficient if the control parameter fluctuates within a region which contains the critical point. The system can be rather fluid, sometimes coming close to the threshold and sometimes wandering far from it. Such a mechanism has been called the 'sweeping of an instability' [754] and intuitively can be well understood. Supposing in our percolation problem that the probability p goes randomly down and up, reaching as far as the percolation threshold p_c; the return distribution will be a weighted mixture of distribution with various values of p. When we are closer to the critical point at p_c, the tails of the return distribution are fatter, because the exponential cutoff is shifted beyond the cluster size about $\sim (p - p_c)^{-2}$. Therefore, the power-law behaviour will dominate on average. The value of the exponent will change, from $\alpha = 3/2$ to $\alpha = 2$, which is good, though, because it brings us even closer to the empirical numbers. We leave the detailed calculations as an exercise for the reader (Problem 3).

The fluidity of the system can also manifest itself in rearranging bonds between agents, thus allowing them to leave old alliances and build new ones. In the original Cont-Bouchaud model we had two possibilities. Either we dissolve all the bonds and create the clusters anew after each trade (annealed case), or keep the clusters fixed during the price evolution but average the statistical properties of the price evolution over all configurations of the percolation clusters (quenched case). Indeed, we were interested mostly in the return distribution, where the two approaches give the same results.

However, when we start investigating the time correlations, especially the volatility clustering, we must decide how fast the cluster configurations are changing on the time scale of individual trades. If the changes in the herding structures are much slower, we can suppose the cluster is fixed. On the contrary, if the herds are formed quickly for each single trade, the cluster structure is annealed. In either case, however, the effect of slowly-decaying volatility correlations is missing.

So, we introduce slow evolution in the cluster structure. After each trading step we randomly choose a pair of agents (or a link on the underlying network, e.g. the Bethe lattice), break the bond connecting them, if it exists, and connect them by a new bond with probability p. The quantity of interest is the autocorrelation of absolute returns. Again assuming linear price impact (3.17) and small activity $a \to 0$, the absolute returns correspond to cluster sizes, so the quantity we need is the autocorrelation of the mean cluster size

$$\overline{S}(t) = \frac{1}{N_{ct}} \sum_{l=1}^{N_{ct}} s_l(t) \qquad (3.21)$$

where N_{ct} is the number of clusters at time t, and $s_l(t)$, $l = 1, \ldots, N_{ct}$ are the sizes of the clusters at that moment. Note that the logic of the Cont-Bouchaud model dictates that the agents who are not connected to anybody else form clusters of size 1, while in the standard formulation of bond percolation a cluster must contain at least two sites connected by a bond. The equation (3.21) is therefore simplified to $\overline{S}(t) = N/N_{ct}$, and instead of the mean cluster size we can examine the number of clusters, which is significantly easier. Of course, as bonds are added and removed, N_{ct} fluctuates and the volatility clustering in the modified Cont-Bouchaud model is indirectly expressed through the autocorrelation of the number of clusters

$$C_c(\Delta t) = \langle N_{ct} N_{ct-\Delta t} \rangle - \langle N_{ct} \rangle \langle N_{ct-\Delta t} \rangle. \tag{3.22}$$

The situation is exceptionally simple on the Bethe lattice with a fixed node degree. Indeed, in the absence of loops any bond decreases the number of clusters by one, and denoting N_{bt} the number of bonds at time t we have $N_{ct} = N - N_{bt}$. The evolution of the number of bonds is a Markov process with transition probabilities

$$\text{Prob}\{N_b \to N_b + 1\} = p\,\frac{N - 1 - N_b}{N - 1}$$
$$\text{Prob}\{N_b \to N_b\} = (1 - p)\,\frac{N - 1 - N_b}{N - 1} + p\,\frac{N_b}{N - 1} \tag{3.23}$$
$$\text{Prob}\{N_b \to N_b - 1\} = (1 - p)\,\frac{N_b}{N - 1}.$$

Those who wonder where the term $N - 1$ comes from should recall that $N - 1$ is the maximum number of bonds on a Bethe lattice with N nodes. We may proceed with the solution by writing the master equation for the distribution of the number of bonds $P_{\text{bond}}(n; t) \equiv \text{Prob}\{N_{bt} = n\}$. One may verify explicitly that the stationary distribution is the binomial one

$$P_{\text{bond}}(n) \equiv \lim_{t \to \infty} P_{\text{bond}}(n; t) = p^n (1 - p)^{N-1-n} \binom{N - 1}{n}. \tag{3.24}$$

To obtain the full time dependence of the probability distribution is a more difficult task. To do that, we write the master equation in the form $P_{\text{bond}}(n; t+1) - P_{\text{bond}}(n; t) = -\sum_{n'} A_{nn'} P_{\text{bond}}(n'; t)$. Then, we can determine, at least in principle, the eigenvalues of the matrix $A_{nn'}$. The lowest eigenvalue is always $\lambda_0 = 0$, and the lowest positive eigenvalue λ_1 determines the rate of asymptotic exponential relaxation toward the stationary state (3.24). The autocorrelation of the number of clusters decays with the same rate

$$C_c(\Delta t) \sim e^{-\lambda_1 \Delta t}, \quad \Delta t \to \infty \tag{3.25}$$

and so does the autocorrelation of absolute returns. We can conclude on a qualitative basis that the volatility clustering in the Cont-Bouchaud model with fluid bonds does exist, but the decay is exponential. Indeed, it was confirmed in numerical simulations [755], showing a slow, although only exponential, decay of the volatility autocorrelation, in contrast to the empirically observed power-law decrease.

The result (3.24) is worth looking at in the limit of a large number of agents $N \to \infty$. With p fixed, the density of bonds $n_b = N_b/(N-1)$ becomes sharply peaked around its average value p and the relative fluctuations in all three quantities N_b, N_c, and $\bar{S}$ decrease as $1/\sqrt{N}$. Thus, the fluctuations in the volatility must decrease in a similar manner when the number of agents increases, and the very existence of the phenomenon of volatility clustering appears related to the finiteness of the number of agents on the stock market. The latter conclusion can surprise us, but we shall see on several additional occasions that many intriguing economic phenomena are most probably 'mere' finite-size effects.

From Eq. (3.24) it also follows that in the limit $N \to \infty$ the concentration of bonds becomes normally distributed. Therefore the distribution of volatility also becomes Gaussian and the Cont-Bouchaud model with fluid bonds can be considered as one of the possible implementations of the simplest stochastic volatility model described by Eq. (2.92).

The Cont-Bouchaud model has been simulated numerically on several types of lattices [756]. It has also been modified in many ways; for example, the trading probability a was coupled by a feedback mechanism to the price movements [757–759], or the clusters were formed through spin-spin interaction instead of the uncorrelated percolation [760, 761]. However, it turned out that the basic features are quite robust when subjected to such variations, which makes the Cont-Bouchaud model a solid starting point for further ramifications [691, 762–770].

3.1.3   Eguíluz-Zimmermann model

Still taking inspiration from the percolation model of Cont and Bouchaud, we can try a substantially different mechanism of herd formation. Consider agents who are hesitating for quite a long time before they sell or buy. They are not waiting passively, though. In the meantime they ask the opinions of the others, assess the odds of various strategies, and form coalitions with those who seem to be the next winners. We can observe the buildup of a complicated social structure, as when a soccer team surrounded by coaches, doctors, and managers prepares for a final match. When the time comes to make a deal, the cluster of agents grown during the formation period acts as a single herd and dissolves afterwards. Therefore, as in the Cont-Bouchaud model, the distribution of returns will reflect that of the cluster sizes, but now the clusters will be dynamical objects instead of static percolation clusters. Note that this mechanism is much more dramatic than the bare fluidity of bonds described at the end of the preceding section. Therefore this model deserves special attention and was also given a special name after its inventors V. M. Eguíluz and M. G. Zimmermann [771].

Coalescing and dissolving herds

We do not assume any a priori social structure among people. Anybody can be paired with anyone else. The configuration of our ensemble of N agents is fully described by a vector $g = [g_1, g_2, \ldots, g_N]$ where g_s is the number of clusters of size s. An obvious normalisation condition $\sum_{s=1}^{N} s\, g_s = N$ restricts the set of allowed configurations.

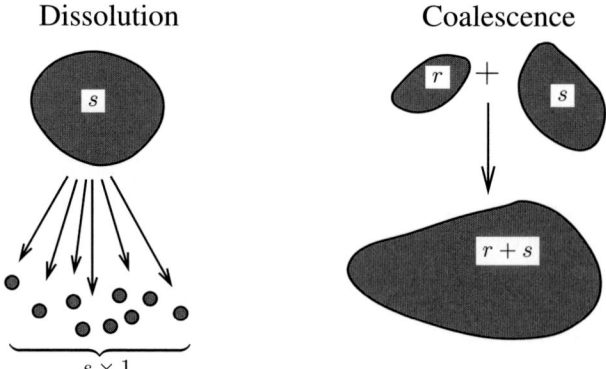

Fig. 3.3 Scheme of the two processes making up the dynamics of the Eguíluz-Zimmermann model. Dissolution occurs with probability $2a$, coalescence with probability $1 - 2a$.

The clusters evolve either by coalescence or by breakup into isolated agents, as shown schematically in Fig. 3.3. The frequency of these two types of events is tuned by the parameter $a \leq 1/2$, as in the Cont-Bouchaud model. With probability $2a$ an agent is chosen at random and then induces all agents within the same cluster to perform the same action. All of them either sell a unit of stock or buy a unit of stock with the same probability. After the trade is completed, the cluster would dissolve into sets of isolated agents. If the state vector was g before the move and the size of the active cluster was s, the new state would be denoted $g^{(-s)}$ with components $g_r^{(-s)} = g_r - \delta_{sr} + s\delta_{1r}$. (The symbol δ_{ij} here is the usual Kronecker delta.) The dynamics is reversible in the sense that, knowing the final configuration and the size s of the cluster affected, we can reconstruct the original configuration. So, we also denote by $g^{(+s)}$ the original configuration which resulted in g after breakup of a cluster of size s. Explicitly we have $g_r^{(+s)} = g_r + \delta_{sr} - s\delta_{1r}$. This property will be useful when we turn to an analytic solution of the Eguíluz-Zimmermann model.

With the complementary probability $1-2a$, a pair of agents is chosen at random and if they belong to different clusters, the two clusters merge. If the two agents share the same cluster, nothing happens. This move is not accompanied by a trade. The stock market remains idle. If the sizes of the two distinct clusters were r and s, the new configuration, denoted $g^{(+rs)}$, would have components $g_k^{(+rs)} = g_k - \delta_{rk} - \delta_{sk} + \delta_{r+s\,k}$. Again, the original configuration can be deduced from the result, provided we know the sizes of the affected clusters. In our notation g was preceded by $g^{(-rs)}$ where $g_k^{(-rs)} = g_k + \delta_{rk} + \delta_{sk} - \delta_{r+s\,k}$.

We could, of course, implement the dynamics of the model on a computer, and indeed, the numerical simulations [771] would show interesting features of the price fluctuations due to herd trading, especially the power-law distribution of returns occurring for small activity a. However, the model is analytically tractable without much difficulty [772–774], and we shall assault it now with the well-tried master equation approach.

Origin of the power-law distribution

The model has no memory beyond the last step, so the dynamics of cluster configurations is a Markov process with transition probabilities

$$\text{Prob}\left\{g \to g^{(+rs)}\right\} = \frac{1 - 2a}{N(N-1)}\, rs\, g_r\, (g_s - \delta_{rs})$$

$$\text{Prob}\left\{g \to g^{(-s)}\right\} = \frac{2a}{N}\, s\, g_s. \tag{3.26}$$

The probability $P(g;t)$ that we find the configuration g at time t evolves according to the master equation

$$P(g;t+1) - P(g;t)$$
$$= \frac{1 - 2a}{N(N-1)} \sum_{r,s=1}^{N} \bigg(rs\, (g_r + 1 + \delta_{rs})\, (g_s + 1)\, P(g^{(-rs)};t)$$
$$- rs\, g_r\, (g_s - \delta_{rs})\, P(g;t) \bigg) \tag{3.27}$$
$$+ \frac{2a}{N} \sum_{s=2}^{N} \Big(s\, (g_s + 1 - \delta_{s1})\, P(g^{(+s)};t) - s\, g_s\, P(g;t) \Big).$$

The full distribution function contains far more information than we could actually use. We shall need the average numbers of clusters of size s and as will be clear a few lines later, as well as the averages of various products of such numbers. For convenience, we normalise them with respect to the total number of agents N. Thus we define

$$n_s(t) = \frac{1}{N} \sum_g g_s\, P(g;t)$$

$$n_{rs}(t) = \frac{1}{N^2} \sum_g g_r\, g_s\, P(g;t) \tag{3.28}$$

$$\cdots$$

$$n_{s_1 s_2 \ldots s_m}(t) = \frac{1}{N^m} \sum_g \left(\prod_{l=1}^{m} g_{s_l} \right) P(g;t).$$

In principle, the full set of functions $n_{s_1 s_2 \ldots s_m}(t)$ at all orders $m = 1, 2, 3, \ldots$ bears as much information as the distribution $P(g;t)$. A chain of equations emerges, reminiscent of the Born-Bogoliubov-Green-Kirkwood-Yvon equations very well known in the physics of fluids, and in order to calculate the functions up to order m, we need the functions at order $m+1$. Fortunately, in the the limit $N \to \infty$ with a kept fixed (more precisely we need $a\sqrt{N} \to \infty$, see [773]), the averages of products are factorised into a product of averages. So, we can write a closed equation for the average number of clusters $n_s(t)$, using $n_{rs}(t) = n_r(t)\, n_s(t)$. Rescaling the time $t \to Nt$ and omitting the terms of order $O(1/N)$, we get the set of equations

Stirling formula Box 3.5

estimates the value of the factorial for large arguments. The exact statement is that [775]
$$x! = \Gamma(x+1) = x^x\,e^{-x}\sqrt{2\pi\,x}\;e^{r(x)}, \text{ with } \tfrac{1}{12x+1} < r(x) < \tfrac{1}{12\,x}.$$
Another good approximation for a factorial is
$$x! \simeq x^x\,e^{-x}\sqrt{(2x+\tfrac{1}{3})\pi},$$
which also gives a reasonable estimate for small values of x, e.g. for $x = 0$ the formula
suggests $0! \simeq \sqrt{\pi/3} = 1.0233267\ldots$.

$$\frac{\mathrm{d}}{\mathrm{d}t}n_s(t) = -2(1-a)\,s\,n_s(t) + (1-2a)\sum_{r=1}^{s-1} r(s-r)\,n_r(t)\,n_{s-r}(t) \text{ for } s > 1 \qquad (3.29)$$

accompanied by the 'boundary' equation

$$\frac{\mathrm{d}}{\mathrm{d}t}n_1(t) = 2a\sum_{r=2}^{\infty} r^2\,n_r(t) - 2(1-2a)\,n_1(t). \qquad (3.30)$$

The stationary state $n_s \equiv \lim_{t\to\infty} n_s(t)$ can be found using the discrete Laplace transform of the function $s\,n_s$. Indeed, from (3.29) it follows that $\hat{y}(z) = \sum_{s=1}^{\infty} z^s\,s\,n_s$ obeys a quadratic equation, implying a square-root singularity, already a good old friend of ours (see Eq. (3.7)). We can easily find that

$$\hat{y}(z) = \frac{1-a}{1-2a}\left(1 - \sqrt{1 - \frac{1-2a}{(1-a)^2}\,z}\right), \qquad (3.31)$$

and expanding in powers of z we have

$$n_s = \frac{1-a}{1-2a}\,\frac{1}{2^{2s-1}}\,\frac{(2s-2)!}{(s!)^2}\left(1 - \left(\frac{a}{1-a}\right)^2\right)^s. \qquad (3.32)$$

For large s, applying the Stirling formula (see Box 3.5) to the result (3.32) gives the exponentially truncated power law

$$n_s \sim s^{-5/2}\,e^{-s/s_0} \qquad (3.33)$$

where the characteristic scale $s_0 = 1/\ln\left(1 + \frac{a^2}{1-2a}\right) \simeq a^{-2}$ diverges for $a \to 0$. For a very small frequency of trades the distribution of cluster sizes becomes power-law with the exponent $\tau = 5/2$, as in the Cont-Bouchaud model just at the percolation threshold. The power-law behaviour is then also translated to the power-law distribution of returns. We again assume a linear price impact as in Eq. (3.17), but, contrary to the Cont-Bouchaud model, only a single cluster is active now; and for any value of a the distribution of returns is $P_X(x) \propto s\,n_s$. We suppose again a linear impact $X \propto s$. Therefore, in the limit of very low activity, $a \to 0$, the returns are power-law distributed with the exponent

$$\alpha = \frac{1}{2}. \qquad (3.34)$$

If you wonder why the exponent in the Eguíluz-Zimmermann model is smaller by one compared to the Cont-Bouchaud model, you should recall that in the Eguíluz-Zimmermann model we randomly choose an agent, while in the Cont-Bouchaud model we choose an entire cluster. Therefore the probability of choosing a cluster of size s and causing a price change proportional to s gets an extra factor s in the Eguíluz-Zimmermann model.

Before assessing the real utility of the Eguíluz-Zimmermann model, we want to raise the question of principal interest. Comparing the Eguíluz-Zimmermann model with models based on percolation, such as the Cont-Bouchaud one, we immediately notice the absence of any conventional critical point. To obtain a power-law distribution of cluster sizes in percolation, we need to tune the probability of establishing a bond so that we are at the percolation threshold. We shall not repeat here the arguments justifying the assumption that reality keeps it close to the critical point. Instead, we should note that in the Eguíluz-Zimmermann model the situation is conceptually much simpler. It is enough to stay at infinitesimally small activity a, or, if you wish, the critical point is exactly at $a = 0$. Now, the question is whether the Eguíluz-Zimmermann model has anything in common with self-organised criticality, where the power-law distributions also arise in the limit of infinitesimally weak external perturbation, be it adding a single grain of sand or kicking a random walker out of the origin and waiting an arbitrarily long time before it returns home.

Indeed, the general scheme of self-organised criticality as a dynamical system with absorbing state(s) which are infinitesimally weakly disturbed can also be adapted for the Eguíluz-Zimmermann model. Frankly, we have not spoken about what happens if we fix $a = 0$ strictly. The reason is that the dynamics is quick and trivial. The clusters never dissolve but always merge. Very soon the state is reached with a single cluster containing all agents, and this is the absorbing state of the dynamics. For any small but non-zero a the absorbing state is broken from time to time, however long we need to wait for such a breakup. Then, the dynamics starts again, wandering randomly in the complicated space of cluster configurations, until the absorbing state is hit again. From this point of view the power-law distribution of cluster sizes corresponds to the long-living, but still non-stationary state of the system dominated by the absorbing state. This makes the connection to more conventional self-organised critical systems more transparent, although it does not seem to bring much practical help for calculation of the properties of the Eguíluz-Zimmermann model.

Now it is time to compare the results of the Eguíluz-Zimmermann model with stock market data. Unfortunately, there is little agreement with real price fluctuations. One problem is that the return exponent (3.34) is much too small to even roughly describe the market dynamics. A more important flaw becomes apparent when we try to figure out what the volatility clustering would look like in the Eguíluz-Zimmermann model. Indeed, there is a memory in the system, but unfortunately the absolute returns, i.e. the sizes of the active clusters are negatively, instead of positively, correlated. To see this, it is enough to consider what happens after a large cluster is selected to be active. After the trade is over, the cluster breaks into isolated agents, and the average cluster size drops significantly. After a big price change one should expect a much smaller change. No extended periods of large volatility are observed.

As a stock-market model the Eguíluz-Zimmermann scheme does not seem to bring much success. But we should not be too strict. First, neither the sandpile nor the Cont-Bouchaud models were much closer to reality in their original formulations. It is only a smart modification that makes them more attractive. One suggestion of how to improve the Eguíluz-Zimmermann model is left to the reader as Problem 4 at the end of this chapter; other variants can be found in the literature [776–782]. Also note that introducing a feedback mechanism relating the value of a with the size of the last dissociated cluster in the form $a \propto s^{-1}$ does lead to positive volatility correlations decaying as a power, as shown in numerical simulations [783].

Second, there may be other, originally unexpected applications. Indeed, it is quite possible that the aggregation models, including the sandpile and Cont-Bouchaud models treated above, can do a better job in modelling the growth, splitting, and merging of companies. Indirectly they can serve to explain the power laws in the return distributions too, if we only ascribe the return distribution to the size distribution of investors acting on the stock market. Let us turn to an example of such a model now.

Sizes of businesses

Let us have N agents who wish to conduct business together. They can aggregate into clusters of various sizes. Any cluster of size $s > 1$ will be called a business. Clusters of size 1 are individual agents with the tendency to cluster with other agents. We do not consider mergers of two businesses of differing sizes, but only growth of businesses by adding individuals. From time to time, a business can go bankrupt and dissolve, leaving a set of separate agents behind.

The evolution of the model proceeds as follows [784]. At each step, an agent is chosen at random. If she belongs to a cluster of size $s > 1$, the cluster dissolves with probability $a(s)$. We let now the probability depend on the cluster size and among possible choices we focus on power laws in the form $a(s) = b\, n_1\, s^{-\beta}$ with $0 < b\, n_1 < 1$ and $\beta \geq 0$. The concentration n_1 of individual agents is incorporated into the factor for further convenience. If the chosen agent is an independent individual, another agent is chosen at random, and the first agent joins the cluster the second agent belongs to. If the second agent is also an individual, a new business of size 2 is created by joining the two.

When solving this model we can proceed in the same way as we did in the case of the Eguíluz-Zimmermann model. The situation is even simpler due to the absence of mergers. The equation for the normalised cluster numbers is

$$\frac{\mathrm{d}}{\mathrm{d}t} n_s = -a(s)\, s\, n_s - n_1\,(s\, n_s - (s-1)\, n_{s-1}) \tag{3.35}$$

for $s > 1$, while for the number of individual agents we have

$$\frac{\mathrm{d}}{\mathrm{d}t} n_1 = \sum_{s=2}^{\infty} a(s)\, s^2\, n_s - n_1\,\Big(\sum_{s=1}^{\infty} s\, n_s + n_1\Big). \tag{3.36}$$

The formal solution of (3.35) for the stationary state is straightforward and, compared with the original Eguíluz-Zimmermann model, we can do well without the discrete Laplace transform. By iteration we obtain

$$s\, n_s = n_1 \prod_{r=2}^{s} \frac{1}{1 + \frac{a(r)}{n_1}} = n_1 \prod_{r=2}^{s} \frac{1}{1 + b\, r^{-\beta}}, \tag{3.37}$$

and for large s we first make a logarithm and then replace the discrete sum by an integral. We thus get

$$
\begin{aligned}
n_s &\sim s^{-1} \exp(-\tfrac{b}{1-\beta}\, s^{1-\beta}) && \text{for } \beta < 1 \\
n_s &\sim s^{-1-b} && \text{for } \beta = 1
\end{aligned}
\tag{3.38}
$$

for $s \to \infty$. Contrary to the Eguíluz-Zimmermann model, the cluster distribution obeys a power law only under special conditions. Moreover, the exponent in the power-law regime is not universal, but depends on the parameter quantifying the tendency of clusters to break into individuals.

The dynamics of businesses and economic aggregation is a vast area on its own, as yet unexplored to a large extent. We stop our journey in this wilderness now, leaving the reader a handful of road signs in Refs. [785–788].

3.2 Agent-based models

While in the preceding section we concentrated mainly on analytically obtained results, now we enter the proper realm of computer simulations. The approaches we will present in the rest of this chapter were often deliberately designed as numerical laboratories, where the researcher may play with values of numerous parameters or even include additional new features. Such an attitude naturally precludes using analytical tools, but on the other hand provides immense intellectual freedom, eventually leading to many discoveries never anticipated by paper-and-pen mathematics. Here the computational science of complex systems plays an invaluable role, which does not fade away even when analytical description of some of the features revealed in artificial computer economies finally becomes available.

3.2.1 Early attempts

Stigler 1964

It is quite exciting to realise that the first numeric experiments with stock-market models appeared soon after the computer became a tool used by the wider scientific community. In 1964 George J. Stigler devised a model which implements a random trading schedule [789], neglecting any strategic reasoning of the buyers and sellers. In his scheme, blind agents place their orders within a specified price interval, and the bids and asks are stored until they are cleared or until they expire. When an agent places her bid at a price larger or equal to any of the existing asks, a trade takes place, and the corresponding bid and ask are removed. And vice versa: a newly placed ask order is cleared in a symmetric manner with an appropriate bid, if it exists.

In Fig. 3.4 we reprint one of the original simulation results published in Ref. [789]. Although the simulations involved only a very few (by our standards) time steps, they showed how it is possible to generate a fluctuating price sequence, much like the one observed in reality, with minimum assumptions. Stigler himself used his simulated price evolution under various conditions as an argument against strict regulation imposed on stock market practices. While nowadays such strong argumentation backed by so little numerical evidence would rather provoke a smile, in the mid-1960s the idea was revolutionary.

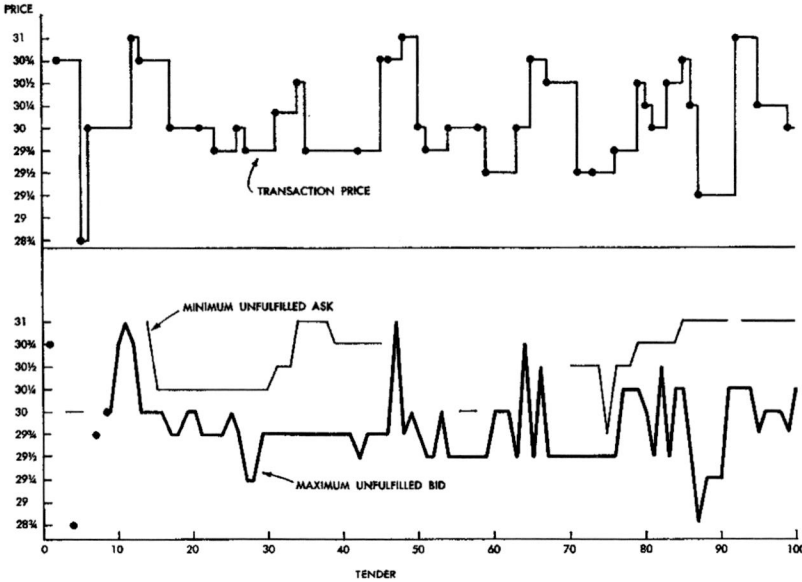

Fig. 3.4 Stigler's original simulations. Reprinted from [789], with permission of the University of Chicago Press.

In fact, the model introduced by Stigler involves on the most basic level all the substantial ingredients a successful stock market model should have. Indeed, its core idea is far more realistic than many other stock-market models cooked up later with powerful PC-workstations at hand.

The Stigler model can be defined as follows. The price logarithm Y_t can fluctuate within fixed bounds; for simplicity let us assume the interval $[0, 1]$. The system evolves in discrete time, and in each step one order of unit volume is placed randomly within the allowed price interval. Buy and sell orders are issued with equal probability. The expiration time of all orders is limited to t_{exp} time steps. Unsatisfied orders older than t_{exp} are removed. Therefore, at each instant there are at a maximum t_{exp} waiting orders. The state of the system is described by the positions and types of all unsatisfied orders. Of course, all waiting sell orders lie above all remaining buy orders.

When a new order arrives, a deal may be transacted, depending on the position of the incoming order. If it is a buy order and happens to be placed below or at the lowest unsatisfied sell order, the two orders are matched and removed from the system. Otherwise the new order is stored in the waiting list. Conversely, if a sell order comes, it is cleared if it is lower than or equal to the highest unsatisfied buy order; otherwise it is added to the actual order list. In our simulation the transaction price is set as the position of the older of the two cleared orders, although other rules (e.g. the mean of the two) can also be used.

The basic properties of the Stigler model can be seen in Fig. 3.5, where we show a typical time series for the prices and returns, defined here as price increments

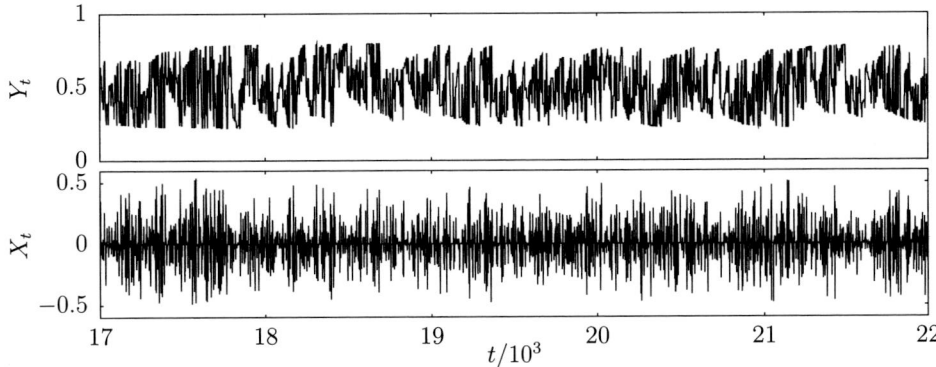

Fig. 3.5 Simulation of the Stigler model. In the upper panel, we show the log-price fluctuations, in the lower panel the corresponding returns. The orders expire after $t_{\mathrm{exp}} = 5000$ steps.

$$X_t = Y_t - Y_{t-1}. \tag{3.39}$$

The price does not depart far from the centre of the allowed interval, which may be interpreted as fluctuations around the constant fundamental price. A glimpse at the return time-series reveals signs of volatility clustering, which is quantified in the autocorrelation functions of the returns and absolute returns defined as

$$\begin{aligned}
\langle X_t\, X_{t-\Delta t}\rangle_c &\equiv \langle X_t\, X_{t-\Delta t}\rangle - \langle X_t\rangle\langle X_{t-\Delta t}\rangle \\
\langle |X_t\, X_{t-\Delta t}|\rangle_c &\equiv \langle |X_t\, X_{t-\Delta t}|\rangle - \langle |X_t|\rangle\langle |X_{t-\Delta t}|\rangle
\end{aligned} \tag{3.40}$$

where $\langle A_t\rangle = \sum_{t=1}^{T} A_t/T$ denotes the time average of a process A_t over the simulation that lasted T time steps.

We can see in Fig. 3.6 that the returns have a short negative autocorrelation, decaying exponentially, while the correlation of absolute returns decays much more slowly, and for large time differences it decreases as a power $\sim (\Delta t)^{-1.3}$. This finding grasps some of the basic features of a real time series: uncorrelated returns and volatility clustering. However, the distribution of the returns does not exhibit fat tails, so from this point of view the Stigler model is unrealistic. However, we will see later in Chap. 4 how Stigler's idea was revived in more sophisticated order-book models.

Kirman's ants

Everybody knows ants: they come in thousands of species and inhabit all corners of the planet. With all their variability they have one important feature in common. When foraging for food, an ant most of the time follows the path marked by pheromones laid by other ants from the same colony. With relatively small probability, the ant sets on an unexplored path. This way the colony struggles to exploit the discovered sources of food to a maximum extent, but at the same time it is able to find an alternative source before the first one is completely exhausted.

In the paper [790] Alan Kirman cited results of experiments with ants choosing two trails for food. The geometry was carefully designed to be symmetric in order to avoid

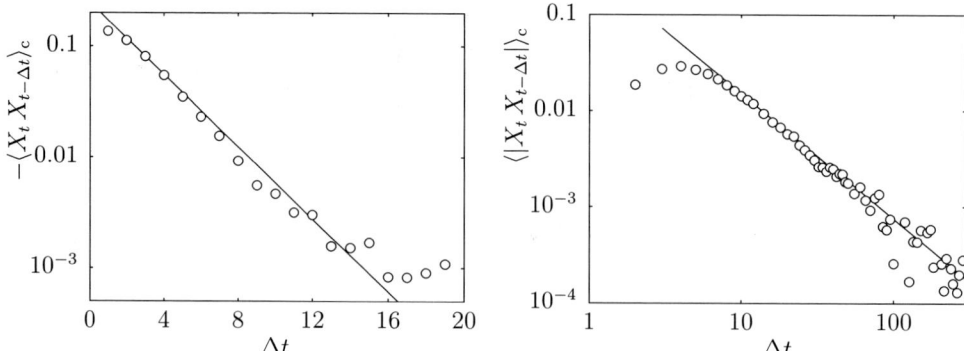

Fig. 3.6 Correlations in price fluctuations in the Stigler model. In the left panel we show the autocorrelation function of returns. The straight line is the exponential fit $\propto \exp(-\Delta t/2.66)$. In the right panel, we show the autocorrelation of absolute returns. The line is the power decay $\propto (\Delta t)^{-1.3}$. The orders expire after $t_{\exp} = 5000$ steps.

any bias toward one of the trails. It was found that the ants preferred one of the trails most of the time but that after some time they swapped paths without any perceptible reason. There is no equilibrium or stationary state characterising the behaviour of the colony. Instead, the colony keeps switching between the two metastable states.

Kirman used the collective intelligence of ants to illustrate his views on the nature of economic processes. In opposition to standard economic theory, concentrated on finding and characterisation of the equilibrium state, Kirman suggested that the apparent equilibria could be mere long-lasting metastable states. The system keeps jumping from one metastable state to another, instead of settling into an equilibrium. That nightmare of the equilibrium theorists, the excessive fluctuations which should not be there but the existence of which is undeniable, can now be attributed to the jump process between several distinct quasi-equilibria.

To investigate the process manifested by the experiment with ants, Kirman introduced a stochastic model of binary choice [790]. There are N agents (ants) choosing one of two options (trails), denoted by $+1$ and -1. The dynamics is driven by two tendencies. First, the agents tend to imitate each other, i.e. choose the same option as some other agent already chose. It corresponds to following the pheromone track. Second, with a small probability ϵ, the choice of the agent changes spontaneously. In the economic context the choices ± 1 can be interpreted as two investment strategies, e.g. holding asset A or asset B, selling vs buying, etc. Therefore, we strip the economic agents of all signs of rationality and let them decide by pure imitation and random swap of preference. The strategies spread through the population by contagion, much like an infectious disease.

Before entering the discussion of the relevance of such a decision-making mechanism for human behaviour in the market, let us see what qualitative and quantitative features can be expected from the model. To this end, we first provide a more formal definition of Kirman's ant model. The state of the agent i at time t is described by

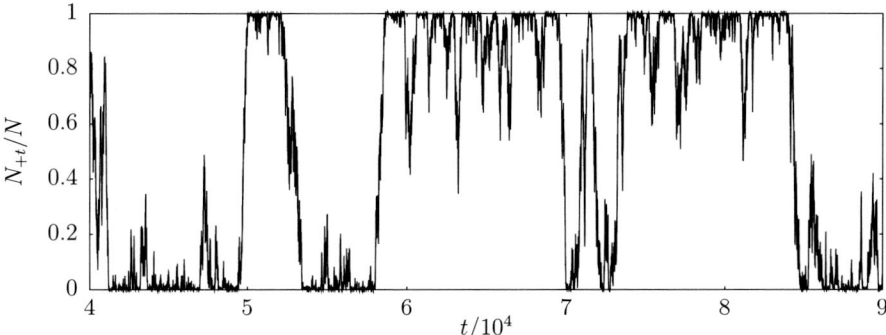

Fig. 3.7 Simulation of Kirman's ant model. We show the time evolution of the fraction of ants choosing the $+1$ option. The parameters of the model are $N = 10^3$, $\epsilon = 2 \cdot 10^{-4}$.

the Ising variable $\sigma_{it} \in \{+1, -1\}$ Denote by N_{+t} the number of agents choosing $+1$ at time t, i.e. $N_{+t} = \sum_{i=1}^{N} \frac{1}{2}(\sigma_{it} + 1)$. The evolution proceeds in discrete time. In each step, we perform an elementary update, defined as follows. First, we randomly choose one agent i and flip her preferred choice with probability ϵ: $\sigma_i \to -\sigma_i$. With the complementary probability $1 - \epsilon$ we also randomly choose a second agent, j, and the first agent adopts the choice of the second agent: $\sigma_i \to \sigma_j$.

We do not suppose any social structure determining the links between the agents. Anybody can be influenced by anybody else with equal probability. Therefore, the state of the system is fully described by the variable N_{+t} and its dynamics is governed by the transition probabilities

$$
\begin{aligned}
\text{Prob}\{N_+ \to N_+ + 1\} &= \frac{N - N_+}{N}\left(\epsilon + (1 - \epsilon)\frac{N_+}{N - 1}\right) \\
\text{Prob}\{N_+ \to N_+ - 1\} &= \frac{N_+}{N}\left(\epsilon + (1 - \epsilon)\frac{N - N_+}{N - 1}\right).
\end{aligned}
\tag{3.41}
$$

For convenience we measure the time t in such a way that an advance from t to $t + 1$ includes N updates according to (3.41). In Figure 3.7, we can see a typical time sequence of the fractions of agents choosing $+1$. We can observe periods in which the fraction is close to either 0 or 1, interrupted by rapid jumps between the two extremes, accompanied by wild fluctuations. The absence of a specific equilibrium is apparent. Closer inspection reveals that the existence of an equilibrium and nature of the fluctuations depend on the frequency of spontaneous change in mind, quantified by the parameter ϵ. For ϵ large compared to $1/N$, the fraction N_{+t}/N fluctuates around the mean value $1/2$, while the behaviour exemplified in Fig. 3.7 occurs in the opposite case, $\epsilon \lesssim 1/N$.

We can also look at the correlations in the time sequence N_{+t}. To this end we extract the autocorrelation function of the differences $\Delta N_{+t} = N_{+t} - N_{+t-1}$ and the absolute difference $|\Delta N_{+t}|$ defined analogously to (3.40). In Fig. 3.8 we can see that the changes in N_{+t} are uncorrelated, while the absolute values of the changes exhibit

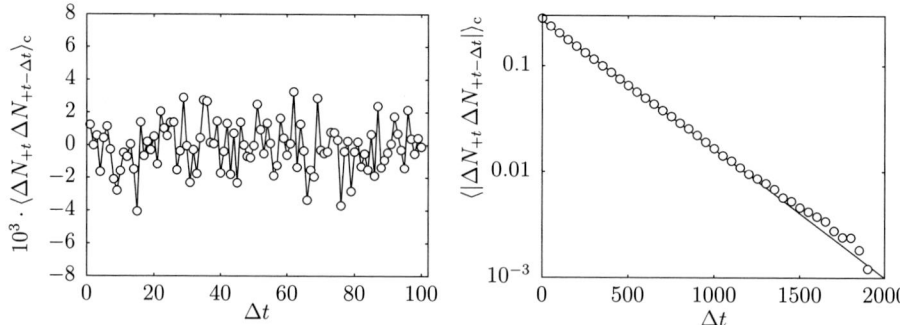

Fig. 3.8 In the left panel, we show the autocorrelation function of the differences ΔN_{+t}; in the right panel we show the autocorrelation function of the absolute differences $|\Delta N_{+t}|$. The parameters of the model are $N = 10^3$, $\epsilon = 2 \cdot 10^{-4}$.

long-time, although exponentially dampened, positive correlations. This is a sign of a weak form of 'volatility clustering'. However, we should be careful with drawing conclusions from the Kirman's ant model, as there is no trading and no price defined in the model. On the other hand, the idea of spreading the trading strategies by simple contagion is implemented in a very transparent way within the Lux-Marchesi model, discussed in Sec. 3.2.3.

Experiments with humans

It is often repeated that the main difference between natural sciences and economics lies in the freedom physicists, biologists, and the like enjoy in carrying out their experiments. Indeed, it is difficult to imagine that economists could do anything comparable to the production of anti-matter in supercolliders. On the other hand, human beings have long been subject to sophisticated experiments, ranging from NMR imaging of active zones corresponding to specific emotions in the brain, to drastic sociological experiments forcing innocent subjects to behave as sadistic guards in an experimental concentration camp.

In fact, experiments with human behaviour in an economic environment do take place quite often, and pedagogic practice at universities, where students have to exchange virtual money for virtual stocks, is a common example of such modelling. Indeed, it is obviously more natural to model the behaviour of humans (e.g. stockbrokers) based on the behaviour of other humans (e.g. students) than modelling an animate human world using inanimate computer machinery. Among various economic experiments, let us focus on one example.

When Jasmina Arifovic tried to explain the persistent fluctuations of foreign exchange rates significantly exceeding any rational reactions to news arrival [791], she used groups of students trading, through a network of computers, three virtual items: goods, currency A, and currency B. The trading scheme was based on the overlapping generations model, defined as follows [792]. In each time step, one round of trading takes place. Every generation lasts two time steps: you might call them youth and adult age, if you like, and we denote them by I and II, respectively. The traders are divided into two equal-sized groups: members of the first group are 'young' at odd time steps and 'adult' at even time steps, while members of the other group are 'young' at even time steps and 'adult' at odd time steps. Hence the name overlapping generations model. Let us first explain the situation with only one currency.

Each trader in each generation receives a certain amount of goods. It is possible to buy or sell goods on the free market. The market price at time t is Z_t. We shall return to the question of establishing the price later. The important point is that the amount received in youth, m_I, is larger than the one received in adult age, m_{II}. Therefore, to ensure a more equilibrated consumption, some goods should be passed from youth to the adult period. But it is supposed that the goods are spoiled after one time step, so the only way to spare something for later use is to sell a certain fraction of goods in the first step and use the money for buying some goods in the second step. So, in the first step, happening at time t, $m_I - m$ units of goods are consumed and m sold at price Z_t, and the trader pockets $m Z_t$ units of the currency, which is then used in the second step, at time $t + 1$, to buy $m Z_t / Z_{t+1}$ goods at the current price Z_{t+1}.

Arifovic assumes that the gained utility is measured as the sum of the logarithms of the goods consumed in the first and second steps, $U = \ln(m_I - m) + \ln(m_{II} + m Z_t / Z_{t+1})$. The experimental subjects are instructed to try to maximise this function.

So far, the traders' only strategy is given by choosing the amount m of the goods sold in the first step. All the rest follows from the choice of this number by all traders. Arifovic enriched the basic scheme of the overlapping generations model by allowing the agents to exchange the goods partially for currency A and partially for currency B. Therefore, the agent has one more degree of freedom, namely the fraction $l \in (0, 1)$ of the goods to be sold in the market of currency A. Let the price of the goods at time t, expressed in currencies A and B, be Z_t^A and Z_t^B, respectively. The exchange rate between the two currencies is $E_t = Z_t^A / Z_t^B$. Then the utility is modified to

$$U(m, l) = \ln(m_I - m) + \ln\left(m_{II} + m\left(l\frac{Z_t^A}{Z_{t+1}^A} + (1 - l)\frac{Z_t^B}{Z_{t+1}^B}\right)\right). \tag{3.42}$$

In each step, the price is set by a formula which mimics the dynamics due to the demand-supply disequilibrium. If there are N traders selling goods at time t, with strategies $[m_i, l_i], i = 1, \ldots, N$, then

$$\begin{aligned} Z_t^A &= MN / \sum_{i=1}^{N} l_i\, m_i \\ Z_t^B &= MN / \sum_{i=1}^{N} (1 - l_i)\, m_i \end{aligned} \tag{3.43}$$

where M is a certain constant which only sets the global price scale and has no effect on the exchange rate. We can note the somewhat unrealistic feature of the model that the price is determined only by the supply and not by the demand for the goods.

The traders can observe the price movements on their computer screens in each generation and decide on the strategy, which is a pair of numbers $[m, l]$. This choice

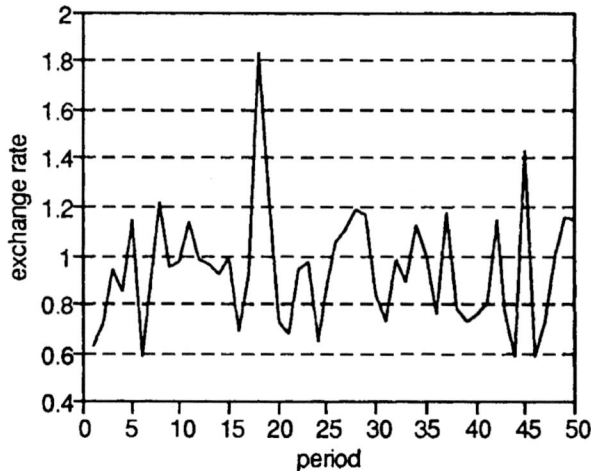

Fig. 3.9 Exchange rate fluctuations observed in the experiments by Arifovic; reprinted from [791], with permission of the University of Chicago Press. The amounts of goods provided in youth and adult age were $m_I = 10$, $m_{II} = 1$, respectively.

relies entirely on the rational (or maybe irrational) reasoning of the human agents. The software then collects the choices of all agents, recalculates the price, and displays the results; the trading then proceeds to the next step. Arifovic conducted many sessions of the experiment and consistently found that while the mean consumption rate in the first step, m, converges toward a specific stationary state with only minor fluctuation, the exchange rate does not exhibit any tendency to approach the equilibrated value. An example of the exchange rate fluctuations observed in the experiment is shown in Fig. 3.9.

. . . and genetic algorithms

In order to better understand the results, Arifovic also modelled the agents' behaviour on a computer, using the genetic algorithm method. The reader will learn more about genetic algorithms in Box 3.7. For now, it is enough to know that genetic algorithm acts much like Darwinian evolution. The strategies are mutated at a small rate, and the agents who possess more successful strategies are replicated with higher probability than the agents whose strategies lead them to failure. Such a mutation-selection process results in self-organised optimisation of the performance of the system. Let us sketch briefly the explanation of persistent fluctuations proposed by Arifovic. In the case of a single currency, an equilibrium is established with only small fluctuations reflecting the spontaneous mutations of the strategies. If we interpret the mutations as rational adaptations to external stimuli, news arrival, etc. the model nicely satisfies the wishes of the equilibrium theory of economics. However, the picture changes completely when two currencies are introduced. Because there is no a priori equilibrium value of the exchange rate, the system is free to fluctuate incessantly along a continuous line of equilibria, instead of settling around one equilibrium point in the space of parameters.

Again, as in the toy example of Kirman's ants, the excessive fluctuations are attributed to the absence of a specific economic equilibrium, or to multiplicity of equilibria. Closer inspection reveals even more features in common with the ants choosing their trail.

The genetic algorithm dynamics drives the system in principle to the state in which all agents share an identical optimal strategy. If that happens, the offers of goods are the same in group I as in group II; therefore the prices are unchanged from one step to the other, and the exchange rate is constant. In this case $\partial U(m,l)/\partial l = 0$, while $\partial U(m,l)/\partial m = 1/(m-m_I) + 1/(m+m_{II})$. Therefore the value of m still tends toward its optimum value $m_{\mathrm{opt}} = (m_I - m_{II})/2$, but l can change arbitrarily, because such a change does not alter the expected utility.

This can be achieved in practice only if all agents change their l simultaneously, which can hardly occur due to mutations alone. However, there is also reproduction in play: if one agent changes its l, its strategy may proliferate before the selection pressure brings the value of l back, thus invading the system with a large sub-population with uniform l, different from the original one. Such a sub-population can eventually prevail by mechanical reproduction, completing the jump of the entire system from one quasi-equilibrium characterised by a certain l to another quasi-equilibrium with a different l. Such jumps were indeed observed in simulations presented in [791].

The dynamics closely resembles the behaviour of Kirman's ants. The role of imitation or pheromone following is now replaced by reproduction of strategies, and instead of two possible choices of the trail we have a continuum of choices for l. A small difference remains: the two ants' trails were symmetric, while strategies with a higher utility are reproduced more often. But as we have already seen, when approaching the theoretical equilibrium state the exchange rate fluctuations are suppressed, implying that all strategies differing by only l are equally good and the agents can arbitrarily switch between them. The ensuing large fluctuations of the exchange rate are purely due to switching between equivalent equilibrium values of l. The theoretical equilibrium state remains purely virtual, and the system never reaches a stationary state. The overlapping generations model with genetic algorithm dynamics was subsequently investigated using more computing power [793–795], and features compatible with empirically observed fat tails and volatility clustering were found. Unfortunately, the data do not seem to show robust power-law tails in the return distribution, so a closer comparison with real data is hardly possible.

More developments

Let us briefly mention several other models suggested for explaining the stylised facts of economic signals. Not all of them lead to fruitful ends, but we want to include here a note for the sake of completeness. Still many more works lie buried in journal repositories.

The EMH says that market dynamics is essentially deterministic. The market reacts to any external influence so that it very quickly returns to a new equilibrium. Then, as already pointed out, a serious problem arises, as prices and other parameters fluctuate much more violently than any external source of perturbation might explain. One way out of this paradox is based on the theory of catastrophes (also called the theory of singularities), which states that continuous change of a control parameter induces a discontinuous jump in the controlled quantity [796]. In the 1970s and 1980s it was indeed a very popular theory, invoked to explain virtually all conceivable abrupt changes in nature, the human body, and society. Naturally, it attracted the economists too.

Zeeman [797] considers a purely deterministic dynamical model with two types of agents. The first type, the naive trader, assumes that the price level reached in the last step of the evolution is the fundamental price of the asset. The second type, the momentum trader, expects that the price trend will essentially be repeated in the next step but modified by a nonlinear signal function, or $Z_{t+1,\text{expected}} - Z_t = S(Z_t - Z_{t-1})$ where a typical realisation of the signal function can be $S(u) = \tanh(u)$. Under these assumptions it is found that the price follows the deterministic dynamics

$$Z_{t+1} = Z_t + F + MS(Z_t - Z_{t-1}) \tag{3.44}$$

characterised by the set of attractors, depending on the two model parameters F and M. The nonlinearity introduced through the signal function causes bi-stable 'catastrophic' behaviour within a certain range of parameters; in such a regime there are two stable fixed points of the dynamics (3.44), and the system discontinuously jumps from the one to the other when the parameters are continuously tuned. The accompanying hysteresis is reminiscent of a first-order phase transition. However intellectually appealing, the Zeeman model does not seem to comply with real observations, as the price sequence in speculative bubbles and crashes does not look very similar to switching between only two potential states of a bi-stable system. Let us also note that the model has been revisited, introducing a stochastic change of parameters F and M [798].

Another way of avoiding the excessive volatility paradox is based on looking for deterministic but chaotic models of economic dynamics. Let us mention only a few examples of results pointing in this direction.

Based on their earlier simulations of economic cycles [799], Youssefmir and Huberman [800] introduced a model in which agents choose between two resources. The utility of the resource depends on the number of agents who are currently using it. The dependence of utility on usage introduces nonlinearity into the system. The agents can have various strategies, taken from a pre-defined set. The probability of choosing a strategy is proportional to the past utility provided by that strategy. This leads to deterministic equations for the concentration of strategies, which were studied by numerical integration, revealing persistent fluctuations in the resource usage and exhibiting volatility clustering. More thorough investigation shows that the turbulent periods correspond to switching between various different but equally efficient strategy mixtures. This is again the recurrent theme already seen in the toy model of Kirman and in the currency exchange model of Arifovic: the volatility is due to switching between multiple optima regardless of whether we work with deterministic or stochastic dynamics.

In a similar spirit, Brock and Hommes [801] tried to reconcile the deterministic view on market dynamics with an underlying stochastic microstructure within a model of heterogeneous agents trading an asset. The authors suppose the population of traders

---

Logistic map | Box 3.6

generates a sequence of numbers $x_t \in [0,1]$, $t \in \mathbb{N}$, according to the equation
$$x_{t+1} = r\, x_t (1 - x_t). \tag{*}$$
Viewed as a dynamical process, the sequence has a point attractor for $r < r_1 = 3$, a cyclic attractor of the length 2 for $r_1 < r < r_2 = 1 + \sqrt{6}$, and generally a cycle of the length 2^n for $r_n < r < r_{n+1}$. The limit of the growing sequence r_n is $r_\infty = 3.569945672\ldots$ and for generic $r > r_\infty$ the process is chaotic. It is perhaps the simplest and most popular example of deterministic chaos. John von Neumann suggested in the late 1940s that the equation (*) with $r = 4$ could be used as a source of computer-generated pseudo-random numbers.

---

is again endowed with a pool of different strategies or predictors of the future price movements. Each strategy in the pool has a certain success rate according to the profit it provides to its users. The strategy with the score U is then used by the agents with probability $\sim e^{\beta U}$, where β is the parameter called by the authors 'intensity of choice'. Of course, a physicist recognises in it the inverse temperature. The probabilities then determine the fraction of agents using that strategy in the next time step, and for a large number of agents we get the set of deterministic equations for concentrations of the strategies in the population. Brock and Hommes examined several simple examples using only two strategies, which may be identified as 'fundamentalist' and 'chartist' (see below). They found that the dynamics is nonlinear with a bifurcation diagram reminiscent of that present in the logistic map (see Box 3.6). Consequently, they argued in favour of deterministic chaos in market dynamics.

A keystone in the construction of the EMH is the assumption of the agents' rationality. As EMH is empirically wrong, one may also try to lift the assumption of full rationality.

Frankel and Froot [802] admitted a deviation from an orthodox rational expectation theory in their attempt to explain the dollar bubble in the early 1980s. They postulated that there are three main types of agents on the market: fundamentalists, chartists, and portfolio managers. All three are completely rational within the bounds of their basic behavioural type. Fundamentalists try to rationally predict the fundamental value of the asset, chartists deduce the trends on a rational basis, and portfolio managers adjust their strategy as a weighted mixture of the expectations of fundamentalists and chartists, depending on which of the two groups is (rationally) perceived as more successful. Yet the whole system does not behave rationally, for none of the three groups takes into account either the mutual influence of the group or the impact of its own decisions.

In the model of Frankel and Froot, both fundamentalists and chartists behave mechanically in the sense that they never change their prediction algorithms. The adaptive element consists in the presence of portfolio managers, who attribute different weights to the first two groups according to their measured performance. However, the adaptation is not fully rational, as the portfolio managers learn about the state of the market more slowly than they are changing it. They do not keep pace with the effect of their own actions. Based on this observation, the model was able to qualitatively reproduce the bubble behaviour seen in the data.

Day and Huang [803] devised a similar model, showing qualitatively realistic features. They also suppose the agents are of two types, called in this context α- and β-investors. The former ones expend effort on finding the fundamental value. They sell if the price is above, and buy if the price is below. In our previous terminology these agents would be called fundamentalists. The latter group of agents behaves differently: as finding the fundamental value is a costly process which may not pay off after all, they simply chase the trend. Day and Huang used a somewhat distorted version of these 'chartists', prescribing them to sell if the price were above a certain value (fixed forever) and buy if it were lower.

Even farther in the direction of heterogeneity and limited rationality goes the attempt to simulate the diversity and evolution of decision-making algorithms using the idea of genetic programming [804–806]. Every agent is equipped with a program for predicting the future price return X_{t+1} based on past returns $X_{t-\tau}$, $\tau = 0, 1, \ldots$. More specifically, the program is represented by a formula assembled in a syntactically correct way from the set of operations $\{+, -, *, /, \ldots\}$, numerical constants, and operands $\{X_t, X_{t-1}, \ldots\}$. An example of such a formula can be the predictor $X_{t+1} = 3\,X_t\,(1 - X_{t-1})$.

Similar in essence to the genetic algorithms mentioned above (see also Box 3.7), genetic programming regards the set of agents' algorithms as an ecosystem subject to Darwinian evolution. Fitness of an algorithm is given by its success in predicting the future, which is reflected by the increase (or decrease) of wealth in agent's pocket. More successful predictors are replicated with larger probability, so the overall performance of the ensemble of agents is improved by adaptation.

In the first simulations [804] the agents were trained on a portion of a real price series. Then it was checked how well the agents predict the rest of the price series. As expected, some of the predictors performed quite well, but the overall capability of learning and predicting was far from satisfactory.

In subsequent work [805, 806] the agents were allowed to play in an artificial market and learn from the other agents' actions. The time series for prices produced within these simulations exhibited excess kurtosis, but the probability distribution for returns was not investigated.

3.2.2 Levy-Levy-Solomon model

Participants in economic life can behave in various ways, depending on the scale of their business, their social and material background, local customs, and many other features we are perhaps not yet aware of. Very often the precise motivation and mechanisms behind their decisions are obscure to the agents themselves. A 'realistic' model of decision-making in the economy remains an unreachable dream and probably needs the intervention of other disciplines such as social psychology or cultural history. We can speculate that the structuralist approach, when taken seriously, can lead to significant progress [28].

For now, we are left with bare hypotheses, and among them the prominent position is occupied by the classical view of people as rational optimising machines, who always know what is good for them, and always select the best one from among all possible choices. Even if we accept a certain possibility of error, and take into account the

Genetic algorithms | Box 3.7 |

were invented in the 1970s by John H. Holland [807]. The aim was to provide new solutions to notoriously difficult optimisation problems, like the travelling salesman problem. Holland's idea originated from observing how well living creatures are adapted to various tasks they face in their lives. The algorithm for such optimisation is well-known: it is Darwinian evolution. Holland implemented Darwinism for evolution of good solutions in optimisation problems. As in nature, the solution is encoded in the genome. Instead of DNA, a string of binary digits is used, each meaning presence (1) or absence (0) of a specific trait characterising the solution. The whole ensemble, or ecosystem, of solutions is then subject to evolution. In each step, the quality of the solutions is calculated and a pair of genomes is selected for sexual reproduction, preferring those which encode better solutions. Some of the poorly performing solutions are replaced by the newborn individuals. The offspring has part of the genome from one parent and the rest from the other parent. This operation, mixing the two genomes, is called crossover. Moreover, the offspring genome can be altered with a small probability by flipping some of the bits. This represents point mutations. The combination of mutation, selection, and crossover leads, after some time, to the emergence of some very good genomes. The best one is then the desired solution of the original problem.

Closely related is the method of **genetic programming**, devised by John R. Koza for finding optimal algorithms, or programs, for performing various calculations [808]. Again, an ensemble of programs is allowed to evolve according to Darwinian rules, and the good ones are expected to emerge spontaneously. Technically it is more complicated than genetic algorithms, as crossover and mutations must preserve syntactically correct code.

definite imprecision in the input information, the general idea stays unchanged. It is supposed that one-dimensional thinking, the search for an extreme of a single quantity, is the driving force of all manoeuvres in the productive lives of all human beings.

However ridiculous it may seem, such a hypothesis must be taken seriously. First, it is evident that avoiding harm and seeking good is hard-wired in our minds. The question is how we know what is good and what is harm in a specific situation. Second, if a person is taught long enough that normality means maximising a predefined quantity, she can hardly resist. People maximise profit because they believe in their hearts that this is the way normal people behave. This belief was conveyed them by social pressure. All apparent exceptions constitute a problem which needs special explanation and in most cases such one is readily found, e.g. within the selfish gene paradigm. It is not our task here to discuss how much this paradigm is descriptive and to what extent it is rather a prescriptive and regulative instrument. Instead, let us present the third and most important argument.

Indeed, it may well be that although individuals never strictly obey the orders the proponents of maximisation impose on them, on average and effectively they behave as if they were maximising something which can be a posteriori read off and identified with the quantity called profit. We cannot help recalling classical mechanics, where massive bodies hardly care about minimising the action functional but their trajectories are still exactly described by the principle of minimum action. The particles behave as rational minimisers of the quantity called action, despite being inanimate pieces of passive matter.

Utility function   Box 3.8

Realistic utility function must have two properties. It is an increasing function, $U(w_1) > U(w_2)$ if $w_1 > w_2$, and moreover, it is concave, i.e. $U\left(\frac{1}{2}(w_1 + w_2)\right) > \frac{1}{2}\left(U(w_1) + U(w_2)\right)$ for any w_1, w_2. The concavity is the mathematical expression of the obvious fact that people try to avoid risk; therefore changes in wealth are more important for them if their starting wealth is low. You can very often see these two conditions stated in the more concise form using the first and the second derivative of the utility function, namely, $U'(w) > 0$ and $U''(w) < 0$. However, such a formulation neglects the fact that there is no guarantee that an empirical $U(w)$ is a differentiable function. On the contrary, if $U(w)$ were measured point-to-point by a smart sociological experiment, the result would most probably be a piecewise linear function, a feature which can only partially be ascribed to bias from the experimental setup, since people often do like to judge the inputs in a jumpy manner. In model situations, the favourite forms of the utility function are logarithmic, $U(w) = \ln w$, exponential, $U(w) = 1 - e^{-aw}$, $a > 0$, or power, $U(w) = w^{1-\nu}/(1 - \nu)$, $0 < \nu < 1$. Lacking empirical guidance, we shall use one or the other according to convenience for the model in question. However, it would be too pessimistic to claim that the form of the utility function is arbitrary. Empirical studies do exist that measure risk aversion in human decisions [809], which is quantitatively expressed by the second derivative $U''(w)$. This enables fitting the free parameter of the utility function, once we have decided on its general shape.

However unproductive the academic debate may be on whether humans really do or do not maximise profit in their everyday lives, it is sensible to investigate the consequences of such an assumption. This brings us back to a more solid scientific ground, even if we remain within a hypothetical sphere. At least we shall know the enemy better.

Optimal investment

It is impossible to give here an account of the theory of optimal investing in the stock market, and after all, it is not the goal of this book. We shall only pick an example of an optimisation strategy which will be useful for our later discussion on a microscopic market model.

Greater wealth will certainly imply a better situation than less wealth. On the other hand, this simple observation does not mean that the quantity to be optimised is necessarily the wealth itself. The only prerequisite is that this quantity is a non-decreasing function of wealth. In fact, various paradoxes were invented to show that wealth cannot be sensibly considered as the quantity to be optimised. Instead, it is supposed that a so-called utility function $U(w)$ exists and every agent struggles to make it as large as possible. (See Box 3.8.) The variable w is usually assumed to be the wealth itself (which is what we will do in this section too), although one can imagine a utility function depending on other quantities, or even on ensembles of variables.

In standard economic theory, people are supposed to rationally maximise their utility function. In practice it means that the economic agents parameterise their behaviour by a set of numbers $f_1, f_2, \ldots, f_n$ on which wealth w depends and then try to tune these parameters so that the expected utility $U(w(f_1, \ldots, f_n))$ is the largest possible. A classical example is a merchant selling a mixture of both good and bad wine. The parameter f is the fraction of the quality ingredient. When f is close to 1, the profit is low, as few people can afford an expensive beverage. If f approaches 0, the profit from selling distasteful swill is also low. The optimum lies somewhere in the middle.

Here we have in mind an optimising strategy for buying, selling, or holding stocks. A very transparent presentation of the problem stems from the work of J. L. Kelly [810–818].

Suppose you have some capital and you want to invest it. You may either buy a riskless asset, e.g. a government bond, or some shares, which promise more gain but you may also lose a lot when the price of the shares drops unexpectedly. Your strategy will be to divide your capital and invest only a fraction $f \in [0, 1]$ into risky shares, while the rest will be kept safe in bonds. To see how the investor's wealth W_t evolves due to the price fluctuations of the shares, we assume that the price Z_t performs a geometric random walk in discrete time t, i.e. the next price will be $Z_{t+1} = (1 + \eta_t)Z_t$ where $\eta_t > -1$ for all times t are random variables which are independent and whose probability distribution does not depend on time. Therefore, in the next time the wealth is

$$W_{t+1} = (1 + f\eta_t)W_t. \tag{3.45}$$

For reasons which originate partially from information theory, Kelly uses the logarithmic utility function. The investor wants to know what fraction f she should keep in the stock in order to gain maximum utility in the long run at time $t \to \infty$. The task is therefore reduced to finding the maximum of the expression

$$L(f) = \langle \ln\left(1 + f\,\eta_t\right)\rangle \tag{3.46}$$

where the angle brackets denote the average over the random factor η_t. Because $L''(f) = -\left\langle\left[\eta_t/(1 + f\,\eta_t)\right]^2\right\rangle < 0$, the maximum can be either inside the interval $(0, 1)$ or at its edges. In Fig. 3.10 we can see how it works. The fluctuations in η_t compete with the trend. If $\langle\eta_t\rangle > 0$ the wealth grows nominally, but it may effectively decrease due to occasionally low value of the random factor η_t. It is important to note the sensitivity of the maximum of $L(f)$ with respect to the value of the average $\langle\eta_t\rangle$. A tiny change may bring the location of the maximum from $f = 0$ to $f = 1$ or vice versa, and only in a narrow range of parameters of the noise η_t is the maximum found inside the interval $(0, 1)$. We shall see in next paragraph that this sensitivity determines the generic behaviour of the Levy-Levy-Solomon model. It was also noticed that such an unrealistic feature of the Kelly method prevents its direct use in everyday practice, as was demonstrated, e.g. on empirical data from the New York Stock Exchange [819]. But even though the Kelly method is ruled out as a means for earning money, it may well serve as a toy example of how true optimisation of an investment works.

Artificial optimisers

In the model of Levy, Levy and Solomon [820–823] the simulated agents perform just the optimisation according to Kelly. The only deviation from the basic Kelly scheme is that it uses a wider range of utility functions in order to allow the agents to differ in their investment preferences. Specifically, we assume the power-law form

$$U(w) = \frac{w^{1-\nu}}{1 - \nu} \tag{3.47}$$

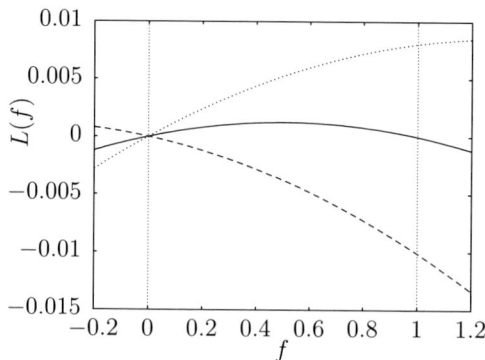

Fig. 3.10 Illustration of the Kelly optimisation. In this example, the noise η_t assumes values $m \pm s$ with equal probability $\frac{1}{2}$. The values of the parameters are $s = 0.1$ and $m = 0.005$ (dotted line), $m = 0.013$ (full line), or $m = -0.005$ (dashed line). We can see how a slight change in m shifts the maximum from the right to the left edge of the interval $f \in [0, 1]$.

and complete the specification postulating that $\nu = 1$ implies $U(w) = \ln w$. Thus we have a one-parametric set of utility functions at our disposal. The quantity ν measures the risk aversion of the agent. Indeed, the larger the value of ν, the more the poor agents feel the fluctuations in their wealth and the less the fluctuations are perceived by the rich agents.

The state of the i-th of the N agents at time t is determined by her wealth W_{it} and by the fraction F_{it} of the wealth kept in the risky asset. The price of the stock at time t is Z_t, so the agent i owns $S_{it} = F_{it}W_{it}/Z_t$ shares. The total number of shares

$$S = \sum_{i=1}^{N} \frac{F_{it}W_{it}}{Z_t} \tag{3.48}$$

is conserved, which plays a decisive rule in calculation of the change in the asset's price.

The wealth of the agents evolves due to changes in the price of the asset and thanks to the dividend d distributed among the shareholders. Suppose the transactions at time $t + 1$ are performed at a hypothetical price z_h, which will be determined in the negotiation according to demand and offer. Then the new, still hypothetical, wealth of the agent i is

$$w_{hi} = W_{it}\left[1 + F_{it}\frac{z_h - Z_t + d}{Z_t}\right]. \tag{3.49}$$

The transactions occur because the agents want to update their investment fractions F_{it} so that their expected utility is maximum. In the Kelly method we tacitly assumed we have full information on the probability distribution of the noise η_t. Here we are in a more complicated situation, for the agents have to guess the probability of future movements only based on their previous experience. The best thing they can do is to watch the price series and estimate the future return, including the dividend

$$\eta_{t+1} = \frac{Z_{t+2} - Z_{t+1} + d}{Z_{t+1}} \tag{3.50}$$

from the returns in M previous steps, $\eta_{t-\tau}$, $\tau = 1, 2, \ldots, M$. In other words, the agents' memory is M steps long, and for the future they predict that any of the M past values may repeat itself with equal probability. Note that the quantities η_t are not random numbers supplied from outside, as in the Kelly optimisation, but result from the dynamics of the Levy-Levy-Solomon model.

Let us denote by angle brackets $\langle \phi(\eta_{t+1}) \rangle_M = \frac{1}{M} \sum_{\tau=1}^{M} \phi(\eta_{t-\tau})$ the average over expected return, for any function $\phi(\eta)$. If the agent i has hypothetical wealth w_{hi} at time $t + 1$ and at the same time she chooses a hypothetical investment fraction f_{hi}, her wealth at the next time $t + 2$ will be $w_{hi}[1 + f_{hi} \eta_{t+1}]$. The optimal strategy should maximise the expected utility $\langle U(w_{hi}[1 + f_{hi} \eta_{t+1}]) \rangle_M$ with respect to the hypothetical fraction f_{hi}. Now we can appreciate the convenient choice of a power-law utility function (3.47), because the expected utility is factorised; and we can find the optimum by maximising the quantity

$$L(f_{hi}) = \left\langle \frac{(1 + f_{hi} \eta_{t+1})^{1-\nu}}{1 - \nu} \right\rangle_M. \tag{3.51}$$

If all of the agents were equal and followed fully rationally the suggestions resulting from the optimisation, there would be no trade, as everybody would insist on holding the same proportion of her wealth in stock. Such a conclusion contradicts the most basic observation of busy economic life. It also casts serious doubts on the very concept of economic agents as rational optimisers. We can see that the very existence of trade relies on heterogeneity of their expectations and strategies. If there was a unique optimum strategy and everybody followed it, the economy would crash. But if there is no global optimum, why should we look for any? We do not undervalue this question, and if we do not venture an answer, it is because we feel it goes much deeper than this book may ever reach.

Of course, there are some more or less standard and more or less expensive ways out of this dilemma. One of them says that it is the beneficial action of pure chance that saves us from an impasse and keeps the trade going. This is the stance adopted in the Levy-Levy-Solomon model. If $f^* \in [0, 1]$ is the value of f_{hi} maximising the expression (3.51), the actual investment fraction the agent i chooses in time $t + 1$ will be

$$F_{it+1} = f^* + \epsilon_{it} \tag{3.52}$$

where the random term ϵ_{it} accounts for uncertainty and subjective error in the decisions of agent i at time t. In practice we take the random variables ϵ_{it} as independent and uniformly distributed in the interval $(-b/2, b/2)$, with the restriction that $f^* + \epsilon_{it} \in [0, 1]$. Knowing the investment fractions, we can compute the actual price at time $t + 1$ from conservation law (3.48). The result is

$$Z_{t+1} = \frac{Z_t \sum_{i=1}^{N} F_{it+1} W_{it} - (Z_t - d) \sum_{i=1}^{N} F_{it+1} F_{it} W_{it}}{Z_t S - \sum_{i=1}^{N} F_{it+1} F_{it} W_{it}}. \tag{3.53}$$

From here the actual return at time t and the new wealth values of the agents are computed

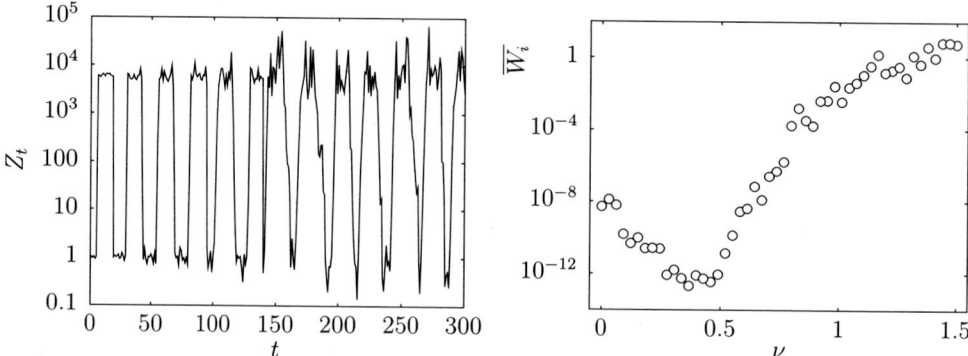

Fig. 3.11 Simulation of the Levy-Levy-Solomon model. In the left panel, time series of the price in a typical run, for $N = 1000$ agents with memory $M = 10$ and aversion exponent $\nu = 0.5$. The dividend is $d = 0.001$ and the noise in the agents' decisions $b = 0.05$. In the right panel, wealth distribution in a typical run with a heterogeneous population, after 500 steps. The $N = 1000$ agents are distributed into 50 groups according to the value of the exponent ν, which ranges from 0 to 1.5. Each point in the graph represents the geometric mean of the wealth values of the agents within one group.

$$\eta_t = \frac{\sum_{i=1}^{N} F_{it+1} W_{it} - (Z_t - d)S}{Z_t S - \sum_{i=1}^{N} F_{it+1} F_{it} W_{it}} \tag{3.54}$$

and

$$W_{it+1} = W_{it}[1 + F_{it}\,\eta_t]. \tag{3.55}$$

Eqs. (3.51) through (3.55) define the dynamics of the Levy-Levy-Solomon model. We can load it onto a computer and look at its behaviour. A typical example of the time evolution of the price is shown in Fig. 3.11. The most characteristic feature of the price series is the periodic oscillation between low and high price levels. The period is determined by the length of the agents' memory M, and the oscillations can be understood if we recall the sensitivity of the Kelly method to the parameters of the noise.

Initially, the sequence of returns kept in the agents' memory is generated randomly, centred around the dividend d. The initial price is set to 1. When the agents start trading, they feel encouraged by the dividend and try to buy the asset. Their investment fractions are close to 1. As a result, the price rises and stays high for some time. But the dividend being constant, it loses importance at a higher price level, and the noise generated by the random component of the agents' decision prevails. The investment fraction drops quickly to a value close to 0; therefore the price falls substantially. In Fig. 3.11 we can see that the higher and lower price levels are as far as four orders of magnitude from each other, which is certainly much exaggerated compared to reality. The negative return perceived in the price drop remains in memory for M steps and screens out the encouragement from the dividend, now important again, as the price is relatively lower. As soon as the bad experience from the drop is forgotten, the investors become optimistic again, the price jumps up to about the same level as before; and this periodic movement keeps going on and on. At later times, the oscillations become less regular due to the random distribution of the agents' wealth, but the general feature of periodicity persists. We can see that the general character of the time series, exhibiting jumps between two well-separated price levels, is caused by the sudden switching between investment fractions close to 0 and 1. The intermediate values are very rare in the Kelly method; therefore the intermediate prices in the Levy-Levy-Solomon model are also rarely visited.

Role of heterogeneity

In its most basic form the Levy-Levy-Solomon model is absolutely unrealistic. Does that mean that it should be discarded, or is there any useful lesson it can teach us? To see what is closer to the truth we can make the agents heterogeneous. We have already discussed the influence of the memory span M on the period of the oscillations. The larger M is, the longer the agents remember the last price jump and keep the investment close to the extreme investment fraction, be it 0 or 1. But if the agents differ in their capacity to remember past returns, various periods interfere in the price series and make it more realistic. If there are several, or even many different memory lengths at play, the price signal would be as irregular as the real stock market fluctuation.

Another kind of heterogeneity may be the varied risk aversion of the agents. Some of them accept the risk more willingly, some are rather reluctant. We quantify the effort to avoid risk by the exponent ν. In the population of agents with several different values of ν the various groups compete among themselves, and we can ask what level of risk aversion results in the highest gain.

We can see a typical result in Fig. 3.11. If we divide the agents into groups according to their value of ν, we observe a non-trivial dependence of the wealth averaged within the group on the risk-aversion exponent ν. If we increase the value from $\nu = 0$, the average wealth first decreases, because risk averters invest less into the asset and therefore gain less dividend. However, the trend reverses around $\nu \simeq 0.5$, and the more careful investors win. The reason for this behaviour, which somewhat contradicts intuition, lies in the details of the investors' reactions to the abrupt drops in price. While the risk-averters sell their shares immediately, contributing to further collapse, the risk-prone agents keep holding their assets for a longer time and suffer a significant loss.

Although the Levy-Levy-Solomon model is too simplistic to be able to reproduce the complexity of stock market fluctuations, some basic mechanisms are present in it. Most importantly, it demonstrates how the imprecision in agents' actions is vital for the liquidity of the market. Second, it shows how sudden price changes, booms and crashes, can result from endogenous sources. In fact, it is the delayed response to past events, kept in the finite memory of only M steps, that makes the market unstable and leads to collective switching from optimistic to pessimistic investments and back. Although no real booms and crashes can be periodic, as this would imply perfect predictability, the basic mechanism, a synchronous response of the investing crowd to certain configurations of past and present information, may be the same in reality as in the model. Moreover, the spurious periodicity can be avoided in more sophisticated versions of the Levy-Levy-Solomon model which were investigated in depth in [822, 824, 825].

3.2.3 Lux-Marchesi model

Soon after economists started to abandon the orthodox EMH view, it became clear that heterogeneity of the agents' behaviour is essential to understand the stylised facts. The immediate question is, how far we must go with the heterogeneity in order to get a realistic-looking model. Thomas Lux came up with the idea that three groups of agents are enough [826–828]. The first of them are the fundamentalists, who observe deviations of the actual price from what they perceive as a fundamental value of the asset. If the price is higher, they sell, as they expect a price drop in the future. And vice versa: if the price is lower, they buy, since their logic predicts the price will rise towards the fundamental value. The second and third group taken together are chartists, people with different talents than fundamentalists. They rely on their ability to predict the price movements in the near future based on the observation of the past price series. We may think they are modern shamans gazing at structureless chaos of rugged lines of turbulent prices, and suddenly—pop!—they know the good or bad fortune. In practice, of course, it is not fortune telling, but implementing sophisticated analytic software that chartists do. In the simplified view of Lux, they are much less smart, though. They are either optimists, who always buy as they constantly expect the price to rise, or pessimists who, on the contrary, keep selling because they think the price must go down. To sum up, the agents are either fundamentalists, or optimists, or pessimists.

So far, the scheme is similar to earlier attempts by Day and Huang [803], Frankel and Froot [802], and Brock and Hommes [801], as explained in Sec. 3.2.1. All of these schemes share the classification of agents as fundamentalists and chartists. However, Lux combined it with Kirman's idea of spreading the behaviour by imitation or contagion. The type of the agent's strategy is not attributed forever; instead, the agents tend to follow other agents and form a herd. The herding effect is even strengthened if the individuals imitate more apparently successful mates. The idea of agents flowing dynamically between three uniformly acting groups was embodied in the model of Lux and Marchesi [826–837]. The three groups can be viewed as three strategies used by the market participants. The sizes of the groups determine the price changes, and the flow of the agents among the three strategies is also influenced by the current price and its trend. But while the price changes are fully deterministic as soon as the group sizes are known, the dynamics of the group sizes themselves is stochastic. Let us first look at the evolution of the prices and later investigate the flow of the strategies.

Price dynamics

The virtual stock market we shall create and animate in the computer will be inhabited by N agents who are offered three possible types of behaviour. The choices are denoted by the symbols 0 for fundamentalists, + for optimists, and − for pessimists. There are N_σ agents in the group $\sigma \in \{-, 0, +\}$. Logarithm of the price of the traded asset at time t is denoted $Y_t = \ln Z_t$. (For brevity, in the following we shall call Y_t 'price', although 'log-price' would be the precise name.) The return is defined simply as the difference $X_t = Y_t - Y_{t-1}$.

The agents' actions lead to price changes, which in turn influence the composition of the strategy mixture in the population. This feedback effect is the main cause of the complex behaviour of the model. What is the response of the price to the agents' actions? To answer that question we need to know how much of the asset the members of the three strategy groups will buy or sell. We assume the ability to buy/sell is not limited either by the financial resources the agents possess or by the amount of the asset available on the market. The price dynamics will be determined purely by the disequilibrium between demand and offer.

The strategies of the three agent types are rather simplistic. The optimists buy in each time step one unit of the asset; the pessimists sell one unit of the asset; the fundamentalists first compare the actual price Y_t with the fundamental level y_f. Then, they buy if the price is higher and sell if the price is lower. We assume that the amount to buy or sell is proportional to the difference $Y_t - y_f$. In principle, we could choose any increasing function of Y_t, equal to zero if $Y_t = y_f$, but we adopted the simplest choice. Altogether, the excess demand is

$$D = N_+ - N_- - N_0 \, c \, (Y_t - y_f) \tag{3.56}$$

where the parameter c quantifies the relative aggressiveness of the fundamentalists with respect to the chartists. The response of the price on the excess demand is deterministic

$$X_{t+1} = Y_{t+1} - Y_t = b \, \frac{D}{N}. \tag{3.57}$$

The price impact depends on the so-called market depth, represented here by the factor b. We can freely choose $b = 1$, thus fixing the unit of price.

Two remarks are due here. First, we entirely neglected the complicated job of the fundamentalists when establishing their most desired quantity, which is the fundamental price. Most of their time, skill, and resources are spent in this effort and properly speaking, their strategies should be modelled in various ways to obtain a reliable estimate for y_f. On the other hand, we admit validity of the EMH to such an extent that the fundamentalists are capable of translating all incoming news and external influence into the fundamental price and do that fully rationally and instantly. As the news is coming randomly, the evolution of the fundamental price is most naturally modelled by a random walk. However, if we want to separate the influence of fluctuations in the fundamental price from the fluctuations induced by the intrinsic agents' dynamics, it is wiser to forget the external influence entirely and keep y_f constant. Because Y_t is interpreted as the logarithm of price, which assumes both positive and negative values, we can choose the fundamental price $y_f = 0$ without loss of generality.

Having discussed the dynamics of price, we must also specify how the sizes of the three groups of agents evolve. Here we apply the idea of opinion spreading by contagion. (We shall discuss opinion spreading in much more detail in Chap. 8.) When two agents meet, the first one can adopt the opinion of the second, like Kirman's ants adopting the trail established by other ants. However, unlike the ant, the agent is able to assess (at least partially) the success of the strategies and to compare her own strategy with the other agent's. This may seem straightforward, but in practice it may prove difficult to say in a given moment whether the fundamentalist or the chartist strategy is superior, because they operate at different time scales. The fundamentalists hope that the deviation from the fundamental price will be rectified and that they will gain the difference when it does so in some uncertain distant future. On the contrary, the chartists bet on the trend followed in the next step, so they live on a very short time horizon.

Strategy flow

Let us see now how the idea of strategy choice by contagion can be implemented in the most simplified manner. If two agents meet, the first can adopt the strategy of the second. The agents can also change their strategy spontaneously and randomly with a small probability ϵ. So far, the dynamics would be a trivial generalisation of the Kirman's ants model to three groups instead of two. But suppose that the agents are not completely blind imitators and take into account the quality of the strategies, as far as they are able to estimate it. The chartists believe in the trend they observe. If the price has risen since the last update, they suppose it will rise again, and vice versa. Therefore, the current optimists, who believed in the price increase and therefore bought a unit of the asset, may convert to pessimists if they observe a price drop.

Thus, switching between optimism and pessimism is based only on the sign of the price change. A change from chartist to fundamentalist strategy and back is based on a slightly more subtle consideration. The fundamentalists sell if the price is above the fundamental value and buy if the price is lower, as they believe that the market forces must sooner or later bring the price back to its fundamental. The catch in it resides in the words 'sooner or later'. In fact, before the price returns, the reward from playing the fundamentalist strategy is not collected, and we have little knowledge about when it will happen. But if a 1 Euro investment pays back only after 10 years, there is something rotten in the state of your finances, isn't there? Contrary to the chartists who gain or lose immediately, depending only on whether their one-step prediction was correct, the fundamentalists must discount the potential gain $|Y_t - y_f|$, taking into account the estimated time before the gain becomes real. In the Lux-Marchesi model, we implement this feature introducing a single discount parameter d. Then the fundamentalist strategy is considered superior to the pessimist if $d|Y_t - y_f| > -X_t$ and superior to the optimist if $d|Y_t - y_f| > X_t$. Thus, we can always order the three strategies according to their estimated success.

Now we are ready to describe one step in updating the strategies of the N agents. To fix the time scale, we assume that the update of a single agent's strategy takes time $1/N$. This update proceeds as follows. We pick one agent at random. She may change strategy spontaneously, with probability ϵ, to any of the remaining 2 possibilities. Otherwise, we pick another agent and compare her strategy with the first. With probability q, the first agent simply copies the strategy from the second, as if she were a pure imitator. But with probability $1 - q$ the first agent compares the quality of her own strategy with the strategy of the second agent and adopts the latter only if it is better.

To put the definition on a more formal basis, let us first denote by g_σ the expected gain of the strategy $\sigma \in \{-, 0, +\}$ on the condition that the current price is Y and the return per unit time is X, as prescribed by (3.57). Using the simplification $y_f = 0$, we have

$$g_\pm = \pm X$$
$$g_0 = d\,|Y|. \tag{3.58}$$

The state of the system is fully described by the price Y and the numbers of agents in each of the three groups N_σ. Only two of them are independent though, and it will be convenient later to describe the state in terms of the number of fundamentalists N_0 and the imbalance between optimists and pessimists $N_\Delta = N_+ - N_-$. The dynamics is a Markov process, so it is fully determined by transition probabilities from state (N_+, N_0, N_-, Y) to state (N'_+, N'_0, N'_-, Y'). In fact, the only allowed transitions correspond to the decrease of a group σ_1 by 1 and the simultaneous increase of another group $\sigma_2 \neq \sigma_1$ by 1; so $N'_{\sigma_1} = N_{\sigma_1} - 1$, $N'_{\sigma_2} = N_{\sigma_2} + 1$, and $N'_{\sigma_3} = N_{\sigma_3}$, where $\sigma_3 = -\sigma_1 - \sigma_2$ is the index of the third, unaffected group. The corresponding transition probability can be expressed as

$$w\Big[(N_{\sigma_1}, N_{\sigma_2}, Y) \to (N_{\sigma_1} - 1, N_{\sigma_2} + 1, Y')\Big]$$
$$= \Big\{\epsilon \frac{N_{\sigma_1}}{2N} + (1 - \epsilon)\frac{N_{\sigma_1} N_{\sigma_2}}{N(N-1)}\big[q + (1-q)\,\theta(g_{\sigma_2} - g_{\sigma_1})\big]\Big\} \tag{3.59}$$
$$\times \delta\Big(Y + \frac{1}{N^2}(N'_+ - N'_- - N'_0\,cY) - Y'\Big)$$

where we used the notation $\theta(g) = 1$ for $g > 0$ and $\theta(g) = 0$ otherwise; $\delta(0) = 1$ and $\delta(y) = 0$ for $y \neq 0$. The first term in the curly brackets comes from a spontaneous change of strategy, while the second comes from contagion. The factor containing the δ-function accounts for the change in price, which is deterministic as soon as the new sizes of the three groups are fixed. Note the extra factor $1/N$ in the price change, compared to Eq. (3.57), which is due to the fact that the elementary update takes time $1/N$.

Simulation results

The reader is certainly eager to see how the model works in practice. First let us look at the time evolution of the number of fundamentalists and both types of chartists. In Fig. 3.12 we can see the concentrations N_0/N and N_Δ/N in a typical run with $N = 1000$ agents. Most of the time, the fundamentalists make up a decisive majority of the population. Their action on the price is reverting to the fundamental value, and comparing the upper and middle panels of Fig. 3.12 we can see that the periods with a high percentage of fundamentalists are marked by fairly small deviations of the price from $y_f = 0$. As a result, the return is also small, as can be seen in the lower panel of Fig. 3.12.

But from time to time, more chartists emerge. While the fundamentalists act as negative feedback on the price, the action of the chartist is just the opposite. The optimists are reinforced if the price rises, which leads to a further immediate increase

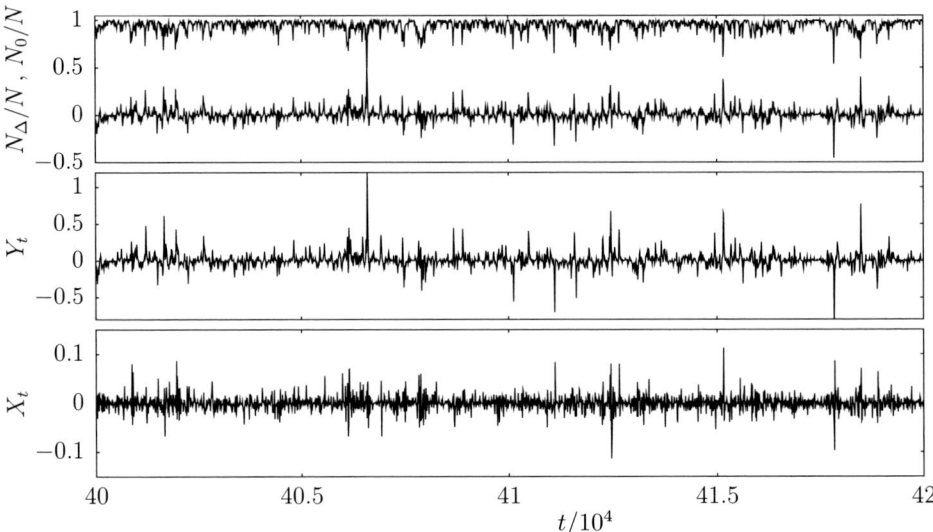

Fig. 3.12 Typical time series for the Lux-Marchesi model. In the upper panel, evolution of the concentration N_0/N of fundamentalists (upper curve) and the relative imbalance N_Δ/N between optimists and pessimists (lower curve). The corresponding price and return are shown in the middle and lower panels, respectively. The number of agents is $N = 1000$, and the parameters of the model are $q = 0.9$, $d = 0.1$, $\epsilon = 0.001$, and $c = 0.8$. Note that the periods of higher volatility coincide with the lower fraction N_0/N of fundamentalists and larger fluctuations in the imbalance N_Δ.

in price. Conversely, the pessimists feed on the decrease, and at the same time they cause a decrease in the price. There is positive feedback induced by the chartists.

For a certain limited time the positive feedback can prevail. In such periods, the number of fundamentalists drops, and the imbalance between optimists and pessimists starts fluctuating violently, which causes high volatility in price, as can be seen from the series of the returns. Such a state lasts for some time before the fundamentalists take over again; so the periods of high volatility punctuate the low-volatility ones. On a qualitative level, we observe the effect of volatility clustering.

Also qualitatively, we observe that the returns are usually small, with relatively rare large spikes. This is rather different from Gaussian noise and suggests fat tails in the return distribution. In a simulation, we measure the cumulative return distribution in the run of length T

$$P^>(x) = \frac{1}{T} \sum_{t=1}^{T} \theta(|X_t| - x). \qquad (3.60)$$

The typical result is shown in Fig. 3.13, and the fat tails are indeed very well visible. However, it is not so clear whether the tail can be characterised by a power law; most probably not, and even if it did obey a power law, the exponent does not seem to be

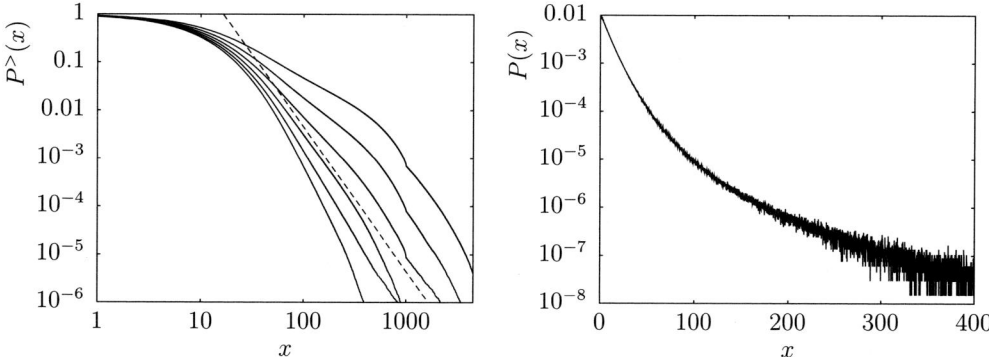

Fig. 3.13 In the left panel, cumulative distribution of returns in the Lux-Marchesi model with $N = 1000$ agents. The statistics were taken in the last two-thirds of a single run taking the time 10^8. The parameters common to all curves are $q = 0.9$, $\epsilon = 0.001$, and $c = 0.8$. The values of d for the seven curves shown are (from top to bottom) $d = 0.06$, 0.07, 0.08, 0.09, 0.1, 0.11, and 0.15. The dashed line is the power law $\sim x^{-3}$. In the right panel, distribution of returns $P(x) = -\frac{\mathrm{d}}{\mathrm{d}r} P^>(x)$ in a semi-logarithmic scale, for $d = 0.09$ with the other parameters the same as in the left panel.

universal. On the other hand, the tail is certainly fatter than an exponential, as the inset in Fig. 3.13 shows. For certain values of the parameters, the distribution is also fairly similar to the empirical one, with tail behaviour close to $P^>(x) \sim x^{-3}$. Note that the decisive parameter for the falloff of the probability at the tail is the discount parameter d, comparing the expected time horizons for the gain of fundamentalists and chartists. The values giving the most realistic results range from $d \simeq 0.09$ to $d \simeq 0.1$, suggesting that fundamentalists are ready to wait about ten times longer for their profit than chartists.

The return and volatility correlations provide us with quantitative information about the predictability and the volatility clustering. In Fig. 3.14 we can see that the returns themselves have a short, exponentially decaying anticorrelation. The volatility, measured as absolute return, is positively correlated on a much longer time scale, but the decay is still exponential, as in Kirman's ant model. In fact, it is not so surprising, as the Lux-Marchesi model is a Markov process, and exponentially fast (or slow) forgetting is a generic feature of Markov processes. Therefore, the effect of volatility clustering is present here, but is not strong enough to reproduce the empirical data.

An interesting feature of the Lux-Marchesi model is weak sensitivity to the values of its parameters. There is no abrupt change in behaviour, and the generic features shown in Figs. 3.12 through 3.14 are observed in more or less clear form for all parameter choices. However, it was noted that increasing the number of agents suppresses the fluctuations and makes the fat tails and other features less pronounced. At first sight this observation may seem disappointing, because it makes difficult the usual procedure of statistical physics, which is the thermodynamic limit $N \to \infty$. On the other hand, this finding is compatible with other approaches to price fluctuations, for example in

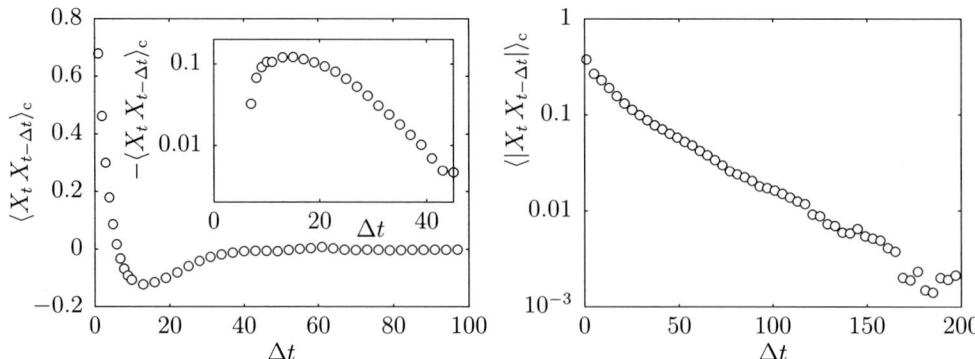

Fig. 3.14 Autocorrelations of return (left panel) and volatility (right panel) in the Lux–Marchesi model are shown, for $N = 1000$ agents. The parameters are $q = 0.9$, $\alpha = 0.1$, $\epsilon = 0.001$, and $\gamma = 0.8$. In the inset, detail of the data shown in the main plot.

the Cont-Bouchaud model discussed in Sec. 3.1.2 or in the models based on order-book dynamics, as we shall see in Chap. 4. Perhaps all the complexity of the stock market is a finite-size effect.

Analytical insight

The transition probabilities (3.59) can be used to derive a master equation for the probability density function $P_{N_0\, N_\Delta\, t}(n_0, n_\Delta)$. (For brevity, we shall drop the subscripts N_0 and N_Δ in the rest of this section.) Then we make the thermodynamic limit $N \to \infty$. In doing so we find that the time derivative on the left-hand side bears the factor $1/N$, as we supposed that one update lasts $1/N$ time units. On the right-hand side we get several drift terms, containing first derivatives of the concentrations. All of them are of the order $1/N$, while the diffusive terms, containing second derivatives, happen to be of the higher order $1/N^2$. So, in the thermodynamic limit the dynamics of the Lux-Marchesi model is effectively deterministic. Explicitly, we get

$$
\begin{aligned}
\frac{\partial}{\partial t} & P_t(n_0, n_\Delta) \\
={}& \frac{\epsilon}{2}\left[\frac{\partial}{\partial n_\Delta}\Big(3 n_\Delta\, P_t(n_0, n_\Delta)\Big) + \frac{\partial}{\partial n_0}\Big((3 n_0 - 1)\, P_t(n_0, n_\Delta)\Big)\right] \\
&+ \frac{(1-\epsilon)(1-q)}{2}\left[\frac{\partial}{\partial n_\Delta}\Big(\big(-((1-n_0)^2 - n_\Delta^2)G_{+-}\right.\\
&\left.\qquad - n_0(1 - n_0 + n_\Delta)G_{+0} + n_0(1 - n_0 - n_\Delta)G_{-0}\big)\, P_t(n_0, n_\Delta)\Big)\right.\\
&+ \frac{\partial}{\partial n_0}\Big((n_0(1 - n_0 + n_\Delta)G_{+0} \\
&\left.\qquad + n_0(1 - n_0 - n_\Delta)G_{-0}\big)\, P_t(n_0, n_\Delta)\Big)\right] + O\!\left(\frac{1}{N}\right)
\end{aligned}
\tag{3.61}
$$

where the factors $G_{\sigma\sigma'} = \text{sign}(g_\sigma - g_{\sigma'})$ determine the orientation of the drift, depending on the expected quality of the three strategies. If the initial condition is nonrandom, i.e. $P_0(n_0, n_\Delta) = \delta(n_0 - n_0(0))\delta(n_\Delta - n_\Delta(0))$, the probability density does not spread out but keeps its δ-function form $P_t(n_0, n_\Delta) = \delta(n_0 - n_0(t))\delta(n_\Delta - n_\Delta(t))$, and no fluctuations appear. We can therefore fully describe the dynamics by differential equations for the trajectories $n_0(t)$ and $n_\Delta(t)$. They are complemented by the equation for the price dynamics, which follows from Eqs. (3.57) and (3.56). The stochastic process Y_t then becomes a non-random function $y(t)$. Altogether we have

$$\frac{d}{dt}n_\Delta(t) = -\frac{3\epsilon}{2}n_\Delta + \frac{(1-\epsilon)(1-q)}{2}\Big[((1-n_0)^2 - n_\Delta^2)G_{+-}$$
$$+ n_0(1 - n_0 + n_\Delta)G_{+0} - n_0(1 - n_0 - n_\Delta)G_{-0}\Big]$$
$$\frac{d}{dt}n_0(t) = -\frac{\epsilon}{2}(3n_0 - 1) - \frac{(1-\epsilon)(1-q)}{2}\Big[n_0(1 - n_0 + n_\Delta)G_{+0} \qquad (3.62)$$
$$+ n_0(1 - n_0 - n_\Delta)G_{-0}\Big]$$
$$\frac{d}{dt}y(t) = n_\Delta - c\,n_0\,y(t).$$

These equations must be completed by three comparisons of the expected gains

$$G_{+-} = \text{sign}(n_\Delta - c\,n_0\,y(t))$$
$$G_{+0} = \text{sign}(n_\Delta - c\,n_0\,y(t) - d\,|y(t)|) \qquad (3.63)$$
$$G_{-0} = \text{sign}(-n_\Delta + c\,n_0\,y(t) - d\,|y(t)|).$$

(For notational simplicity, we dropped the explicit time dependence of $n_0(t)$ and $n_\Delta(t)$ everywhere on the right-hand side.) The above equations (3.62) and (3.63) fully describe the dynamics of the Lux-Marchesi model for $N = \infty$. Of course, it is impossible to solve this set of nonlinear equations analytically. However, there are some special cases in which the solution is feasible and provides useful insight into the complexity of the model.

Provided we know the densities $n_0(t)$ and $n_\Delta(t)$, we can write the evolution of the price explicitly as

$$y(t) = \int_0^t n_\Delta(t')\,e^{-c\int_{t'}^t n_0(t'')\,dt''}\,dt' + y(0)\,e^{-c\int_0^t n_0(t')\,dt'}. \qquad (3.64)$$

Now we should ask under what conditions we are able to compute the desired densities analytically.

Let us forbid the spontaneous changes of strategy with $\epsilon = 0$. If the concentration of fundamentalists is zero at the beginning, there is no mechanism which could turn it non-zero at any later time. (And this would remain true even if we kept the diffusive terms of order $1/N$). So, we get a single equation for the quantity $n_\Delta(t)$, which can be easily solved as it does not depend on the price $y(t)$. There is a trivial stationary solution $n_\Delta(t) = 0$ with a constant price. However, if we perturb the initial state $n_\Delta(t) = 0$ by an infinitesimal amount, the perturbation amplifies and $n_\Delta(t)$

steadily grows towards 1 or decreases towards -1, depending on the sign of the initial perturbation. The explicit solution is

$$n_\Delta(t) = \pm \tanh\left(\frac{1-q}{2}\, t\right), \tag{3.65}$$

and when we insert it into Eq. (3.64), we can see that the price grows or decreases linearly without bound. This solution shows that the system is hopelessly unstable without fundamentalists. Therefore, it is impossible to build a sensible market model with less than three strategies.

To get something sensible, we must start with non-zero $n_0(0)$. Then we get into trouble, because the equations for $n_0(t)$ and $n_\Delta(t)$ depend on the price through the factors (3.63). But these factors can assume only discrete values ± 1, so our strategy now might be to find specific solutions with these factors fixed and then 'glue' these solutions together at the points where the factors change sign. In fact, there are only four possible combinations of the values (G_{+-}, G_{+0}, G_{-0}), namely the following: I $(1, 1, -1)$, II $(1, -1, -1)$, III $(-1, -1, -1)$, and IV $(-1, -1, 1)$. Closer inspection of Eqs. (3.62) reveals that the behaviour of the model in region III differs from II just by inverting the sign of $n_\Delta(t)$, and the same holds true for the mutual relationship between regions IV and I. So, it is sufficient to solve Eqs. (3.62) in regions I and II, and fortunately it is possible to write these solutions explicitly. However, this is not a full victory, because gluing the solutions at the borders between regions I to IV is beyond the reach of analytical methods and must be done numerically.

Let us now list the already announced solutions. For simplicity, we set $q = 0$. Solutions with arbitrary q differ only by rescaling the time t by the factor $1 - q$. In the region I we get

$$
\begin{aligned}
n_0(t) &= \frac{4A\,\mathrm{e}^t}{\Big((B-2)\,\mathrm{e}^t + A\Big)\Big((B+2)\,\mathrm{e}^t + A\Big)} \\[2mm]
n_\Delta(t) &= \frac{(B^2 - 4)\,\mathrm{e}^{2t} - A^2}{\Big((B-2)\,\mathrm{e}^t + A\Big)\Big((B+2)\,\mathrm{e}^t + A\Big)}
\end{aligned}
\tag{3.66}
$$

and in region II the solution is

$$
\begin{aligned}
n_0(t) &= \frac{\mathrm{e}^t}{C + \mathrm{e}^t} \\[2mm]
n_\Delta(t) &= -\frac{(2D - C)\,\mathrm{e}^t + CD}{(\mathrm{e}^t + C)(\mathrm{e}^t + D)}
\end{aligned}
\tag{3.67}
$$

where A, B, C, and D are integration constants to be determined from the initial and gluing conditions.

These partial solutions can provide some qualitative yet exact information on the dynamics of the Lux-Marchesi model, as described by Eqs. (3.62). Suppose we start from region I. We can see that within the solution (3.66) the concentration of fundamentalists decreases to zero, while the disequilibrium between optimists and pessimists is saturated at its maximum value $n_\Delta = 1$. This also means that the price grows without limits. Therefore, at a certain time the factor G_{+0} must become negative, and we enter region II, where solution (3.67) applies. Here, on the contrary, the fundamentalists take over, and their concentration eventually saturates at value $n_0 = 1$. At the same time, the disequilibrium vanishes. From the general formula for price evolution (3.64), we can see that for $n_\Delta = 0$ the price relaxes to zero from any initial condition and irrespective of the detailed form of $n_0(t)$. This means that a time must come when the increase of price reverts and we pass from region II to region III, where the pessimist strategy is taken as superior to the optimist one, but fundamentalists are still the best, as the price is high enough. But the drop in price is further strengthened by the growing pessimists, and soon we find ourselves entering region IV. The further evolution is then a mirror image of what we have just described, and we end with a picture of quasi-periodic oscillations of price and the concentrations $n_0(t)$ and $n_\Delta(t)$. The complexity is implied by the nonlinear character of these oscillations, which very probably have a chaotic nature [827].

Thus we arrive at a somewhat paradoxical situation. We stressed earlier that the evolution of the stock market is much closer to a purely stochastic process than to deterministic chaos, as was believed some 20 years ago. Now we can see that the stochastic model we developed brings us back to a deterministic evolution with chaotic features. The stochasticity is recovered if we take into account the effect of a finite number of agents N, but its role consists rather in resetting, from time to time, the initial conditions for otherwise deterministic dynamics. It is highly probable that in reality the finite-size effects are so strong that the deterministic nature of the Lux-Marchesi model is buried deep under the noise. This is another example of the peculiar nature of econophysical models, which consists in the fact that the interesting phenomena are very often finite-size effects, and introducing limits of infinite number of agents is methodologically misleading. Even though the Lux-Marchesi model deviates from empirical data in some important aspects (most notably, it underestimates volatility clustering), generally it seems to grasp the coarse-grained mechanism of stock-market fluctuations.

3.3   What remains

It is better to stop here and conclude this chapter by mentioning several interesting approaches the reader can study from the (immense!) journal literature on microscopic market models.

First, we should note that the Cont-Bouchaud model is not the only scheme in which the concept of percolation was applied. There are also very interesting models of social percolation introduced in Refs. [838, 839].

One of the first significant attempts in the direction of agent models was the Santa Fe artificial market [840–845], developed by the joint group of economists and physicists at the Santa Fe Institute. The model is rather involved, and its bare simplification later gained much more fame under the label of minority game. We shall devote our entire Chap. 5 to minority game, so we limit ourselves here to a few basic ideas of the Santa Fe artificial market model.

The agents observe the price movements and react according to their respective strategies. There are three possible actions: to buy a unit of stock, to sell, or to do nothing. The strategy of an agent consists in comparing the sequence of price movements in M last steps with a set of patterns. Each pattern has an action attached to it, and if the agent finds that a pattern matches the actual price history, she performs the action attached to that pattern. The set of patterns and actions is different for each of the agents, and in the computer simulation it is chosen randomly at the beginning. Some of the agents have better strategies than the others and their wealth grows. On the other hand, the worst-performing agents are periodically wiped out from the system and are replaced by fresh agents with new randomly generated strategies. The price fluctuations emerging in this model are leptokurtic, i.e. the tails are heavier than Gaussian. Unfortunately, a detailed comparison with empirical data is not available.

The model introduced by Caldarelli, Marsili, and Zhang [846] is close in spirit to the Santa Fe artificial market but the implementation is substantially different. The agents are given certain predictors, the function of which is to guess future price change on the basis of past prices. In contrast to the Santa Fe market, the Caldarelli-Marsili-Zhang model uses various combinations of first, second or even higher derivatives of the price and their powers to predict the future. Again, the worst agents are replaced by randomly created newcomers. It was found that the price fluctuations exhibit scaling and power-law tails. The Hurst exponent $H = 0.62$ is close to the empirical value. However, the results were considered poorly reproducible by some authors [32].

Another model worth mentioning studies the market ecology of producers and speculators [847]. In a heat engine the energy comes in at a higher temperature and flows out at a lower one, producing useful work and increasing entropy. Similarly, in the stock market the producers must follow the natural rhythm of their work, injecting a large amount of information into the system. Indeed, everybody knows the seasons of wheat, grapes, or olives, and the speculator can count on her fingers when it pays to buy and when to sell the agricultural commodities. Part of the profit ends in the pockets of those who did not actually work in the fields, but the farmers are happy that somebody is ready to buy their goods immediately. By buying at a high offer and selling at high demand, the speculators themselves dampen the oscillations of the price. We might say that they feed on the information contained in the price signal, thus increasing entropy. Producers and speculators can be compared to the parts of the heat engine attached to high- and low-temperature reservoirs, respectively. The machine can work because it is an open system and therefore can steadily produce entropy.

The model implemented in [847] is close to the Santa Fe or Caldarelli-Marsili-Zhang market models, as it again assumes that the agents decide according to their strategies, or predictors. Here the worst performing agents in both speculators' and producers' groups are also replaced by fresh blood. The new feature is the possibility that the agents take no action if they feel that the conditions on the market are not favourable. This can be compared to the grand-canonical ensemble, where the particles can also go in and out of the system. In the context of a stock-market model, this idea was first advocated by Zhang [75], and the two-component model [847] was subsequently purified and elaborated into a very elegant and rich model called the grand-canonical minority game, which will be discussed at length in Chap. 5.

Another very interesting model was developed by Donangelo and Sneppen [848, 849] showing how money emerges spontaneously from mutual exchange of goods. Finally, we leave the model of Iori [850–853], the models seeking Nash equilibria [854, 855], the Langevin approach [856], as well as a number of others [854, 857–882] to the self-study of the interested reader.

Problems

1. Investigate the following variant of the supply network. The agents are divided into layers $i = 1, 2, \ldots, L$, containing M agents each. The agents in the first layer receive orders randomly from the outside. An agent at level i becomes active and topples if she has $z \geq 2$ orders to deal with. If that happens, she chooses randomly two agents on the lower level $i + 1$ and sends each of them 1 order. Her own heap of orders therefore decreases by 2. The orders sent off by the agents at the lowest layer are lost. Find the distribution of avalanche sizes if L and M are large. What happens if $M = 2$? And what if the orders from the lowest layer are not lost but return back to the system at the top layer?

2. Write your own computer code implementing the Cont-Bouchaud model on additional special social networks: 2D square lattice, 3D cubic lattice, randomly connected graph with a fixed node degree k (use small values, e.g. $k = 2$ or $k = 3$), etc. For more examples of networks, look at Chap. 6. An efficient algorithm for counting the size of clusters arising in the simulations can be found in [751].

3. Distribution of sizes of the percolation clusters in the Cont-Bouchaud model close to the percolation threshold is given by the general formula

$$P_{\text{clu}}(s) \sim s^{-\tau} e^{-b\, s\, |p - p_c|^{1/\sigma}}, \quad s \to \infty \qquad (3.68)$$

with a certain constant b. Assume a linear price impact (3.17) and very low activity $a \to 0$. Find the distribution of returns if the percolation probability p fluctuates randomly within the interval $[0, p_c]$ with smooth distribution $P_p(p)$.

4. Consider the following modification of the Eguíluz-Zimmermann model. With probability $2a$ an agent is picked at random. Provided the agent is contained in a cluster of size s, the cluster dissociates with probability $s^{-\delta}$, $\delta \geq 0$, or remains intact with probability $1 - s^{-\delta}$. With the complementary probability $1 - 2a$ two agents are chosen randomly, and if their clusters have sizes r and s, the two clusters merge with the probability $(rs)^{-\delta}$ or remain separated with the probability $1 - (rs)^{-\delta}$. Find the stationary distribution of cluster sizes n_s. For an inspiration, you can look at [778].

5. Find the stationary distribution for Kirman's ants, i.e. asymptotic probability distribution $P(k)$ for the process defined by transition probabilities (3.41). Hint: in the limit $N \to \infty$, introduce the continuous variable $x = k/N$ and write the Fokker-Planck equation for probability density $P(x)$.

4
Order-book dynamics

When your naive curiosity in the ways people gain and lose wealth crosses the first level of initiation, you probably glimpse the vertiginous hierarchy of insiders, not unlike Dante's circles within circles of the Inferno. Or else you may think of multiple skins of an onion and believe, unless you are a devout post-modernist, that there is something in the centre which dominates the subordinate peripheral layers.

Indeed, if you decide to invest your idle savings into shares of a company, you must visit an agency which will do the job for you. The time when you met the brokers in coffee-houses has been long gone. Your agent will probably contact other specialists, and your desired transaction will proceed through several intermediaries, get transformed, sliced into chunks of a convenient size, and mingled with other transactions, until it reaches the very person whose occupation it is to issue, register, and execute elementary orders to buy and sell. After all those steps have been done, the result goes back through the same chain of middle-persons to get decoded and reassembled so that eventually you pay the agreed-on amount and become the share-owner.

It would be very difficult to model this process in complete detail. Too many inessential features appear there. Every single securities market has some unique mechanisms. By including them in the model we would lose the clear view of generic functions common to all places and times. Moreover, the tendency to make trading more and more computer-based eliminates some of the intermediaries and attributes new roles to the others.

Already in the 1960s Osborne [883] suggested handing the execution of trades over to an electronic calculator. He depicts the stock market as a black box, whose inputs are the issued orders and output is the price. As a next step, he formulates an equation connecting the input and output. It is not a fundamental problem to program that formula on a computer, so one human layer (perhaps even more) can be replaced by a machine. There are arguments for why automated trading via limit order books is the inevitable future (which is quickly becoming the present) of stock markets [884].

Our main objective is to formulate models which can reproduce reasonably well the stylised facts on stock market fluctuations. But which of the layers at work are responsible? Is it the most external community of speculators, freely using the wildest strategies they could invent? Is it the layer below, still pursuing their own goals and procedures but limited to the assignments from their clients? Or, is it the most internal core of those who execute the elementary orders? Or, perhaps it is a combination of all? We do not yet know. The typical scheme investigated in Chap. 3 was the Lux-Marchesi model. The driving forces are the speculators, i.e. the peripheral layer, and the whole rest of the stock-market machinery is considered a mechanism responding

---

Reaction-diffusion processes ┌─────────┐ Box 4.1 └─────────┘

are abstract models for chemical reactions between substances which are not stirred, e.g. within a gel. In the absence of convection the reaction is diffusion-limited, for the molecules must diffuse individually within the close vicinity of each other to react. Perhaps the first important experimental discovery pertaining to reaction-diffusion processes appeared in 1896, the observation of Liesegang rings formed when silver nitrate is dropped in the centre of a Petri dish containing a gel with the solution of potassium dichromate. Concentric patterns of precipitated silver are observed after a few hours. In a model, we have a certain number of particles of different species, A, B, C, $\ldots$, performing a random walk. The diffusion constant may also be species-dependent. Various reactions may occur, for example, a binary one written as $A + B \to C$, which means that when two particles of type A and B happen to be at the same site, they produce a new particle C with a certain reaction probability k. Other examples of reactions include catalysis $A + D \to C + D$, annihilation $A + A \to \emptyset$, replication $A \to 2A$, autocatalysis $A + B \to 2A$, and many more.

The reaction-diffusion processes were most intensively studied in the context of pattern formation, where they help us understand where the zebra's stripes and leopard's spots come from, or why insects' bodies as well as our backbone are segmented [888].

---

in a simple deterministic way to the speculators' activity. The complexity is produced at the outer shell.

In this chapter we invert the perspective. What if the fat tails, scaling, etc. emerge from the very mechanism of placing and clearing the most elementary orders, whatever random flux of orders the speculators send towards the centre? If that were true, strategic thinking would be irrelevant, at least for the universal phenomenology of the market, and the only thing to investigate would be the innermost machinery. For example, one should study Osborne's equation or the behaviour of the software used now in electronic markets. Humans are largely excluded.

At least we do not count on human intelligence. Indeed, experiments were performed [885–887], where 'zero-intelligence' agents implemented by simple computer programs with no strategy traded together with humans (business students, as usual in this kind of experiments). Surprisingly, zero-intelligence agents showed as much efficiency as living people. Individual rationality of the agents had no effect. This finding provides another argument against the universal validity of the efficient market hypothesis. Indeed, if rationality has no observable impact on the market data, why do we need the assumption that people behave like rational optimisers of their individual profits? The hypothesis of rationality may be as superfluous in the economy as the hypothesis of ether in the theory of electromagnetic waves.

We shall see in this chapter that such an extreme formulation goes much too far. However, it is very instructive to see how much of the empirical findings can be explained supposing that people have no soul and no brain.

4.1 Reaction-diffusion and related processes

We have seen basic empirical facts about order books in Chap. 1. Supposing for simplicity that all orders have the same volume, or that larger orders are effectively composed of several orders of unit size, we may think of the orders as particles placed on a line, i.e. on the price axis. The most elementary fact is that we must distinguish between

---

Reaction-diffusion processes: mean-field approach | Box 4.2

On a basic level the reaction-diffusion processes are described by partial differential equations describing the space-time variations of the concentrations c_A, c_B, ... of all species involved. E.g. for one species with a diffusion constant D subject to the annihilation $A + A \to \emptyset$ with a reaction constant k, the equation is

$$\tfrac{\partial}{\partial t}\, c_A = D \tfrac{\partial^2}{\partial x^2}\, c_A - k\, c_A^2. \qquad (*)$$

The complexity originates from the nonlinear terms occurring generically in equations describing the reaction-diffusion processes, as is clearly seen in the example of equation (*). However, a full description is even harder than that, because the differential equations provide only a mean-field approximation of the process. As they do not trace individual particles, the information about the fluctuations around the average concentration is lost. To proceed further, more advanced techniques are necessary, which fall far beyond the scope of this book [889].

---

buying and selling orders. Accordingly, we have also two types of particles, asks (A) and bids (B). A trade occurs when A and B happen to be in the same place. The two orders of opposite type are executed and removed from the book. We may represent such an event as an annihilation reaction

$$A + B \to \emptyset. \qquad (4.1)$$

To complete the model, we must specify how the particles can come to the same place to interact and how new particles are supplied, replacing the annihilated ones.

4.1.1 Instantaneous execution

Stigler again

Perhaps the simplest variant is embodied in the model of Stigler [789], investigated in Sec. 3.2.1. Let us translate it into the language of the reaction-annihilation process [890]. There is a fixed interval of allowed prices. The particles are deposited one by one at random points within this interval. Immediately after deposition the particle tries to annihilate another particle of the opposite type. So, when A is deposited at price z, we check if there is any B at price $z' \geq z$. When found, the new A reacts according to Eq. (4.1) with B placed at the highest price of those exceeding z. The position of such B is then the actual price. If no such B can be found, A remains where it was. We may also imagine the process in an alternative way. The particles do not fall on the interval from above, but rather they are pushed horizontally from the edges of the interval. So, the newly deposited A moves swiftly from the right edge of the allowed interval to the left, towards its random but predestined position. If it meets any B on its way, these A and B annihilate each other. If no B is met, the new A reaches its destination and gets stuck there.

We have seen in Sec. 3.2.1 that the Stigler model quite reasonably reproduces the phenomenon of volatility clustering. The autocorrelation of absolute values of price changes decays as a power law, although the exponent is much higher than the empirical value. On the other hand, when we look at the distribution of price changes, no sign of power-law tails is observed. This is a serious flaw of the Stigler model and we must look for alternatives.

Genoa artificial market

A rather involved modification of the Stigler model appeared much later under the name of the Genoa artificial market [890–898]. The model contains many ingredients, and therefore it is very plastic. We show here a reduced version, keeping only the most significant (in our view) new features.

First, the new orders, either bids or asks, will not be placed at an arbitrary position within the allowed price interval, but close to the current price. We introduce two parameters into the model. The first parameter d is the width of the price interval into which the new order is deposited. The second parameter s measures the distance of the centre of this interval from the current price. This way we allow the price to fluctuate without any a priori bounds. To avoid the obvious problem that the price may sometimes wander into negative numbers, we interpret the coordinate on which orders are placed as the logarithm of price $y = \ln z$, instead of the price itself. Then, if the position of the last transaction was y_0, we deposit a new ask at a point chosen arbitrarily within the interval $[y_0 + s - d/2, y_0 + s + d/2]$, or a new bid within the interval $[y_0 - s - d/2, y_0 - s + d/2]$. We can see that the asks are, on average, put higher than the current price, while bids are put below it on average. Indeed this is the behaviour to be expected from sensible investors. However, the allowed intervals for bids and asks may overlap as long as $s < d/2$, so transactions are possible.

The second and most important improvement over the Stigler model is the feedback from the past volatility to the actual value of the parameters s and d. Principally, there is no clue how large these parameters should be. However, a posteriori their values can be compared with the scale of typical price fluctuations, measured by the volatility. Denoting by y_t the price logarithm at (discrete) time t and $x_t = y_t - y_{t-1}$ the return, we define the instantaneous volatility as absolute price changes averaged using an exponentially decaying kernel

$$v_t = \lambda \sum_{t'=0}^{\infty} (1 - \lambda)^{t'} |x_{t-t'}|. \tag{4.2}$$

Then, the feedback means introducing time-dependent parameters s_t, d_t, instead of fixed s and d, and keeping them proportional to the instantaneous volatility:

$$\begin{aligned} s_t &= \frac{g}{b} v_t \\ d_t &= g \, v_t. \end{aligned} \tag{4.3}$$

The newly introduced constants b and g are parameters of the simplified Genoa market model. The ratio between the width of the allowed interval and its shift off the current price is kept fixed and equal to b. Thus, the pattern of order placement is zoomed in and out according to the actual volatility, with proportionality factor g.

At each step, one order is placed and, depending on its position relative to other orders, it is either executed immediately or remains in the order book. As in the Stigler model, we discard orders sitting in the book for too long. So, at each step we first check whether there is an order placed N steps ago and if we find one, we remove it. Therefore, the maximum number of orders of both types simultaneously

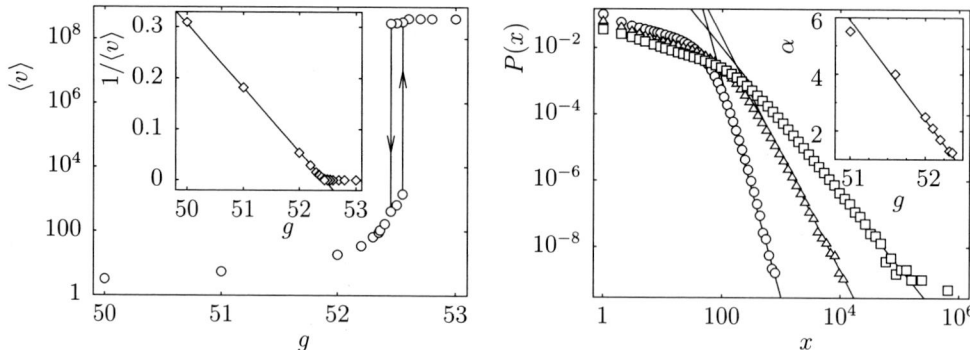

Fig. 4.1 Genoa market model. All data correspond to $N = 1000$, $b = 7$, and $\lambda = 0.001$. In the left panel, the dependence is shown of the average volatility on the width-to-volatility ratio g. The points for $\langle v \rangle$ exceeding about 10^8 should be considered as effectively infinite. The finite value is due to the hard limits on volatility imposed by computer implementation. The hysteresis curve, indicated by lines with arrows, which apparently suggests first-order phase transition, is a subtle effect due to fluctuations in a finite-size system. It should vanish in the thermodynamic limit; actually the transition is of a second order. In the inset, the same data are plotted differently to confirm the type of behaviour predicted by Eq. (4.5). The solid line is the dependence $\propto (52.4 - g)$. In the right panel, we show the distribution of returns. Different symbols correspond to $g = 51$ ($\bigcirc$), 52 ($\triangle$) and 52.36 ($\square$). The lines are the power laws $\propto x^{-1-\alpha}$ with the exponents (from left to right) $\alpha = 5.5$, 2.5, and 1.2. In the inset, the values of exponents extracted from the power-law tails of price-change distributions are shown for various values of g. The solid line is the dependence $(\alpha - 1) \propto (52.4 - g)$. Both panels show consistently that $g_c \simeq 52.4$ for this choice of parameters.

present in the book is N. Clearly, the behaviour of the model is sensitive to the value of N, because it essentially determines the density of orders. Because the trade occurs at the place of an extremal order, i.e. the lowest ask or the highest bid, the typical spacing between adjacent orders determines the scale of the one-step price changes and therefore the volatility. Higher N implies lower volatility and vice versa.

Suppose for a while that we switch off the feedback from the volatility to the parameters s and d. Instead, we fix their ratio b and look how the average volatility $\langle v \rangle$ in the simulation depends on the value of d. For large d this parameter selects just the scale on the axis of log prices y, so it is natural to expect that the volatility is proportional to d. For smaller d the maximum number of orders N is decisive for the volatility, and the linear dependence on d breaks down. Therefore, we can suppose

$$\langle v \rangle \simeq \frac{1}{g_c} d + A \qquad (4.4)$$

with some constants A and g_c, depending on b and N.

Now, if the feedback is effective, the average volatility in a stationary state should be compatible with Eq. (4.4), which produces a divergence when g approaches to g_c

$$\langle v \rangle \simeq \frac{g_c\, A}{g_c - g}, \quad g \to g_c. \tag{4.5}$$

Such a behaviour is confirmed in the simulations, as documented in Fig. 4.1. A second-order phase transition occurs at $g = g_c$, and we can consider the inverse average volatility $\langle v \rangle^{-1}$ as an order parameter, vanishing linearly at the critical point.

The distribution of returns is of central importance, and we can see in Fig. 4.1 its shape for several choices of parameters. The power-law tail $P(x) \sim x^{-1-\alpha}$ develops when we approach the critical point. This seems to be good news, but there are features which calm down our optimism. First of all, the power-law behaviour is not limited to the critical point or its close vicinity. So, it does not seem to be related to critical behaviour but rather results from the generic mechanisms of power-law distributions produced in multiplicative processes [899, 900]. And most importantly, the tail exponent α is not universal. It decreases continuously when we approach g_c from below, until it reaches the value $\alpha = 1$ at the critical point.

On the other hand, such a behaviour helps us understand the origin of the phase transition. The feedback between order placement and volatility, prescribed by (4.3), results in a power-law distribution of returns, with a parameter-dependent exponent. If we suppose that the system is ergodic in the sense that the initial conditions are not remembered forever, the time-average of volatility $\langle v \rangle$ should coincide with the average computed from the return distribution

$$\bar{v} = \int_0^\infty x\, P(x)\, \mathrm{d}x. \tag{4.6}$$

Now, $\bar{v}$ is finite only if $\alpha < 1$ and diverges as the exponent α approaches 1. Therefore, the critical point is reached for the combination of control parameters b and g that leads to the exponent $\alpha = 1$ in the price-change distribution. The phase transition is the consequence of a special type of power-law distribution, not the other way round.

Tuning the parameters, we can easily get distributions of price changes with exponents close to the empirical ones. Too easy, alas! We do not know why the parameters should have this rather than that value. However, the observation that the feedback between actual volatility and order placement leads to power-law distribution has great value, for it proves to be a generic feature common to a wide range of other models too.

4.1.2 Bak-Paczuski-Shubik model

Diffusion

Another possibility of how A and B can be allowed to meet is to let them walk randomly. We can imagine that the trader places an order at a certain price, but after a while she feels that the price was not quite good. She takes the order out and redeposits it somewhere else. In the model of Bak, Paczuski and Shubik [901], the new position is next to the old one, shifted randomly up or down. The movement of the order can be viewed as diffusion. The model then comes close to the very thoroughly studied class of reaction-diffusion processes.

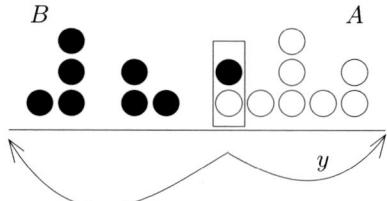

Fig. 4.2 Scheme of the Bak-Paczuski-Shubik model. Particles representing asks (A) and bids (B) perform a random walk. When they meet, a trade occurs and the orders are removed. To keep the number of orders constant, new particles are immediately supplied at the ends of the allowed price segment.

Let us first look at the computer simulation of the model. The particles are placed on discrete points within the segment of allowed log-prices $y = 1, 2, \ldots, L$, and in each time step they can move one unit up or down. There are $N/2$ particles of either species. When two particles A and B meet, they are taken out and redeposited, as illustrated in Fig. 4.2. The point where the reaction took place determines the current price. The simplest rule for the redeposition of the particles is to move them to the ends of the segment, i.e. A is placed at $y = L$ and B at $y = 1$. In Fig. 4.3 we can see the typical evolution of the configuration of orders, as well as the price changes. The first impression is that the transactions occur rather rarely, so the price remains unchanged for a relatively long time and then makes a rather long jump. This way, the Bak-Paczuski-Shubik model produces a stochastic process for the price, described by transaction times T_i, and transaction prices y_i, numbered by discrete indices $i = 1, 2, \ldots$ Alternatively, we can describe it by waiting times $\Delta T_i = T_i - T_{i-1}$ and price jump lengths $x_i = y_i - y_{i-1}$. When we discussed the continuous-time random walks in Sec. 2.1.3, the waiting times and jump lengths were postulated. Here we have a mechanism which generates them. Now, let us look at the properties of the process, as revealed in numerical simulations. We denote M the number of price jumps in the simulation.

The distribution $P(x)$ of the price jumps is shown in Fig. 4.4. We do not observe any signs of fat tails. Instead, the distribution can be fitted very well on a slightly distorted Gaussian, $P(x) \simeq a\mathrm{e}^{-b\,x-c\,x^2}$, with appropriate constants a, b, and c. As we already noted, quite a long time may pass from one transaction to another, while the order book undergoes many internal reorganisations. We can measure the distribution of the waiting times

$$P^>(\Delta t) = \frac{1}{M} \sum_{i=1}^{M} \theta(\Delta T_i - \Delta t), \tag{4.7}$$

and as shown in Fig. 4.4, the distribution is exponential. This suggests that the times of transaction constitute a Poisson point process, at least approximately. We should also ask how much the waiting times and price jumps are correlated. To this end, we measure the average jump conditioned to the waiting time

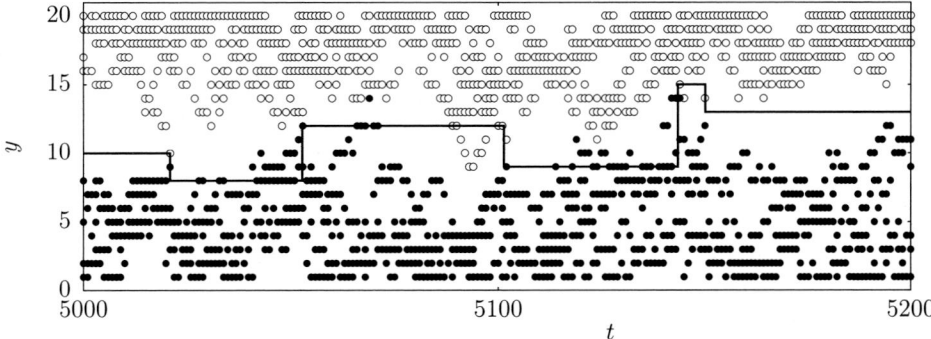

Fig. 4.3 Typical evolution of the Bak-Paczuski-Shubik model. The total number of particles is $N = 10$, and the width of allowed prices $L = 20$. The time is measured so that each particle moves on average once per unit time. Full circles represent bids, empty circles asks. The line traces the current price.

$$\langle x | \Delta t \rangle = \frac{\sum_{i=1}^{M} |x_i| \, \delta(\Delta T_i - \Delta t)}{\sum_{i=1}^{M} \delta(\Delta T_i - \Delta t)}. \tag{4.8}$$

As we can see in Fig. 4.4, on average, a longer wait is followed by a longer jump. This finding is quite natural, as longer reorganisation of the order book may finally bring the new price much farther from the original one.

In order to see the scaling of the price fluctuations and determine the Hurst exponent, we usually draw the quantity $R(\Delta t)$, as defined in Box 4.3. However, in the Bak-Paczuski-Shubik model it is more convenient to use a simpler quantity, measuring the amplitude of the price fluctuations during time interval Δt as

$$\langle |\Delta y|_{\max} \rangle = \left\langle \max_{T_i, T_j \in [t, t+\Delta t]} |y_i - y_j| \right\rangle_t \tag{4.9}$$

where $\langle \ldots \rangle_t$ means the average over all times t in the course of the simulation. In principle, both quantities should behave the same, $\langle |\Delta y|_{\max} \rangle \sim R(\Delta t) \sim (\Delta t)^H$ in the asymptotic regime $\Delta t \to \infty$. However, in practice factors like statistical noise and the finiteness of the system make the former or the latter quantity more suitable, depending on the model under consideration.

In the Bak-Paczuski-Shubik model, we can observe two regimes, as shown in Fig. 4.5. For short time distances Δt, the price moves ballistically, $\langle |\Delta y|_{\max} \rangle \sim \Delta t$. There is a crossover at a typical scale Δt_c, and for larger time distances the movement is subdiffusive, $\langle |\Delta y|_{\max} \rangle \sim (\Delta t)^{1/4}$, so that the Hurst exponent is

$$H = \frac{1}{4}. \tag{4.10}$$

The initial transient regime with ballistic motion of price can be understood taking into account the distribution of waiting times shown in Fig. 4.4. If the time distance

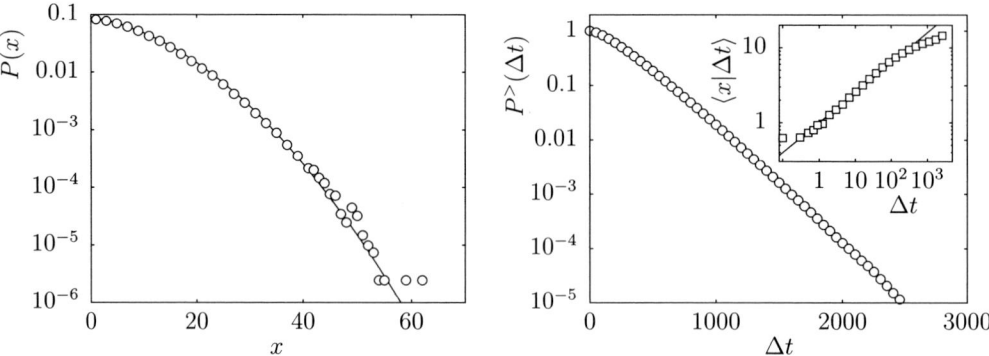

Fig. 4.4 Simulation results for the Bak-Paczuski-Shubik model. In the left panel, distribution of price jumps is shown for $N = 100$, $L = 250$. The solid line is the fit $0.09 \exp\left[-x/28 - (x/19)^2\right]$. The right panel shows the distribution of waiting times between trades. In the inset we can see the average price jump conditioned to the waiting time just before the jump. The straight line is the power dependence $\propto (\Delta t)^{0.4}$.

Δt is shorter than the average waiting time, it is not likely that more than one jump occurs during the interval Δt. Supposing that the jumps have a typical length, the quantity (4.9) is proportional to the probability that a jump does occur, which is in turn proportional to Δt. Hence the observed ballistic behaviour of prices.

For extremely long times, the amplitude of the price fluctuations is determined by the system size L rather than by the process itself. Figure 4.5 shows that the quantity $\langle |\Delta y|_{\max} \rangle$ gets saturated and approaches a constant when $\Delta t \to \infty$. It is especially well visible for a small system size L, where the ballistic regime is directly followed by the saturated regime. Therefore, the relevant regime characterised by the Hurst exponent $H = 1/4$ can be observed only for L large enough to open a time window between the initial ballistic transient and the final saturation. This is a point we must keep in mind when making our simulations.

Density of orders

The reaction-diffusion character of the Bak-Paczuski-Shubik model suggests that we could directly employ the approximate analytical tools which are successfully used in many other models of this kind. For basic information, see Boxes 4.1 and 4.2. The reaction described by (4.1) leads to a set of coupled partial differential equations for the concentrations c_A, c_B of the particles A and B, respectively. This way we get the average densities of bids and asks within the order book.

Assume for the moment that the reaction (4.1) may not occur always and immediately after the particles meet. In this case the reaction constant k is finite and the equations describing the evolution of the concentrations are

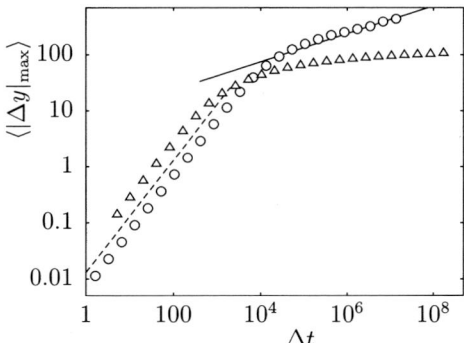

Fig. 4.5 Dependence of the amplitude of price fluctuations on the time interval. The parameters are $N = 4 \cdot 10^4$, $L = 2 \cdot 10^4$ ($\bigcirc$) and $N = 100$, $L = 250$ ($\triangle$). The straight lines are the powers $\propto (\Delta t)^{1/4}$ (solid) and $\propto \Delta t$ (dashed).

$$\frac{\partial}{\partial t} c_A = D \frac{\partial^2}{\partial y^2} c_A - k\, c_A\, c_B$$
$$\frac{\partial}{\partial t} c_B = D \frac{\partial^2}{\partial y^2} c_B - k\, c_A\, c_B. \tag{4.11}$$

They must be supplemented by appropriate boundary conditions reflecting the way the new particles are added to the system. The orders keep flowing in, replacing the annihilated ones, and it is natural to assume that their average number is kept constant. In our case we suppose that the reaction occurs on a segment of prices $[0, L]$ and there are as many bids as there are asks, namely

$$\int_0^L c_A(y)\mathrm{d}y = \int_0^L c_B(y)\mathrm{d}y = \frac{N}{2}. \tag{4.12}$$

Furthermore, we require that the concentrations are non-negative $c_{A,B} \geq 0$. It is convenient to express the concentrations in terms of the difference, $c_- = c_A - c_B$ and the sum, $c_+ = c_A + c_B$. For the difference, we immediately have the equation

$$\frac{\partial}{\partial t} c_- = D \frac{\partial^2}{\partial y^2} c_- \tag{4.13}$$

which is linear and therefore easily solvable for any value of the reaction constant. The sum of the concentrations obeys the equation

$$\frac{\partial}{\partial t} c_+ = D \frac{\partial^2}{\partial y^2} c_+ - \frac{k}{2}\big(c_+^2 - c_-^2\big) \tag{4.14}$$

which is nonlinear; but we can bypass this difficulty by assuming that the reaction constant is very large. Indeed, in the Bak-Paczuski-Shubik model the particles react and annihilate whenever they meet. This exactly corresponds to the limit $k \to \infty$, simplifying the equation (4.14) into the bare relation

$$|c_+| = |c_-| \tag{4.15}$$

ensuring that the regions where c_A and c_B are non-zero do not overlap. Indeed, the infinite reaction rate implies that the particles are annihilated immediately upon encounter and we cannot find a finite concentration of both A and B at the same place. Therefore, we only need to solve the equation for the difference c_- and equate the density of asks and bids with its positive and negative parts, respectively, i.e. $c_A = c_- \, \theta(c_-)$ and $c_B = -c_- \, \theta(-c_-)$.

It is easy to find the stationary solution of Eq. (4.13) subject to condition (4.12). We have

$$c_-(y) = \frac{4N}{L^2}\left(y - \frac{L}{2}\right) \tag{4.16}$$

and the resulting particle concentrations, i.e. the average densities of bids and asks are shown in Fig. 4.6.

Price fluctuations

To see how the price fluctuates we must go beyond the description of the process using particle densities $c_{A,B}$. The actual price is determined by the position where the last reaction took place. The sequences of reaction times $t_1, t_2, \ldots$ and prices $y_1, y_2, \ldots$ are random processes reflecting the (much more complicated) dynamics of reactions between diffusing bids and asks. Unfortunately, little is known analytically about the price fluctuations produced this way [902]. However, we can learn quite enough from a closely related problem, which does have an analytical solution. Imagine that initially the particles of the two types reside in opposite halves of the segment $[0, L]$, with uniform density. We suppose now that the reaction constant k is finite, so the particles of type A and B can mix. The reaction will happen around the centre of the segment. The product of the concentrations measures the intensity of the reaction under way, and we describe the profile of the reaction front using the quantity $\chi(y) = c_A(y) \, c_B(y)$. The time dependence of the width of the reaction front $w = \left[\int (y - L/2)^2 \chi(y) \, dy / \int \chi(y) \, dy\right]^{1/2}$ will bear some information about the fluctuations. It was found [903] that it grows rather slowly,

$$w(t) \sim t^{1/4} \sqrt{\ln \frac{t}{t_0}} \tag{4.17}$$

with parameter t_0 depending on the details of the model. The critical exponent $1/4$ occurring in this formula was obtained by several independent paths [904–907], so the value is quite well established.

Translating this result into our original problem of price fluctuations, we conclude that the typical amplitude of price fluctuations within a time interval of length Δt grows asymptotically as $(\Delta t)^{1/4}$. Thus, the Hurst exponent in the Bak-Paczuski-Shubik model is expected to be

$$H = \frac{1}{4}. \tag{4.18}$$

In this consideration we disregarded the logarithmic factor, which may turn out to be quite important when we want to check the result (4.18) with numerical simulations.

How to guess the Hurst exponent from the book profile

The value $1/4$ of the Hurst exponent can be obtained by a simple heuristic argument. However poorly justified from a mathematical point of view, it elucidates why the value $1/4$ is so general that many other models investigated further in this chapter will also share the same Hurst exponent as the Bak-Paczuski-Shubik model.

Change of price Δy realised over the time interval Δt is due to a temporary imbalance between the supply of bids and asks within that time interval. If we assume that bids and asks arrive randomly, the actual imbalance, measured from a starting time instant, follows a random walk. Therefore, the magnitude of the imbalance Δv is scaled with time as

$$\Delta v \sim (\Delta t)^{1/2}. \tag{4.19}$$

On the other hand, suppose $n(y)$ is the average order-book profile, i.e. the average density of bids or asks. Here the variable y is the distance from the current price. The change in price Δy due to the imbalance of orders is related to the volume of the imbalance Δv as $\Delta v = \int_0^{\Delta y} n(y) \mathrm{d}y$. Coming back to the Bak-Paczuski-Shubik model, we identify $n(y)$ with the concentration of either bids or asks

$$n(y) = c_+(y + \frac{L}{2}) = \frac{4N}{L^2} |y|. \tag{4.20}$$

The vital ingredient is the linear dependence $n(y) \sim y$, which implies that the volume eaten off the book profile is $\Delta v \sim (\Delta y)^2$; together with the random-walk assumption (4.19) it gives

$$\Delta y \sim (\Delta t)^{1/4}. \tag{4.21}$$

Hence the value $H = 1/4$ for the Hurst exponent. We can see immediately that this value is related only to the linearity of the order-book profile in the vicinity of the price. Other details of the profile are irrelevant, and the Hurst exponent is universal. More generally, if the order-book profile close to the current price behaves like $n(y) \sim y^{\beta-1}$, then the Hurst exponent becomes

$$H = \frac{1}{2\beta}. \tag{4.22}$$

Note, however, that this argument may hold only for short enough times Δt. When the volume imbalance is too big, the profile can no longer be considered linear and the Hurst plot ceases to be universal.

Imitation and drift

Let us consider an example of a modification which significantly changes the book profile, but without a change in the Hurst exponent. In the original setup, the particles-orders were allowed to diffuse freely left or right without any bias. In reality, the change

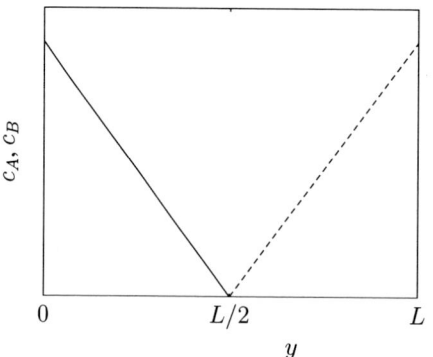 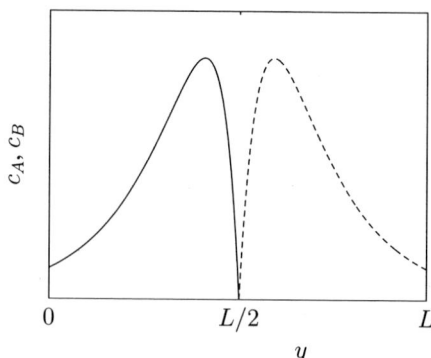

Fig. 4.6 Profiles of the order book in the Bak-Paczuski-Shubik model (left panel) and in the same model but with drift and imitation included (right panel) are shown. The full line denotes the concentration of bids, the dashed line the concentration of asks.

in the order position, which we describe here as diffusion, is due to the change in the investor's strategy. The price is perhaps inadequate, so let us shift the order lower or higher. Naturally, the asks will be mostly shifted low, while the bids are shifted high. This induces a drift in the particles' diffusion. The particles A will drift downwards, the particles B upwards.

At the same time, we shall also introduce another unrelated ingredient, which is imitation [908]. The investors may look where other investors place their orders and put their own close to the imitated ones. In the language of reaction-diffusion processes it corresponds to proliferation:

$$A \to A + A$$
$$B \to B + B. \tag{4.23}$$

The partial differential equations describing the evolution of the concentrations of the particles A and B are now

$$\frac{\partial}{\partial t} c_A = D \frac{\partial^2}{\partial y^2} c_A + b \frac{\partial}{\partial y} c_A + a\, c_A - k\, c_A\, c_B$$
$$\frac{\partial}{\partial t} c_B = D \frac{\partial^2}{\partial y^2} c_B - b \frac{\partial}{\partial y} c_B + a\, c_B - k\, c_A\, c_B. \tag{4.24}$$

The parameters a and b quantify the rates of imitation and the drift, respectively. Again, it is possible to introduce the sum and difference of the concentrations, $c_+ = c_A + c_B$ and $c_- = c_A - c_B$. The equation we obtain is slightly more complicated than (4.13), because c_- does not decouple completely from c_+. However, for instant annihilation, i.e. $k \to \infty$, we can again use the relation (4.15) to get

$$\frac{\partial}{\partial t} c_- = D \frac{\partial^2}{\partial y^2} c_- + b \frac{\partial}{\partial y} |c_-| + a\, c_-. \tag{4.25}$$

The stationary solution is then found independently in the regions where $c_- > 0$ (only particles A are present there) and where $c_- < 0$ (just particles B). As a result, we finally get

$$c_-(y) = R \exp\left(-\frac{b}{2D}\left|y - \frac{L}{2}\right|\right) \sinh\left[\frac{\sqrt{b^2 - 4aD}}{2D}\left(y - \frac{L}{2}\right)\right]. \qquad (4.26)$$

Again, the concentrations of A and B are extracted as positive and negative parts of c_-, respectively. The order-book profile is just $n(y) = |c_-(y + L/2)|$. The parameter R is fixed by the condition that there are on average $N/2$ particles of either type, which is the same as before, Eq. (4.12).

In Fig. 4.6 we can compare the order-book profiles in the original Bak-Paczuski-Shubik model and in the same model with imitation and drift. The latter is clearly more realistic, because the empirical data (see Sec. 1.3) show decaying density of orders when we get farther from the actual price. On the other hand, the behaviour close to the price is qualitatively the same in both variants of the Bak-Paczuski-Shubik model. The density of orders is proportional to the distance from price, which implies the universal value of the Hurst exponent $H = 1/4$. Modifications of the model have no impact on the value of the Hurst exponent as long as the essence of the Bak-Paczuski-Shubik model remains unchanged.

4.2 Reactions of limit and market orders

In the models discussed so far we have not distinguished between limit and market orders. In reality, this distinction is a fundamental issue in all computerised stock markets. Limit orders are placed in the book at a fixed price and wait there until they are executed or cancelled. Cancellations occur when the owner of the order decides that the price is not adequate, or it has been waiting too long already, or for any other subjective reason. Therefore, the limit orders never move, contrary to the Bak-Paczuski-Shubik model. They just come in and go out.

Market orders do not specify any price. The incoming market order to buy matches with the limit order to sell that is placed at the lowest price. (By analogy, we see how it works for sell orders.) This results in immediate execution of both orders involved. This transaction fixes the actual price.

Because the orders never move but are just placed and removed, we can classify the models of this type under the label of deposition-evaporation processes. Placements of orders amount to deposition of particles, while evaporation corresponds to both cancellation of orders (evaporation from the bulk of the material) and their execution (evaporation from the extreme edge of the material).

There is a wide variety of models differing in the way the limit orders are placed, in the rules for cancellation of orders, and so on. We shall show only two basic variants. The first is the Maslov model, and the second stems from several works of the group of Farmer and his collaborators. To fix simple names, we call the latter the uniform deposition model. In all the models described here, all orders have the same volume, considered unity.

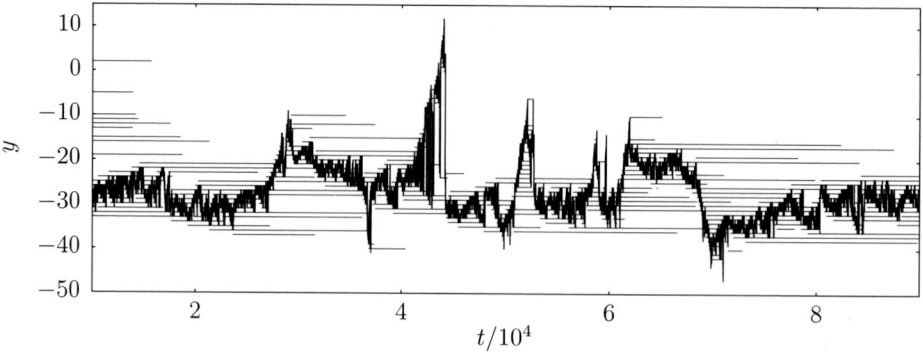

Fig. 4.7 Example of the evolution of the Maslov model with cancellations. Each horizontal segment represents an order, placed where the segment starts and cancelled or executed where the segment ends. The rugged line is the time dependence of the price. The parameters of the model for this plot are $\overline{N} = 100$ and $q = 0.05$.

4.2.1 Maslov model

Deposition, execution, cancellation

The limit orders are preferably placed near the price. As an extreme situation, we can imagine the (logarithm of the) price as an integer number and always place the new orders at distance 1 from the price. All the orders have the same unit volume. This is the principal feature of the model introduced by Maslov [909].

The model evolves in discrete time t. The order book contains N_{At} asks and N_{Bt} bids. The state of the book is described by the positions of the asks $a_{1t} \le a_{2t} \le \ldots$ and bids $b_{1t} \ge b_{2t} \ge \ldots$ and by the current price y_t. The lowest ask always stays above the highest bid, $b_{1t} < a_{1t}$ and the price lies between them, $b_{1t} \le y_t \le a_{1t}$.

In the simplest situation, the only actions we can take at each step are deposition and execution. We can deposit a single bid one unit below the current price, i.e. at position $y_t - 1$, or we can deposit one ask at the position $y_t + 1$. Another possibility is to issue one market order. It can be an order to buy, resulting in execution of the highest bid at position b_{1t}. The executed order is removed from the book, and the new price is set to the position of the executed order, $y_{t+1} = b_{1t}$. Similarly, we can issue an order to sell. Then, the lowest ask is removed and the new price is $y_{t+1} = a_{1t}$. Each of these four actions occurs with identical probability $1/4$.

If we allow cancellation of orders, the process becomes more complicated. Two new parameters enter the definition of the model. The probability of cancellation will be quantified by the parameter $q \in [0, 1]$, and the average number of orders present in the order book will be tuned by the parameter $\overline{N}$. The actual probability that cancellation of an order will happen at time t is proportional to the number of existing orders $N_t = N_{At} + N_{Bt}$.

At each step of the process, we first decide which of the three types of events will occur. The probabilities that, at time t, deposition, execution, and cancellation of an order takes place are

Hurst plot $\boxed{\text{Box 4.3}}$

provides one of the ways to quantify the fluctuations in a time series. In econophysics, we have in mind the time series of price, either empirical or obtained in numerical simulations. Suppose the price at time t is y_t. We specify some discrete unit of time δt. For simplicity, let $\delta t = 1$. We define the one-step price changes $x_t = y_t - y_{t-1}$. In the Hurst plot, we draw the dependence of the quantity

$$R(\Delta t) = \left\langle \frac{\max_{t',t'' \in [t,t+\Delta t]} |y_{t'} - y_{t''}|}{\sqrt{\langle x_{t'}^2 \rangle_{t'} - \langle x_{t'} \rangle_{t'}^2}} \right\rangle_t \qquad (*)$$

on the time difference Δt. The average $\langle \ldots \rangle_t$ is taken over all times t in the course of the time series, while the average $\langle \ldots \rangle_{t'}$ is taken over the interval $t' \in (t, t + \Delta t)$.

The definition $(*)$ looks rather complex and needs some explanation. In the numerator, we put the maximum price difference observed within the time interval $[t, t + \Delta t]$. It quantifies the amplitude of price fluctuations, and its dependence on the length of the time interval Δt should yield the Hurst exponent. However, in practice the dependence is obscured by the effect of volatility clustering. Indeed, the maximum price difference must be averaged over all times t in the empirical or simulated data. Time periods with high volatility dominate this average, while periods with low volatility are suppressed. If we had as long a time series as we would please, this would have little effect eventually, because the long-time correlations in absolute value of price changes would be superseded by even longer times than our data cover. If that were the case, we could easily use the simpler quantity

$$\langle |\Delta y|_{\max} \rangle = \left\langle \max_{t',t'' \in [t,t+\Delta t]} |y_{t'} - y_{t''}| \right\rangle_t. \qquad (**)$$

In reality the correlation time is not negligible compared to the length of the time series. Therefore, to counterbalance the volatility clustering, we divide the maximum price difference in the numerator of $(*)$ by the average volatility within the same interval $[t, t + \Delta t]$. That is the quantity in the denominator. Only after that do we average over all times t. For long time differences Δt, both quantities $(*)$ and $(**)$ should behave in the same manner $\langle |\Delta y|_{\max} \rangle \sim R(\Delta t) \sim (\Delta t)^H$. Here, H is the Hurst exponent, i.e. the principal quantity we want to extract from the Hurst plot analysis.

$$W_t^{\text{dep}} = \frac{1}{2 + q\left(\frac{N_t}{N} - 1\right)}$$

$$W_t^{\text{exe}} = \frac{1 - q}{2 + q\left(\frac{N_t}{N} - 1\right)} \qquad (4.27)$$

$$W_t^{\text{can}} = \frac{q\,\frac{N_t}{N}}{2 + q\left(\frac{N_t}{N} - 1\right)},$$

respectively. Furthermore, we must decide if the event touches bids or asks. In the first two of the three cases, we select bid or ask with equal probability. In the case of cancellation, we choose any of the existing orders with equal probability; therefore the probability to cancel a bid is proportional to the number of bids, and the probability to cancel an ask is proportional to the number of asks. In deposition and cancellation events, the current price remains unchanged, $y_{t+1} = y_t$. At execution, the price changes according to the same rule as described above.

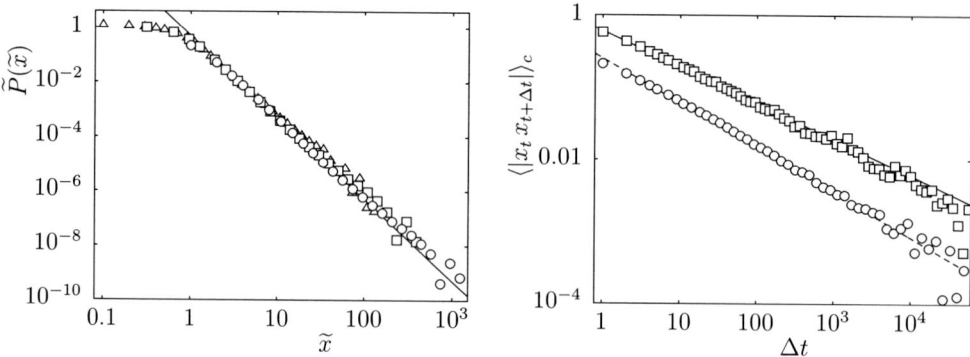

Fig. 4.8 Properties of the Maslov model. In the left panel, the rescaled distribution of price changes is shown, according to Eq. (4.28). The symbols correspond to time distances $\delta t = 1$ (O), 100 (□), and 10^4 (△). The straight line is the power dependence $\propto \tilde{x}^{-3}$. In the right panel we can see autocorrelation of the absolute values of one-step price changes without cancellations (□) and with cancellations, using parameters $q = 0.01$ and $\overline{N} = 1000$ (O). The straight lines are power laws $\propto (\Delta t)^{-0.5}$ (solid) and $\propto (\Delta t)^{-0.62}$ (dashed).

We can see a typical space-time diagram of the evolution of the Maslov model in Fig. 4.7. The price wanders erratically, surrounded by a cloud of orders. Periods of low volatility alternate with periods of high volatility, suggesting that volatility clustering is present here. As we shall soon see, it can be confirmed quantitatively.

Price changes and scaling

The first quantity to investigate is the distribution of price changes $x_t = y_t - y_{t-\delta t}$ during a fixed time interval δt. This distribution will be denoted $P(\delta t; x)$.

Let us first look at the properties of the Maslov model without cancellations. This case is special, as it exhibits scaling behaviour. The distribution of price changes can be rescaled so that

$$P(\delta t; x) = (\delta t)^{-1/4} \widetilde{P}(\tilde{x}) \tag{4.28}$$

where the rescaled price change is $\tilde{x} = (\delta t)^{-1/4} x$. This behaviour is documented in Fig. 4.8. The exponent $-1/4$ in this scaling law implies that the Hurst exponent is $H = \frac{1}{4}$. The simulation data in Fig. 4.8 also show that the scaling function has a power-law tail $\widetilde{P}(\tilde{x}) \sim \tilde{x}^{-1-\alpha}$ with exponent $\alpha = 2$.

If we allow cancellations of orders, the situation changes [890]. In Fig. 4.9 we can see that the distribution of one-step price changes, i.e. $P(\delta t; x)$ for $\delta t = 1$, behaves differently. For a small value of the evaporation probability, the tail of the distribution is still close to a power law. For example, the special values of the parameters $q = 0.01$ and $\overline{N} = 100$ lead to the distribution with a power-law tail characterised by exponent $\alpha = 3$. However, this behaviour is highly non-universal. Not only does the effective exponent in the tail of the distribution depend on the model parameters, but for larger q it loses the power-law character completely. In the model with cancellations, the power-law-looking tail for small but positive q is a coincidence, a remnant of the

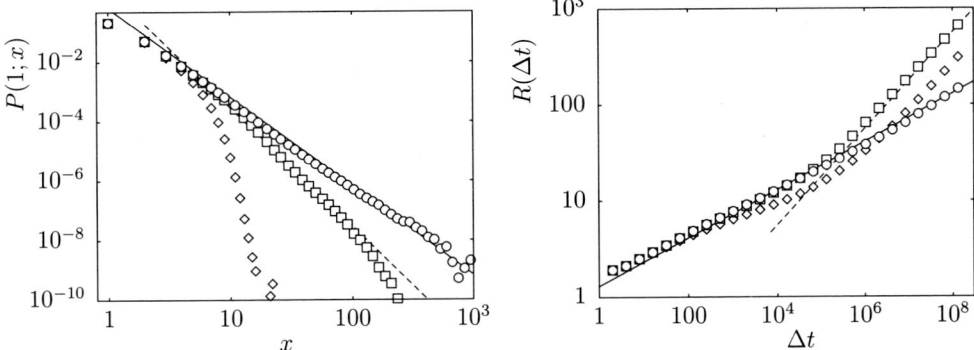

Fig. 4.9 Properties of the Maslov model without cancellations (○) and with cancellations, using parameters $\overline{N} = 100$ and $q = 0.01$ (□), $q = 0.05$ (◇) are shown. In the left panel, the distribution of one-step price changes is shown. The straight lines are power laws $\propto x^{-3}$ (solid) and $\propto x^{-4}$ (dashed). In the right panel, the Hurst plot. The straight lines are power laws $\propto (\Delta t)^{1/4}$ (solid) and $\propto (\Delta t)^{1/2}$ (dashed).

power-law behaviour observed for $q = 0$ rather than a generic feature. This is also supported by the loss of scaling as soon as we allow cancellations. While for $q = 0$ the scaling (4.28) is confirmed very well, for $q > 0$ it breaks down.

Autocorrelations

Besides the distribution of price changes we are eager to see the autocorrelation of absolute values of price changes, as it measures the effect of volatility clustering, already glimpsed within the sample evolution in Fig. 4.7. We measure the autocorrelation function of one-step price changes $x_t = y_t - y_{t-1}$ defined as

$$\langle |x_t\, x_{t+\Delta t}| \rangle_c = \langle |x_t\, x_{t+\Delta t}| \rangle_t - \langle |x_t| \rangle_t \langle |x_{t+\Delta t}| \rangle_t. \tag{4.29}$$

where again, $\langle \ldots \rangle_t$ means average over all times t in the course of the simulation.

The result is shown in Fig. 4.8. We can clearly see the power-law decay of the autocorrelations

$$\langle |x_t\, x_{t+\Delta t}| \rangle_c \sim (\Delta t)^{-\tau} \tag{4.30}$$

with the exponent very close to the value $\tau = 1/2$ in the case without cancellations. Contrary to the difference in the price-change distribution, the autocorrelations change only very weakly if we allow cancellations of orders. As is also seen in Fig. 4.8, the only effect of cancellations is a mild increase in the value of the exponent τ. Cancellations have nearly no influence on volatility clustering.

Hurst exponent

We have already noticed the value of the Hurst exponent $H = 1/4$ which results from the scaling (4.28). It is instructive to look at the Hurst plot, where the quantity $R(\Delta t)$, defined in Box 4.3 is drawn. Without cancellations, the dependence $R(\Delta t) \sim (\Delta t)^{1/4}$

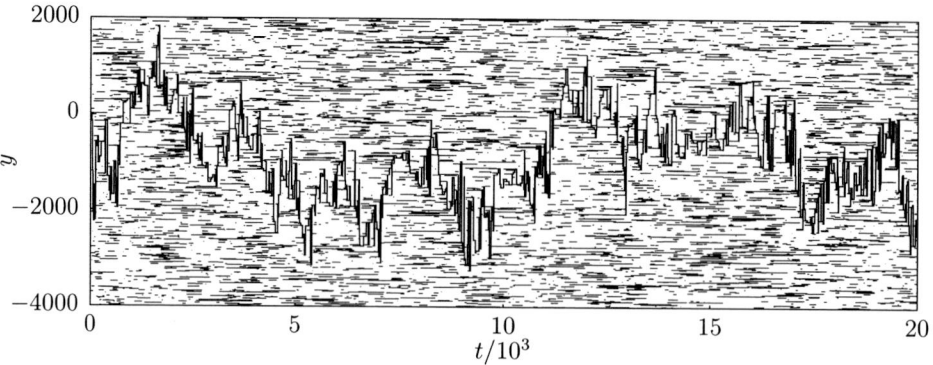

Fig. 4.10 Example of the evolution of the uniform deposition model. Each horizontal segment represents an order, placed where the segment starts and cancelled or executed where the segment ends. The rugged line is the time dependence of the price. The parameters of the model for this plot are $\overline{N} = 100$ and $q = 0.9$. The width of the range of allowed prices is $L = 10^4$.

is extended over the whole range of time differences, confirming the value $H = 1/4$ for the Hurst exponent. The cancellations of orders change the situation dramatically. For short time differences Δt shorter than the average time, an order survives without being cancelled; the behaviour seen in the Hurst plot is still compatible with the power law $R(\Delta t) \sim (\Delta t)^{1/4}$. For longer times it breaks down, and for Δt much longer than the lifetime of orders a different power law sets on, namely $R(\Delta t) \sim (\Delta t)^{1/2}$, which corresponds to the price wandering like a random walk.

An important quantity to compare with the empirical data is the order-book profile $n(y)$, i.e. the average number of limit orders found at the distance y from the current price. We can see in Fig. 4.12 what it looks like both in the case with cancellations and without them. In the latter case, the profile is a decreasing function for $y > 0$, while in the former case there is a pronounced maximum at a certain distance from the price, while close to the price the density of orders is small. This result is somewhat counter-intuitive. Naively, one would expect that cancellations affect mostly the orders far from the price; therefore close to the price the profile would be the same as without cancellations. But the simulations show just the opposite. Comparing the results with the empirical findings of Sec. 1.3, we conclude that the introduction of cancellations makes the model much closer to reality.

Mutually comparing the Maslov model with and without cancellations, we certainly perceive the aesthetic qualities of the latter case. No cancellations imply absence of further free parameters q and $\overline{N}$; the price-change distribution is scaled perfectly according to (4.28); the power-law tail in the scaled distribution is governed by a nice integer exponent $\alpha = 2$; and a clear-cut Hurst plot confirms the simple value $H = 1/4$ of the Hurst exponent already known from the scaling analysis. All of it is lost as soon as we allow cancellations of orders. On the other hand, the loss of beauty is compensated by the fact that the Maslov model with cancellations is much

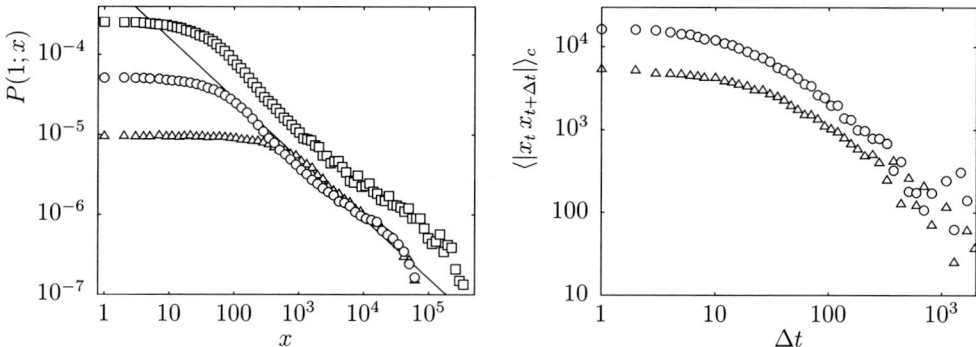

Fig. 4.11 In the left panel, one-step price changes in the uniform deposition model are shown for parameters $\overline{N} = 10^4$ and $q = 0.5$ ($\square$); $\overline{N} = 10^4$ and $q = 0.9$ ($\bigcirc$); $\overline{N} = 10^3$ and $q = 0.9$ ($\triangle$). The allowed price range is $L = 10^6$. The straight line is the power decay $\propto x^{-3/4}$ In the right panel, the autocorrelation of absolute values of one-step price changes. The parameters are $L = 10^3$, $q = 0.9$, $\overline{N} = 100$ ($\bigcirc$) and $\overline{N} = 10^3$ ($\triangle$).

more realistic. First, cancellations are part of reality. We cannot simply ignore them. Furthermore, the exponent found in empirical data is much closer to the value $\alpha = 3$, which may be achieved by tuning the parameters of the cancellations. The Hurst exponent $H = 1/4$ is completely wrong, as we have seen in Sec. 1.2.2. Therefore, insisting on its value $H = 1/4$ would be somewhat perverse. Also the order-book profile becomes much more realistic as soon as we introduce cancellations. Finally, the scaling found in the empirical price-change distribution does not extend to arbitrarily long times. For longer times it breaks down, just as it breaks down in the Maslov model with cancellations. All of this makes the Maslov model with cancellations a good candidate for order-book dynamics.

4.2.2 Uniform deposition model

Deposition arbitrarily far from price

The model, which is, to a certain extent, complementary to the Maslov model, but also shares a great many common features with it, was developed in a series of works by Farmer and collaborators [910–914]. Many variants were investigated from several points of view. The plasticity of the scheme makes it very attractive for experimenting with various features of the model. Here, we shall look at the most fundamental core of the whole family of models, and we shall especially stress its differences from the Maslov model treated in the last section. To set the names, we shall call the core model we describe here the uniform deposition model.

At first sight, the only difference between the uniform deposition model and the Maslov model consists in the deposition rule for incoming orders. In the Maslov model, new orders are always placed within a fixed distance δy from the current price. We used just the distance $\delta y = 1$, but any variant in which the orders are placed randomly but not farther than δy from the current price shares the same universal features.

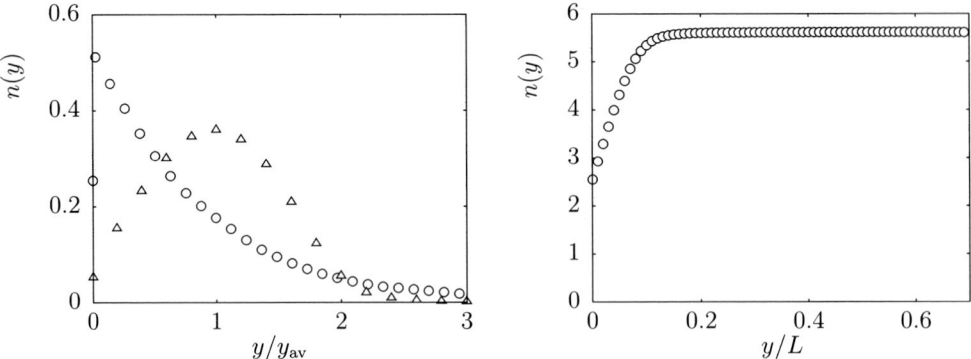

Fig. 4.12 Order-book profile in the Maslov model (left panel) and uniform deposition model (right panel). For the Maslov model, we compare the case without cancellations ($\bigcirc$) and with cancellations, with parameters $\overline{N} = 10^5$, $q = 0.05$ ($\triangle$). For convenience, the distance from the current price x is plotted relative to its average value $y_{av} = \int_0 y\, n(y)\mathrm{d}y / \int_0 n(y)\mathrm{d}y$. For the uniform deposition model, the parameters are $q = 0.1$, $\overline{N} = 10^4$, and $L = 10^5$.

On the contrary, if we allow the orders to be deposited arbitrarily far from the price, the situation changes. In the uniform deposition model, the deposition probability is uniform on the whole range of prices. The only exception is that we do not allow deposition exactly at the current price. Formally, the probabilities of deposition, execution, and cancellation events are given by the same formulae (4.27) as in the case of the Maslov model. There are differences, though. First, in a simulation the price axis cannot extend to infinity. Therefore, we limit the range of allowed prices to the segment of length L. We suppose L is even. As we already said, deposition at the current price y_t is forbidden. Thus, the new orders are deposited at any of the L points $-L/2, -L/2 + 1, \ldots, y_t - 1, y_t + 1, \ldots, L/2$ with equal probability. The order deposited at a point below y_t is interpreted as a bid, and the order deposited above y_t is interpreted as an ask.

Execution of orders works exactly the same as in the Maslov model, so it is unnecessary to discuss it further. But in cancellations, there is an important conceptual difference. The probability given by (4.27) is the same as in the Maslov model, but in the Maslov model the orders are clustered around the price, and the cancellation events are somehow a complement or correction to the natural execution of the limit orders by incoming market orders. Thus, in the Maslov model, q is typically a small number compared to 1. In contrast, in the uniform deposition model cancellations are essential, because orders are deposited in the whole allowed segment and ought to also be removed from areas where the price rarely wanders; otherwise the density of orders would quickly blow up. Therefore, q is comparable to, although smaller than, 1. Typically, the simulations are performed in a regime in which $1 - q$ is much smaller than 1.

An example of the typical evolution of the state of the order book in uniform deposition model is shown in Fig. 4.10. It looks as if the price 'crawled' in a homogeneous sea of orders. A closer look again reveals traces of volatility clustering, as large price changes are often followed by other large price changes, and vice versa. Superficially, the price fluctuations are not much different from the Maslov model.

Price changes and volatility clustering

Quantitative analysis shown in Fig. 4.11 corrects this view. We can first look at the distribution of one-step price changes $P(1; x)$. It is interesting that, in contrast to the Maslov model, the tail of the distribution is completely insensitive to the probability of cancellations q. It also looks nice that the tail follows a power-law decay $P(1; x) \sim x^{-1-\alpha}$. However, the value of the exponent $\alpha = -1/4$ is completely wrong compared to the empirical data.

   Also in Fig. 4.11 we can see the autocorrelation function of absolute values of one-step price changes defined in Eq. (4.29). Compared to the Maslov model, the memory seen in the autocorrelations is shorter and the decay faster than a power law, although at the same time it is slower than exponential, as can be checked by plotting the same data as in Fig. 4.11 in a semi-logarithmic scale instead of a double-logarithmic one. This means that in uniform deposition model we also observe strong volatility clustering, but it is still weaker than in the Maslov model.

Book profile

As we did with the Maslov model, we also look at the average density of orders in uniform deposition model. The order-book profile is shown in Fig. 4.12, in parallel with the results for the Maslov model. In fact, the only important difference from the Maslov model is that the order density approaches a constant when we go very far from the current price. Of course, this is dictated by the uniform probability of deposition. Clearly, if we postulated a deposition probability which vanishes far away from the price, we would get a much more realistic order density. The problematic point of such an approach is that we do not a priori know how the deposition probability should decrease as a function of distance from the price. In such a situation, it is straightforward to take just the probability found in the empirical studies. This is a successful strategy and has been applied, e.g. in Refs. [261, 915].

4.3 Schematic models

Every model attempts to grasp important features of reality. Big differences may arise according to what 'important' means. We can try to fit the empirical data as close as possible and become satisfied only after all exponents, distributions, etc., agree up to a level of statistical errors. Alternatively, we can try to concentrate on generic and universal features and investigate a model which may look odd when compared to reality but is able to mimic just these generic features. These are the kind of models we call schematic. They do not aim at a thorough description, but rather they try to elucidate one single feature separately, isolated from the rest of the complex reality. That is why schematic models can hardly be used for the prediction of the future in the stock market. Instead, they provide some insight into what and why.

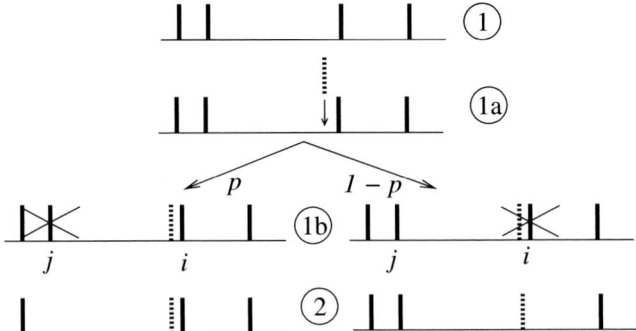

Fig. 4.13 Illustration of the interacting gaps model. From an original configuration in step 1, two ways can lead to a final configuration at step 2. In the intermediate step 1a, a new order (dashed) is deposited at distance 1 from an existing order (index i). In intermediate step 1b, one of the previously existing orders is executed. This way the total number of orders is kept unchanged. The executed order is either the order i next to which the new order is placed (this happens with probability $1 - p$) or the order j situated on the other side with respect to the new order (with probability p).

4.3.1 Interacting gaps

Gaps instead of particles

When a physicist looks at an order book, she almost immediately translates it into a system of particles on a line. The particles are added and removed according to rules which are specific to this or that order-book model. The state of the book is described by the positions of the particles, accompanied by the current price. However, the same information is also contained in the difference variables. Instead of coordinates of the particles, we can describe the state using the distances between them. We shall call these distances gaps. Some care must be taken in order to correctly translate the information on the price position, but essentially, this is the idea behind the gap models of order-book dynamics.

When orders are deposited, executed, or cancelled, the lengths of the gaps are changed, but such change affects only at most two neighbouring gaps, as if the two gaps in contact underwent a reaction producing one or two new gaps in place of the old ones. Indeed, if an order is executed or evaporated, two adjacent gaps collapse into one. When an order is placed, one gap is split into a pair of adjacent gaps. To mimic such dynamics, Solomon introduced the interacting gaps model, which was first investigated in [916] and then in [917].

The model is built upon three gross simplifications. First, we do not make any difference between bids and asks. Therefore, all gaps are treated on an equal basis. In reality, the gap between the highest bid and lowest ask, i.e. the spread, is certainly different from all other gaps. Similarly, the gap between the lowest and second lowest ask has different properties than the gap between the second and third lowest ask, and so on. None of these subtleties are considered in the interacting gaps model. We also insist that at most one order can lie at one place. Therefore, all gaps have positive lengths.

The second simplification consists in imposing strict conservation of orders. In reality, the number of orders fluctuates around an average value. In the interacting gaps model, the number of orders is always constant. This is guaranteed by two things. First, we do not have any cancellations. Second, deposition of an order is immediately followed by execution of another order. Both actions are united in a single step of the dynamics.

The third simplification concerns the position of the price. We would like the price to coincide with the position of one of the existing orders. The correct place for setting the price would be the position of the order which was executed in the last step. However, in this way the price would lie somewhere in the empty space between a pair of orders. Therefore, if an order is executed, the price is set at either the right or the left neighbouring order. We take left and right randomly with equal probability $1/2$. Thus, the price change is always equal to the length of one of the gaps. This definition anticipates that we shall later equate the distribution of gap lengths with the distribution of price changes.

The dynamics is illustrated in Fig. 4.13, where one step is sketched in detail, including the intermediate states. The configuration 1 is the initial state. In 1a, a new order is placed next to the order i representing the position of the current price. On the side opposite from than i with respect to the new order, there is another order, denoted j. One of the old orders is then executed. There are two possibilities, shown at the intermediate state 1b. With probability p, the order j is executed, while i is executed with probability $1 - p$. In the language of gaps, two reactions are possible. The neighbouring gaps can change so that one of them shrinks by one unit and the second becomes longer by one unit. This happens with probability $1 - p$. Or, one of the gaps shrinks up to the minimum length 1, while the second is lengthened by as much as the first one was shortened. Such a collapse has probability p. We shall not go into details, for example about what we should do if the place where we want to put the new order is occupied. Instead, we define now the model more formally in terms of gap dynamics.

Gap dynamics

We have $N + 1$ orders of unit volumes placed at integer positions. The state of the order book is described by the lengths of the gaps $g_i > 0$, $i = 1, 2, \ldots, N$ separating the orders. In order to specify which pair of gaps is affected at time t, we introduce the index k_t, so that the gap number k_t interacts with the gap number $k_t + 1$. The extreme values $k_t = 0$ and $k_t = N$ are special, and what happens there will be explained later. As for the dynamics of the index k_t, we take the simplest assumption that it performs a random walk, i.e. $k_{t+1} = k_t \pm 1$ with equal probability $1/2$, only keeping in mind that there are bounds $0 \le k_t \le N$.

The interaction goes as follows. For brevity, we denote $k_t = j$. First, we generate a random number $\sigma = \pm 1$, representing the left or the right side, with equal probability $1/2$ for both signs. Then, with probability p the collapse of one of the gaps occurs. The gap lengths are updated according to the rule

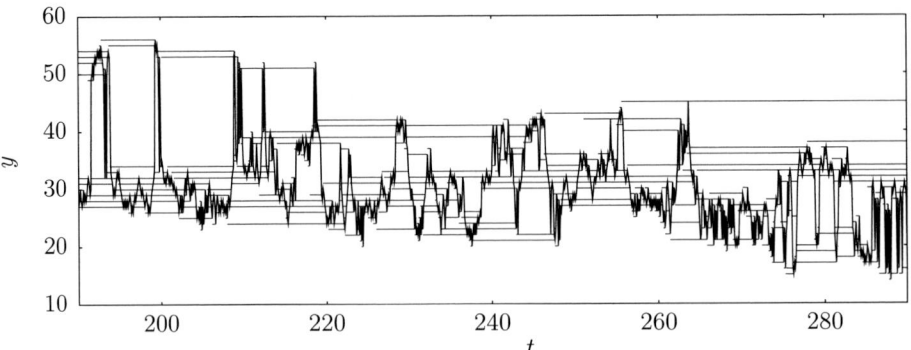

Fig. 4.14 Example of the evolution of the interacting gaps model. The horizontal lines represent the positions of the orders, the rugged line is the time dependence of the price. The parameters of the model are $N = 10$, $\bar{g} = 5$, $p = 0.3$.

$$(g_j, g_{j+1})(t+1) = \begin{cases} (1, g_j(t) + g_{j+1}(t) - 1) & \text{for } \sigma = +1 \\ (g_j(t) + g_{j+1}(t) - 1, 1) & \text{for } \sigma = -1. \end{cases} \quad (4.31)$$

Otherwise, there is a shift in the gap lengths, which occurs with complementary probability $1 - p$ and the gaps are updated as

$$(g_j, g_{j+1})(t+1) = (g_j(t) + \sigma, g_{j+1}(t) - \sigma) \quad (4.32)$$

on condition that both $g_j(t) + \sigma \geq 1$ and $g_{j+1}(t) - \sigma \geq 1$. If either of these conditions is not satisfied, the gaps do not change.

Some modifications must be employed at the left and right ends of the sequence of gaps, i.e. for $j = 0$, or $j = N$. If, for example, $j = N$, the gap g_N has no neighbour gap to interact with according to rules (4.31) and (4.32). Then, we choose $\sigma = \pm 1$ as before. The collapse occurs only if $\sigma = +1$ and it affects only one gap

$$g_j(t+1) = 1. \quad (4.33)$$

For $\sigma = -1$, nothing happens. The shift also affect only one gap, but otherwise the rule is the same as before, i.e.

$$g_j(t+1) = g_j(t) + \sigma \quad (4.34)$$

only if $g_j(t) + \sigma \geq 1$.

The remaining piece to be specified is the movement of price. In fact, it is enough to specify the price change at each step of the dynamics. The most realistic prescription for the location of the current price y_t is the position of the order separating the two gaps after their interaction, i.e. at order number k_t. In the previous step, the interacting pair was specified by the index k_{t-1}. Because k_t performs a simple random walk, k_{t-1} was either one unit above or below k_t. Accordingly, the previous price y_{t-1}

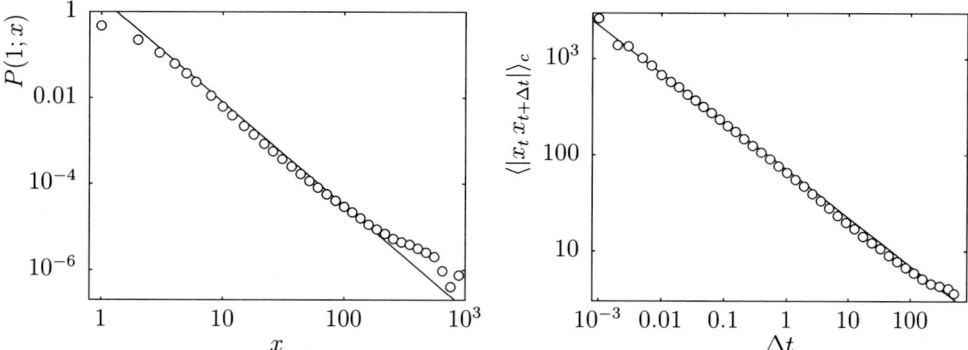

Fig. 4.15 Properties of the interacting gaps model for parameters $N = 1000$, $\overline{g} = 50$, and $p = 0.1$. In the left panel, distribution of one-step price changes is shown. The straight line is the power dependence $\propto x^{-5/2}$. In the right panel, autocorrelation of absolute values of one-step price changes is shown. The straight line is the power law $\propto (\Delta t)^{-1/2}$.

was either at the order number $k_t + 1$ or at the order number $k_t + 1$. In the former case the price goes down, in the latter it goes up. Therefore, the one-step price change is

$$x_{t+1} = y_{t+1} - x_t = \begin{cases} g_{k_t-1}(t+1) & \text{for } k_t > k_{t-1} \\ -g_{k_t}(t+1) & \text{for } k_t < k_{t-1}. \end{cases} \tag{4.35}$$

We can see that the absolute value of the price change equals one of the gaps. This is just the feature we wanted to incorporate into the model, because the statistics of price changes is translated into the statistics of gap lengths.

Simulations

In practical implementations, one more parameter enters the model. It is the initial value of the average length of the gaps $\overline{g}$. Usually, the initial condition is chosen such that all gaps are equal, $g_i(0) = \overline{g}$ for all i. Thorough studies show that in the long-time limit, the initial condition is forgotten, and moreover, in the stationary state, the results are independent of the value of $\overline{g}$. Nevertheless, in the results we shall show, we also specify the value of $\overline{g}$ used.

Let us first look at an example of the evolution of the order book in the interacting gaps model, as shown in Fig. 4.14. At first sight, the evolution looks very similar to the Maslov model. The volatility clustering seems to be present as well. The main difference in the visual impression is that the price is much less surrounded by orders. In the interacting gaps model, we found much too often most of the orders on the same side of the price. This is certainly not a realistic feature.

Quantitatively, we can easily measure distribution of gap lengths and distribution of one-step price changes. It is very satisfactory that both of these quantities follow a power law with the same exponent

$$\begin{aligned} P(g) &\sim g^{-1-\alpha} \\ P(1;x) &\sim x^{-1-\alpha} \end{aligned} \tag{4.36}$$

where $\alpha = 3/2$. The price-change distribution is shown in Fig. 4.15. This way we are able to mimic the coincidence of price-change and first gap distribution found in empirical data [290]. The power-law tail in the distribution is quite satisfactory itself, as is the coincidence of gap and price-change distributions. In fact, this coincidence is deliberately built in into the model. On the worse side, as in the Maslov model, we must admit that the numerical value of the exponent α is significantly smaller than the empirical one. Moreover, contrary to the Maslov model, the interacting gaps model is so rigid that the change of the exponent towards the correct value cannot be achieved by tuning an appropriate parameter.

We can also look at the volatility clustering. The autocorrelation function of one-step price changes is shown in Fig. 4.15. We can see the power-law decay which closely matches the result for the Maslov model without cancellations, including the same value of the exponent $\tau = 1/2$. Thus, we observe a certain universality in this value. On the other hand, if we make the Hurst plot for the interacting gaps model, we extract the Hurst exponent $H = 1/2$, equal to the random-walk value and strikingly different from the Maslov model. The reason for such behaviour is that at long times the price dynamics is dominated by the random-walk movement of the index k_t. There is an area open for experimenting with more sophisticated prescriptions for the process k_t.

Mean-field solution

One of the features which make the interacting gaps model attractive, despite its weaknesses, is the possibility of solving it within the mean-field approximation. Unlike other models, here the mean-field solution provides miraculously accurate results.

In the one-dimensional version, only adjacent gaps can interact. The essence of the mean-field approximation is to allow interaction between any pair of gaps, chosen at random, in the same spirit as in Refs. [918–921]. Thus, in each step we first choose two gaps j and l. There is no reason to consider j first and l second or the converse; therefore the random choice of $\sigma = \pm 1$ is now superfluous. The update rules (4.31) and (4.32) are simplified so that the collapse

$$(g_j, g_l)(t+1) = (1, g_j(t) + g_l(t) - 1) \tag{4.37}$$

occurs with probability p and shift

$$(g_j, g_l)(t+1) = (g_j(t) + 1, g_l(t) - 1) \tag{4.38}$$

with probability $1 - p$, on the condition that $g_l(t) > 1$. There is a subtle but fundamental difference from the original interacting gaps model. In the one-dimensional case, the sum of the lengths of interacting gaps is conserved in all events except those occurring at the edges. This leads to slowly but inevitably forgetting the initial value of the average gap length $\bar{g}$. On the contrary, in the mean-field version the sum of gap lengths is strictly conserved, so $\bar{g}$ is a parameter of the model which does not change. We shall soon see how this fact somewhat complicates the solution. Note, however, that the conservation of the sum of gaps can also be imposed on the one-dimensional version if we change the boundary conditions. Indeed, putting the model on a ring of a finite length, instead of an infinite line, the first gap would interact with the last one, and the conservation would hold. However, it would look rather strange if the coordinate along the ring were interpreted as a price. That is why we prefer to avoid the periodic boundary conditions. There is yet another consequence. The realistic price axis has no strict bounds from either above or below (recall that actually we always speak of the logarithm of price, not the price itself, so that there is indeed no lower bound). In the simulations of the interacting gaps model we have seen that all the orders are concentrated in a very narrow interval compared to the potentially infinite price axis. If we translate this finding into the mean-field version of the model, it means that the average gap length $\bar{g}$ should be very large, because the average is dominated by a single but enormously large 'gap', measured over an imaginary circle: going from the highest ask to $+\infty$ then jumping to $-\infty$, thus closing the circle, and then reaching the lowest bid. Later we shall see that this fact is important for interpretation of the results of the mean-field approximation.

Let us proceed with the solution of the mean-field variant of the model. All information on the state of the process at a specific time t is contained in the distribution function of all configurations of gap lengths $P(\{g_i\};t)$. We shall need only the distribution of one gap (for example the first one) because all of them are statistically equivalent: $P_1(g;t) = \sum_{\{g_i\}} \delta(g_1 - g)P(\{g_i\};t)$. As an auxiliary quantity we shall also need the joint distribution of two gaps (for example the first and the second one) $P_{12}(g,g';t) = \sum_{\{g_i\}} \delta(g_1 - g)\delta(g_2 - g')P(\{g_i\};t)$.

From the elementary moves (4.37) and (4.38), we derive the following exact equation for the evolution of the one-gap distribution

$$
\begin{aligned}
N\Big[P_1(g;t+1) - P_1(g;t)\Big] = &-2P_1(g;t) \\
&+ (1-p)\Big[\delta(g-1)P_1(1;t) + P_1(g+1;t) + P_1(g-1;t) \\
&+ P_{12}(g,1;t) - P_{12}(g-1,1;t)\Big] \\
&+ p\Big[\delta(g-1) + \sum_{u=1}^{g} P_{12}(u,g+1-u;t)\Big].
\end{aligned}
\tag{4.39}
$$

We are interested in the stationary distribution $P_1(g) = \lim_{t\to\infty} P_1(g;t)$. It is possible to show that in the limit of large number of gaps, $N \to \infty$, the stationary two-gap distribution is factorised

$$
\lim_{t\to\infty} P_{12}(g,g';t) = P_1(g)P_1(g').
\tag{4.40}
$$

However, when taking the thermodynamic limit $N \to \infty$ we miss certain essential features of the distribution. Later we shall discuss how to bring these features back, for otherwise we fall into annoying inconsistencies.

In the stationary state the left-hand side of (4.39) vanishes, and using the factorisation (4.40) we get a closed equation for the distribution of gaps $P_1(g)$. It can be easily solved using the discrete Laplace transform $\widehat{P_1}(z) = \sum_{g=1}^{\infty} z^g P_1(g)$. The point is that the difference equation (4.39) turns into a trivially solvable quadratic equation

$$-2\widehat{P}_1(z)+p\left[z+\frac{1}{z}\left(\widehat{P}_1(z)\right)^2\right]$$
$$+(1-p)\left[\left(z+\frac{1}{z}\right)\widehat{P}_1(z)-(1-z)(1-\widehat{P}_1(z))P_1(1)\right]=0. \tag{4.41}$$

Once we have $\widehat{P}_1(z)$, we can deduce the behaviour of $P_1(g)$ for large g, because it is related to the behaviour of $\widehat{P}_1(z)$ at $z \to 1$. If $\widehat{P}_1(z)$ is analytical, i.e. it has all derivatives and they are finite at $z=1$, the distribution $P_1(g)$ has an exponential tail. A power-law singularity in $\widehat{P}_1(z)$ at $z=1$ implies a power-law tail in $P_1(g)$. This is exactly what happens in the solution of (4.41), as we shall now show.

The missing piece in the puzzle is the still unknown number $P_1(1)$, the probability that the gap has the minimum size 1. It should be established from the condition that the initial value of the average gap length $\bar{g}$ is conserved forever; therefore it can be calculated from the stationary distribution. This leads to the relation between $\bar{g}$ and $P_1(1)$ in the form

$$\bar{g}=\lim_{z\to1^-}\frac{\mathrm{d}}{\mathrm{d}z}\widehat{P}_1(z)$$
$$=1+\frac{1-p}{2p}P_1(1)-\sqrt{\left[\frac{1-p}{2p}P_1(1)\right]^2-\frac{1-p}{p}(1-P_1(1))}. \tag{4.42}$$

As a function of $P_1(1)$ this expression reaches maximum at $P_1(1) = P_1^c(1) \equiv \frac{2p}{1-p}\left(1/\sqrt{p}-1\right)$. The maximum value of the average gap size is then

$$\bar{g}_{\max}=\frac{1}{\sqrt{p}}. \tag{4.43}$$

We must distinguish three regimes. For $\bar{g} < p^{-1/2}$ the one-gap distribution has an exponential tail, because $\widehat{P}_1(z)$ is analytic at $z \to 1^-$. The second regime occurs exactly at the critical value $\bar{g} = p^{-1/2}$. We denote $P_1^c(g)$ the one-gap distribution at this critical point. If we insert the value $P_1^c(1)$ into the solution of the equation (4.41), we obtain

$$\widehat{P}_1^c(z)=1-\frac{1}{\sqrt{p}}(1-z)-\frac{(1-\sqrt{p})^2}{2p}(1-z)^2$$
$$+\frac{(1-\sqrt{p})}{p^{3/4}}|1-z|^{3/2}\sqrt{1+\frac{(1-\sqrt{p})^2}{4\sqrt{p}}(1-z)}. \tag{4.44}$$

The $(1-z)^{3/2}$ singularity at $z \to 1$ is the fingerprint of the power-law tail in the distribution of gap sizes, with exponent $5/2$, so

$$P_1^c(g)\sim g^{-5/2},\ \text{for}\ g\to\infty. \tag{4.45}$$

The third regime with $\bar{g} > p^{-1/2}$ is the most subtle, but it is precisely the relevant one to be compared with the simulations. In an infinite system, there is no stationary state. However, it is easy to understand what type of non-stationarity we face. As the time goes on, the gap distribution is split into two substantially different contributions. First, there is a portion proportional to the critical gap distribution $P_1^c(g)$. Its Laplace transform is given by (4.44). Second, there is a δ-function part, which shifts to larger and larger values, but its weight is shrinking. This gives us a hint for the behaviour of a finite system. For very large but finite N, the ultimate stationary state is essentially equal to the combination of the critical gap distribution $P_1^c(g)$ and a delta-function which comes from a single huge gap. This is a kind of condensation phenomenon. The total sum of gap lengths is conserved in the dynamics, but in the stationary state most of the sum is gathered by one single gap. It is worth noting that a similar condensation phenomenon was observed in quite different branches of econophysics, for example in the theory of wealth distribution [922].

We already explained why the single huge gap is irrelevant for the comparison with either simulations or the empirical data on order books. Therefore, we disregard it and take as our result the critical distribution

$$P_1(g) = P_1^c(g) \sim g^{-5/2}, \text{ for } g \to \infty. \tag{4.46}$$

This also means that this distribution is universal, independent of p and the number of gaps N. The agreement with the simulation results (4.36) is excellent. This is one of the few but important examples of the systems where the results for one-dimensional and mean-field variants of the model coincide [748, 923–925]. Finally, let us note that models of this kind have been investigated for a long time in various branches in physics, starting from the classical works of Smoluchowski on coagulation [926] and continuing to the current day [919–921, 924, 927–932]. That is why it deserves to be studied on its own.

4.3.2 Crippled order book

One side only

The order book has two sides. There are bids on the left from the current price and there are asks on the right from the price. The two sides evolve to some extent independently. Certainly, the deposition and cancellation on one side does not influence the other, at least in the models like Maslov and uniform deposition model. If that is also the case in reality is another question, and we shall not touch upon it here. The independence, however partial and approximate, suggests studying the dynamics of one side only, regardless of what is happening with the other side.

Such a crippled order book, deprived of one of its legs, cannot reveal all of the information hidden in the full order book, but it still keeps many interesting structures which are helpful in studying the order dynamics. The orders reside on the half-line of non-negative numbers. We can interpret it so that we look only at asks, and the origin of the half-line denotes the position of the highest bid, which never changes. The distance of the lowest ask from the origin is the spread. The orders are deposited with an uniform density and they can be cancelled in the same way as in the uniform deposition model. The arrival of a market order results in execution of the lowest ask, i.e. the order closest to the origin.

We cannot expect to get the fluctuations of price in this model. However, we can obtain the average density of orders and the probability distribution of the spread. The calculation is quite easy, as we shall see now.

Orders in a bin

The trick consists in counting the number of orders present in the interval $[0, L]$. Let us denote their number at time t by N_t. During the infinitesimal time interval dt this number can increase by 1 due to deposition, with probability $\alpha L\, dt$. The deposition rate is proportional to the length of the interval, because the deposition is uniform. We could also postulate a non-trivial spatial dependence of the deposition rate. The results we shall show can easily be generalised to that case.

The number N_t can also decrease by 1, due to two mechanisms. With probability $\beta\, N_t\, dt$ one order is cancelled and with probability $\gamma\, dt$ one order (the lowest one) is executed, on the condition that there is something to be executed, i.e. if $N_t > 0$. The dynamics is equivalent to the random walk on a semi-infinite linear chain. The sites are indexed by non-negative integer numbers n. The hopping rates to neighbouring sites are

$$
\begin{aligned}
w_{n \to n+1} &= \alpha\, L \\
w_{n+1 \to n} &= (n+1)\beta + \gamma.
\end{aligned}
\tag{4.47}
$$

The information on this process is contained in the probability distribution $P(n; t) = \text{Prob}\{N_t = n\}$. It obeys the master equation

$$
\begin{aligned}
\frac{d}{dt} P(n; t) = {}& \alpha L\, P(n-1; t) + \Big((n+1)\beta + \gamma \Big) P(n+1; t) \\
& - \big(n\beta + \gamma + \alpha L \big) P(n; t)
\end{aligned}
\tag{4.48}
$$

which holds for all $n \geq 0$, with the condition at the boundary $P(-1; t) = 0$ for all times t. All we need is the stationary solution $P(n) = \lim_{t \to \infty} P(n; t)$ of Eq. (4.48). As the problem is one-dimensional, the solution can be found by a recursive application of Eq. (4.48) with zero on the left-hand side. The result is

$$
P(n) = \frac{(\alpha L)^n}{Z\, (\gamma + \beta)(\gamma + 2\beta) \dots (\gamma + n\beta)}.
\tag{4.49}
$$

The normalisation factor Z ensures that $\sum_{n=0}^{\infty} = 1$ and can be expressed using the confluent hypergeometric function

$$
Z = {}_1F_1\left(1; \frac{\gamma}{\beta} + 1; \frac{\alpha L}{\beta} \right)
\tag{4.50}
$$

(see Box 4.4 if unclear).

The average density of orders, or the order-book profile, can be computed by differentiation of the average number of orders inside the interval $[0, L]$ with respect of its length L. Thus

$$
n(y) = \frac{d}{dL} \sum_{n=0}^{\infty} n P(n) \bigg|_{L=y}.
\tag{4.51}
$$

> **Hypergeometric and confluent hypergeometric functions** | Box 4.4 |
>
> are some of the most frequently used special functions in physics and applied mathematics [933]. The hypergeometric function of variable z and parameters a, b, and c is defined by the series
> $$F(a, b; c; z) = 1 + \sum_{k=1}^{\infty} \frac{a(a+1)\ldots(a+k-1)b(b+1)\ldots(b+k-1)}{c(c+1)\ldots(c+k-1)\,k!} z^k.$$
> Similar but slightly less complicated is the series defining the confluent hypergeometric function of variable z and parameters a and c
> $$_1F_1(a; c; z) = 1 + \sum_{k=1}^{\infty} \frac{a(a+1)\ldots(a+k-1)}{c(c+1)\ldots(c+k-1)\,k!} z^k.$$
> For some special values of the parameters the hypergeometric and confluent hypergeometric functions are reduced to expressions containing only elementary functions. For example
> $$(1-z)^{\nu} = F(-\nu, 1; 1; z)$$
> $$\ln(1-z) = -zF(1, 1; 2; z)$$
> $$_1F_1(1; 2; z) = (e^z - 1)/z.$$
> Other special values of the parameters lead to various special functions. For example, the Bessel function of an imaginary argument, which we use in Sec. 8.2.1 and Sec. 8.2.3, is
> $$I_{\nu}(z) = \frac{(z/2)^{\nu}}{\Gamma(\nu+1)} e^{-z} \,_1F_1(\nu + \tfrac{1}{2}; 2\nu + 1; 2z),$$
> and the elliptic integral we encounter in Sec. 8.2.1 is
> $$K(z) = \frac{\pi}{2} F(\tfrac{1}{2}, \tfrac{1}{2}; 1; z^2).$$
> There are various identities connecting these functions. For example, the differentiation is performed according to
> $$\frac{d}{dz} \,_1F_1(a; c; z) = \frac{a}{c} \,_1F_1(a+1; c+1; z).$$

Using the formula for differentiation of confluent hypergeometric functions we obtain a seemingly complicated result. To make the formula simpler, we introduce the notation

$$f_k(y) = \,_1F_1\left(k; \frac{\gamma}{\beta} + k; \frac{\alpha y}{\beta}\right). \tag{4.52}$$

Using that, we can write

$$n(y) = \frac{\alpha}{\beta + \gamma} \frac{f_2(y)}{f_1(y)} + \frac{\alpha^2 y}{\gamma + \beta} \frac{\frac{2}{\gamma+2\beta} f_3(y) f_1(y) - \frac{1}{\gamma+\beta} f_2^2(y)}{f_1^2(y)}. \tag{4.53}$$

An example of the order-book profile calculated in this way is shown in Fig. 4.16. We can see that it agrees qualitatively very well with the density of orders in the uniform deposition model as shown in Fig. 4.12. Far from the current price, the profile is constant, in contrast to the empirical data; but this feature can easily be cured if we suppose that the deposition rate decays with the distance from the origin. Another quantity the crippled order-book model offers is the probability distribution of the spread b. It is related to the probability that the interval $[0, L]$ is empty. The probability distribution for the spread is again obtained by differentiation

$$P(b) = -\frac{d}{dL} P(0)\bigg|_{L=b}. \tag{4.54}$$

A calculation similar to that of the order-book profile leads us to the result

$$P(b) = \frac{\alpha}{\beta + \gamma} \frac{f_2(b)}{f_1^2(b)}. \tag{4.55}$$

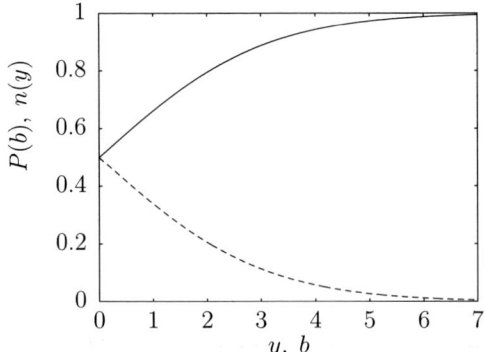

Fig. 4.16 Properties of the crippled order book model are shown for $\alpha = \beta = \gamma$. The full line is the average density of orders $n(y)$; the dashed line is the probability density of the spread, $P(b)$.

In Fig. 4.16 we can see the results for the special values of the parameters, $\alpha = \beta = \gamma$. In this case, the confluent hypergeometric functions arising in the formulae (4.53) and (4.55) can be expressed using exponentials. However, the main qualitative features are generic and hold for all values of the parameters. The density of orders is finite at $y = 0$ and increases with increasing y, eventually reaching saturation at a non-zero value for $y \to \infty$. This is just the behaviour we observed in uniform deposition model. As we already stressed, the model can be made more realistic assuming that the deposition rate decreases with y. In that case, the density of orders would exhibit a maximum and approach zero for large y, in accordance with the empirical data. The probability distribution for the spread is always a decreasing function.

Two bins

To conclude the discussion of the crippled order book, let us note that the model can be developed further considering two intervals instead of one. We can trace the time evolution of the number N_{1t} of orders in the interval $[0, L_1]$, and the number N_{2t} of orders in the interval $[L_1, L_2]$. The process in question would be a two-dimensional random walk on a set of pairs of non-negative integers. The complication which makes this process less trivial is the dependence of transition rates on the position of the walker. The jumps in the variable N_2 are influenced by the value of the variable N_1. This precludes separating the movements in the two Cartesian directions.

The information we could gain in this analysis would tell us the distribution of distances between the lowest and second lowest asks. We have already pointed out several times that this is the quantity which essentially copies the distribution of the price changes. It could be extracted from the probability that in the stationary state we find $N_1 = 1$ and $N_2 = 0$. Unfortunately, a compact analytic formula for the distribution of occupation in the two intervals is not available, but the problem can be solved numerically, imposing hard finite bounds on the numbers N_1 and N_2.

4.3.3 Order-book dynamics as a matrix multiplicative process

Simplified configurations

The most complicated feature of all order-book models we have discussed so far is the fluctuating number of orders at all distances from the current price. It would be much simpler if the instantaneous configuration of the order book was parameterised by a small number of variables. As an example we can imagine that the configuration of bids, i.e. orders lying below the current price, is represented by their average density ρ_-, and similarly, the configuration of all asks by the density ρ_+. It is clearly an oversimplification, because the configuration is wildly fluctuating, and, moreover, the average density depends on the distance from the current price.

One can expect that this simple model would be adequate if the average order density satisfactorily reflects the instantaneous configuration of the orders on the book. This would hold if the spatial fluctuations in the instantaneous configuration were small, or, alternatively, if the empty spaces separating the points at which orders are concentrated are not very wide. This means that the granularity of the order book can be neglected, and the configuration can be replaced with a smooth function.

Let us check how far we can go in assuming such a gross simplification [934]. The configuration of the order book is given by two numbers. Instead of average densities $\rho_\pm$, it will be more convenient to work with potential shifts of price $x_\pm$, defined simply as

$$x_\pm = \frac{1}{\rho_\pm}. \tag{4.56}$$

How to understand this definition? Suppose the density of orders is ρ_+ above the current price, and a market order arrives to buy one unit of volume. Then, if the density is truly uniform, the executed limit orders lie up from the current price as far as the point lying $1/\rho_+$ above it. And this is just the position of asks below the price, hence the definition (4.56). Thus, we simplified the state of the order book to the two-component vector

$$X = \begin{pmatrix} x_+ \\ x_- \end{pmatrix}. \tag{4.57}$$

To specify the dynamics, we must establish how this vector is changed upon the arrival of market and limit orders. The method will always be the same. After the arrival of an order O_1, we try to calculate anew the potential change in price that would occur if a market order O_2 arrived immediately after the order O_1. This will be the effect of the order O_1 on the vector (4.57).

Let us first take the market orders. If the order O_1 is an order to buy a unit volume, the price is shifted upwards by x_+. Then, if the order O_2 is again an order to buy, the second price shift is again x_+; while if it is an order to sell, the downward price movement can be decomposed into two parts. First, the shift due to the order O_1 has left an empty space of width x_+. The price must travel this void. Second, on the left side of this void there are orders with density ρ_-; therefore price is shifted leftwards by x_-. In total, the price shift is $x_+ + x_-$. In terms of the state vector X the effect of a market order to buy can be expressed as the matrix multiplication

$$X \to X' = T_+ X \tag{4.58}$$

where

$$T_+ = \begin{pmatrix} 1 & 0 \\ 1 & 1 \end{pmatrix}. \tag{4.59}$$

Similarly, a market order to sell transforms the state as $X \to X' = T_- X$, where

$$T_- = \begin{pmatrix} 1 & 1 \\ 0 & 1 \end{pmatrix}. \tag{4.60}$$

The effect of limit orders is more tricky. We assume that the limit and market orders may not be of the same size. One of them may be taken as the unit of measurement, so let the market orders have unit volume, while the limit orders have volume $v < 1$. In order to keep the average number of orders constant, we tune the probability to put a limit order to $p = 1/(1 + v)$ and the probability to issue a market order to $1 - p = v/(1 + v)$.

We also suppose that the limit orders are placed at a small but finite distance from the current price. Let this distance be d. The model will work well in the regime of potential price changes $x_\pm$ larger than d. When $x_\pm$ and d become comparable, the details of order placement become relevant, and a more detailed model would be necessary.

So, suppose there is a buy limit order (a bid) placed at distance d below the current price. This is the order O_1. Now, if O_2 is a buy market order, it does not feel the presence of the new order O_1, and the price change is x_+ as before. If O_2 is a sell market order, the situation is somewhat more complicated. The market order has volume 1 and is partially matched by the volume v of the new limit order. The rest, volume $1 - v$, is satisfied via the orders which were present before. It is supposed that these orders have density $\rho_- = 1/x_-$, so that the shift in the price due to the order O_2 is $(1 - v)/\rho_- = (1 - v)x_-$. To sum up, the effect of the new bid is expressed by the matrix multiplication $X \to S_- X$, where

$$S_- = \begin{pmatrix} 1 & 0 \\ 0 & 2 - \frac{1}{p} \end{pmatrix}. \tag{4.61}$$

In this expression we used the relation $p = 1/(1 + v)$. Similarly, the effect of a new limit order to sell (an ask) is $X \to S_+ X$, where

$$S_+ = \begin{pmatrix} 2 - \frac{1}{p} & 0 \\ 0 & 1 \end{pmatrix}. \tag{4.62}$$

Therefore, the dynamics of the order book is modelled by a matrix multiplicative process

$$X_{t+1} = M_t X_t \tag{4.63}$$

where at each time t the matrix M_t is one of the four matrices S_+, S_-, T_+, and T_-. The first two possibilities are taken with the same probability $p/2$, and the remaining

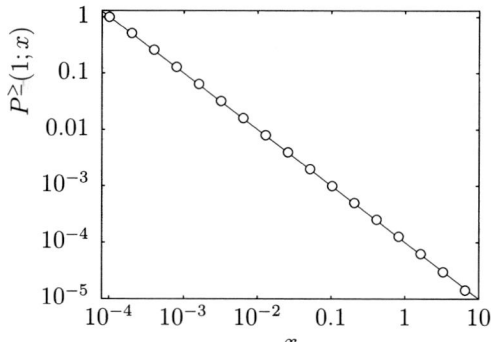

Fig. 4.17 Cumulative distribution of one-step price changes in the matrix multiplicative process for order-book dynamics. The line is the power x^{-1}.

two occur with probability $(1-p)/2$ each. Along with the evolution of the state vector, we can define the changes of the current price y_t as

$$y_t = y_{t-1} + x_t \qquad (4.64)$$

where the one-step increments are $x_t = 0$ if $M_t = S_\pm$ and $x_t = \pm x_\pm$ if $M_t = T_\pm$. The components $x_\pm$ of the state vector X_t are positive at all times. They never decrease when a market order is issued, i.e. when the state vector is multiplied by one of the matrices $T_\pm$. On the other hand, arrival of a limit order leads to a decrease of one of the components of the state vector. However, we stressed in advance that the orders are placed at minimum distance d. This is the remnant of the inherent discreteness of the price axis. Price changes cannot be indefinitely small. Therefore, we require that neither x_+ nor x_- can be smaller than d. If multiplication by the matrices $S_\pm$ results in the decrease of x_+ or x_- below d, it is set to the value d by definition.

In this way we implement the feature of repulsion from zero. The multiplicative processes repelled from zero have been studied thoroughly [170, 173, 676–679, 682, 690, 691, 899, 900, 935, 936], and the generic feature which emerges is the power-law tail in the distribution of the values of the process. That will also be the case here.

Distribution of price changes

The process (4.63) can be very easily simulated on computer. We show in Fig. 4.17 the cumulative distribution of one-step price changes. As we can see, it clearly has a power-law tail

$$P^{\geq}(1;x) \equiv \mathrm{Prob}\{x_+ \geq x\} \sim x^{-\alpha} \qquad (4.65)$$

with exponent $\alpha = 1$. The value of the exponent is too small, both in comparison with the Maslov model ($\alpha = 2$) and with the empirical data ($\alpha \simeq 3$), but this discrepancy can be cured. It turns out that the value $\alpha = 1$ is related to the assumption that the order density is uniform on average, so that the average order-book profile does not depend on the distance from the price. But we can do better. For example, we can assume that the average profile is linear, or even that the density is proportional to the distance from the price. The model remains essentially unchanged; only the matrices $T_\pm$ and $S_\pm$ will be given by expressions other than (4.59) to (4.62). With such freedom in the model definition, the exponent α can be tuned within wide range. We shall not proceed further in this direction. The simulations are easy enough. Instead, we shall show purely analytical arguments which lead to the value $\alpha = 1$ we obtained numerically.

How to find the exponent analytically

Mathematically, the problem is reduced to the investigation of the product of random matrices

$$\prod_{t=0}^{T} M_t. \tag{4.66}$$

If we had scalars instead of matrices, we could take the logarithm of the product and apply the well-known theory of sums of random variables [775], including the central limit theorem and other results. Here, the problem is much more complicated and despite much effort, few results are known rigorously [937, 938]. The complexity stems from the fact that the matrices M_t do not commute, and the order in which they appear in the product (4.66) matters. There is a formal similarity to time-ordered products of operators, which arise in quantum mechanics and field theory [939].

We shall proceed by an approximation, which nevertheless gives correctly the value of the exponent. The first step is the change of variables. Instead of the potential price changes in positive and negative directions $x_\pm$, we shall use the combinations

$$
\begin{aligned}
\widetilde{x}_+ &= \frac{1}{2}(x_+ + x_-) \\
\widetilde{x}_- &= \frac{1}{2}(x_+ - x_-).
\end{aligned}
\tag{4.67}
$$

The interesting variable will be the first one, $\widetilde{x}_+$, which is the average potential price change. The other variable, $\widetilde{x}_-$, measures the instantaneous disequilibrium between asks and bids in the order book. The transformed state vector

$$\widetilde{X} = \begin{pmatrix} \widetilde{x}_+ \\ \widetilde{x}_- \end{pmatrix} \tag{4.68}$$

evolves in the same way as X, except the matrices $T_\pm$ and $S_\pm$ are replaced with their transformed counterparts:

$$
\begin{aligned}
\widetilde{T}_+ &= \frac{1}{2}\begin{pmatrix} 3 & 1 \\ -1 & 1 \end{pmatrix} \\
\widetilde{T}_- &= \frac{1}{2}\begin{pmatrix} 3 & -1 \\ 1 & 1 \end{pmatrix} \\
\widetilde{S}_+ &= \frac{1}{2}\begin{pmatrix} 3 - \frac{1}{p} & 1 - \frac{1}{p} \\ 1 - \frac{1}{p} & 3 - \frac{1}{p} \end{pmatrix} \\
\widetilde{S}_- &= \frac{1}{2}\begin{pmatrix} 3 - \frac{1}{p} & \frac{1}{p} - 1 \\ \frac{1}{p} - 1 & 3 - \frac{1}{p} \end{pmatrix}.
\end{aligned}
\tag{4.69}
$$

Now we apply an approximation. The change of the variable $\widetilde{x}_+$ depends on this variable itself and on the value of the other variable $\widetilde{x}_-$. The approximation consists in replacing the actual value of $\widetilde{x}_-$ with its average over all realisations. We thus neglect correlations between $\widetilde{x}_+$ and $\widetilde{x}_-$. This is the same idea as that of the mean-field approximation, used in various disguises throughout all disciplines of physics. Here the mean-field approximation makes things even simpler than usual. Due to the symmetry between bids and asks in the matrix multiplicative process, the average of the difference variable $\widetilde{x}_-$ must be zero. The approximate process for the variable $\widetilde{x}_+$ is then

$$\widetilde{x}_+(t+1) = m_t\,\widetilde{x}_+(t) \tag{4.70}$$

where the multiplicative factors are given by the upper left corners of the matrices $\widetilde{T}_\pm$ and $\widetilde{S}_\pm$; therefore

$$m_t = \begin{cases} \frac{3}{2} - \frac{1}{2p} & \text{with probability} & p \\[2mm] \frac{3}{2} & \text{with probability} & 1-p\,. \end{cases} \tag{4.71}$$

The process $\widetilde{x}_+(t)$ is now the usual scalar multiplicative process repelled from zero, which we mentioned earlier. We can directly apply the results known to be valid for such processes. First, it can be shown that the precise way the variable is repelled from zero is immaterial as long as we are only interested in the tail of the distribution for $\widetilde{x}_+$. The stationary distribution $P(\widetilde{x}_+)$ for a large enough $\widetilde{x}_+$ obeys the equation

$$P(\widetilde{x}_+) = \frac{2p}{3 - \frac{1}{p}} P\left(\frac{2\,\widetilde{x}_+}{3 - \frac{1}{p}}\right) + \frac{2(1-p)}{3} P\left(\frac{2\,\widetilde{x}_+}{3}\right). \tag{4.72}$$

Second, the tail has the form of a power law. If we insert

$$P(\widetilde{x}_+) \simeq A\,(\widetilde{x}_+)^{-1-\alpha}, \qquad \widetilde{x}_+ \to \infty \tag{4.73}$$

into Eq. (4.72), we get the equation for the exponent α in the form

$$1 = p\left(\frac{3}{2} - \frac{1}{2p}\right)^\alpha + (1-p)\left(\frac{3}{2}\right)^\alpha. \tag{4.74}$$

There is always the trivial solution $\alpha = 0$, which follows from the conservation of probability in the process (4.70). The second, non-trivial solution of the equation (4.74) is

$$\alpha = 1 \tag{4.75}$$

as can be easily checked. The mean-field approximation gives the same result we have seen in the numerical simulations.

It is interesting to note that the same consideration cannot be used for the difference variable $\widetilde{x}_-$. The argument we used above now fails because the average of $\widetilde{x}_+$ is not zero. Even worse, the average of $\widetilde{x}_+$ is not well defined, because its distribution decays too slowly for $\widetilde{x}_+ \to \infty$.

The matrix multiplication process is indeed a very rudimentary model of the order book. Amazingly, it provides a very simple explanation of the power-law tail in the distribution of price changes. Quantitatively, the exponent is not very realistic, but, as we already said, this aspect can be considerably improved by taking more realistic functions for the average order density instead of the homogeneous one.

4.4 What remains

The modelling of order-book dynamics follows the line of studies on market microstructure [940–944]. However, the older works were concerned mostly with the existence of equilibrium and in choosing the optimal strategy. The agents are classified into three groups, the uninformed (or noise) traders, the informed traders (the insiders), and the market makers (or specialists). The agents should decide whether to issue a market order or to place a limit order and, if the latter is chosen, at what price. Given the rules of the market, the optimal action for informed traders and market makers is looked for [331, 945–949]. The heterogeneity in the agents' expectations plays a crucial role. The equilibrium is reached assuming that the agents think strategically and optimise their actions [950–955]. The question which is often asked is what the optimal strategy is for execution of orders, especially the large ones [956–965]. On the other hand, physical thinking advocates description in terms of inert bodies. That is why we did not spend time with these theories, although we acknowledge their importance.

There are many other works we do not have space to describe in detail. The computer simulations of order books were performed as early as 1976 by Garman [940]. Among the 'modern' approaches, we should especially mention the models based on queueing theory [575]. The price axis is supposed to be discrete, and at each site there is a queue of orders. These orders arrive and are cancelled or executed as usual. In the simplest setting, we consider only two queues, one for bids and one for asks. These queues are those closest to the actual price. New orders are deposited only in these queues. If one of the queues is emptied, say, on the ask side, the price is shifted one unit upwards (or downwards, if it were bids). Suddenly, on the bid side there is an empty queue and on the ask side there is a queue of a random length, which has been hidden deep in the book until now. Within this model, many quantities can be calculated analytically [84, 966, 967], including the distribution of waiting times between successive price changes and the diffusive behaviour of price at very large times. The model was further improved by considering deposition farther from the price, according to empirical data [968] and by considering correlations in the order flow [969]. Alternative queueing models also exist [970–972].

As we have seen, the dynamics of the full order book is so complex that analytic results are scarce. One of the attempts to tackle the problem mathematically is based on a set of coupled stochastic differential equations [973]. The information on the equilibrium state is thus obtained. We also amply discussed the schematic models, which do provide us with some analytical insight. A very nice approach in this direction is the application of the asymmetric exclusion process much studied in physics in the context of surface growth and transport phenomena very far from equilibrium [974]. Adapted to the dynamics of the order book, it gives the value of the Hurst exponent $H = 2/3$ exactly [975]. Another approach, analogous to the matrix multiplicative process of Sec. 4.3.3 was suggested, using stochastic differential equations with additive noise [324].

We have already mentioned that the uniform deposition model can be made more realistic and in fact connected with the Maslov model by taking deposition probability as a function of the distance from the price. Such studies have actually been done, for example supposing exponential [976–978] or power-law [261] decay of the deposition probability. Even closer to reality is the 'empirical behavioural' model of Mike and Farmer [915, 979] in which the deposition probability is adjusted according to empirical data and has a power-law dependence on the distance. Similarly, the cancellation rates are also chosen according to empirics. Moreover, the correlations in the signs of incoming orders (say, buy is sign +1, sell is sign −1), which are known to have very long memory, are implemented in the model, also to closely follow the empirical data. The model reproduces reality very well and was further developed [980]. One of the most notable improvements is the feedback from the volume present at the best bid or ask to the volume of incoming orders of the opposite type [289]. Another kind of feedback is obtained if the limit orders are deposited at a distance proportional to the spread [981, 982]. Interesting features emerge if the rates of deposition, cancellation, and execution are themselves fluctuating quantities [983].

The question why the deposition probability decreases as a power when we go farther from the price remains open. The mechanism responsible for this power law is probably related to the optimisation of investments performed by agents working at widely dispersed time horizons [263]. Actually it is reasonable to expect that the distribution of time horizons and (related to it) distribution of distances is maintained by equilibration, so that all agents expect just the same average gain, irrespective of the time horizon on which they act. This idea would certainly deserve a better formalisation.

Although we mentioned the opposition of agents thinking strategically vs zero-intelligence agents, these two approaches can be combined. For example, we can mix fundamentalist and chartist behaviour types in agents that compute where to place the limit order based on observation of past prices [984–986]. The agents can also learn from their past experience [986] and imitate each other [987].

The order book can also be modelled rather phenomenologically, introducing (directly unobservable) response functions [988] or assuming probability distributions behind the scene, which are then calibrated with respect to empirical data [312].

We shall leave the study of various other models [976, 989–1001] to the reader. She may find help in some review articles [890, 1002, 1003] or in a proceedings volume [47].

Problems

1. Consider, as a toy model, the one-queue order book. Suppose that only one side, with respect to the price, can evolve, for example the asks. Denote N_t the number of orders in the queue at time t. New orders arrive at rate α per unit time. If $N_t > 0$, one order is executed at rate γ or cancelled at rate βN_t. The execution of the last remaining order results in resetting the queue. Simultaneously, the price changes by one unit. The sign of the change will be clarified soon.
   With probability $1/2$ the number of orders in the queue after such an event remains zero. It means that from now on, the opposite side of the order book will evolve, e.g. whereas the queue represented asks before, now it will represent bids. It also implies that the sign of the next price move will be the opposite of the sign of the change which has just occurred.
   Or, with a complementary probability $1/2$, the length of the queue is set to a random number from a Poisson probability distribution $P(n) = e^{-\lambda} \lambda^n / n!$. In this case, bids remain bids and asks remain asks. Also the sign of the next price change remains the same as the sign now.
   What is the distribution of waiting times between price changes? Imagine that you wait a fixed time interval after a price change. You wait until another price change occurs or until the time expires, whatever comes first. Thus, three cases can happen, namely no price change, price change of the same sign as the first one, and price change of the opposite sign. What are the probabilities of these cases? For inspiration, look at Ref. [575], Chap. 5, and at papers [966, 967].

2. Modify the matrix multiplicative process so that, instead of constant order density on either side of the price, it assumes linear dependence. What is the value of exponent α?

5
Minority game

5.1 Rules of the game

Imagine you make a winter trip to the Alps with two of your friends who are economists. The weather is terrible, falling snow prevents you from seeing more than a few steps ahead, and the wind and frost are exhausting. Finally you reach the chalet, happy to be alive, and soon the fire and hot, thick soup make you feel like you are in heaven. But if the foul weather continues for some days, you will need to invent another form of excitement.

Soon you recourse to your eternal theme, the origin of stock market fluctuations. You start playing a tiny Wall Street next to your fireplace. In each round, each of you three decides in secret whether she wishes to buy or sell a single share to or from an inexhaustible share-owning institution. Then the decisions are revealed, and virtual trading takes place. The price of the shares evolves, depending on whether there are more of you wishing to buy than to sell, or vice versa. For example, you can agree that the price rises by 1 per cent if the majority buys, and drops by 1 per cent if the majority sells.

Obviously, if one of you manages to do the opposite action than the remaining two, she gains an advantage. Indeed, either she sells at a higher price, because the other two caused excess demand, or she buys at lower price if the majority decided to sell. You soon realise that the winning strategy is to be in the minority, and you simplify the rules of the game you have just invented so that those in the majority group lose a point while the one in the minority receives a point. On average, you are always losing, but the one of you who is smart enough can gain at the expense of the other two. You will probably repeat the game again and again and try to guess the actions of the others by observing the outcomes of the past rounds. As time passes, you will probably learn some skills, or at least you should not lose as much as in the beginning. When the sun reappears, you shall wake up as experts in this bad-weather entertainment.

The story I have just told you should be sufficient for you to capture everything about how the game, now called minority game, is played. As with all good models in physics and economics, it is extremely simply formulated and gives extremely rich and unexpected behaviour. This chapter will reveal some of its mysteries.

5.1.1 Inductive thinking

To believe in a human guided by totally rational reasoning requires a good deal of abstraction. Fully rational decisions simply cannot be found in reality. Yet the idea is so deeply rooted in orthodox economic theory that it is assumed at least as a benchmark or a starting point of developing a corrected theory.

Types of irrationality

Very often the limits imposed by Nature on our rational capabilities are described as deficient information. If we knew what the governor of the central bank decided just an hour ago, we would be better off than those of our colleagues who have to wait until tomorrow morning for the official communication. Indeed, more transparency in the business environment would likely be beneficial. The analysts, including empirical econophysicists, would certainly be happy if they knew all the details about the myriads of transactions, from huge to tiny, which are now veiled by business secrecy.

Sometimes the information is missing because no one has ever had the idea to store it. Historical commodity prices from times before newspapers with regular records were established have been lost. Even now lots of communication goes by phone (but note that mobile calls are archived, at least temporarily, in case the police need them later!), and the precise timing of many events escapes human attention.

Even when the information is there, it may not be enough to imply a reliable statement on, e.g. the probability that a business will fail next month. Of course, human actions are also influenced by random noise, weather, car accidents, falling rocks, and the like. In short, the sources of irrational aspects in our decisions are multiple. But we can still hope that in principle we can improve our ability of deduction by accumulating more and more new information. Such an attitude relies on an essentially static worldview, as if there were an objective, universally valid optimal decision, and a supernatural being possessing all the information about the system would be able to rationally deduce the best solution. Our corrupted terrestrial bodies can never be so smart, but we can still approach the ideal state almost indefinitely. This caricature depicts the optimistic programme of the Enlightenment. We can classify it as boundedly rational deductive thinking. Modelling such a situation implies 'simply' adding an extra stochastic element into an optimisation problem corresponding to a fully rational choice. However, the fact that the rationality in our decisions is bounded does not mean mere noise in our decisions. The lack of information makes some decisions impossible and changes the economic setup, as shown by Akerlof in his famous article on the market of 'lemons' [127] and also stressed by other important economists [854].

On the other hand, there is a deeper hindrance to our rational aspiration. Everyone must agree that all economic decisions are made for the future, while all the information they are based upon resides in the past. Inferring future events from past ones is an inductive rather than a deductive procedure.

However the induction may seem equivalent to deduction with limited information, especially to those who strive to describe dynamics as statics in one more dimension, we believe such a view misses an essential distinction.

Indeed, while in 'space' dimensions there may be multiple instances of the same process, e.g. many customers buying cars, and we can build statistics on such an ensemble, the time course is always unique. Surely, nobody can average over several alternative French Revolutions to set bounds on the likelihood of Napoleon's takeover. Back to our example of a car market, the customers viewed as realisations of a stochastic process are so strongly correlated in the time domain (virtually none will buy a Porsche before 1900!) that the story of adding time as a mere extra dimension is hardly productive.

History never repeats itself exactly; and if we want to relate causes with consequences, we face a much more severe shortage of data than if we, for example, statistically analyse customer preferences at a given period of time. Moreover, if we are lucky and succeed in distilling a causal chain in the course of economic events, such knowledge obviously changes the future, further weakening the effect as more people know of its existence. On the contrary, knowing the distribution of products sold in one country can by no means change the distribution in another country measured at the same time.

To sum up, inductive thinking is a much more difficult endeavour than mere deduction with limited information. That is why induction requires specific approaches and specially designed models.

Looking more closely at how inductive thinking works in real situations, we soon notice that the decisions are based on pattern recognition. Humans possess certain ensembles of 'images' stored in their memories, associating certain outcomes with preceding situations. The images are not collected systematically nor are they rationally classified. We cannot be too far wrong if we consider them random. The images, or patterns, are then used to make decisions by matching them with reality and choosing among those which are compatible with the current situation. This implies that we can attribute a quality to the images, highlighting those which suggested a beneficial decision in the past. The important point in inductive thinking is that we constantly assess and re-assess the mental images we carry in our heads, and this way we continuously react to the changing rules of the world around us. Even if the teacher at a high school taught us a virtually random collection of facts, as soon as we master the art of selecting among them what is most likely useful now, and even more importantly, if we are capable of constant updating of the usefulness of what we know, we cannot fail totally in our adaptation for life.

El Farol bar

W. Brian Arthur deserves merit for bringing inductive thinking into the economy in terms of a simple model [1004]. It was inspired by the El Farol bar in Santa Fe, which attracted fans of live Irish music every Thursday night. Suppose there are 100 people who consider going to the bar, but there are only 60 seats available. It is annoying to go to the bar if it is overcrowded, and it is also a pity to stay home while vacant chairs are longing for us there. Should we stay or should we go?

But how can the individuals coordinate so that the bar attendance is close to its capacity if they cannot communicate with each other before they decide about their night programme? We would expect that the number of visitors will be completely random, and the source of entertainment will be used ineffectively. But if the people are given the record of the number of visitors in the past weeks, they may infer inductively how many are likely to be there next time. If the prediction is below 60, the decision is to go; if it exceeds 60, it is better to stay.

Arthur's pioneering idea was that we can model the situation supposing each person has a certain fixed, and relatively small, number of predictors. Based on the past sequence of attendances, say ... , 51, 36, 82, 45, 66, 49, the predictors compute the next attendance. Predictors can process the information of the past attendances in many ways. For example they can say that next time the attendance will be a) the same as the last one, or b) take a rounded average of the last three attendances, or c) the same as p weeks ago (detecting cycles of period p), and d) many more rules the reader can surely invent herself. If the predictor expects attendance below 60, it suggests that the agent go to the bar; otherwise it suggests she should stay home.

An important point is that the agents are given several such predictors and in each step attribute plus or minus points to the predictors, depending on their success in suggesting the correct action. The agent then decides to use the predictor which has maximum points. Such a mechanism emulates the bounded intelligence of the music fans. They will never know for sure what to do, but they will adapt to the conditions of the world. Let us stress again that the predictors are random and do not contain anything that could be a priori useful to guess the correct answer.

In a computer simulation, Arthur showed that the agents soon self-organise so that the attendance fluctuates around the optimal value 60. It is a good result for computer boxes who do not know anything about the optimum and whose 'education' is in fact randomly acquired garbage!

The equilibrium is not found by a predefined algorithm, and the imprecision in the inference is not due to an external noise. The correct answer emerges spontaneously from the collective action of the agents. The setup is very close in spirit to solving problems by genetic algorithms or genetic programming (see Box 3.7). Indeed, the genetic algorithm selects good answers from a pool of arbitrary shots by an artificial selection process, while in the El Farol problem the solution emerges among random and individually senseless predictors.

Moreover, the agents themselves provide both the signal to be deciphered and the noise obscuring it. Thus, they constitute a complete self-sustained ecology, much the same as humans themselves provide the environment and determine conditions of life for humankind. The microcosm of El Farol is a distillate of the whole human society.

5.1.2   The algorithm

Arthur's analysis of the El Farol bar problem was a lucid demonstration of how the agents find the equilibrium by induction. When we turn to the investigation of the fluctuations around the stationary state, things become much more complicated. The complexity stems from the intrinsic frustration (see Box 5.1 for a precise definition). Indeed, it is impossible to find a general optimum strategy for the El Farol bar visitor, as this would imply that all people would take the same action, and the attendance would be either zero or full. Surely that is not close to optimum resource utilisation. The idea of a representative agent is of no use here.

Instead, we need to know how the agents adapt to each other, keeping in mind that their strategies are better or worse only conditioned to other agents' actions. This way we can investigate the crucial question of how close the system can approach the optimum or how large are the fluctuations are around the equilibrium.

Frustration $\boxed{\text{Box 5.1}}$

Let us explain the concept of frustration using a simple example of a spin system, as shown in the following figure.

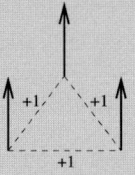 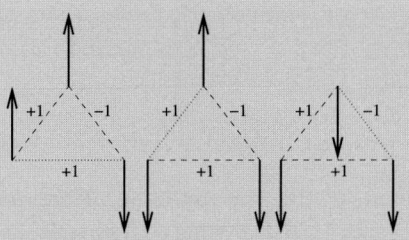

Ferromagnetic case: $1 \times 2 = 2$ groundstates. All bonds satisfied.

Frustrated case: $3 \times 2 = 6$ groundstates. Unsatisfied bond is shown as a dotted line.

If three Ising spins are bound together by ferromagnetic interaction so that the Hamiltonian is $H = s_1 s_2 + s_2 s_3 + s_3 s_1$, the ground state is obviously the uniform configuration $s_1 = s_2 = s_3$. All three bonds can be considered 'satisfied', because all three pairs of spins individually are in their lowest-energy configuration. The groundstate is twice degenerate due to the global symmetry; the energy does not change if we flip all spins simultaneously. On the other hand, if one of the bonds is antiferromagnetic, so that $H = s_1 s_2 + s_2 s_3 - s_3 s_1$, there is no configuration in which all three bonds are satisfied. Such a situation is called frustration. Since there are three equivalent choices for the unsatisfied bond, there are three different groundstate spin configurations (taking into account the spin-flip symmetry, we have altogether six groundstates).

Frustration is the source of complex behaviour of many systems, especially disordered ones. As already seen in our example, frustration leads to proliferation of equilibrium states. For example in spin glasses the number of equilibria increases exponentially with the number of spins.

Minority wins

To this end, in the work of Challet and Zhang the rules were further simplified [1005, 1006]. To keep the optimum state trivial, we assume that the agents can take one of two possibilities. Just as the bar visitors could go out or stay home, here we let the agents go one step up or down. To keep in touch with reality we can imagine the step up to be an order to buy a unit of a commodity, and conversely the step down to be an equal-sized sell order.

Having N agents, denote by $a_i(t) \in \{-1, +1\}$ the action of the i-th agent at time t. The aggregate movement of the whole ensemble of agents

$$A(t) = \sum_{i=1}^{N} a_i(t) \tag{5.1}$$

will be called attendance, in analogy with the El Farol bar model, although now it is rather related to the shift in the commodity price. If the attendance is positive, the price rises and those who decided to sell, i.e. agents who acted as $a_i(t) = -1$, can feel rewarded, as they get a better price than one step before. On the contrary, buyers, $a_i(t) = +1$, suffer a loss, as they spent more money than they would if they had bought one step earlier. Conversely, negative attendance rewards buyers and punishes sellers. Therefore, it is always beneficial to go against the trend and stay in the minority. This is why the model was named 'minority game'. To avoid ambiguity in telling which group makes up the minority, we always expect that the number of agents N is odd. Obviously, the optimal situation is reached when the size of the minority group is as large as possible, or, equivalently, when the attendance is as close as possible to zero.

The minority reward rule is expressed formally as the prescription for the update of agents' wealth $W_i(t)$ as follows

$$W_i(t) - W_i(t-1) = -a_i(t) \operatorname{sign} A(t). \tag{5.2}$$

Different variants of the rule can be implemented, replacing the function $\operatorname{sign} A(t)$ by a more general form $G(A(t))$, where $G(x)$ is an arbitrary anti-symmetric non-decreasing function, $G(x) = -G(-x)$. It can be interpreted as the impact of the demand-offer disequilibrium on the commodity price. A very natural choice is the linear price impact $G(x) = x$, and we will use it in the analytical solution of the minority game. As will be stressed later, the precise form of $G(x)$ is of little importance, and one may choose one or another depending on the particular question asked.

The reward rule in the minority game is very transparent and simple to implement because of its immediacy. The players collect their points at the same moment (or, to be more precise, an infinitesimally short time after) they make their actions. This is perfectly appropriate for bar attendance; but if we want to interpret the minority game as a trading model, a complication arises. We should note that the gain expressed by Eq. (5.2) is potential, rather than actual. Indeed, in order to know how much the seller really earned we would need to know at which price she acquired the commodity in a more or less distant past. Similarly, the buyer will know the outcome only after she sells the commodity back in the future. To compute the financial effect of an action at time t, it is necessary to know the price movements within a longer time span, not just the immediate change suggested by the attendance $A(t)$ right at the same step as the action was taken. In the minority game, such multitemporal nature of stock-market activity is drastically simplified.

Multiple strategies

So far we have dealt with the gain the agents receive or the loss they must suffer for their actions. Obviously, they want to act so that they get rewarded, but how to choose what to do? To this end, each agent is given a set of S strategies (mostly $S = 2$ will be enough) predicting the right action in the next game, based on the outcomes in the last M steps. The parameter M measures the length of the agent's memory. As the game goes on, the agents learn which of the S strategies is worth using. We shall soon explain how they manage that, but now let us see how the strategies are devised.

To reduce the information content to a manageable amount, we shall not store the full sequence of attendances $A(t')$, but only the series telling us which choices were the profitable ones

$$\chi(t') = -\operatorname{sign} A(t') \tag{5.3}$$

for $t' = t-1, t-2, \ldots, t-M$. We shall call $\chi(t)$ the outcome of the game at time t. A strategy is a prescription which predicts the next profitable action from the binary string of M past outcomes

$$\mu(t) = [\chi(t-1), \dots, \chi(t-M)] \tag{5.4}$$

kept in the agents' memory. Therefore, the strategy is a map $\{-1,+1\}^M \to \{-1,+1\}$. Distinguishing among the S strategies of the agent i by index s (we can especially choose $s \in \{-1,+1\}$ for $S = 2$) we denote $a_{s,i}^\mu$ the action suggested by s-th strategy of agent i, provided the sequence of past outcomes is μ.

The number of possible sequences of outcomes is $P = 2^M$, and the number of all possible strategies is a very large number 2^{2^M} already for a moderate memory length M. This means that the collection of agents is indeed very heterogeneous, and it is practically impossible to find two agents equipped with the same strategies. On the other hand, the strategies, albeit different, may be significantly correlated. The strategies in a pair can, for example, differ only in a few cases of the past outcomes μ, which may never even occur in the course of the game. If that happens, the agent effectively has only one strategy. If we imagine each a strategy as a P-component vector, there can be at most P mutually orthogonal, i.e. uncorrelated strategies. The effective size of the strategy space P should then be compared with the number of agents N among whom the strategies are distributed. So, we can anticipate that the properties of the minority game will depend on the scaling parameter

$$\alpha = \frac{P}{N} = \frac{2^M}{N} \tag{5.5}$$

when both the number of agents and the memory length go to infinity.

Dynamics of scores

We have just understood that the strategies take responsibility for the agents' behaviour, and every agent has to choose somehow which of her S strategies is the best at that moment. Now we must decide how to measure the quality of the strategies. A very simple method is comparing the outcomes of the game with the predictions of all strategies. Every strategy of each agent keeps a record of its score $S_{s,i}$ and receives $+1$ or -1 points depending on whether it suggested the right or wrong action, respectively. The strategy with maximum points is then actually used by the agent. Another important thing is that the strategies which are not currently used are continuously tested, so that the player learns which of her patterns of behaviour is better suited to current circumstances.

We can formalise this rule for $S = 2$ by introducing the difference $q_i(t) = (S_{+,i} - S_{-,i})/2$ in scores of the two strategies of the agent i. (The factor $1/2$ is used for future convenience). First, we introduce some notation which will be useful throughout the rest of this chapter. We denote

$$\omega_i^\mu = \frac{1}{2}\left(a_{+,i}^\mu + a_{-,i}^\mu\right)$$
$$\xi_i^\mu = \frac{1}{2}\left(a_{+,i}^\mu - a_{-,i}^\mu\right) \tag{5.6}$$
$$\Omega^\mu = \sum_{i=1}^N \omega_i^\mu.$$

Ergodicity breaking   Box 5.2

A dynamical system is considered ergodic if, starting from a random initial configuration, sooner or later it visits all configurations allowed a priori. In that case, the time-averaged quantities do not depend on the initial condition. Knowing the time spent at individual configurations, we can replace the time average by the statistical average governed by an appropriate measure on the configuration space. The measure assigns to each set of states a probability that is proportional to the time spent in these states during the evolution of the system.

If, on the contrary, there are configurations which are a priori allowed but dynamically inaccessible from a given starting point, the ergodicity is broken. In a non-ergodic system, the time averages do depend on the initial conditions.

Given the scores' differences, the agents choose from among their strategies according to

$$s_i(t) = \operatorname{sign} q_i(t) \tag{5.7}$$

and their actions are

$$a_i(t) = a_{s_i(t)}^{\mu(t)} = \omega_i^{\mu(t)} + s_i(t)\xi_i^{\mu(t)}. \tag{5.8}$$

The attendance follows immediately from (5.8)

$$A(t) = \Omega^{\mu(t)} + \sum_{i=1}^{N} \xi_i^{\mu(t)} s_i(t). \tag{5.9}$$

The score differences are updated in each step according to

$$q_i(t+1) - q_i(t) = -\xi_i^{\mu(t)} \operatorname{sign} A(t) \tag{5.10}$$

which can be written, using (5.7), and (5.9), in a compact form

$$q_i(t+1) - q_i(t) = -\xi_i^{\mu(t)} \operatorname{sign}\left(\Omega^{\mu(t)} + \sum_{j=1}^{N} \xi_j^{\mu(t)} \operatorname{sign} q_j(t)\right). \tag{5.11}$$

Eq. (5.11), together with the prescriptions (5.3) and (5.4) for the memories $\mu(t)$, fully describes the minority game.

To summarise, the minority game consists of the coupled dynamical processes $q_i(t)$ and $\mu(t)$ defined above. Note that the actual wealth of the agents collected according to (5.2) does not enter the dynamics. It is rather a secondary by-product, while the virtual points attributed to the strategies play the primary role.

Initial conditions

The last piece completing the picture of the minority game is the question of its initial conditions. The question is less trivial than one might think. Indeed, we shall soon see that the most intriguing feature of the basic minority game is the presence of a phase transition from an efficient but non-ergodic regime to an inefficient ergodic phase. The ergodicity breaking implies dependence on an initial condition which persists for an infinitely long time (see Box 5.2).

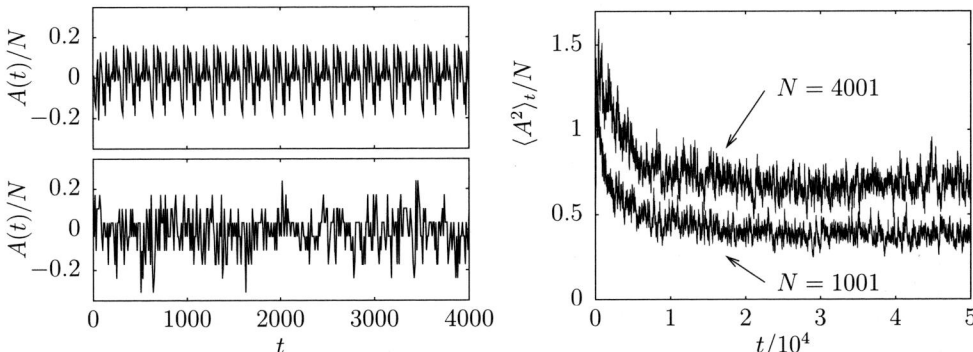

Fig. 5.1 Time series for the attendance (left two panels) and local time average of the square attendance (right panel). In the upper left panel, the memory is $M = 5$, and the number of agents is $N = 1001$; while in the lower left panel $M = 5$ and $N = 29$. In the right panel, the memory is $M = 10$ and the number of agents $N = 1001$ (lower line) and $N = 4001$ (upper line). The parameter of the time average is $\lambda = 0.01$.

The canonic choice of the initial conditions is randomly drawn $\mu(0)$ with uniform probability distribution, and all strategies have the same (e.g. zero) score. It means $q_i(0) = 0$ for all agents i. All the results of numerical simulations presented in the next section will use these initial conditions. We shall briefly comment on the influence of non-zero initial values for q_i in due course.

Given the assignment of strategies $a^{\mu}_{s,i}$ and initial conditions, the dynamics of the minority game is fully deterministic. In actual simulations we must average the results over a very small subset of all possible strategy assignments and choices of initial memory string $\mu(0)$. This introduces a casual element into the otherwise non-random dynamics. We shall see later that adding some stochasticity into the minority game rules, especially to the strategy choice (5.7), makes the game 'softer' and more amenable to analytic investigation.

5.1.3 Phase transition

It is indeed very easy and straightforward to embody the algorithm of the canonic minority game, as expressed by Eqs. (5.3), (5.4), and (5.11), in computer code and observe the results. The minority game owes much of its appeal to the simplicity of the code and the ease with which a newcomer can touch and feel the behaviour of the agents.

Adaptation

The first thing to observe is the evolution of attendance $A(t)$. Already, on a very qualitative level we can distinguish between two types of behaviour, as seen in Fig. 5.1. Let us fix the memory length M and vary the number of agents N. If N is large enough, the time series sweeps through an initial transient state and then becomes periodic. On the contrary, if N is small, the attendance follows a rather chaotic course.

The most important quantitative measure in the minority game is the volatility

$$\sigma^2 = \lim_{T \to \infty} \frac{1}{T} \sum_{t=1}^{T} A^2(t). \tag{5.12}$$

It tells us how efficiently the agents utilise the available resources. Indeed, if exactly half of the population is in the minority, the average gain of all agents is maximum (zero), and the volatility reaches its minimum value, which is zero. In the simulations, we always observe positive volatility, and it is easy to check that σ^2 grows when the average gain decreases.

The principal question is whether the agents are able to adapt online to the behaviour of other agents so that the volatility is kept as small as possible. To examine that question, we introduce a time-local version of the volatility, computed during the simulations, as $\langle A^2 \rangle_t = \lambda A^2(t) + (1 - \lambda) \langle A^2 \rangle_{t-1}$. This means averaging the volatility over roughly $(1 - \lambda)^{-1}$ last steps, with the weight decreasing exponentially when we go deeper into the past. In the example shown in Fig. 5.1 we can clearly see how the initially large volatility decreases, until it is saturated at a certain level, which depends on the parameters N and M. The suppression of the volatility is clear evidence of self-organisation due to individual learning of the agents.

Minimum of volatility

The greatest surprise of minority game comes when we plot the dependence of stationary volatility on the memory length M. We can see a pronounced minimum, indicating that there is a certain optimum size of the memory, beyond which the agent does not become more 'intelligent'. On the contrary, too much information leads to confusion [1007, 1008].

Interestingly, if we plot the volatility per agent σ^2/N against the scaling variable $\alpha = 2^M/N$, all results fall onto a single curve, as shown in Fig. 5.2. The minimum sharpens as we increase the system size and eventually approaches a singularity at the critical value $\alpha_c \simeq 0.34$. This marks a dynamic phase transition of a rather unusual character. To learn more about it, we must first determine the corresponding order parameter.

Efficiency and order parameters

The market is considered efficient if there is no information left in the price signal. This means that you cannot use some freely accessible information and make a profit from it. In a similar spirit, we can investigate the efficiency of the agents in extracting and destroying the information stored in the sequence of attendance. More precisely, we shall ask what the average sign of the attendance is, with fixed memory pattern μ

$$\langle \text{sign} A | \mu \rangle = \lim_{T \to \infty} \frac{\sum_{t=1}^{T} \delta_{\mu\,\mu(t)} \text{sign} A(t)}{\sum_{t=1}^{T} \delta_{\mu\,\mu(t)}}. \tag{5.13}$$

If for a pattern μ the average is significantly different from zero, it is possible to predict which will be the most probable winning side. If the non-zero average persists, it means

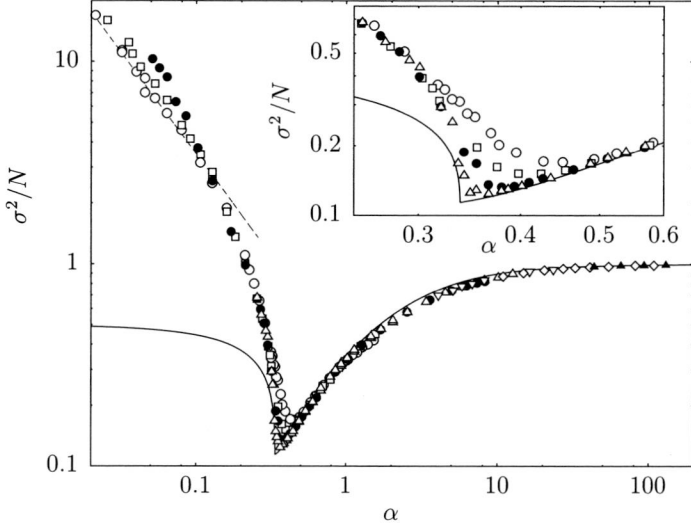

Fig. 5.2 Dependence of volatility in the minority game on the scaling parameter $\alpha = 2^M/N$, for several combinations of memory M and number of agents N. In the inset, detail of the same data close to the transition point. The symbols correspond to memory length $M = 5$ ($\bigcirc$), 6 ($\square$), 8 ($\bullet$), 10 ($\triangle$), 12 ($\triangledown$), 14 ($\diamondsuit$), and 16 ($\blacktriangle$). The solid line is the result of analytic calculations using the replica method. The dashed line is the dependence $\sim \alpha^{-1}$.

that the agents are not able to use this information, and the system as a whole is not efficient [1008]. The global measure of efficiency is the average over all 2^M memory patterns

$$\theta^2 = \frac{1}{2^M} \sum_{\mu=0}^{2^M-1} \langle \operatorname{sign} A | \mu \rangle^2. \tag{5.14}$$

It may be appropriately called predictability, as it measures the amount of remaining information in the time series [1009].

The data shown in Fig. 5.3 demonstrate that in the crowded phase, $\alpha < \alpha_c$, the system is efficient, as all information is eliminated from the signal. We can see it qualitatively, observing rather large deviations from zero in $\langle \operatorname{sign} A | \mu \rangle$ when the number of agents is relatively small, and also quantitatively, in the dependence of θ^2 on the scaling parameter α. In the crowded and efficient phase, the predictability is virtually zero, while for $\alpha > \alpha_c$ it continuously grows. Therefore, θ^2 can be considered as an order parameter.

Frozen agents

It often happens in dynamic phase transitions that the order parameter is not unique. It is also the case here. Let us look at the way an agent uses her two strategies. She can alter them more or less regularly, so that both strategies are used equally often.

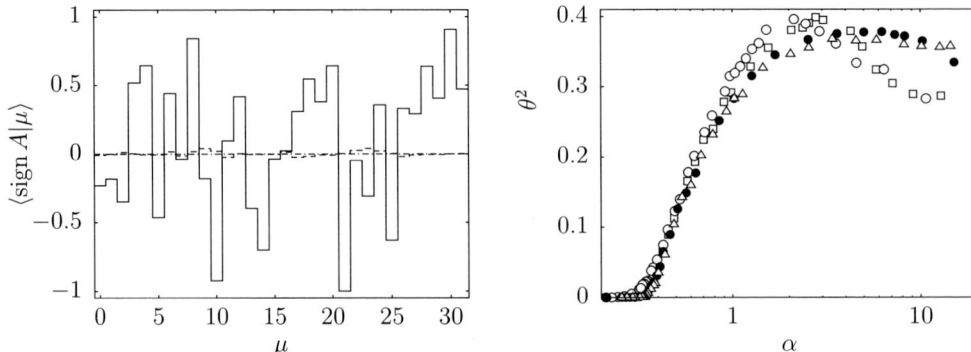

Fig. 5.3 In the left panel, the average sign of the attendance depending on the actual memory pattern μ is shown, for $M = 5$. The number of agents was $N = 51$ (solid line), 81 (dashed line), and 101 (dotted line). In the right panel, we can see the average predictability of the time series, measured according to (5.14), for $M = 5$ ($\circ$), 6 ($\square$), 8 ($\bullet$), and 10 ($\triangle$).

But it may also happen that one of the strategies is chosen more often than the other. After a long enough time the preference for one strategy should be clearly visible in the statistics. We can call this effect polarisation of the agents. The quantitative measure of the polarisation of agent i is the time-averaged difference in the scores of the two strategies

$$v_i = \lim_{t \to \infty} \frac{1}{t} |S_{+,i}(t) - S_{-,i}(t)|. \tag{5.15}$$

In Fig. 5.4 we can see a histogram of the agents' polarisation. For $\alpha > \alpha_c$ it has two pronounced peaks, showing that there are indeed two types of agents, the first switching strategies all the time, the other having one preferred strategy that is used all the time. The pressure induced by the difference in scores need not be very high, but in the histogram it is clearly visible. The latter agents are called frozen, and we should ask how many of them are there. Looking again at Fig. 5.4 we observe that the higher peak in the histogram vanishes in the efficient phase $\alpha < \alpha_c$, so there are no frozen agents. Quantitative statistics are shown in the right panel of Fig. 5.4. Surprisingly, the fraction of frozen agents ϕ grows when we decrease α, reaching its maximum at $\alpha = \alpha_c$, and then drops discontinuously to zero.

The fraction of frozen agents, too, can be chosen as an order parameter, as it vanishes in one phase and stays non-zero in the other. And we should not be confused by the fact that the transition looks continuous, i.e. second order, from the point of view of order parameter θ^2, while the second order parameter ϕ indicates a discontinuous, i.e. first-order transition. The classification pertinent to equilibrium order transitions has limited applicability when the transition is a dynamic one and should not be taken too seriously.

How many patterns occur?

There is also a third characteristics that distinguishes between the efficient and inefficient phases. If we measure the frequency with which a certain memory pattern

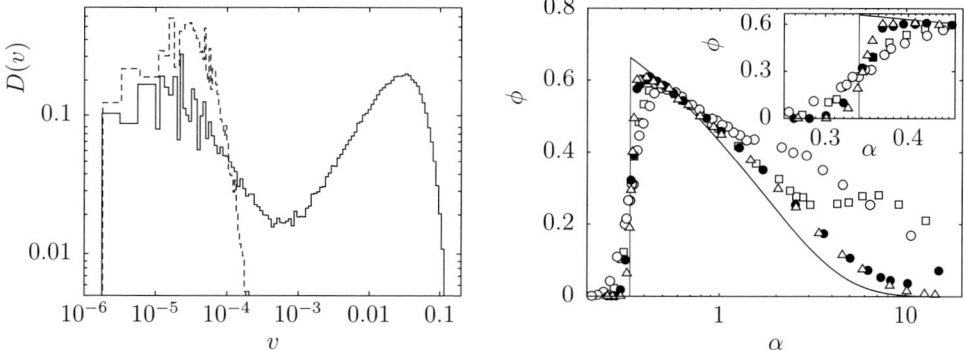

Fig. 5.4 In the left panel, a histogram of the polarisation of the agents, for $M = 8$ and number of agents $N = 251$ (solid line) and $N = 901$ (dashed line). In the right panel, the fraction of frozen agents for $M = 5$ ($\bigcirc$), 6 ($\square$), 8 ($\bullet$), and 10 ($\triangle$). The line is the analytic result from the replica approach. In the inset, detail of the transition region.

appears, we find that it is quite homogeneous in the efficient phase, while in the inefficient one it is increasingly uneven.

Denoting $p_\mu = \lim_{T \to \infty} \sum_{t=1}^{T} \delta_{\mu(t)\mu}/T$ the relative frequency with which μ is found in true dynamics, we can measure the entropy of the distribution established in the course of the dynamics

$$\Sigma = -\sum_{\mu} p_\mu \ln p_\mu \qquad (5.16)$$

and compare it with the entropy $\Sigma_0 = M \ln 2$ of the uniform distribution. We can see in Fig. 5.5 that the difference $\Sigma_0 - \Sigma$ is indeed negligible in the symmetric phase, $\alpha < \alpha_c$, while in the asymmetric phase it is positive and grows with α. It means that the space of memories is visited non-homogeneously. For very long memory M, the effective number of actually occurring μ's can even be rather small, as shown in the inset of Fig. 5.5. The data suggest that for fixed N the effectively visited volume is scaled as $e^\Sigma \sim 2^{\gamma M}$ with $\gamma \simeq \frac{1}{2}$; so the relative fraction of visited μ's shrinks to zero as $2^{-(1-\gamma)M}$ for $M \to \infty$.

5.2 Towards an analytical solution

While numerical experiments with the minority game are readily accessible to anybody who has elementary skills in computer programming, some of the principal questions can be answered only by analytical approaches. Are the efficient and inefficient phases separated by a true phase transition, or is it only a crossover phenomenon? Is the efficiency in the low-α phase perfect or just very high? Can we give an exact proof of the scaling property, i.e. that in the thermodynamic limit $M, N \to \infty$, all relevant quantities depend on the memory length and the number of agents only through the parameter $\alpha = 2^M/N$?

It turns out that tackling the minority game analytically requires special and rather sophisticated techniques. Despite the big challenges posed by the novelty of the dynamic process underlying minority game, a large part of the physics behind it is now understood. There are essentially two approaches being used in the analytical study of minority game. The first one is the replica method (see Boxes 5.4 and 5.5; for more information, see Ref. [1010]), starting from a mapping of the dynamical process on an effective equilibrium problem. The other one fully accounts for the dynamics using the generating functional method [1011, 1012]. Here we deal with the replica method only. As for the generating functional method, we recommend the book [34], devoted to its use in the minority game.

But before proceeding further, we must introduce some simplifications and modifications of the canonical minority game, which do not change the essence of the model, but make it more amenable to solution.

5.2.1 A few modifications

Quasi-irrelevance of memory

Historically, the first important step forward was the observation that in numerical simulations the results remain (nearly) the same if we replace the true dynamics of the memorised outcomes $\mu(t)$ as in (5.4), by memories μ drawn randomly from the set of all $P = 2^M$ possibilities [1013, 1014]:

$$\text{Prob}\{\mu(t) = \mu\} = \frac{1}{P}. \tag{5.17}$$

This seemingly formal step constitutes a fundamental change of view. We may interpret $\mu(t)$ as an external information, unrelated to the previous results of the game and essentially random. The only important point is that all agents are given identical information. We can also consider $\mu(t)$ as a task or requirement the agents have to fulfil. The complexity of the game stems from the fact that all agents must solve the same task simultaneously, individual agents' solutions interfere with each other, and there is no general solution available for all. That is the built-in frustration in minority game, as we alluded to previously.

The independence of $\mu(t)$ on the real history is indeed a crucial simplification, because the processes $q_i(t)$ are now decoupled from the process $\mu(t)$. The original dynamics turns into a Markov process, and all memory effects are washed out. When we find it convenient, we shall call this modification the Markovian minority game.

As the original minority game gives nearly the same results as the Markovian version, we can indeed say that the memory in the minority game is (nearly) irrelevant. However, claims about the irrelevance of memory cannot go too far, since the true dynamics of memories does not visit all points in the space of μ's with equal probability, as we saw in the previous paragraph.

Nevertheless, replacing the true dynamics $\mu(t)$ by randomly drawn μ's, we keep all essential complexity of the minority game in place. The Markovian process we get is quantitatively slightly different from the original minority game, but we gain much better access to analytical tools.

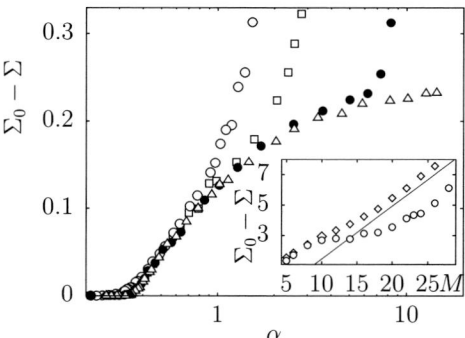

Fig. 5.5 Entropy of the distribution p_μ of the frequency of the memory patterns μ, relative to the entropy of the uniform distribution Σ_0. In the main figure, the memory length is $M = 5$ ($\bigcirc$), 6 ($\square$), 8 ($\bullet$), and 10 ($\triangle$). In the inset, the same quantity for a small number of agents, $N = 3$ ($\lozenge$) and $N = 5$ ($\square$). The straight line has the slope $\ln \sqrt{2}$, indicating the behaviour $e^\Sigma \sim 2^{M/2}$ for large M.

Besides the technical advantage, there is an interesting lesson we can learn immediately. At the beginning we assumed the agents learn inductively how to predict future attendance. In the Markovian minority game there is no future to predict, or more precisely, there is no past on which such a prediction could be based. Instead, the agents simply adapt to each other under various external circumstances, embodied in the binary strings μ. We started with inductively thinking individuals, but we find that their behaviour looks very much like they were optimising their position within an environment made up of all the remaining population. Deductive thinking regains a part of its credit.

Thermal minority game with linear payoff

So far, the strategy was chosen in a deterministic way according to the difference in strategy scores (5.7). A technically useful modification consists in allowing probabilistic choice of the strategy, giving smaller or larger preference to the one with a higher score. Similarly as in Monte Carlo simulations of equilibrium systems, we introduce a parameter Γ, analogous to the inverse temperature [1019–1026]. If the difference of the strategies' scores for the agent i is $q_i(t)$, she chooses the strategy s with probability

$$\text{Prob}\{s_i(t) = s\} = \frac{e^{s\,\Gamma\,q_i(t)}}{2\cosh\Gamma\,q_i(t)}. \tag{5.18}$$

The value of parameter Γ measures the level of randomness in the choices. For $\Gamma \to \infty$ the dynamics becomes deterministic, and we recover the original minority game prescription. If the strategy was selected many times with the same score difference, we would get the following average choice

$$\langle s_i(t)\rangle = \tanh\Gamma q_i(t). \tag{5.19}$$

Spin glasses and neural networks Box 5.3

are now classical examples of systems where quenched randomness plays a fundamental role [1010]. Spin glasses as real systems are typically noble metals (Au, Ag, or Cu) alloyed with a small percentage (one to five per cent) of a magnetic impurity, like Fe or Mn. They exhibit a peculiar magnetic state below a critical temperature, where ergodicity is broken in a large number of different stable states unrelated to each other by any obvious transformation. The most studied model of a spin glass is the Sherrington-Kirkpatrick model [1015] described by the Hamiltonian

$$H = -\sum_{i,j} J_{ij} S_i S_j \qquad (*)$$

where $S_i \in \{-1, +1\}$ are Ising spins, and J_{ij} are quenched random couplings drawn from a normal distribution.

Neural network models represent a drastic simplification of the function of the brain, but can serve as an effective technical basis for devising artificial neural circuits. The neurons are described by Ising spins, where the value -1 is interpreted as an inactive and $+1$ as an active state of the neuron. In the Hopfield model [1016] of the neural network we can identify the Hamiltonian, which has a formally identical form as $(*)$, but the couplings have the following form

$$J_{ij} = \sum_{\mu} \xi_i^{\mu} \xi_j^{\mu} \qquad (**)$$

where $\xi_j^{\mu} \in \{-1, +1\}$ are chosen randomly. The canonical interpretation says that ξ_j^{μ} are patterns stored in the memory of the neural network which can be retrieved by suitable dynamics imposed on the neurons [1017].

Comparing the Hamiltonian $(*)$ with (5.27) and the expression for the couplings $(**)$ with (5.24), we discover a close analogy with the minority game. The replica solution of the minority game took direct inspiration from the paper [1018] on neural networks.

We shall soon see how and why this average becomes the central quantity in the replica approach to minority game.

Further simplification can be achieved if we change the payoff rule. Instead of adding or subtracting one point depending on the bare fact that the agent was in the minority or majority, we can provide a payoff proportional to the departure from the ideal equilibrium half-to-half state. Therefore, the payoff will not be proportional to the sign of the attendance as in (5.10), but to the attendance $A(t)$ itself. The formula for the update of the scores' differences is now linear and has the form

$$q_i(t+1) - q_i(t) = -\xi_i^{\mu(t)} \left(\Omega^{\mu(t)} + \sum_{j=1}^{N} \xi_j^{\mu(t)} s_j(t) \right) \qquad (5.20)$$

where the Markov processes $\mu(t)$ and $s_i(t)$ are governed by the probabilities (5.17) and (5.18). Note that the expression in parentheses in Eq. (5.20) is just the attendance $A(t)$.

Batch minority game

Both in the standard minority game and in the modifications described above, the scores, i.e. variables $q_i(t)$, are changed in every step, after each choice of the external information $\mu(t)$. However, we can expect that for large systems the evolution of $q_i(t)$ is rather slow, and significant changes occur only on the timescale comparable with the total number of possible μ's, which is $P = 2^M$. This observation suggests a modification which is called the batch minority game.

Instead of updating the scores in each step, the variables q_i are changed only after the agents are given all of the P possible μ's (in arbitrary order). The agents react according to the rules of the thermal minority game explained above, but during such round of P steps the choice of the strategy is always governed by the same distribution (5.18). The change in q_i from the round l to $l+1$, i.e. from the time $t = Pl$ to $t' = Pl+P$ is

$$q_i(l+1) - q_i(l) = -\frac{1}{P} \sum_{\mu=0}^{P-1} \xi_i^\mu \left(\Omega^\mu + \sum_{j=1}^{N} \xi_j^\mu s_j(t) \right). \tag{5.21}$$

The factor $1/P$ was introduced for further convenience. This expression can be further simplified, because in thermal minority game the choice of $s_j(t)$ is independent of μ, and for large P we can write

$$\sum_{\mu=0}^{P-1} \xi_j^\mu s_j(t) \simeq \sum_{\mu=0}^{P-1} \xi_j^\mu \langle s_j(t) \rangle, \tag{5.22}$$

and the thermal average $\langle s_j(t) \rangle$ is given by Eq. (5.19). So, we conclude that the dynamics is given in a compact form by

$$q_i(l+1) - q_i(l) = -h_i - \sum_{j=1}^{N} J_{ij} \tanh \Gamma\, q_j(l) \tag{5.23}$$

where we denote

$$h_i = \frac{1}{P} \sum_{\mu=0}^{P-1} \xi_i^\mu \, \Omega^\mu$$

$$J_{ij} = \frac{1}{P} \sum_{\mu=0}^{P-1} \xi_i^\mu \, \xi_j^\mu. \tag{5.24}$$

5.2.2 Replica solution

Dynamics of magnetisation

For replica calculations [1027–1030] it is more convenient to express the dynamics in terms of the magnetisation values, $m_i = \tanh \Gamma q_i$, instead of the score differences q_i. The time evolution is particularly simple in the limit for small Γ. This is what we shall develop in the following.

In the limit $\Gamma \to 0$ we can expand the change of the magnetisation from one round to the other as

$$m_i(l+1) - m_i(l) = \Gamma\,(1 - m_i^2(l))(q_i(l+1) - q(l))$$
$$- \Gamma^2\, m_i(l)(1 - m_i^2(l))(q_i(l+1) - q(l))^2 + O(\Gamma^3) \tag{5.25}$$

and, keeping only the lowest order in Γ, we get the following dynamics of the magnetisation values,

$$m_i(l+1) - m_i(l) \simeq -\Gamma\left(1 - m_i^2(l)\right)\left[h_i + \sum_{j=1}^{N} J_{ij} m_j\right]$$

$$= -\Gamma\left(1 - m_i^2(l)\right)\frac{1}{2}\frac{\partial}{\partial m_i} H(m_1, m_2, \ldots, m_N) \tag{5.26}$$

where we have denoted

$$H(m_1, m_2, \ldots, m_N)$$

$$= \frac{1}{P}\sum_{\mu=0}^{P-1}(\Omega^\mu)^2 + 2\sum_{i=1}^{N} h_i m_i + \sum_{i,j=1}^{N} J_{ij} m_i m_j \tag{5.27}$$

$$= \frac{1}{P}\sum_{\mu=0}^{P-1}\left[\sum_{i=1}^{N}\left(\frac{1}{2}(a_{+,i}^\mu + a_{-,i}^\mu) + \frac{1}{2}(a_{+,i}^\mu - a_{-,i}^\mu)\, m_i\right)\right]^2,$$

a function which plays a central role in the subsequent calculations. Indeed, the evolution according to (5.26) closely follows a gradient descent in the 'potential' H. The stationary state corresponds to the minimum of H, which can be considered as a Lyapunov function for the dynamics (5.26).

Before we proceed to solving the minimisation problem, we should clarify the relationship of the Hamiltonian H with observable quantities. In the original minority game the agents were rewarded according to the sign of the attendance, while now the gain is proportional to the attendance itself. So, the definition of the predictability according to Eq. (5.14) is less appropriate if we want to be consistent, and we should rather use the quantity

$$\theta_A^2 = \frac{1}{P}\sum_{\mu=0}^{P-1}\langle A|\mu\rangle^2 \tag{5.28}$$

as a measure of the information content. We can see in Fig. 5.6 that it behaves qualitatively very similar to θ^2, shown above in Fig. 5.3. Most notably, both θ^2 and θ_A^2 vanish in the symmetric phase $\alpha < \alpha_c$, indicating that all information contained in the signal was used by the agents and thus disappeared.

The attendance provided the external information μ is given by Eq. (5.9). Averaging the variables $s_i(t)$ in a stationary state is performed according to the probability distribution (5.18), so $\langle s_i(t)\rangle = m_i$, and, inserting the definitions (5.6), we can see that the expression for predictability coincides with the Hamiltonian (5.27). This means that the measured predictability is equal to the minimum of the Hamiltonian because the stationary magnetisation values m_i are exactly those which minimise H.

The relation between the Hamiltonian and the volatility σ^2 is less straightforward. We can write

$$\sigma^2 = \frac{1}{P}\sum_{\mu=0}^{P-1}\langle A^2|\mu\rangle \tag{5.29}$$

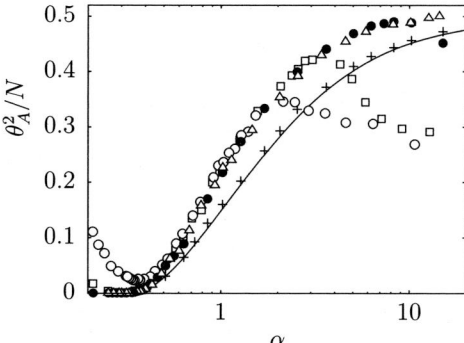

Fig. 5.6 Average predictability of the time series, measured according to (5.28), for $M = 5$ (○), 6 (□), 8 (●), 10 (△). The line is the analytical result from the replica approach, the symbols + denote results of the simulation of the minority game with uniformly sampled memories μ, for $M = 8$.

and express the attendance, again using Eq. (5.9). The problem arises when we come to averaging the products of $s_i(t)$, because in general they cannot be factorised, $\langle s_i(t)s_j(t)\rangle \neq \langle s_i(t)\rangle\langle s_j(t)\rangle = m_i m_j$. We obtain

$$
\sigma^2 = \frac{1}{P}\sum_{\mu=0}^{P-1}\left(\sum_{i=1}^{N}(\omega_i^\mu + \xi_i^\mu m_i)\right)^2 + \frac{1}{P}\sum_{\mu=0}^{P-1}\sum_{i=1}^{N}\left(\xi_i^\mu\right)^2(1 - m_i^2)
$$
$$
+ \frac{1}{P}\sum_{\mu=0}^{P-1}\sum_{i,j=1}^{N}(1 - \delta_{ij})\xi_i^\mu\xi_j^\mu\langle(s_i(t) - m_i)(s_j(t) - m_j)\rangle.
$$
(5.30)

In the first term we recognise the Hamiltonian (5.27), and the second term can easily be computed once we obtain the magnetisation values, by minimisation of H. However, the third term cannot be obtained only from the dynamics of magnetisation according to Eq. (5.26). More detailed studies [1031] show that in the ergodic phase $\alpha > \alpha_c$, the product averages can be factorised, and therefore the third term in Eq. (5.30) can safely be neglected. On the other hand, in the non-ergodic phase the factorisation breaks down, and the value of the third term depends on the thermal parameter Γ. However, in the limit $\Gamma \to 0$ it vanishes again, so for infinitesimally small Γ we can calculate σ^2 for all values of the parameter α by studying only the static properties of the Hamiltonian H. The price to pay will be the quantitative disagreement between σ^2 as calculated in the non-ergodic regime and numerical simulations of the original minority game, because the latter corresponds to the opposite limit $\Gamma \to \infty$. To placate the nervous reader we can say that, qualitatively, the analytic results from the minimisation of the Hamiltonian H give a reasonably true picture of what is seen in the simulations, not only in the ergodic phase but also in the low-α phase, where ergodicity is broken. For the exact treatment we would need to resort to the generating functional method [34].

Replica trick I: generalities | Box 5.4

The replica trick is a technique for computing averaged properties of a random system in thermodynamic equilibrium. Let us have a Hamiltonian $H(s, a)$ depending on state variables s (e.g. spins) and moreover on random parameters a (e.g. randomly placed impurities). As a are random variables, so are the Hamiltonian, the partition function $Z = \sum_s e^{-H}$, the free energy $F = -\ln Z$, and all other thermodynamic quantities (for brevity we fix the inverse temperature at $\beta = 1$). Physically relevant results are the averages over the disorder, most notably the mean free energy $\overline{F} \equiv \int P(a)F\mathrm{d}a$. From probability theory we know that all information on a random variable F is contained in its characteristic function $Z(n) = \overline{e^{-nF}} = \overline{Z^n}$, depending on the complex variable $n \in \mathbb{C}$. For example, the average $\overline{F} = -\lim_{n\to 0} \frac{\mathrm{d}}{\mathrm{d}n} Z(n)$ and the fluctuations related to the second moment $\overline{F^2} = \lim_{n\to 0} \frac{\mathrm{d}^2}{\mathrm{d}n^2} Z(n)$ can readily be obtained.

The difficulty arises when we want to actually compute $Z(n)$ for a given model. It turns out that we are unable to get it for a general n, except for positive integer values $n \in \mathbb{N}$. We are saying that we have n 'replicas' of the original system. To calculate the derivatives and the limit $n \to 0$, we must first extend the analytic continuation of the function $Z(n)$ from $\mathbb{N}$ to the rest of the complex plane, or at least to a certain neighbourhood of the point $n = 0$. To prove the existence and uniqueness of this continuation is one of the hardest open problems in contemporary mathematical physics. The best we can do here is to assume that the continuation does exist and that it is unique. Some general features can be stated, though. The continuation must always be extended around an accumulation point of the set on which $Z(n)$ is known. Here the only accumulation point is $n = \infty$, so we should effectively work with infinitely many replicas. The continuation in fact goes from the neighbourhood of $n = \infty$ to the neighbourhood of $n = 0$.

Finally, we can also learn an important general feature of the agents' behaviour in minority game. Each of them tries hard to maximise her individual profit, and that is why she assesses the quality of the strategies and chooses the best one. However, we have just seen that it is not the overall loss, measured by the volatility σ^2, but the information content, or H, that is minimised in the dynamics. In short, the agents are not fully aware of the collective effect of their actions. Instead of optimising the global performance, they merely devour as much information as they can.

Effective spin model

We have seen that we can formulate the problem as finding the ground state for the system of soft spins $m_i \in [-1, 1]$ described by the Hamiltonian H. To this end we shall first investigate its behaviour at finite temperature β and eventually determine the ground-state properties by sending $\beta \to \infty$. To pursue such a program we need to overcome a serious hindrance. The Hamiltonian depends on the strategies $a_{s,i}^\mu$, which are selected randomly at the beginning and introduce the quenched disorder into the Hamiltonian.

In fact, such a problem is nothing really new in statistical physics. Spin glasses and neural networks (see Box 5.3) are very well described by very similar Hamiltonians. The presence of quenched randomness is the feature all of these models have in common, and it can be very effectively tackled using the replica method. (See Boxes 5.4 and 5.5.) The quantity of prime interest is

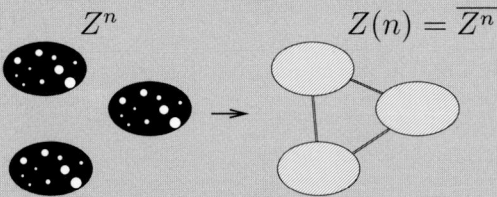

Replica trick II: a toy example $\boxed{\text{Box 5.5}}$

One can hardly imagine a simpler system than one Ising spin in a random magnetic field. The Hamiltonian is
$$H = -h\,s$$
with $s \in \{-1, +1\}$, and the random variable h has normal distribution $P(h) = (2\pi\sigma^2)^{-1/2} \exp(-h^2/2\sigma^2)$. For a positive integer n we have $Z(n) = \overline{Z^n} = \overline{\left(\sum_s e^{hs}\right)^n} = \overline{\sum_{s_1}\sum_{s_1}\dots\sum_{s_n} e^{h\sum_{a=1}^n s_a}}$. The disorder average is elementary and yields $Z(n) = \sum_{\{s_a\}} e^{\frac{1}{2}\sigma^2(\sum_{a=1}^n s_a)^2} = \sum_{\{s_a\}} e^{\sum_{a<b}^n \sigma^2 s_a s_b + n\sigma^2/2}$.
We can see that our disordered system is effectively described by the Hamiltonian $H_n = -\sigma^2 \sum_{a<b}^n s_a s_b - n\sigma^2/2$ representing n interacting replicas of the original model. This is a generic property of all the replica calculations: n non-interacting replicas of the disordered system are transformed to n interacting replicas, but without disorder. The following figure is a sketch of that transform.

The minus sign in the Hamiltonian H_n means that the interaction between replicas is effectively attractive. This is also a generic feature. The replicas always attract each other and the strength of the attraction, here measured by the parameter σ, is larger as the disorder becomes stronger, i.e. as the random ingredients in the original Hamiltonian fluctuate more.
We may proceed with calculating $Z(n)$ with the aid of the Hubbard-Stratonovich transform, introducing an auxiliary field Q. We get $Z(n) = \sum_{\{s_a\}} \int \frac{dQ}{\sqrt{2\pi}} \exp\left(-\frac{1}{2}Q^2 + Q\sigma \sum_{a=1}^n s_a\right) = \int \frac{dQ}{\sqrt{2\pi}} e^{-\frac{1}{2}Q^2} (2\cosh Q\sigma)^n$,
hence the free energy has the mean value $\overline{F} = -\int \frac{dQ}{\sqrt{2\pi}} e^{-\frac{1}{2}Q^2} \ln(2\cosh Q\sigma)$.

$$Z(n) = \overline{\left(\int_{-1}^1 d[m_i]\, e^{-\beta H}\right)^n} \tag{5.31}$$

where the overbar denotes the disorder average, i.e. the average over realisations of the strategies $a_{s,i}^\mu$, and we introduced a shortcut notation for multiple integrations $\int_{-1}^1 d[m_i] \equiv \int_{-1}^1 dm_1 \int_{-1}^1 dm_2 \dots \int_{-1}^1 dm_N$ which will be used in various modifications throughout the calculation.

To perform the disorder average in Eq. (5.31) for a positive integer n we formally introduce n replicas of the system with state variables m_i^a, $a = 1, 2, \dots, n$. Then we can see that the Hamiltonian (5.27) consists of $P = 2^M$ terms, each of them being the sum of n squares. The squares in exponents are conveniently simplified using the Hubbard-Stratonovich transform (see Box 5.6), introducing in return nP new auxiliary fields z_a^μ. But it turns out that all contributions for different μ are disorder-averaged independently of the others, and we end with a product of P identical factors, each of them containing only n auxiliary fields z_a. Explicitly, we find

---

Hubbard-Stratonovich transform | Box 5.6

is based on the identity
$$\int_{-\infty}^{\infty} \exp\left(-\tfrac{1}{2}x^2 + Ax\right) \frac{\mathrm{d}x}{\sqrt{2\pi}} = \exp\left(\tfrac{1}{2}A^2\right).$$
It is used quite often to convert four-spin interactions into two-spin ones and terms which are quadratic in state variables into linear terms, at the expense of introducing a new auxiliary continuous variable. A posteriori, such a variable is usually identified with certain equilibrium characteristics of the model.

For example, the fully connected Ising model with Hamiltonian $H = -\frac{1}{N}\sum_{i<j}^{N} J s_i s_j$, where $s_i \in \{-1, +1\}$, is solved as follows. The partition function is $Z = \sum_{\{s_i\}} \mathrm{e}^{-H} = \sum_{\{s_i\}} \exp\left(\frac{J}{2N}\left(\sum_{i=1}^{N} s_i\right)^2 - \frac{J}{2}\right)$ and, introducing an auxiliary variable $\tilde{m}$ we have $Z = \sum_{\{s_i\}} \mathrm{e}^{-J/2} \int \frac{\mathrm{d}\tilde{m}}{\sqrt{2\pi}} \exp\left(-\frac{1}{2}\tilde{m}^2 + \tilde{m}\sqrt{J/N} \sum_{i=1}^{N} s_i\right) = \mathrm{e}^{-J/2} \int \frac{\mathrm{d}\tilde{m}}{\sqrt{2\pi}} \mathrm{e}^{-\frac{1}{2}\tilde{m}^2} \left(2\cosh \tilde{m}\sqrt{J/N}\right)^N$.

Changing the variable to $m = \tilde{m}/\sqrt{JN}$, we get $Z = \mathrm{e}^{-J/2} \int \mathrm{e}^{-N\mathcal{F}(m)} \sqrt{\frac{JN}{2\pi}}\, \mathrm{d}m$ with $\mathcal{F}(m) = \frac{1}{2}Jm^2 - \ln\left(2\cosh Jm\right)$. In the thermodynamic limit $N \to \infty$ we apply the saddle-point method for the evaluation of the integral and arrive at the following expression for the free energy density $f = -\lim_{N\to\infty}(\ln Z)/N = \mathcal{F}(m^*)$, where m^* is determined by the minimality condition $\frac{\mathrm{d}}{\mathrm{d}m}\mathcal{F}(m)|_{m=m^*} = 0$. Explicitly, it satisfies the equation $m^* = \tanh Jm^*$ which leads to the interpretation of the value m^* as average magnetisation, $m^* = \langle s_i \rangle$.

$$Z(n) = \int_{-1}^{1} \mathrm{d}[m_i^a] \left\{ \int_{-\infty}^{\infty} \mathrm{d}\left[\frac{z_a}{\sqrt{2\pi}}\right] \mathrm{e}^{-\frac{1}{2}\sum_{a=1}^{n} z_a^2} \right.$$

$$\times \prod_{i=1}^{N} \overline{\exp\left(\mathrm{i}\sqrt{\frac{\beta}{2P}} \sum_{a=1}^{n} z_a \left(1 + m_i^a\right) a_{+,i}\right)}$$

$$\left. \times \overline{\exp\left(\mathrm{i}\sqrt{\frac{\beta}{2P}} \sum_{a=1}^{n} z_a \left(1 - m_i^a\right) a_{-,i}\right)} \right\}^P . \tag{5.32}$$

The averages over the variables $a_{s,i}$ are performed using the following simple trick, which works for $P \to \infty$

$$\overline{\exp\left(\mathrm{i}\frac{C}{\sqrt{P}} a_{s,i}\right)} = \frac{1}{2}\sum_{a=\pm 1} \mathrm{e}^{\mathrm{i}Ca/\sqrt{P}} = \cos\left(\frac{C}{\sqrt{P}}\right) \simeq 1 - \frac{1}{2P}C^2 \simeq \exp\left(-\frac{C^2}{2P}\right). \tag{5.33}$$

This way we get in the exponent terms of the type $m_i^a m_i^b$ indicating interaction between replicas. A short exercise in algebra yields

$$Z(n) = \int_{-1}^{1} \mathrm{d}[m_i^a] \left\{ \int_{-\infty}^{\infty} \mathrm{d}\left[\frac{z_a}{\sqrt{2\pi}}\right] \mathrm{e}^{-\frac{1}{2}\sum_{a=1}^{n} z_a^2} \right.$$

$$\left. \times \exp\left(-\frac{1}{2}\frac{\beta}{\alpha} \sum_{a,b=1}^{n} z_a z_b \left(1 + \frac{1}{N}\sum_{i=1}^{N} m_i^a m_i^b\right)\right) \right\}^P . \tag{5.34}$$

The pivotal parameter $\alpha = P/N$ appears for the first time here. For the integration over magnetisation values, m_i^a it is very inconvenient that the term $\sum_{i=1}^{N} m_i^a m_i^b$ appears inside the bracket $\{\ldots\}^P$. To take it out, we introduce another set of $n(n+1)/2$ variables q_{ab}, $a \leq b$, with δ-functions guaranteeing that $q_{ab} = \frac{1}{N}\sum_{i=1}^{N} m_i^a m_i^b$. Thus

$$
Z(n) = \int_{-1}^{1} \mathrm{d}[m_i^a] \int_{-\infty}^{\infty} \mathrm{d}[q_{ab}] \left(\prod_{a \leq b}^{n} \delta\left(q_{ab} - \frac{1}{N}\sum_{i=1}^{N} m_i^a m_i^b\right) \right)
$$
$$
\times \left\{ \int_{-\infty}^{\infty} \mathrm{d}\left[\frac{z_a}{\sqrt{2\pi}}\right] \mathrm{e}^{-\frac{1}{2}\sum_{a=1}^{n} z_a^2} \exp\left(-\frac{1}{2}\frac{\beta}{\alpha}\sum_{a,b=1}^{n} z_a z_b (1 + q_{ab})\right) \right\}^P. \tag{5.35}
$$

The δ-functions are then expressed by integral representation $\delta(x) = \int_{-\infty}^{\infty} \frac{\mathrm{d}y}{2\pi}\mathrm{e}^{\mathrm{i}xy}$. Furthermore, in the exponent we recognise a quadratic form in the variables z_a, so the integration over these variables is straightforward in principle. We complete the definition of quantities q_{ab} by symmetrisation, $q_{ab} = q_{ba}$, and define a $n \times n$ matrix M with elements $M_{ab} = \delta_{ab} + \frac{\beta}{\alpha}(1 + q_{ab})$. The Gaussian integration over z_a gives $(\det M)^{-1/2}$. Therefore

$$
Z(n) = \int_{-\infty}^{\infty} \mathrm{d}[q_{ab}] \int_{-\mathrm{i}\infty}^{\mathrm{i}\infty} \mathrm{d}\left[\frac{-\mathrm{i}N\,r_{ab}}{2\pi}\right] \exp\left(-N\sum_{a \leq b}^{n} r_{ab}\,q_{ab}\right)
$$
$$
\times (\det M)^{-P/2} \left[\int_{-1}^{1} \mathrm{d}[m^a] \exp\left(\sum_{a \leq b} r_{ab}\,m^a m^b\right)\right]^N \tag{5.36}
$$
$$
= \int_{-\infty}^{\infty} \mathrm{d}[q_{ab}] \int_{-\mathrm{i}\infty}^{\mathrm{i}\infty} \mathrm{d}\left[\frac{-\mathrm{i}N\,r_{ab}}{2\pi}\right] \exp\left(-N\beta n\mathcal{F}\right)
$$

with effective replicated free energy

$$
\mathcal{F}(q_{ab}, r_{ab}) = \frac{1}{n\beta}\sum_{a \leq b}^{n} r_{ab}\,q_{ab} + \frac{\alpha}{2n\beta}\ln\det M
$$
$$
- \frac{1}{n\beta}\ln\int_{-1}^{1} \mathrm{d}[m^a] \exp\left(\sum_{a \leq b} r_{ab}\,m^a m^b\right). \tag{5.37}
$$

The last integral in Eq. (5.36) can be taken by the saddle-point method (see Box 2.9 if unclear), as in the thermodynamic limit $N \to \infty$ we have

$$
Z(n) \simeq \exp\left(-N\beta n\mathcal{F}(q_{ab}^*, r_{ab}^*)\right), \tag{5.38}
$$

and the calculation reduces to finding the position q_{ab}^*, r_{ab}^* of the minimum of $\mathcal{F}$, i.e. solving the set of $n(n+1)$ equations

$$
\frac{\partial}{\partial q_{ab}}\mathcal{F}(q_{ab}^*, r_{ab}^*) = \frac{\partial}{\partial r_{ab}}\mathcal{F}(q_{ab}^*, r_{ab}^*) = 0. \tag{5.39}
$$

Fortunately, we can look for the solution in a very simple replica-symmetric form

$$
q_{ab}^* = q + (Q - q)\delta_{ab}
$$
$$
r_{ab}^* = r + (R - r)\delta_{ab}, \tag{5.40}
$$

reducing the number of free parameters to only four. Replica symmetry means that a matrix does not change under any permutation of its indices. The most general form of such matrices is just given by (5.40). It is possible to prove that this solution is thermodynamically stable, which fact justifies the replica symmetry [1032]. The effective free energy becomes

$$\mathcal{F} = \frac{1}{\beta}\left(RQ - \frac{1}{2}rq\right) + \frac{\alpha}{2\beta}\ln\left(1 + \frac{\beta(Q-q)}{\alpha}\right) + \frac{\alpha(1+q)}{2(\alpha + \beta(Q-q))}$$
$$- \frac{1}{\beta}\int_{-\infty}^{\infty}\frac{dz}{\sqrt{2\pi}}e^{-\frac{1}{2}z^2}\ln\int_{-1}^{1}e^{\left(R-\frac{1}{2}r\right)m^2 + z\sqrt{r}\,m}dm + O(n). \tag{5.41}$$

The last term, containing integrals over m and z, looks like the free energy of a particle in potential $V_z(m) = -\left(R - \frac{1}{2}r\right)m^2 - z\sqrt{r}\,m$, averaged over Gaussian-distributed random external field z. We shall use that analogy in practical calculations, introducing z-dependent averages with respect to the potential $V_z(m)$ as $\langle\ldots\rangle_z = \int_{-1}^{1}\ldots e^{-V_z(m)}dm/\int_{-1}^{1}e^{-V_z(m)}dm$. Note also that the dependence on n occurs only in a term of higher order $O(n)$, so we can safely perform the limit $n \to 0$ now. The minimum of the effective free energy is found by differentiating $\mathcal{F}$ with respect to the four free parameters. After a short calculation we get four coupled transcendental equations

$$R - \frac{r}{2} = -\frac{\alpha\beta}{2}\frac{1}{\alpha + \beta(Q-q)}$$
$$r = \alpha\beta^2\frac{1+q}{(\alpha + \beta(Q-q))^2}$$
$$Q - q = \frac{1}{\sqrt{r}}\int_{-\infty}^{\infty}\frac{dz}{\sqrt{2\pi}}e^{-\frac{1}{2}z^2}z\langle m\rangle_z \tag{5.42}$$
$$Q = \int_{-\infty}^{\infty}\frac{dz}{\sqrt{2\pi}}e^{-\frac{1}{2}z^2}\langle m^2\rangle_z.$$

Let us recall that we are looking for the minimum of the Hamiltonian (5.27), so we need the solution of (5.42) in the limit $\beta \to \infty$. We should distinguish between two cases. First, the difference $Q - q$ can approach a finite limit, so that the quantity

$$\chi \equiv \beta(Q - q) \tag{5.43}$$

diverges. This is the case of the symmetric, non-ergodic phase with $\alpha < \alpha_c$. Indeed, χ can be interpreted as susceptibility, measuring sensitivity to initial conditions. Diverging susceptibility implies that an arbitrarily small change in initial conditions persists infinitely long, marking the non-ergodic behaviour.

The second solution is characterised by finite susceptibility, so $Q - q \to 0$ for $\beta \to \infty$. This is the asymmetric phase, $\alpha > \alpha_c$ and we shall now spend some time analysing the results we can infer from the solution of (5.42).

Properties of the ergodic phase

The most technically difficult part of the set (5.42) is calculation of the averages $\langle m\rangle_z$ and $\langle m^2\rangle_z$ and then integrating over z. However, in the limit $\beta \to \infty$ the algebraic manipulations are significantly simplified, as we can see from the first two equations in set (5.42) that $r \sim \beta^2$ and $R - \frac{1}{2}r \sim \beta$; so the potential $V_z(m) \sim \beta$ for $\beta \to \infty$. The averages with respect to such potential are dominated by its single minimum at a point $m = m^*$, which can be readily obtained. To simplify the notation we write $V_z(m) = \sqrt{r}\left(\frac{1}{2}\zeta m^2 - z m\right)$, where $\zeta = -\frac{2}{\sqrt{r}}\left(R - \frac{1}{2}r\right)$. For $\beta \to \infty$ the parameter ζ approaches a finite limit,

$$\zeta = \sqrt{\frac{\alpha}{1 + Q}}. \tag{5.44}$$

According to the value of z the minimum lies either inside the interval $(-1, 1)$ or at the endpoints -1 or 1. We list the results in the following table, showing the values of the averages in each of the three ranges of z, together with the results of the integration over z in the corresponding intervals.

$z \in$		$(-\zeta, \zeta)$	$(-\infty, -\zeta]$	$[\zeta, \infty)$
m^*		z/ζ	-1	1
$\langle m \rangle_z$		z/ζ	-1	1
$\langle m^2 \rangle_z$		z^2/ζ^2	1	1
$\int \frac{dz\, e^{-z^2/2}}{\sqrt{2\pi}} z \langle m \rangle_z$	$\frac{1}{\zeta}\operatorname{erf}\left(\frac{\zeta}{\sqrt{2}}\right) - \sqrt{\frac{2}{\pi}} e^{-\frac{\zeta^2}{2}}$			$\sqrt{\frac{2}{\pi}} e^{-\frac{\zeta^2}{2}}$
$\int \frac{dz\, e^{-z^2/2}}{\sqrt{2\pi}} \langle m^2 \rangle_z$	$\frac{1}{\zeta^2}\operatorname{erf}\left(\frac{\zeta}{\sqrt{2}}\right) - \sqrt{\frac{2}{\pi}}\frac{1}{\zeta} e^{-\frac{\zeta^2}{2}}$			$1 - \operatorname{erf}\left(\frac{\zeta}{\sqrt{2}}\right)$

$$\tag{5.45}$$

Putting together the pieces contained in (5.45) and the relation between ζ, α and Q given by (5.44), we obtain the following expressions for Q and the susceptibility χ, parameterised by ζ.

$$\alpha = 2\zeta^2 + (1 - \zeta^2)\operatorname{erf}\left(\frac{\zeta}{\sqrt{2}}\right) - \sqrt{\frac{2}{\pi}}\,\zeta\, e^{-\frac{\zeta^2}{2}}$$

$$Q = 1 - \operatorname{erf}\left(\frac{\zeta}{\sqrt{2}}\right) + \frac{1}{\zeta^2}\operatorname{erf}\left(\frac{\zeta}{\sqrt{2}}\right) - \sqrt{\frac{2}{\pi}}\frac{1}{\zeta} e^{-\frac{\zeta^2}{2}} \tag{5.46}$$

$$\chi = \frac{\alpha\operatorname{erf}\left(\frac{\zeta}{\sqrt{2}}\right)}{\alpha - \operatorname{erf}\left(\frac{\zeta}{\sqrt{2}}\right)}.$$

This solution breaks down when the susceptibility diverges, i.e. for $\alpha \to \operatorname{erf}\left(\frac{\zeta}{\sqrt{2}}\right)$. This condition fixes the critical value of the control parameter α through the pair of equations

$$2\zeta_c = \zeta_c \operatorname{erf}\left(\frac{\zeta_c}{\sqrt{2}}\right) + \sqrt{\frac{2}{\pi}} e^{-\frac{\zeta_c^2}{2}}$$

$$\alpha_c = \operatorname{erf}\left(\frac{\zeta_c}{\sqrt{2}}\right). \tag{5.47}$$

The numerical solution gives

$$\zeta_c = 0.43632656\ldots, \quad \alpha_c = 0.33740018\ldots \tag{5.48}$$

which agrees very well with the value observed in simulations, $\alpha_c \simeq 0.34$.

Now we are ready to compare the analytical results with simulations in the ergodic regime, $\alpha > \alpha_c$. The predictability (5.28), corresponds to the minimum of the Hamiltonian (5.27), and in our calculation it is just the minimum of the free energy (5.41) in the limit $\beta \to \infty$. We find

$$\theta_A^2 = \frac{\alpha^2}{2} \frac{1 + Q}{(\alpha + \chi)^2}.$$
(5.49)

From Eq. (5.30) it follows that the volatility, averaged over the realisations of the strategies, is related to the predictability as

$$\sigma^2 = \theta_A^2 + \frac{1 - Q}{2}.$$
(5.50)

As a brief comparison of the analytical results with numerical data, look at Fig. 5.6. Clearly, the original minority game is slightly off the analytical prediction, which is due to the fact that we assumed all memory patterns to be equally probable. If we implement this assumption in simulations, the agreement is excellent.

We can also look at the volatility in Fig. 5.2. The agreement is rather satisfactory, and even more, we can see that it is better the closer we are to the critical point. Indeed, this is due to the fact that the distribution of memory patterns becomes uniform at α_c. This leads us to conjecture that the analytically-found critical value (5.48) of α_c is in fact an exact result.

5.3 Minority game as a market model

The original ambition of the minority game, and its predecessor, the El Farol bar problem, was to explore the capabilities of inductive reasoning. The scope is much broader than mere market modelling and touches virtually on all aspects of human life. The task is to survive in a dangerous and nearly unpredictable world. We have seen that the method incorporated in the minority game is to identify and partially utilise the fragments of information contained in past data. Extraction of the hidden information is an essentially collective effect, and the principles of the minority game do not allow reduction of the dynamics to the behaviour of a representative agent. An economically thinking person must ask immediately if the minority game can reproduce the observed behaviour of market and if it can be used for predicting future market movements, or, in short, to earn money. Several remarks are worth making before we set out on this track. First, one must always bear in mind that predicting the future price and understanding how the market works are two different, perhaps even mutually exclusive, things. The minority game may function very well as a heart of a software package, someone can even think of patenting it, but the contribution to our knowledge can be close to nil. On the other hand, minority game can teach us a lot of things about agents' coordination and information ecology which may never be translated into net financial profit.

Besides these general observations, one should also notice several practical problems in interpreting the minority game as a market model. Despite the often used pedagogical motivation of minority game as a toy model of stock trading, minority game as such went much too far with abstraction, and we must return some steps back. First, we must reintroduce price, money, and capital. Then, we run into a far-reaching problem of calculating the payoff. The immediate rewarding in the minority game algorithm was soon judged unrealistic and multitemporality has to be taken into account. And most importantly, the standard minority game does not exhibit any signs of the basic stylised facts, which is a minimum requirement for a market model to be taken seriously. Now we turn to several variants of the minority game introduced to cope with the just-mentioned imperfections.

5.3.1   Producers and speculators

In the standard minority game, all agents are essentially equal. They differ only in their randomly chosen strategies. An important step towards reality is to include two types of behaviour, which we call producers and speculators [847]. Suppose there are N_p and N_s agents of each type, respectively.

Contrary to speculators, producers are quite limited in their speculation ability. They largely follow the circumstances induced by the logic of production. Any change in their behaviour occurs at a much longer time interval than the decisions of speculators. In the minority game, we can model this fact by giving just one strategy to each producer, while the speculators will have two, as ordinary minority game agents have.

This formalism closely follows the scheme of ordinary minority game [1033–1035]. Speculators indexed by $i = 1, 2, \ldots, n_s$ are endowed with pairs of strategies $a_{s,i}^\mu$, $s = \pm 1$, depending on P memory patterns μ. Similarly, the unique strategies of producers $j = 1, 2, \ldots, N_p$ are $a_{\mathrm{prod},j}^\mu$. We have already seen that the only dynamic variables of the game are differences in the scores of the strategies. Now, these differences are attributed only to speculators, but the formula for the differences also contains the strategies of producers. This formula is formally identical to (5.20) with a small change in the definition

$$\xi_i^\mu = \frac{1}{2}\left(a_{+,i}^\mu - a_{-,i}^\mu\right)$$
$$\Omega^\mu = \sum_{j=1}^{N_p} a_{\mathrm{prod},j}^\mu + \frac{1}{2}\sum_{i=1}^{N_s}\left(a_{+,i}^\mu + a_{-,i}^\mu\right). \tag{5.51}$$

The solution of this version of minority game is straightforward using the replica method. We can write the Hamiltonian

$$H(m_1, m_2, \ldots, m_{N_s})$$
$$= \frac{1}{P}\sum_{\mu=0}^{P-1}\left[\sum_{j=1}^{N_p} a_{\mathrm{prod},j}^\mu \right.$$
$$\left. + \sum_{i=1}^{N_s}\left(\frac{1}{2}(a_{+,i}^\mu + a_{-,i}^\mu) + \frac{1}{2}(a_{+,i}^\mu - a_{-,i}^\mu)\, m_i\right)\right]^2 \tag{5.52}$$

and closely follow the steps described in Sec. 5.2.2.

It turns out [1033] that the main feature, which is the presence of phase transition, also remains valid in the presence of speculators, but the location of the critical points

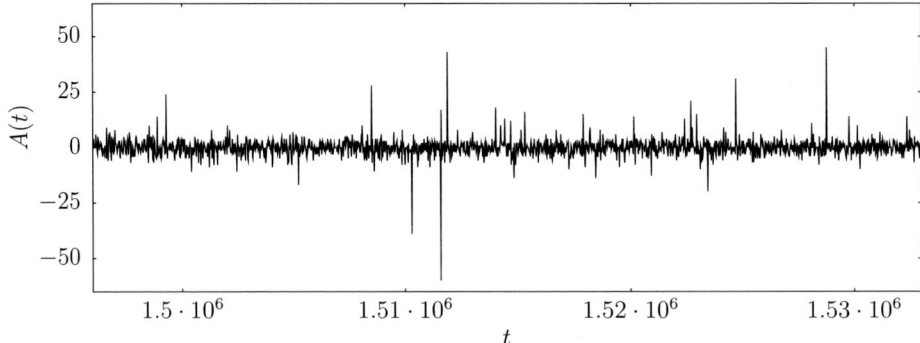

Fig. 5.7 Typical realisation of the time series of attendance in the grand-canonical minority game. The parameters are $N_p = 16$, $N_s = 481$, $M = 4$, and $\epsilon = 0.01$.

is shifted. The ruling parameter is the ratio $\alpha = P/N_s$. As producers increase the information content, the critical value of the parameter α decreases with increasing N_p. It is interesting to observe how the gain of the two groups changes when the number of speculators and producers is varied. In the symmetric phase the producers always have zero gain, and the gain of speculators is large and negative. We can cross the critical point in two ways. First, we can keep the number of speculators constant and increase the number of producers. This way the gain of producers decreases deeper and deeper below zero with increasing N_p, while the gain of speculators increases and eventually becomes positive. Second, we can fix the number of producers and decrease the number of speculators. In so doing, the gain of producers always decreases. On the other hand, the speculators' gain depends on the number of producers. If N_p is large enough, the speculators' gain increases when their number decreases, but if N_p is small, there is a maximum in the dependence of the speculators' gain on their number. What is small/large N_p depends on the number of memory patterns P among which the agents are able to distinguish. Therefore, we can consider this model an example of a market ecosystem consisting of two species of agents, namely producers and speculators. The speculators are like plants, producing information instead of glucose. The speculators can be compared to herbivorous animals, feeding on the information created. However superficial this analogy might be, it helps us understand the coexistence of various types of market participants.

5.3.2 Grand-canonical minority game

Fluctuating number of participants

The classical minority game is a negative-sum game. The agents are forced to play even if they lose constantly. We can modify the rules so that the losers can leave the game, at least for a while, if the losses are too severe. The point is that generically in minority game the strategies which are not used have overall better performance than those which are used. We expect that the absence of a player affects the adaptation of other agents in such a manner that after a while, the strategy which was absent

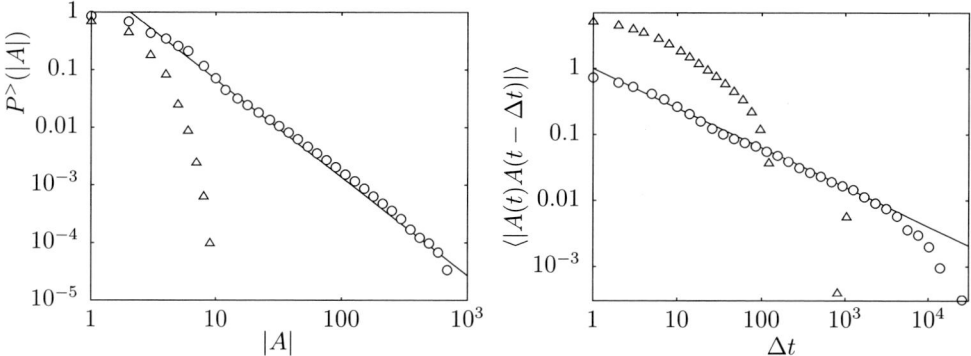

Fig. 5.8 Properties of the time series of attendance in the grand-canonical minority game. In the left panel, cumulative distribution of the absolute attendance. The symbols distinguish between two typical situations: realisation with high kurtosis ($\bigcirc$) and with low kurtosis ($\triangle$). The line is the power law $\propto |A|^{-1.7}$. In the right panel, the autocorrelation of absolute attendance is shown. Again, we show typical realisation with strong ($\bigcirc$) and weak ($\triangle$) volatility clustering. The line is the power law $\propto (\Delta t)^{-1/2}$. In both panels, the parameters are $N_p = 16$, $M = 4$, and $\epsilon = 0.01$. In the left panel, we have $N_s = 1601$, while in the right panel $N_s = 481$.

becomes profitable again. The player joins the game, and adaptation changes again. In analogy with the grand-canonical ensemble of statistical physics [1036], where the particles may enter and leave the system freely, this variant of the game is called the grand-canonical minority game [75].

Again, the agents are divided into producers and speculators. The producers have one strategy and always play. The speculators differ slightly from those of Sec. 5.3.1. They have only one strategy, but they can choose whether to use it or not to play at all. The decision of the speculator i at time t is denoted $d_i(t) \in 0, 1$. Let us denote $a^{\mu}_{\text{prod},j}$ and $a^{\mu}_{\text{spec},i}$ the strategies of producers and speculators, respectively. The attendance is

$$A(t) = \sum_{j=1}^{N_p} a^{\mu(t)}_{\text{prod},j} + \sum_{i=1}^{N_s} a^{\mu(t)}_{\text{spec},i} d_i(t). \tag{5.53}$$

Instead of a difference in scores for a speculator, we have a single score of the only strategy the speculator owns. We shall denote it by $q_i(t)$, as it plays the same role as the scores' difference in ordinary minority game. The score is updated as

$$q_i(t+1) - q_i(t) = -a^{\mu(t)}_i A(t) - \epsilon \tag{5.54}$$

where the positive parameter ϵ represents the threshold for agents' participating or not. As in the thermal minority game, the choice of action the agent takes is probabilistic, rather than deterministic. A parameter Γ, playing the role of inverse temperature, is introduced. The probability of participation is

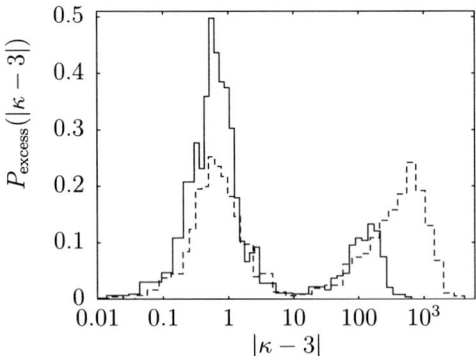

Fig. 5.9 Histogram of the excess kurtosis of the attendance distribution in grand-canonical minority game for $N_p = 16$, $M = 4$, $\epsilon = 0.01$, $N_s = 801$ (solid line), and $N_s = 321$ (dashed line). The data are collected from 1000 independent realisations.

$$\text{Prob}\{d_i(t) = 1\} = \left(1 + e^{-\Gamma\, q_i(t)}\right)^{-1}. \tag{5.55}$$

It is possible to work out the same procedure using the replica trick, as used for the standard minority game. The behaviour of the model is governed by three parameters. Two of them generalise the parameter $\alpha = P/N$ of the usual minority game and are defined as $n_s = N_s/P$ and $n_p = N_p/P$. As usual, $P = 2^M$, and M is the memory length. The remaining parameter is the threshold ϵ. The role of magnetisation is taken by the averages $m_i = \langle d_i(t)\rangle \in [0,1]$. The Hamiltonian to be minimised is

$$
\begin{aligned}
&H(m_1, m_2, \ldots, m_{N_s}) \\
&= \frac{1}{P}\sum_{\mu=0}^{P-1}\left[\sum_{j=1}^{N_p} a_{\text{prod},j}^\mu + \sum_{i=1}^{N_s} a_{\text{spec},i}^\mu\, m_i\right]^2 + 2\epsilon\sum_{i=1}^{N_s} m_i.
\end{aligned}
\tag{5.56}
$$

It turns out that the behaviour is substantially different for positive and negative values of ϵ. There is a first-order transition exactly at $\epsilon = 0$. The more interesting region is $\epsilon > 0$. We can look at the number $N_{s,\text{act}}$ of active speculators, i.e. those who have $m_i > 0$ in the state that minimises the Hamiltonian (5.56). When we increase n_s, the number $n_{\text{act}} = N_{s,\text{act}}/P$ first increases, then reaches a maximum, and then decreases to a finite limit when $n_s \to \infty$. This means that adding more speculators does not increase the number of active ones. The newcomers remain out of business anyway, at least in the statistical sense.

Note that all of these results hold in the thermodynamic limit $P \to \infty$, with n_s and n_p fixed. It is possible to show that in this limit the distribution of attendance is Gaussian [1037]. This is not good news. We wanted a model which would give us fat tails, but none are obtained. However, the model is more complex than the replica analysis in thermodynamic limit reveals. For an alternative view on the grand-canonical minority game, see Refs. [1038–1040].

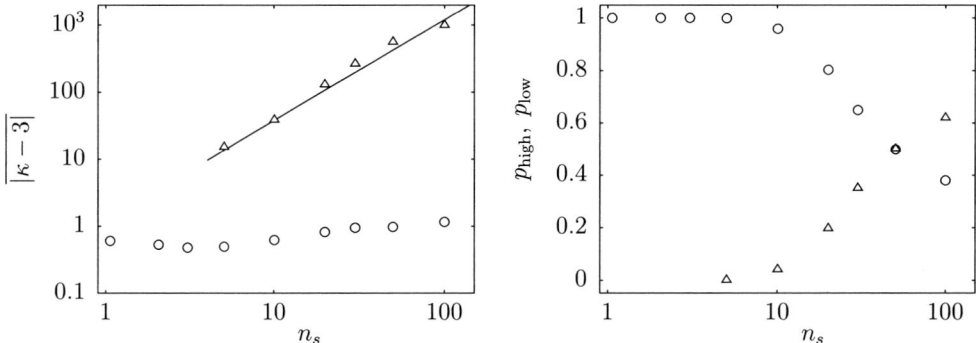

Fig. 5.10 Kurtosis of the attendance distribution in the grand-canonical minority game. In the left panel, average excess kurtosis in the low-kurtosis ($\bigcirc$) and high-kurtosis ($\triangle$) realisations. The line is the power dependence $\propto n_s^{1.5}$. In the right panel, the fractions of low-kurtosis ($\bigcirc$) and high-kurtosis ($\triangle$) realisations. The parameters are $N_p = 16$, $M = 4$, $\epsilon = 0.01$.

Are stylised facts compatible with minority game?

It comes as sort of a surprise that the behaviour seen in the simulations is not quite the same as predicted by the replica treatment. As for the average volatility, fraction of active agents and similar quantities, the agreement is good. However, when we start to look at the time series of the attendance, more complex features emerge. We can glimpse some of it in the sample time series of attendance shown in Fig. 5.7. Due to speculators coming in and out, the attendance fluctuates much more than in the classical minority game. The occasional spikes seen in the figure should warn us that the fluctuations are far from Gaussian as predicted by replica calculation. It turns out [1037, 1041, 1042] that the inconsistency is due to finiteness of the set of agents used in the simulations. This is an intriguing finding. All the interesting features of grand-canonical minority game are in fact finite-size effects.

To assess the properties of the time series quantitatively, we plot in Fig. 5.8 the cumulative distribution of the attendance

$$P^>(|A|) = \lim_{T \to \infty} \frac{1}{T} \sum_{t=1}^{T} \theta\left(\left|\sum_{i=1}^{N} a_i(t)\right| - |A|\right) \qquad (5.57)$$

where $\theta(x) = 1$ for $x > 0$ and $\theta(x) = 0$ for $x \le 0$.

There is a surprise in the data. In fact, we observe two types of behaviour: either the attendance is distributed as power law, $P^>(|A|) \sim |A|^{-\alpha}$ with exponent $\alpha \simeq 1.7$, or the distribution is exponential. Which of these two cases is realised in a particular time series depends on the specific choice of the strategies. Similarly, we also investigate the autocorrelation function of absolute values of the attendance. It is also shown in Fig. 5.8. Again, we observe that the time series fall into two groups. In some realisations the autocorrelation function falls off exponentially and in others it decreases slowly as a power law $\langle |A(t)A(t-\Delta t)| \rangle \sim (\Delta t)^{-\tau}$ with the exponent close to $\tau \simeq 0.5$. So, what is the generic behaviour of the game?

As a partial but well-defined quantitative measure, we can calculate the kurtosis of the distribution if the attendances

$$\kappa = \frac{\left\langle \left(A(t) - \langle A(t)\rangle\right)^4\right\rangle}{\left\langle \left(A(t) - \langle A(t)\rangle\right)^2\right\rangle^2}. \tag{5.58}$$

If the distribution were Gaussian, the kurtosis would be $\kappa = 3$, so the excess kurtosis, defined as $|\kappa - 3|$, tells us how much the distribution deviates from the Gaussian shape. We expect that the distribution which is close to a power law will have a very large excess kurtosis, while slight modifications of the Gaussian will exhibit a much smaller value.

In Fig. 5.9 we plot the histogram of the excess kurtosis found in many realisations of the grand-canonical minority game. We can clearly observe two distinct peaks, proving that there are indeed two qualitatively different types of behaviour. The key parameter for the appearance of the peak with high kurtosis is the reduced number of speculators n_s. The peak starts to emerge at about $n_s \simeq 10$, and for larger n_s the position of the peak shifts to a fairly high kurtosis value. Simultaneously, the high-kurtosis peak gains in weight.

In Fig. 5.10 we plot the average position of the low- and high-kurtosis peaks through the excess kurtosis averaged over low- and high-kurtosis realisations separately. We find that in the high-kurtosis samples the kurtosis grows as a power $|\kappa - 3| \sim n_S^\gamma$ with exponent $\gamma \simeq 1.5$. In the same Fig. 5.10 we also plot the fractions of high-kurtosis and low-kurtosis realisations. It shows that for large enough n_s most of the realisations have very large kurtosis and therefore can be characterised by fat tails in the distribution of attendance.

Deeper analysis of the coexistence of the two types of realisations traces the effect back to the presence of the first-order transition at $\epsilon = 0$. This is an analogy of the coexistence of liquid and solid phases (ice and water) occurring just at the melting temperature. In finite systems, there is a region around the point $\epsilon = 0$ where the coexistence can be observed. When the size of the system goes to infinity, the width of this region shrinks to zero. It is unrealistic to assume that the threshold ϵ would be precisely equal to zero. That is why the interesting phenomena are observed just as a finite-size effects.

We can conclude this section with a positive answer to the question raised in its title. Yes, there is a quite natural way to reproduce basic stylised facts, namely power-law return distribution and power-law absolute return autocorrelation, in a variant of the minority game. The unexpected observation that all the stylised facts observed in grand-canonical minority game are in fact finite-size effects is not a flaw. Just the opposite is true. It reveals that the finite and discrete natures of the real stock market might be vital characteristics behind its complex features. Recall that we arrived to similar conclusions also when studying the Cont-Bouchaud and Lux-Marchesi models in Chap. 3. This observation also puts in doubt various attempts to establish a continuum limit in models of economic behaviour. The fluctuations are substantially influenced by unavoidable granularity of the economic systems. We can also hardly be expect that a kind of macroscopic thermodynamics of economy can be developed, as it always involves a limit to the infinite number of particles or agents. The proper theory should work on a mesoscopic, rather than macroscopic scale, in the sense that the number of agents is neither very small nor very large.

5.4 What remains

Since its introduction in 1997, the minority game has grown into a diversified branch of non-equilibrium statistical physics. As such, it has started its independent life, only partially related to its original purpose of a model for agents' behaviour in socio-economic systems. There are good reviews [1043, 1044] and books [22, 33, 34] which cover most of the developments. Let us briefly comment on further important developments which we could not include in this chapter.

One of the fundamental issues of the theory of minority game is the full account of the coupled dynamics of memories and scores. This is achieved using the generating functional method [1045–1051]. We omitted it in this chapter because the approach is, in our opinion, too technical to be exposed here. But the reader does not lose much, because there is an excellent book [34] entirely devoted to this method.

Besides the replica and generating functional methods, we should also mention the analytical procedure based on crowd-anticrowd theory, developed by Johnson et al in an ample series of papers [778, 1052–1077] investigating in depth various aspects of the minority game and its ramifications.

An original approach avoiding both analytical work and numerical simulations was applied in the interactive minority game [1078, 1079] . Instead of formulae and computer codes, people are used as testing animals. A web page (www3.unifr.ch/econophysics/minority/game/) was created, where anybody can come and play as an agent against a group of other players embodied in a Java applet. Quite surprisingly, people differ very much in their ability to compete with the program. Some users outperform the computerised agents quite a lot, other users lose constantly. One day, perhaps, the interactive minority game will be used as a unified test of intelligence. Who knows?

The minority rule is so intimately tied to the whole ensemble of problems treated using minority game that it seems paradoxical for one to switch and study the majority game instead. Yet the majority game reveals a surprisingly rich structure [1080]. Formally, the only difference consists in replacing the minus signs on the right-hand sides of equations (5.11) and (5.20) with plus signs. The most interesting setup is a mixture of majority and minority games. The players differ according to the rule used for updating the scores of their strategies. Those using the minority rule can be compared with market fundamentalists, believing that there is a correct market price and, whenever the actual price deviates, it should come back soon. On the contrary, the players using the majority rule are like trend followers, doing mostly what the majority does. Depending on the relative fractions of fundamentalists and chartists, the global features of the game look different. The most interesting finding is that there is an intermediate range of the ratio of the fraction in which autocorrelations of the attendance are extremely close to zero. This means that the time series of attendance behaves very much like the price fluctuations in real markets [1081]. On the other hand, both pure majority and minority games exhibit strong correlations or anticorrelations, respectively, in the attendance. Analytical solution of the mixed majority-minority game can be found in Refs. [1082, 1083] and further simulations in Refs. [1084–1086].

Including the majority rule is not the only possible way to change the reward attributed to the strategies. A very interesting idea is implemented in the so-called $-game [1087, 1088]. It is based on the observation that the agent in a stock market must predict the price two steps ahead, not just one, as assumed in standard minority game. Indeed, if we decide to buy a stock at time t, it will influence the price at the next time $t + 1$. We shall assume that from the time $t + 1$ it will gain value until a certain future time $t' > t + 1$. This introduces the feature of multitemporality. The second time t' can lie in an a priori unknown instant in future. This is the time the investor will sell back the asset and pocket the financial gain. The simplest hypothesis on the occurrence of this second time instant will be $t' = t + 2$. Based on such reasoning, it was suggested that, instead of the prescription (5.11), the scores of the strategies should be updated as

$$q_i(t + 1) - q_i(t) = \xi_i^{\mu(t-1)} A(t). \tag{5.59}$$

It turns out [1087, 1088] that the $-game behaves like a mixture of majority and minority agents, thus accomplishing something similar to what was set up artificially in the mixed majority-minority game.

A similar multi-temporal structure is also present in the so-called escape game [1089]. At each time, the agents can decide to hold $(a_i(t) = 1)$ or not to hold $(a_i(t) = 0)$ an asset. It is assumed that the value of the asset is proportional to the attendance $A(t) = \sum_i a_i(t)$, i.e. to the number of agents holding the asset. We can say that the asset is as valuable as it is popular among the agents. Those who hold the asset are rewarded each time the others decide to acquire it and punished if the others decide to get rid of the asset. The strategy is easy to formulate here: be first to buy, and be first to sell. Those who are at the tail of the herd are losers. In order to stimulate the players, a premium is divided among those who hold the asset. If the premium is sufficiently large, the behaviour is very similar to the ordinary minority game. But if the premium is small, the dynamics becomes increasingly complex. A multitude of metastable states occurs, and the system performs intermittent hops from one to the other. In this sense, the dynamics is reminiscent of spin glasses [1010], or glasses in general.

The classical minority game can also be developed in other directions. For example, the agents can take into account their own impact on the system [1032]. In this case, the replica solution reveals that the replica symmetry is broken and many stable equilibrium states coexist. The agents can undergo an evolutionary process inspired by natural selection [1090–1092]. The agents can also trade in several assets, i.e. several sets of memories $\mu_1(t), \mu_2(t), \dots$ are supplied to the agents simultaneously [1093]. The agents can have more than two choices [1094, 1095]. The agents can live in a non-stationary environment [1096]. The scores of strategies do not remember their performance forever but fade away [1097]. Surprisingly, it was found that both replica and generating-functional methods fail to account for this fadeout [1098]. New analytical tools are required here.

In the standard minority game, all agents are equal. In reality, however, the traders act at different time scales and with different weights. These features can be relatively easily incorporated into the mechanism of minority game, and analytical solutions for such situations are available [1035, 1099–1102]. Under certain conditions, it can be shown that the heterogeneity is beneficial for the system as a whole [1102].

Another way to introduce inequality is the following. In the standard minority game, every agent interacts with all others via the common memory pattern μ. What if we place the agents into a space and allow them interact locally? This is the basic idea of various species of local minority games [1103–1106]. The agents may also copy each other [1107]. Here we arrive at the problem of the emergence of leadership in the community of economic subjects. This was studied within the framework of minority game by allowing the agents either to use the best of their own strategies or to look at their local neighbourhood and imitate the best of their neighbours [1108]. The basic topology under the agents' community can be either a simple chain [1108, 1109] or a random network [1110, 1111]. This substrate network tells us which agents can communicate with each other. It was found that the imitation structures create a second-layer network of dependence among the agents, on top of the substrate network. The imitation network falls generically into the class of scale-free networks [1110, 1111]. This is in agreement with empirical studies of networks of dependence among economic entities [364]. (See Chap. 6 for more information on networks). The idea was further developed in Refs. [1112–1123].

We do not have space here to go deeper in the various literature [1124–1163]. Instead, as a dessert, the reader can taste how introducing quantum strategies [1164, 1165] leads to the even richer and more wonderful world of quantum minority game [1166–1168].

Problems

1. The case $M = 1$, $N = 3$, $S = 2$ is just small enough to be investigated by complete enumeration. Write your own program to accomplish that task and look at the periodicity of the attendance, and distributions of the quantities σ^2, θ^2 and ϕ. How fast does the number of possible assignments of strategy pairs to agents grow with M and N?

2. Imagine what happens with minority game if there are no memories at all, and each agent has the same pair of strategies $a_{+,i} = +1$, $a_{-,i} = -1$. The dynamics of the scores' differences (5.23) defines a dynamical system. Show that a critical value Γ_c exists so that for $\Gamma < \Gamma_c$ the attractor is a point, while for $\Gamma > \Gamma_c$ it is a limit cycle of length 2 steps. What is the average σ^2 in the first and second cases? How does it depend on the number of agents N? For a hint, look at [1160].

3. The agents in minority game can have different weights [1035, 1102]. Divide the set of agents into G groups, g-th group containing N_g agents acting with weight I_g, so that the attendance is $A(t) = \sum_g I_g \sum_{i_g=1}^{N_g} a_{g,i_g}(t)$. Solve this variant using the replica method. Using for simplicity just two groups, show that the sum of gains of all agents is still negative as in the standard minority game, but the average gain of the group with lower weight can be positive.

6
Network economy

When physicists think of elementary entities in their theories, be it atoms, spins, molecules or elementary particles, they like to place them regularly in space or let them move freely with no preference for any place or direction. This intuition is embodied into the assumption of translational and rotational invariance, either continuous (shifting all particles by any distance does not matter) or discrete (displacement by an integer multiple of an elementary distance does not make any change), and mathematically expressed in terms of symmetry groups.

The mental inclination toward highly symmetric spaces and arrangements seems to be supported by beautiful structures of crystals as well as by the achievements of fundamental physics, which to a large extent consists in looking for the appropriate symmetry group underlying the observed phenomena. Indeed, the crystal symmetry allows us to speak of energy bands in solids and provides a sound basis for the notion of quasi-particles. It is the symmetry of space-time that renders us the images of distant galaxies through freely travelling photons, for no symmetry would mean no photons, and no photons would mean no vision whatsoever. All of this is true despite the indisputable fact that the world surrounding us is full of structures lacking any trace of symmetry. Think of the long knotted molecule of DNA in every cell, a fragment of pumice, or hair on your head when you wake up early in the morning. To describe these structures we need more advanced techniques, and most of the problems related to them still await an effective tool for solution. Although it is rather inappropriate to recourse here into the realms of science fiction, one may imagine creatures adapted to living in spaces with no symmetry at all, which are able to process information transmitted by complex elementary excitations replacing the ordinary photons. This idea is very relevant in some contemporary technological applications, and the theory of random networks is one of valuable tools used there.

A complementary attitude seems to have pervaded the social sciences and economy. Here, the 'elementary particles' are human beings or companies, contracts, or pieces of information. Recognising the intractable complexity of the tangled web of mutual contacts and influences, we often forget it completely and think in terms of representative agents and aggregate quantities. A physicist would perhaps recognise in such approach to reality a kind of 'mean-field approximation', a beloved instrument of physicists themselves.

So, we can see that the two approaches are not so distant from each other as one may think at first glance. Either we constrain the universe into a regular, highly symmetric framework or forget any structure at all. Both ways are extremely efficient and successful, but answer only some specific questions. As we shall see, sometimes they completely miss the real phenomena, for example, if we ask how measles is spread. To go further, we need to know details on the disordered structure of our universe. And as the material subject of study in this book is human society in its various aspects, we have to collect information on how the society is actually shaped and investigate how the peculiarities of the social structure affect its functioning. The basic assumption throughout this chapter will be that the numerous interdependences we find in society can be expressed in terms of a collection of networks, each of them mapping a certain aspect of pairwise interactions among humans or human collectives, or even products of human activity, for example, books like the one you are currently reading.

6.1   Random graphs

6.1.1   About graphs

The mathematical fabric for the scientific description of social and other networks is provided by the graph theory [1169]. Let us quickly review its basic concepts.

A graph consists of vertices and edges. Every edge connects two (not necessarily distinct) vertices. We may or may not distinguish the order in which we take the two vertices connected by an edge. Accordingly, we say that the edge is oriented or unoriented, respectively. An oriented graph has all edges oriented, and similarly the unoriented graph consists only of unoriented edges. (We do not consider the case in which both oriented and unoriented edges are present in one graph, although it would be quite possible.) Formally, a graph is a pair of sets, $G = (\mathcal{V}, \mathcal{E})$, where $\mathcal{V}$ is the set of vertices and $\mathcal{E} \subset \mathcal{V} \times \mathcal{V}$ the set of edges. For an unoriented graph we require that for any $v, w \in \mathcal{V}$ such that $(v, w) \in \mathcal{E}$ we also have $(w, v) \in \mathcal{E}$, but the two pairs are considered as one edge. If not stated differently, throughout this chapter N_G will stand for the number of vertices in the graph G. We shall often omit the index G if the graph in question will be evident from the context. An example of an unoriented graph is given in Fig. 6.1.

The structure of a graph can be conveniently described by the adjacency matrix. Let us number the vertices $v_i \in \mathcal{V}$ by indices $i = 1, 2, \ldots, N$. We define the $N \times N$ matrix with elements

$$a_{ij}(G) = \begin{cases} 1 & \text{if } v_i, v_j \in \mathcal{V} \text{ and } (v_i, v_j) \in \mathcal{E} \\ 0 & \text{elsewhere} \end{cases}. \tag{6.1}$$

The adjacency matrix of an unoriented graph is symmetric. Any oriented graph G can be converted into the corresponding unoriented graph G^S by symmetrisation of the edges. The adjacency matrix of the symmetrised graph is $a_{ij}(G^S) = 1 - (1 - a_{ij}(G))(1 - a_{ji}(G))$. Note that we can allow loops, i.e. edges starting and ending at the same vertex. This corresponds to a 1 on the diagonal of the adjacency matrix. The definition of a graph can be further generalised if we want to model multiply connected vertices. In this case $a_{ij}(G)$ is the number of edges going from vertex v_i to v_j.

Graphs are characterised in a great many ways. The first characteristics we shall work with will be related to the degrees of the vertices in a graph. The degree of a vertex v is the number of edges attached to that vertex. Note that if the edge makes a loop, it is counted twice, taking contribution from both ends of the edge. With this

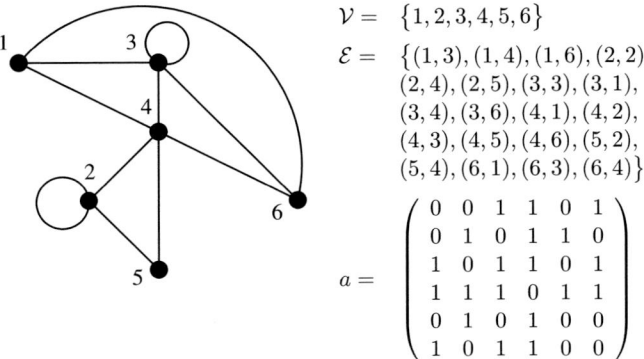

$$\mathcal{V} = \{1,2,3,4,5,6\}$$

$$\mathcal{E} = \{(1,3),(1,4),(1,6),(2,2),$$
$$(2,4),(2,5),(3,3),(3,1),$$
$$(3,4),(3,6),(4,1),(4,2),$$
$$(4,3),(4,5),(4,6),(5,2),$$
$$(5,4),(6,1),(6,3),(6,4)\}$$

$$a = \begin{pmatrix} 0 & 0 & 1 & 1 & 0 & 1 \\ 0 & 1 & 0 & 1 & 1 & 0 \\ 1 & 0 & 1 & 1 & 0 & 1 \\ 1 & 1 & 1 & 0 & 1 & 1 \\ 0 & 1 & 0 & 1 & 0 & 0 \\ 1 & 0 & 1 & 1 & 0 & 0 \end{pmatrix}$$

Fig. 6.1 An example of an unoriented graph with 6 vertices and 11 edges. The sets of vertices and edges are shown, as well as the adjacency matrix a. Degrees of the vertices are, for example, $d(1) = d(6) = 3$ or $d(3) = 5$. The distances are $d(1,6) = 1$, $d(1,5) = 2$ etc. The degree sequence of this graph is $[5,5,4,3,3,2]$.

definition the sum of degrees of all vertices is twice the number of all edges. Formally, in an unoriented graph $G = (\mathcal{V}, \mathcal{E})$ we define

$$d_G(v) = |\{v' \in \mathcal{V} : (v,v') \in \mathcal{E}\}| + |\{(v,v)\} \cap \mathcal{E}| \tag{6.2}$$

or, in terms of the corresponding adjacency matrix

$$d_G(v_i) = \sum_j (1 + \delta_{ij}) a_{ij}(G). \tag{6.3}$$

The second term in Eq. (6.2) and the Kronecker δ appearing in Eq. (6.3) account for the double counting of edges in loops, as explained above.

The reason for that double counting becomes clear when we introduce analogous quantities for oriented graphs. As the order of vertices adjacent to an edge matters, we distinguish the in-degree, i.e. the number of edges coming toward the vertex, from the out-degree, which is the number of edges going away from the vertex. We give the definitions in terms of the adjacency matrix; the reader can easily formulate it herself in a way analogous to (6.2). Thus

$$d_G^{in}(v_i) = \sum_j a_{ji}(G)$$
$$d_G^{out}(v_i) = \sum_j a_{ij}(G). \tag{6.4}$$

We can see that the edge which is a loop is counted as both in-degree and out-degree. Therefore, the sum of in-degree and out-degree is just the (ordinary) degree of the vertex in the symmetrised graph, $d_{G^s}(v) = d_G^{in}(v) + d_G^{out}(v)$.

Although a graph may be a very intricate object, a good deal of important information is already provided by the degree sequence, which is the ordered list of degrees of all vertices $[d_1, d_2, \ldots, d_N]$, $d_i \geq d_{i+1}$.

For example, all vertices can have the same order k. Such graphs are called k-regular, or simply regular, if we do not care about the specific value of k. In the special case of a 3-regular graph we speak of a cubic graph. A simple example are the corners of a cube connected by the adjacent edges, hence the name.

Not every N-tuple of natural numbers makes a degree sequence of a graph. For example $[1, 0]$ is obvious nonsense, because the sum of all degrees must be an even number (each edge has two ends!). A slightly less trivial example is the sequence $[4, 2]$, which is impossible if we forbid multiple edges connecting the same pair of vertices.

The degree sequence provides a very crude description of the graph. If we need more detailed information, we need more sophisticated means. One of the typical approaches used in graph theory is to look for specific types of subgraphs in a graph. The meaning of the word is self-explaining. $G' = (\mathcal{V}', \mathcal{E}')$ is a subgraph of $G = (\mathcal{V}, \mathcal{E})$, denoted $G' \subset G$, if $\mathcal{V}' \subset \mathcal{V}$ and $\mathcal{E}' \subset \mathcal{E}$. Less formally, we also say that G contains G'.

Which kinds of subgraphs will be of interest for us? To know some, we define several special types of graphs. A complete graph is an unoriented graph containing all $(N - 1)N/2$ possible edges connecting its N vertices. Complete subgraphs of a larger graph are often called cliques. A path of length l connecting vertices u and v is a graph which contains just vertices $v_0 = u, v_1, v_2, \ldots, v_l = v$ and edges $(v_0, v_1), (v_1, v_2), \ldots, (v_{l-1}, v_l)$. A graph G is called connected, if for all pairs of its vertices u and v we find a path connecting u and v and contained in G. There may be more such paths, with different lengths. The minimum of these lengths is called the distance of vertices u and v in the graph G, and denoted $d_G(u, v)$. It can be expressed in terms of the powers of the adjacency matrix

$$d_G(u_i, u_j) = \min\{n : (a^n)_{ij} \neq 0\}. \tag{6.5}$$

A cycle of length l contains l vertices and as many edges. Numbering the vertices $1, 2, \ldots l$, the edges are $(1, 2), (2, 3), \ldots (l-1, l), (l, 1)$. Very often we ask if a given graph contains cycles, and if it does, how large they are and how many of them there are. A connected graph which does not contain any cycle is called a tree. We can see that a path is a special case of a tree. Note that all trees have one less edge than there are vertices. Interestingly, the opposite is also true: any connected graph with N vertices and $N - 1$ edges is a tree. If a graph does contain cycles, they can have various lengths. We shall be especially interested in triangles, i.e. the cycles of length 3, which is the minimum possible. If there are only a few short cycles, we can consider the graph as locally tree-like and use this property in various approximate calculations. We shall see later that this is often the case.

Another important class of graphs is that of bipartite graphs. Their set of vertices can be decomposed into two disjoint sets, $\mathcal{V} = \mathcal{V}_1 \cup \mathcal{V}_2$, $\mathcal{V}_1 \cap \mathcal{V}_2 = 0$, so that all edges have their ends in different sets, i.e. if $(u, v) \in \mathcal{E}$, then either $u \in \mathcal{V}_1$ and $v \in \mathcal{V}_2$ or $u \in \mathcal{V}_2$ and $v \in \mathcal{V}_1$. Real examples of bipartite graphs abound. For example, $\mathcal{V}_1$ can be the set of articles in scientific journals, $\mathcal{V}_2$ the set of their authors and the edges denote authorships. For a given graph, it may not be obvious at first glance whether or not it is bipartite. Fortunately, they are characterised by the absence of cycles of odd lengths. For example, every tree is a bipartite graph, because it does not contain any cycles at all. To see it easily, pick any vertex of the tree as its root and put all vertices at odd distance from the root in the set $\mathcal{V}_1$ and all vertices at even distances (including the root itself) in the set $\mathcal{V}_2$. If a bipartite graph contains all allowed edges, so that any vertex from $\mathcal{V}_1$ is connected by an edge to each vertex in $\mathcal{V}_2$, the graph is called a complete bipartite graph. If the numbers of vertices in the two subsets are $m = |\mathcal{V}_1|$ and $n = |\mathcal{V}_2|$, we denote the corresponding complete bipartite graph as $K_{m,n}$.

Quantitative measures of some subgraphs of a graph G are reflected in several important numbers characterising the graph. The diameter of the graph G is the maximum of distances between all pairs of vertices in G, $\mathrm{diam}_G = \max\{d_G(u,v) : u, v \in G\}$. If the graph is not connected, the diameter is considered infinite. For a connected graph, we also define the average distance,

$$\overline{l_G} = \frac{1}{N(N-1)} \sum_{u,v \in \mathcal{V}, u \neq v} d_G(u,v). \tag{6.6}$$

In generic situations the diameter and the average distance behave similarly as functions of the number of vertices. The length of the shortest cycle contained in the graph is called its girth. Unlike the diameter, the girth bears less relevant information, because we are more interested in how many cycles there are of a certain length, or how probable it is that a given vertex belongs to a cycle of a certain length. So, instead of girth, we shall try to establish a more vague but at the same time more useful quantity, which is the typical cycle length.

As we already said, the cycles of length 3, or triangles, are especially interesting. Their concentration tells us how many vertices connected to a chosen vertex are also connected among themselves. Locally, it is measured by the clustering coefficient of a vertex v, which is the number of edges between neighbours of v, (there are $d_G(v)$ of them), divided by the number of all possible edges between these neighbours. Using the adjacency matrix, we express the clustering coefficient of vertex v_i in the graph G as

$$C_G(v_i) = \frac{\sum'_{j,k} a_{ij} a_{ik} a_{jk}}{\sum'_{j,k} a_{ij} a_{ik}} \tag{6.7}$$

where the prime in the summation symbol means that the sum is restricted to indices satisfying $i \neq j, i \neq k, j \neq k$. As a global characteristic, we introduce the clustering coefficient of the graph

$$C_G = \frac{\sum'_{i,j,k} a_{ij} a_{ik} a_{jk}}{\sum'_{i,j,k} a_{ij} a_{ik}} \tag{6.8}$$

which compares the number of triangles in the graph with one third of the number of 'vees', i.e. pairs of edges which have exactly one vertex in common. Indeed, each triangle contains three 'vees', and the numerator in (6.8) counts each triangle 6 times (number of permutations of the three vertices) while the denominator counts each 'vee' twice. Note that (6.8) is not the vertex clustering coefficient (6.7) averaged over all vertices v_i, as one might naively think.

6.1.2 Graph ensembles

Erdős and Rényi

Graphs are tricky enough, but the objects we shall study in this chapter are even more complicated. What we have in mind are random graphs, and in mathematical language they are sets of graphs endowed with a probability measure [753]. The first construction of random graphs we mention here is based on graph ensembles, which are simply sets of graphs $\mathcal{G}$ with at most N vertices, together with probability $\mathcal{P}(G)$ defined for any element $G \in \mathcal{G}$. They were first introduced in the late 1950s by Hungarian mathematicians Pál Erdős and Alfréd Rényi [1170, 1171], who also proved most of the essential theorems on them, although some important forerunners and people working simultaneously along similar lines should not be forgotten [1172–1175].

The first graph ensemble, studied by Erdős and Rényi, is denoted $\mathcal{G}_{N,M}$ and consists of all graphs with N vertices and M edges, each of them taken with the same probability

$$\mathcal{P}(G) = \binom{N(N-1)/2}{M}^{-1}. \tag{6.9}$$

The second graph ensemble is perhaps even simpler, consisting again of graphs with N vertices, where each of the possible $N(N-1)/2$ edges is present with probability $p \in [0, 1]$ and absent with probability $1 - p$. The ensemble is denoted $\mathcal{G}_{N,p}$. The choice $p = 1$ corresponds to the complete graph, while for $0 < p < 1$ the ensemble contains all possible graphs with N vertices (there are $2^{N(N-1)/2}$ of them), with a probability depending on the actual number of edges. If there are M edges in graph $G \in \mathcal{G}_{N,p}$, then $\mathcal{P}(G) = p^M (1 - p)^{N(N-1)/2-M}$. In informal colloquial jargon, the graphs from both ensembles $\mathcal{G}_{N,M}$ and $\mathcal{G}_{N,p}$ are referred to as Erdős-Rényi graphs, and indeed the properties of the two are very similar.

Another prominent graph ensemble $\mathcal{G}_{N,k-\mathrm{reg}}$ consists of all k-regular graphs with N vertices, each of them with equal probability. Seemingly simple, such random graphs turn out to be much harder to study than the Erdős-Rényi graphs.

Most of the properties of the graph ensembles refer to the limit of large size, $N \to \infty$. As all statements on a property P of random graphs have probabilistic content, we shall always understand them as saying that the probability that property P is satisfied goes to 1 in the limit $N \to \infty$.

In this sense we ask the first question about Erdős-Rényi graphs, namely how large the connected components are in such a graph. Let us concentrate on the ensemble $\mathcal{G}_{N,p}$. Intuitively, if p is very small, the vertices will rarely be connected at all, and the graph will consist of many small connected components. On the contrary, if p is close to 1, nearly all possible edges will be present, and there will be one big component comprising all vertices, except for maybe a few of them. There is a strong rigorous result on the size of the largest component in a graph from the ensemble $\mathcal{G}_{N,p}$, stated as follows [1176]. Suppose $c = pN$ stays constant as $N \to \infty$. Then for $c < 1$ every component has $O(\ln N)$ vertices, while for $c > 1$ there is a number $\alpha(c) > 0$ so that the largest component has $\alpha(c)N + o(N)$ vertices, and all other components have $O(\ln N)$ vertices. In the latter case the largest component contains a finite fraction of all the vertices and it is called the giant component.

How should we understand this result? For given $p = c/N$ there will be on average $pN(N-1)/2$ edges out of the total $N(N-1)/2$. So, there is on average $c(1 - 1/N)/2$ edges per vertex, and as each edge has two ends, and for large N, the average degree of a vertex in such random graph is c. The above theorem says that if the average degree is larger than 1, there is a giant component containing a non-negligible fraction of the vertices, while if the average degree drops below the critical value 1 the graph breaks up into many parts, each containing an infinitesimal fraction of the size of the graph. We used the word 'critical' on purpose. Indeed, the problem is closely related to the mean-field version of a bond percolation problem, which was mentioned in Sec. 3.1.2 in the context of the Cont-Bouchaud model of the stock market. The value $c = 1$ marks the percolation threshold, which is the critical point of a second-order phase transition.

Another important result for the Erdős-Rényi graphs concerns their degree sequence. Quite trivially, the average degree is

$$\overline{d_G} = (N - 1)p. \tag{6.10}$$

Selecting one vertex at random, there are $N - 1$ potential edges leading from it, and each of them is present with probability p. Consequently the vertex has degree k with probability given by the binomial distribution

$$P_{\text{deg}}(k) = \binom{N-1}{k} p^k (1 - p)^{N-1-k}, \tag{6.11}$$

which goes to the Poisson distribution if $N \to \infty$ with fixed average degree $c = pN$. The result can be formulated in a stronger way. Indeed, we spoke about only a single randomly chosen vertex, which tells us little about the entire degree sequence. Instead, we shall now ask how many vertices in a given random graph have degree k. We denote this (random) number N_k. The degree distribution (6.11) bears the information on the average, $\langle N_k/N \rangle = P_{\text{deg}}(k)$, but the point is that we can say more. Let us formulate the theorem first and discuss it afterwards.

If N_k is the number of vertices of degree k in a graph from the ensemble $\mathcal{G}_{N,p}$, $c = pN$ is constant as $N \to \infty$, and $f_k = c^k e^{-c}/k!$, $k \geq 0$ is the mass function of the Poisson distribution, then for any $\epsilon > 0$

$$\lim_{N \to \infty} \text{Prob}\left\{ (1 - \epsilon)f_k \leq \frac{N_k}{N} \leq (1 + \epsilon)f_k \right\} = 1. \tag{6.12}$$

This result means that a sufficiently large graph has almost surely $N f_k$ vertices of degree k, not more and not less. Not only do we know the average properties of the graph, but in the limit $N \to \infty$ all Erdős-Rényi graphs look the same, at least with regard to their degree sequence. Later we shall see that many more graph properties reveal the same finding: if the number of vertices grows to infinity, all graphs in the ensemble $\mathcal{G}_{N,p}$ are equal.

Now, how is the structure of Erdős-Rényi graphs described in terms of various subgraphs they contain? The most important subgraphs are cycles, and we should ask how many of them there are. The answer is that the average number of cycles of length k in a graph from $\mathcal{G}_{N,p}$ is

$$\langle N_{k-\text{cyc}} \rangle = \frac{N!}{2k\,(N-k)!}\,p^k, \tag{6.13}$$

which can easily be seen. Indeed, if we choose a sequence of k vertices, then the probability that they are connected by a cycle is p^k, because each of the k edges is placed independently of the others with probability p. How many of such sequences are there? After choosing one vertex out of N, we have $N - 1$ possibilities to choose the next, etc. giving the total $N(N - 1)(N - 2)\ldots(N - k + 1)$. Not all of them are different, though. There are k possibilities for where to start the sequence for a given cycle and two directions in which to proceed along it. Combining the pieces, we get the result (6.13).

In many calculations we consider the presence of short loops as an undesired nuisance. To assess how serious it is, we would like to estimate the length of the shortest cycle we must take into account. For a given vertex v in the graph G the relevant quantity will be the length of the shortest cycle going through the vertex v, or

$$g_G(v) = \min\{k : \text{there is a cycle } G_k \subset G \text{ of length } k \text{ and } v \in G_k\}. \tag{6.14}$$

The girth of the graph G is the minimum of $g_G(v)$ over all $v \in G$, but as we already noted, this is not the quantity which interests us here. Instead, we would like to know the average $\overline{g_G} = \sum_v g_G(v)/N$. The formula (6.13) will help us to estimate its value. As each cycle of length k goes through k vertices, the average number of k-cycles which go through a given vertex is $k\langle N_{k-\mathrm{cyc}}\rangle/N$. For large N and $c = pN > 1$ it is a growing function of k, so larger cycles are more frequent. The length l of the shortest cycle can be estimated by requiring that there is on average only one cycle of the length l, i.e. $l\langle N_{l-\mathrm{cyc}}\rangle/N = 1$. The value obtained this way will serve as an approximation for the average shortest cycle, $\overline{g_G} \simeq l$. Using (6.13) and assuming that $l \ll N$ we get the estimate

$$\overline{g_G} \simeq \frac{\ln N}{\ln c}. \tag{6.15}$$

We can see that the cycle length grows with N, so that we can claim that locally the graph looks like a tree if N is large enough. On the other hand, the growth is logarithmic, i.e. rather slow, and a graph with a thousand vertices and $c \simeq 2$ can have cycles of typical lengths of about ten. We can also see that if c decreases to the critical value, equal to one, the length of the shortest cycles diverges, $\lim_{c \to 1+} \overline{g_G} = \infty$, which is yet another sign of the fact that the graph breaks into many small disconnected pieces.

A similar line of thought will lead us to the estimate of the average distance between vertices in the Erdős-Rényi graph. Choosing vertices $u, v \in G$, a path of length l connecting u and v requires l edges and $l - 1$ intermediate vertices. The latter can be chosen in $(N - 2)(N - 3)\ldots(N - l)$ ways while the set of l necessary edges is here with probability p^l. From here we can deduce a formula analogous to (6.13) for the average number of paths between u and v having length l. Fixing again $c = pN > 1$ and requiring that the average number of shortest paths is approximately 1 we get the result

$$\overline{l_G} \simeq \frac{\ln N}{\ln c} \tag{6.16}$$

showing that the shortest cycles and shortest paths have about the same length!

It is instructive to compare these findings for the Erdős-Rényi graphs with the results we can obtain in a similar manner for the regular graphs. As the average degree in the Erdős-Rényi graph is $\overline{d_G} = c$, we should compare it with the c-regular graph for integer c. If we try to compute the probability that edges exist among the prescribed $l-1$ intermediaries $v_1, v_2, \ldots, v_{l-1}$, we place the edges one by one starting from the vertex u. Suppose that the first edge (u, v_1) is placed with probability $p = c/N$. Then, the second edge (v_1, v_2), going from the first to the second intermediary, is not placed with the same probability p, because the vertex v_1 has a definite degree c, and the fact that it already has one edge attached to it changes the probability for the other edges. A simple consideration shows that the second edge is present with probability $\frac{c-1}{c}p$, and the same will hold for all other edges in the path from u to v. This leads us to the estimate $\overline{l_G} \simeq \frac{\ln N}{\ln(c-1)}$, which looks similar to that for the Erdős-Rényi graph if we are interested only in the dependence on the number of vertices; but the dependence on c is different. In terms of cycles and paths, the structures of Erdős-Rényi and c-regular graphs are mutually similar for large c, but definitely different if c is small. In fact, speaking of diameter instead of average distance, one can prove rigorously that for the c-regular graph, $c \geq 3$, $\mathrm{diam}_G = \ln N / \ln(c-1)$ in the limit $N \to \infty$ [1177].

To compute the clustering coefficient, it is sufficient to consider one triple of vertices. They form a triangle if all three edges are present, i.e. with probability p^3. There are three possibilities to have a 'vee' within the triple, each of them with probability p^2. So, the average number of triangles in the graph is $\binom{N}{3}p^3$, and the average number of 'vees' correspondingly $3\binom{N}{3}p^2$. Hence

$$C_G = p = \frac{c}{N}, \tag{6.17}$$

and we can see that in the Erdős-Rényi graphs with a fixed average degree, the clustering coefficient vanishes as the size of the graph grows.

The Erdős-Rényi graphs are frequently said to have the 'small-world' property together with low clustering. Vaguely, this means the distances in the graph are short, but the vertices do not form densely connected cliques, characterised by many triangles. More precisely, the graph is considered a 'small-world' one if the average distance grows logarithmically with the number of vertices; and indeed, this is the case of the Erdős-Rényi graph, as demonstrated in (6.16). The motivation for this concept is provided by hypercubic lattices in d dimensions. Considered as graphs, these lattices have an average distance which grows with the number of vertices as power $\sim N^{1/d}$; thus the typical distances in the small-world graphs are shorter than on a hypercube in however large a dimension. If the vertices were people placed on a two-dimensional regular grid, reminiscent of Manhattan, they would be separated by distances of $\sim \sqrt{N}$ steps; but if they are dispersed over an Erdős-Rényi graph, they are much closer to each other, only $\sim \ln N$ steps from each other.

There is a certain inconsistency in considering the logarithmic increase with the number of vertices. If we look at the average distance, we say that it is $\sim \ln N$, i.e. rather small. At the same time, computing the shortest cycle going through a given vertex, we find that it grows as $\sim \ln N$, i.e. it is large. This short-though-long paradox reveals that we are looking at the graph from different perspectives. Having in mind local properties like the average number of first or second neighbours, $\ln N$ is a large number, and we say that the graph is tree-like. If we investigate global properties such as distance, $\ln N$ is considered small, and we speak of small worlds.

Graphs with prescribed properties

The random graphs of the Erdős-Rényi ensembles $\mathcal{G}_{N,M}$ and $\mathcal{G}_{N,p}$, as well as the regular graphs, are intriguing mathematical objects, but they provide too little versatility to be used for modelling real situations. Most notably, the degree sequence is always the same, Poissonian for Erdős-Rényi, or constant for regular graphs. Networks we shall encounter for which the random graphs serve as models, exhibit various types of degree sequences, but rarely, if ever, is Poissonian found. This leads us to the question of how to construct a graph ensemble with a prescribed degree sequence which could be taken, e.g. from some concrete empirical data. Such an ensemble would be the model of the specific real situation.

We have already noted that not every degree sequence $D = [d_1, d_2, \ldots, d_N]$ can be realised in a graph. However, if we allow for loops and multiple edges between pairs of vertices, the only requirement is that the sum of degrees $\sum_i d_i$ is even. There is a construction which shows that a graph corresponds to any degree sequence satisfying this small constraint. Simultaneously it provides a clue for calculating probabilities of graphs in the corresponding ensemble, which we denote $\mathcal{G}_D$. A graph from this ensemble is obtained as follows [1178]. Imagine that each vertex puts out as many hands as the degree sequence dictates, i.e. d_1 for vertex 1, d_2 for vertex 2, etc. Now randomly choose a pair of hands and let them join. Then, choose another pair out of the remaining free hands and repeat the procedure until no uncoupled hands remain. For a more formal description, look at Box 6.1.

To see the basic properties of the graphs from the ensemble $\mathcal{G}_D$, we proceed in similar way as with Erdős-Rényi graphs. The results are, to a large extent, qualitatively equal, but modifications arise due to the heterogeneity in the degrees. Given the degree sequence D, denote by N_k the number of vertices with degree k and $2L = \sum_k k\, N_k = \sum_{i=1}^{N} d_i$. The simplest quantities are the averages

$$\overline{k} = \frac{1}{N} \sum_k k\, N_k$$
$$\overline{k^2} = \frac{1}{N} \sum_k k^2\, N_k,$$

$$(6.18)$$

and the rest mostly relies on them. The average distance in the graph is estimated by calculating the average number of paths of length l connecting a pair of vertices $v_0, v_l \in \mathcal{V}$ having degrees k_0 and k_l, respectively. A path of length l has $l-1$ intermediaries $v_1, v_2, \ldots, v_{l-1}$. Let us denote by k_i the degree of the vertex v_i. With these degrees fixed, start by looking at the probability that v_0 and v_1 are connected by an edge. There are $k_0\, k_1$ possible ways to join one of the k_0 hands of the vertex v_0 with any of the k_1 hands stemming from the vertex v_1. The remaining hands can be paired in $1 \cdot 3 \ldots (2L - 3)$ ways out of the total $1 \cdot 3 \ldots (2L - 1)$ possible pairings. This gives the probability

$$\text{Prob}\{v_0 \text{ and } v_1 \text{ connected by an edge}\} = \frac{k_0\, k_1}{2L - 1}. \qquad (6.19)$$

Supposing now that v_0 and v_1 are mutually connected, what is the probability that v_2 is also connected to v_1? The same consideration can be applied, with one simple

---

Graphs with prescribed degree sequence | Box 6.1

A graph with multiple edges connecting the same pair of vertices is a slightly more complicated object $G = (\mathcal{V}, \mathcal{E}, \varepsilon)$. In addition to the vertex and edge sets $\mathcal{V}$ and $\mathcal{E}$, respectively, there is an incidence function $\varepsilon : \mathcal{E} \to \mathcal{V} \times \mathcal{V}$ saying which vertices are connected by which edge. We consider identical two graphs which differ only in the permutation of the elements of $\mathcal{E}$. Let $D = [d_1, d_2, \ldots, d_N]$ be the desired degree sequence, and $2L = \sum_i d_i$. For a graph from the ensemble $\mathcal{G}_D$, we use $\mathcal{V} = \{1, 2, \ldots, N\}$ and $\mathcal{E} = \{1, 2, \ldots, L\}$. To construct the incidence function, consider the set $\mathcal{S} = \{1, 2, \ldots, 2L\}$, and the map $f(a)$ from $\mathcal{S}$ to $\mathcal{V}$ such that $f(a) = v$ if $\sum_{i=1}^{v-1} d_i < a \leq \sum_{i=1}^{v} d_i$. Now let $\{\{a_1, b_1\}, \{a_2, b_2\}, \ldots \{a_L, b_L\}\}$ be the partitioning of $\mathcal{S}$ into disjunct two-element subsets, $a_i \neq b_i$, $\{a_i, b_i\} \cap \{a_j, b_j\} = \emptyset$ and $\bigcup_{i=1}^{L} \{a_i, b_i\} = \mathcal{S}$. For each such partitioning we set $\varepsilon(i) = (f(a_i), f(b_i))$. It is interesting to note that the number of ways the partitioning can be done is $1 \cdot 3 \cdots (2L - 1)$, which is exactly the number of all Feynman diagrams containing L propagators for a scalar field theory. This correspondence can be pursued in depth to formulate a field theory of random graphs, see, e.g. Refs. [1179–1184].

---

but far-reaching modification that there are only $k_1 - 1$ hands available to pair with k_2 hands of the vertex v_2. Hence,

$$\text{Prob}\{v_1 \text{ and } v_2 \text{ connected} \,|\, v_0 \text{ and } v_1 \text{ connected}\} = \frac{(k_1 - 1)\, k_2}{2L - 3}, \qquad (6.20)$$

and the same procedure can be repeated until the end of the path. Now we need to take into account how many choices of the intermediaries there are. We assume the graph is large and so is the number of vertices of each degree in question, $N_{k_i} \gg l$, $i = 0, 1, \ldots, l$ (otherwise the combinatorics would be much harder). Then, the number of choices is $\simeq \prod_{i=1}^{l-1} N_{k_i}$. Eventually we sum over all choices of the degrees of all vertices in the path, i.e. the intermediaries as well as the ends v_0 and v_l, and we get a closed expression for the average number of paths of length l between two randomly chosen vertices

$$\langle N_{l-\text{path}} \rangle = \frac{\overline{k}}{N} \left(\frac{\overline{k^2} - \overline{k}}{\overline{k}} \right)^{l-1}. \qquad (6.21)$$

The average distance follows immediately:

$$\overline{l_G} \simeq 1 + \left(\ln \frac{N}{\overline{k}} \right) \left(\ln \frac{\overline{k^2} - \overline{k}}{\overline{k}} \right)^{-1} \simeq \left(\ln \frac{\overline{k^2} - \overline{k}}{\overline{k}} \right)^{-1} \ln N. \qquad (6.22)$$

At first sight it is only a slight modification of the result (6.16) for the Erdős-Rényi graphs, as the dependence on the graph size is again logarithmic. For nearly homogeneous networks, where both $\overline{k}$ and $\overline{k^2}$ converge to a finite limit when $N \to \infty$, it is indeed so. However, if the degree sequence contains many vertices with a large degree, things can differ substantially. For example, as we shall soon see, the degree distribution for many real-world graphs has a power-law tail $N_k \sim k^{-\gamma}$, and for $\gamma = 2$ we have $\overline{k} \sim \ln N$, $\overline{k^2} \sim N$, and the average distance approaches a finite limit, instead of diverging for increasing graph size.

The clustering coefficient can be obtained by repeating the same consideration for the case of triangles and 'vees'. The result is

$$C_G \simeq \frac{\left(\overline{k^2} - \overline{k}\right)^2}{N\overline{k}^3}.\tag{6.23}$$

This result is again similar to but not quite the same as the expression (6.17) for the Erdős-Rényi graph. In a generic situation with finite $\overline{k^2}$ the clustering coefficient decreases as N^{-1} with the graph size, but taking once more the example of power-law distributed degrees, now with exponent $\gamma = 3$, we have $\overline{k^2} \sim \ln N$ and $C_G \sim N^{-1} \ln^2 N$, so that the decrease of the clustering coefficient is significantly slower than for the Erdős-Rényi graph. When we use the formula (6.23) for the case of exponent $\gamma = 2$, we find, to our great surprise, that the clustering coefficient increases as $\sim N$, which is obviously wrong because C_G cannot exceed 1 by definition. The contradiction is clarified when we realise that Eq. (6.19) for the probability of two vertices to be connected tacitly assumes that $k_0 k_1 \ll \overline{k}N$. To see it on a simple example, let us imagine a graph with degree sequence $[N-1, N-1, 2, 2, \ldots, 2]$. The first two vertices are connected to all remaining ones, while the third, fourth, etc. vertices are connected just to the first two. The formula (6.19) estimates the probability of the first two being connected as $(N-1)^2/(4N-5)$, which is larger than 1. This is absurd.

If there are many vertices with a degree comparable to N, as happens for power-law distributions with exponent $\gamma < 3$, the whole line of thought leading to (6.23) must be rectified to give sensible results. The reader is encouraged to try it herself (Problem 1). Nevertheless, the general conclusion can be drawn that the random graphs with a given degree sequence can have a large clustering coefficient, if only the tail of the degree distribution is fat enough. This makes them significantly different from Erdős-Rényi graphs.

Small worlds of Watts and Strogatz

There is a sound reason for the quest for graph models with a large clustering. It was found empirically, for example in the pioneering work of Duncan J. Watts and Steven H. Strogatz [1185], that the clustering in real-world networks is high—much higher than in the Erdős-Rényi graphs of comparable size. At the same time, the small-world property, i.e. logarithmic growth of the average distance, which is typical of random graphs, is also well documented empirically, at least on a qualitative level.

We have already seen that the small-world property and high clustering coexist in graphs with power-law degree distribution if the exponent is small enough. A more transparent recipe for high-clustered small-world networks was suggested in the previously mentioned work [1185]. The idea is to merge a regular lattice, for example a cycle with second-nearest-neighbour connections, with an Erdős-Rényi random graph. The hybrid will inherit the clustering from the lattice and the average distance from the Erdős-Rényi graph.

There are several ways to implement this scheme. Let us present the most typical one. We have N vertices on a ring and each of them is connected by an edge to its $2k$ neighbours, from the nearest up to the k-th nearest ones. Moreover, other edges are added, with probability p_{WS} per each edge in the ring, i.e. randomly chosen pairs of

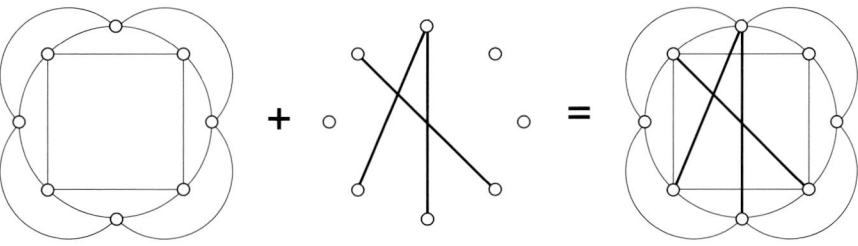

Fig. 6.2 Illustration of the Watts-Strogatz construction. The ring with $N = 8$ and $k = 2$ is combined with an Erdős-Rényi random graph with 3 edges; thus $p_{WS} = 3/16$.

vertices are mutually connected with probability $p = Nkp_{WS}/\binom{N}{2}$, making shortcuts between vertices which would have been quite distant from each other. The additional edges correspond to an Erdős-Rényi graph from the ensemble $\mathcal{G}_{N,p}$ superimposed on the regular ring we started with. Therefore, there are on average $Nk(1 + p_{WS})$ edges in the graph and the average degree is $\langle d_G \rangle = 2k(1 + p_{WS})$. More formally, the vertex set is $\mathcal{V} = \{0, 1, \ldots, N - 1\}$, and the edge set is the union of the edge set of an Erdős-Rényi graph from the ensemble $\mathcal{G}_{N,p}$ and the edges in the ring $\mathcal{E}_{\text{ring}} = \bigcup_{i \in \mathcal{V}} \{(i, i-k), (i, i+k), \ldots, (i, i-1), (i, i+1)\}$. In the latter expression, periodic boundary conditions are assumed, so $i < 0$ is identified with $i + N$, etc. An illustration of the construction is shown in Fig. 6.2.

As with the other graph ensembles, we ask again what the average distance is on the Watts-Strogatz graph. It turns out that the relevant control parameter is the total number of shortcuts $N_s = Nkp_{WS}$. Let us see what happens if N_s is either very small or very large.

Clearly, the distance between vertices i and j on the ring is $\min\{\lceil |i-j|/k \rceil, \lceil |i-j-N|/k \rceil\}$. (For those who are not familiar with the notation we recall that $\lceil x \rceil$ denotes the ceiling function, i.e. the least integer which is $\geq x$.) Therefore, the average distance is $\overline{l_G} = N/(4k)$ for a large enough N. This is the case of $p_{WS} = 0$, or $N_s = 0$.

On the other hand, for large N_s the shortcuts determine the distance in a decisive way. However, it would be a mistake to simply copy the expression (6.16) for the Erdős-Rényi graph, as the path is composed of alternating shortcuts with the passages along the ring. The latter have an average length $\xi/2k$, where $\xi = 1/(kp_{WS})$, also called correlation length, is the inverse of the concentration of the shortcuts on the ring.

Imagine now that the graph is collapsed in such a manner that only $2N_s$ vertices which are ends of the shortcuts are left, and the passages connecting the shortcut ends along the ring are replaced by single edges. The result is a random graph with an average degree equal to $c = 3$. On such a graph, the average distance is $\simeq \ln 2N_s / \ln 3 \simeq \ln 2N_s$, according to (6.16). When we translate this finding to the original Watts-Strogatz graph, we should take into account that every second step corresponds to a passage of length $\xi/2k$, so we find that its average distance is $\overline{l_G} \simeq \frac{1}{4k}\xi \ln 2N_s = \frac{N}{4k}(Nkp_{WS})^{-1}\ln(2Nkp_{WS})$.

The expressions we have just found in the opposite limits of small and large numbers of shortcuts suggest a scaling formula for the average distance

$$\overline{l_G} = \frac{N}{k} f(Nkp_{ws}) \tag{6.24}$$

where the scaling function $f(x)$ has the following limits

$$f(x) \simeq \begin{cases} \frac{1}{4} & \text{for } x \ll 1 \\ \frac{1}{4x} \ln 2x & \text{for } x \gg 1. \end{cases} \tag{6.25}$$

The scaling behaviour is indeed confirmed by numerical simulations and analytical approximations. We shall quote without proof the result of a mean-field approximation [1186]

$$f(x) = \frac{1}{2\sqrt{x^2 + 2x}} \operatorname{arctanh} \frac{x}{\sqrt{x^2 + 2x}}, \tag{6.26}$$

and it can be easily checked that the limit behaviour (6.26) indeed holds. In practice, however, the expression (6.26) does not provide an agreement with numerical data which could be deemed satisfactory. One can do better using computer-assisted enumeration of graphs with a fixed small number of shortcuts. Hence, we can construct Padé approximants and the third-order one [1187]

$$f(x) = \frac{1}{4} \frac{1 + 1.825x}{1 + 1.991x + 0.301x^2} \tag{6.27}$$

works quite well, although it does not have the exact asymptotic behaviour for $x \to \infty$.

6.1.3 Graph processes

Instead of considering the graph ensembles as static objects, we can think of building a graph by adding its constitutive elements, vertices, edges or even larger pieces, one by one. In some cases it is just a suitable way of looking at graphs which are otherwise fairly static, as Erdős-Rényi graph ensembles are, but sometimes the graph process emulates creation of the real network we are modelling.

Adding edges

Let us start with the definition. Suppose we have the vertex set $\mathcal{V}$ of N. The graph process is a sequence G_t, $t = 0, 1, 2, \ldots$, in the space of graphs with the same vertex set $G_t = (\mathcal{V}, \mathcal{E}_t)$, so that the edge sets grow monotonously, $\mathcal{E}_t \subset \mathcal{E}_{t+1}$. This means that the edges are added in the course of the process, but never removed or replaced. Strictly speaking, the sequence G_t is just a single realisation of the graph process, and to have complete description of it we must prescribe a probability to each realisation.

If we start with an empty graph, $\mathcal{E}_0 = \emptyset$ and a single edge is added in each step so that $\mathcal{E}_t$ contains t edges, and furthermore, if all such realisations are equally probable, then the state of the process at time M is exactly equivalent to the Erdős-Rényi graph ensemble $\mathcal{G}_{N,M}$.

How can the other basic ensemble $\mathcal{G}_{N,p}$ be regarded as a graph process? There are $N(N-1)/2$ pairs of vertices which can be potentially joined by an edge. Assign to all these pairs independent random variables with a uniform distribution in interval $[0, 1]$, so that $p_{(i,j)}$ belongs to the pair (i, j). Given a realisation of these random numbers,

Threshold functions | Box 6.2

We consider only monotone properties $\mathcal{P}$ of graphs, which means that if $\mathcal{P}$ holds for a graph G_t, then it also holds for any $G_{t'}$ if $t' > t$. Intuitively, a property is monotone if it cannot be spoilt by the addition of edges. For example if there exists a giant component containing a finite fraction of nodes, it can only be larger when an edge is added, but never smaller. In physical terms, it is related to a phase transition, namely crossing the percolation threshold. If we look at other monotone properties of graphs, we find many other phase transitions, occurring at different times. It is important to note that the critical points of these transitions depend on the number of vertices N. To describe this dependence we introduce for each property $\mathcal{P}$ a threshold function $\tau(N)$ such that

$$\lim_{N \to \infty} \text{Prob}\{G_t \text{ has property } \mathcal{P}\} = \begin{cases} 1 & \text{if} \quad t/\tau(N) \to \infty \quad \text{as } N \to \infty \\ 0 & \text{if} \quad t/\tau(N) \to 0 \quad \text{as } N \to \infty \end{cases}$$

which means, translated into human language, that if the graph is large enough, then the property almost surely holds for graphs with more than $\sim \tau(N)$ edges while it is almost surely violated for graphs with less than $\sim \tau(N)$ edges. In terms of the parameter p, the property $\mathcal{P}$ first emerges at the (size dependent) critical value $p_c = 2\tau(N)/N^2$. Obviously, different graph properties can have different threshold functions and different critical values p_c.

We have seen that the threshold function for the existence of the giant component is $\tau(N) = \frac{1}{2}N$. Some other examples follow. For a graph being connected it is $\tau(N) = \frac{1}{2}N \ln N$; for existence of a cycle of length k in the graph it is $\tau(N) = N$, for any $k \geq 3$; for existence of a complete subgraph with k vertices it is $\tau(N) = N^{2(k-2)/(k-1)}$; for existence of a tree of k vertices within the graph it is $\tau(N) = N^{(k-2)/(k-1)}$. The notion of the threshold function can be even more refined, introducing the hitting time for a graph property. While $c\tau(N)$ can serve as a threshold function as well as $\tau(N)$, independently of the constant factor $c > 0$, the hitting times contain the information on the proper value of the factor c. In fact, the functions $\tau(N)$ given above for the emergence of a giant component and for connectedness are hitting times, with the correct factors $1/2$ in front.

put them in an ascending order up to the value p, so that $0 \leq p_{(i_1,j_1)} \leq p_{(i_2,j_2)} \leq \cdots \leq p_{(i_K,j_K)} \leq p$. Then, at time $t \leq K$ the edge set will consist of edges (i_1, j_1) through (i_t, j_t), and then the evolution stops, $\mathcal{E}_t = \mathcal{E}_K$ for $t > K$.

We can also invert the logic and consider the parameter p as time-dependent. Again ordering all the random numbers assigned to vertex pairs, at time t we add the edge (i_t, j_t) and have $p(t) = p_{(i_t,j_t)}$. The motivation comes from a surprising observation made by Erdős and Rényi, namely, that many properties of the graphs change abruptly at a certain value of p, or, in the language of graph processes, at a certain time t. We have already seen an example, when we explained the emergence of the giant component at $p \simeq 1/N$ or at time $t \simeq N/2$. Similarly, it was proved that first cycles appear in the graph around time $\sim N$, independently of their length, and the graph becomes connected around time $\simeq \frac{1}{2}N \ln N$. For more information see Box 6.2.

Barabási and Albert

When modelling the properties of the World Wide Web, Albert-László Barabási and Réka Albert introduced a simple though incredibly rich model, which marked a small revolution in the field of random graphs. The Barabási-Albert graph process differs

from what we have seen up to now, as it adds not only edges but also vertices. Therefore, it describes a growing network [1188–1191].

First we expose this idea on an intuitive level. In each step we add a web page
with a fixed number m of links on it. The links point to already existing pages, and we
suppose that the probability that a new link refers to a specific old page is proportional
to the number of links already leading to that page. The principle that 'popular' pages
gain even more popularity in this way is called preferential attachment. This model
was devised in order to reproduce the power-law distribution of degrees in the graph
corresponding the structure of the WWW, and we shall now show that growth and
preferential attachment are indeed sufficient to produce these power laws.

When we want to formalise the above-outlined idea in a well-defined graph process,
we must first answer a few technical questions. Perhaps the simplest is what vertex
set we should choose. As we think of graphs growing without limitation, it is natural
to take the vertex set infinite; we use the set of all natural numbers, $\mathcal{V} = \{0, 1, 2, \ldots\}$.
When calculating any properties of the graph process at time t, we shall consider only
the properties of the vertices 0 through t. Another problematic issue is the proper
initial condition. If there are no edges whatsoever at the beginning, and all degrees
are zero, how could we proceed by adding edges with a probability proportional to the
degree? Moreover, what should the m edges going from the first vertex be attached to
if no other vertices are present? There are several ways to go round this problem, and
each of them leads to slightly different graphs. Unfortunately, the choice cannot be
deemed irrelevant, and we must be careful to specify which one we are using. Before
stating the precise initial condition we consider here, we mention the last remaining
problem. When more than one new edge is added with one new vertex, i.e. $m \geq$
2, the preferential attachment principle formulated above does not tell us what the
correlation is among degrees of the m old vertices. The level and type of this correlation
matters a lot. Although the most natural assumption is that of maximum independence
among the m vertices chosen, we shall see that graphs where the dependence is strong
are equally important and interesting.

Let us describe our preferred version of the Barabási-Albert graph process. Note
that the edges in the Barabási-Albert graph are naturally ordered, as they always
join the 'new' vertex t with some of the 'older' vertices $s < t$; so we shall distinguish
between the out-degree, which is always m, and the in-degree, which has non-trivial
behaviour. We shall use a slightly generalised form of preferential attachment, where
the probability of joining a vertex by a new edge is a linear function of its actual
in-degree. The absolute term of this dependence is a crucial parameter determining
the structure of the network, besides the number m of edges added in one step. It
is convenient to express the absolute term as bm, with $b > 0$. The vertex set being
fixed, the process implies only growth of the edge sets; so at time t we have graph
$G_t = (\mathcal{V}, \mathcal{E}_t)$, with edge sets increasing in time, $\mathcal{E}_t \subset \mathcal{E}_{t+1}$.

Suppose at first that $m = 1$. At time $t = 0$ there are no edges, $\mathcal{E}_0 = \emptyset$. At later times,
$t \geq 1$, one edge connecting vertex $t \in \mathcal{V}$ with another vertex which was connected to
the rest of the graph at some earlier time. Let us denote e_t the edge added at time
t, so that $\mathcal{E}_t = \mathcal{E}_{t-1} \cup \{e_t\}$. The choice of which edge is to be added is based on the
in-degrees of already connected vertices. We denote $k_{t-1}(s) = d_{G_{t-1}}^{\text{in}}(s)$ the in-degree

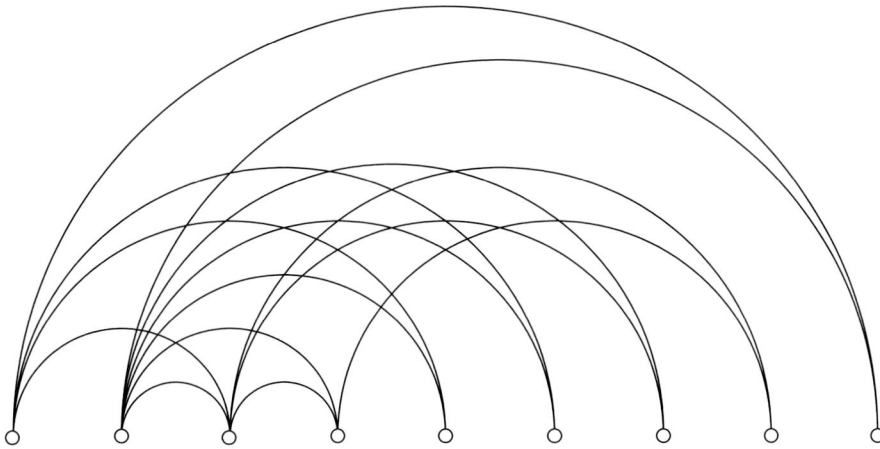

Fig. 6.3 Illustration of the Barabási-Albert graph process. The vertices are added, proceeding from the left to the right. Every new vertex contributes by $m = 2$ edges going from itself to 2 distinct older vertices, according to the preferential attachment principle. We can clearly see that older vertices have on average a higher degree.

of the vertex s at time $t-1$. According to the principle of preferential attachment, the probability that the vertex t is joined to vertex $s < t$ is a linear function of $k_{t-1}(s)$, so

$$\text{Prob}\{e_t = (t, s)\} = \frac{k_{t-1}(s) + b}{(b+1)t - 1}. \tag{6.28}$$

In the denominator we recognise the sum $\sum_{s'=0}^{t-1} \left(k_{t-1}(s') + b\right)$, ensuring the proper normalisation of the probability.

The graphs produced in this way are trees. It is relatively easy to calculate properties of their degree sequences. We can write the master equation for the probability $P_{s,t}(k) = \text{Prob}\{d_{G_t}^{\text{in}}(s) = k\}$ that at time t the vertex s has in-degree k,

$$P_{s,t+1}(k) = \frac{k-1+b}{(1+b)t - 1} P_{s,t}(k-1) + \left[1 - \frac{k+b}{(1+b)t - 1}\right] P_{s,t}(k) \tag{6.29}$$

where we completed the definition for negative k by requiring $P_{s,t}(k) = 0$ for $k < 0$. The boundary condition for the equation (6.29) is $P_{t,t}(k) = \delta_{0k}$, expressing the fact that the 'newest' vertex has no incoming edges yet.

It is easier to work with averaged quantities, namely, to calculate the global degree distribution at time t, which is $P_t(k) = \frac{1}{t+1} \sum_{s=0}^{t} P_{s,t}(k)$. We obtain the following difference equation

$$\begin{aligned}
(t+1)\Big[P_{t+1}(k) - P_t(k)\Big] &= -P_t(k) \\
&+ \frac{t}{(1+b)t - 1}\Big[(k-1+b)P_t(k-1) - (k+b)P_t(k)\Big] + \delta_{0k}.
\end{aligned} \tag{6.30}$$

If there is a stationary distribution $P_\infty(k) = \lim_{t\to\infty} P_t(k)$, it must obey the recurrence formula

$$\frac{k+2b+1}{1+b}P_\infty(k) = \delta_{0k} + \frac{k-1+b}{1+b}P_\infty(k-1) \tag{6.31}$$

with the obvious condition $P_\infty(k) = 0$ for $k < 0$. We can easily solve Eq. (6.31) by iteration, so

$$P_\infty(0) = (1+b)\frac{1}{2b+1}$$

$$P_\infty(1) = (1+b)\frac{b}{(2b+1)(2b+2))} \tag{6.32}$$

$$P_\infty(2) = (1+b)\frac{b(b+1)}{(2b+1)(2b+2)(2b+3))}$$

$$\vdots$$

and generalising the pattern you have surely spotted in (6.32), we obtain the final formula for the degree distribution in the Barabási-Albert graph with $m = 1$:

$$P_\infty(k) = (1+b)\frac{\Gamma(2b+1)\Gamma(k+b)}{\Gamma(b)\Gamma(k+2b+2)}. \tag{6.33}$$

Using the Stirling formula (see Box 3.5 if unclear) we can deduce that at large k the degree distribution obeys the power law

$$P_\infty(k) \sim k^{-2-b}, \quad k \to \infty. \tag{6.34}$$

Having solved the case $m = 1$ we turn to a slightly more complex situation with $m \geq 2$. At least two new edges emanate from every new vertex, and the graph is no longer a tree. There are several recipes for where to send the m new edges, all of them compatible with the principle of preferential attachment. Let us first show three recipes, in which the edges are as independent as possible.

The first two of them are very similar, but differ in whether we do or do not allow multiple edges joining the same pair of vertices. If we do allow multiple connections, then all of the m edges added at time t are connected to some of the $t-1$ older vertices according to the probability

$$p_t(s) = \frac{k_{t-1}(s) + mb}{m((b+1)t - 1)}, \tag{6.35}$$

analogous to the expression (6.28). The initial condition is the same as in the $m = 1$ case, i.e. $\mathcal{E}_0 = \emptyset$; and at later times the probability that the vertex $s < t$ is joined to vertex t by l edges, out of the total m, is

$$\text{Prob}\{k_t(s) - k_{t-1}(s) = l\} = \binom{m}{l}(p_t(s))^l(1 - p_t(s))^{m-l}. \tag{6.36}$$

A similar line of thought as in the case $m = 1$ leads us to the expression for the stationary distribution of in-degrees [1190]

$$P_\infty(k) = (1+b)\frac{\Gamma\big((m+1)b+1\big)\,\Gamma(k+mb)}{\Gamma(mb)\,\Gamma\big(k+(m+1)b+2\big)}. \tag{6.37}$$

We can see that the tail of the distribution again has a power-law form with the same exponent as in the $m = 1$ case. Thus, the asymptotic behaviour (6.34) holds independently of the value of m and the only crucial parameter is b, the 'initial attractiveness' of the vertices. To sum up, the exponent describing the power-law of the in-degree distribution in the Barabási-Albert graph is

$$\gamma_{\mathrm{in}} = 2 + b. \tag{6.38}$$

As we shall see later, the empirically-found values of this exponent usually lie in the interval $[2,3]$; therefore we can adjust the Barabási-Albert graph to the particular situation by appropriate choice of the parameter $b > 0$.

The fact that we allowed multiple edges facilitated the solution and ensured complete independence of the m edges added at the same time. If we forbid that multiplicity, the edges are no longer independent, but the correlation becomes very weak at larger times. So, in the limit $t \to \infty$ it does not matter if we allow multiple edges or not, and the properties of the two variants of the Barabási-Albert graphs are identical. To see an example of a graph produced in this way look at Fig. 6.3.

To see precisely how we proceed, we first specify the initial condition. The first m-tuple of edges must be connected to at least m different vertices, so in the first $m - 1$ steps no edges are added, and the edge sets are empty, $\mathcal{E}_t = \emptyset$ for $t = 0, 1, \ldots, m-1$.

In the following steps, $t \geq m$, the vertex t is iteratively connected to m older vertices, by edges $e_t^1, e_t^2, \ldots, e_t^m$. For the first edge, the attachment probability is a direct generalisation of the formula (6.28), but the placement of any further edge must take into account where the other edges have already been attached. In any case, however, $\mathrm{Prob}\{e_t^l = (t,s)\} \propto k_{t-1}(s) + mb$, although the proper normalisation of the probability is not as simple as in Eq. (6.28).

The third way to construct Barabási-Albert graphs with $m > 1$ is also possible. We can first create a realisation of the Barabási-Albert graph process with $m = 1$, stop at some time $t = mu - 1$, $u = 1, 2, \ldots$, and let every m-tuple of vertices $\{ml, ml + 1, \ldots, m(l-1) - 1\}$, $l = 0, 1, \ldots, u-1$, coalesce into one 'bigger' vertex. The edges are brought together, so that each bigger vertex has exactly m edges going from it. The advantage of such an approach is that all properties of these graphs can be, in principle, deduced directly from the Barabási-Albert graph process with $m = 1$. The disadvantage, on the other hand, is that the m edges going from one vertex are not completely equivalent. Moreover, loops can be created when coalescing the vertices, which sometimes is not desirable.

Adding triangles

The m simultaneously added edges in the Barabási-Albert graphs being independent, they do not contribute much to the clustering coefficient of the graph. We shall quote without proof the estimate for the clustering in the Barabási-Albert graphs defined in the last paragraph [1176]

$$C_G \simeq \frac{m-1}{8}\frac{(\ln t)^2}{t} \tag{6.39}$$

showing that the clustering is nearly as small as for the Erdős-Rényi graphs. Fortunately, it is relatively easy to tune the Barabási-Albert graph process so that it produces graphs with larger or smaller clustering, according to our wishes. Of course, there are some strict bounds we cannot surpass, and one of the purposes of this paragraph will be to estimate these bounds.

Let us see how it works for $m = 2$. We also forbid multiple edges now. Clearly, adding two new edges at time t connected to vertices s_1 and s_2 can produce a triangle or not, depending on whether there was an edge connecting s_1 with s_2. It is possible to create a triangle in each step, starting with two vertices connected by an edge, so $\mathcal{E}_1 = \{(1,0)\}$. Then, in each subsequent step $t \geq 2$, we choose an existing edge $e \in \mathcal{E}_{t-1}$ with uniform probability, and connect the two new edges to the endpoints of e, i.e. if $e = (s_1, s_2)$, $s_1 > s_2$, then $\mathcal{E}_t = \mathcal{E}_{t-1} \cup \{(t, s_1), (t, s_2)\}$. This construction satisfies the preferential attachment principle, although in a specific form. The edges are naturally oriented, even if we do not take the orientation into account explicitly. The newer of the two vertices, s_1, is chosen uniformly from among the $t - 1$ vertices available (obviously, the vertex 0 must be excluded), but the older vertex, s_2, is chosen with probability proportional to its in-degree, just due to the random choice of the edge. Therefore, the probability that a vertex s will be joined to t by an edge is equal to $\text{Prob}\{k_t(s) - k_{t-1}(s) = 1\} = 1/(t-1) + k_{t-1}(s)/(2t-1)$, so that it is linear function of the in-degree of s, as required. The preferential attachment emerges automatically. Moreover, we can see that the corresponding parameter is $b = 1$; hence the exponent in the power-law tail is $\gamma_{\text{in}} = 3$.

We are now ready to estimate the clustering coefficient. The number of triangles at time t is exactly $t - 1$, and we only need to calculate the number of 'vees'. In this respect we shall ignore the orientation of the edges. The edge (t, s_1) creates on average as many 'vees' as is the average total degree, which is $2(2t - 1)/t \simeq 4$ (for large t we neglected the fact that $s_1 \neq 0$). The other edge, when connected to a vertex with in-degree k, which happens with probability $k/(2t - 1)$, creates $k + 2$ 'vees'. Summing over all vertices, we have on average $\sum_{s=0}^{t-1} k_{t-1}(s)\big(k_{t-1}(s) + 2\big)/(2t - 1)$ new 'vees'.

Now we can see that the dominant contribution comes from the sum in which the square of the in-degree is not bounded as $t \to \infty$. Instead, the power-law tail of the in-degree distribution $P_{\text{in}}(k) \sim k^{-3}$ leads to logarithmic divergence. Therefore, the total number of 'vees' added at step t behaves like $\ln t$ and the clustering coefficient decreases as

$$C_G \sim \frac{1}{\ln t} \tag{6.40}$$

i.e. much more slowly than for the Erdős-Rényi graph, although it still does not approach a finite limit when the size of the graph increases. We cannot add more than one triangle at a time, and the number of 'vees' added also cannot be made smaller than $\ln t$ as it follows from the preferential attachment, which we want to keep at any cost. So, we conclude that (6.40) represents the highest clustering we can ever attain in a Barabási-Albert graph of this sort. One possibility to get a finite clustering in a graph process with preferential attachment would be to choose the parameter b in such a way that the sum $\sum_s^{t-1} k^2(s)$ would converge for $t \to \infty$, i.e. prescribe $b > 1$ by hand. We leave the discussion of this possibility to the reader.

Instead, we shall mention the opposite situation. Up to now, we have tried to add as many triangles as possible, to ensure high clustering. But it is also possible to avoid adding triangles at all, keeping the clustering at zero by definition. One simple method for making the avoidance of triangles compatible with preferential attachment is just to invert the procedure of adding triangles. Again, we choose one vertex $s_1 < t$ with uniform probability among those already present, exclude the two vertices which are connected to s_1 by an edge, and choose among the remaining $t - 3$ vertices according to the preferential attachment rule like that of Eq. (6.28). The probability that a randomly chosen vertex s receives an edge from t is then a linear function of its in-degree, which is all we want. Thus, the clustering coefficient of Barabási-Albert graphs has a natural upper bound, but the lower bound is zero. An interesting alternative to get highly-clustered graphs with power-law degree distribution is described in Refs. [1192, 1193].

We have seen that the Barabási-Albert graph process can be modified and tuned in many ways, producing graphs with variable properties. This very fruitful ground has been explored from many points of view [1194–1215], but we stop here and turn to even simpler processes generating graphs with power-law degree distributions.

Adding copies of the graph itself

Most of this chapter is devoted to random graphs and their real embodiments, random networks. However, some of their properties can be well demonstrated on graphs which are deterministic, with no random element. These graphs are very easy to handle; therefore, they provide a helpful tool for various analytical calculations, although the lack of randomness also brings about some unnatural features.

One of the main motivations is to model strongly heterogeneous graphs, namely those with power-law degree distribution. The Barabási-Albert graph process is the prominent model, but still too complicated for some purposes. It is possible to construct a graph with power-law degree distribution iteratively, at each step making several identical copies of the entire graph and then joining them in a specific way. In fact, there is experimental evidence that metabolic networks, describing the chemical reactions in the cells of our body, look very much like that. They consist of hierarchically assembled modules [1216–1218]. Something similar was also observed in the structure of the internet [1219].

Let us see one possible way to make the model in practice [1220]. At the time $t = 0$ the graph consists of a single vertex and no edges. At the time $t = 1$ we add p new vertices and join them to the first vertex. In what follows, we shall call the first vertex of the graph its root. At the next time, $t = 2$, we make additional p copies of the existing graph and join all vertices in the copies to the root in the original. At time t the graph contains $N_t = (p+1)^t$ vertices. We can see an illustration of the construction for $p = 2$ in Fig. 6.4. The graph thus obtained is usually called a deterministic scale-free graph. A formal description of the process can be conveniently formulated in terms of the adjacency matrix. Denote by $A(t)$ the adjacency matrix at time t, with initial condition $A_{11}(0) = 0$. At time $t + 1$ we make a direct sum of $p + 1$ times the matrix $A(t)$ itself and then add the new edges to the root. We denote $B(t + 1)$ the matrix corresponding to the new edges, $B_{mn}(t + 1) = B_{nm}(t + 1) = 1$ for $n = 1$,

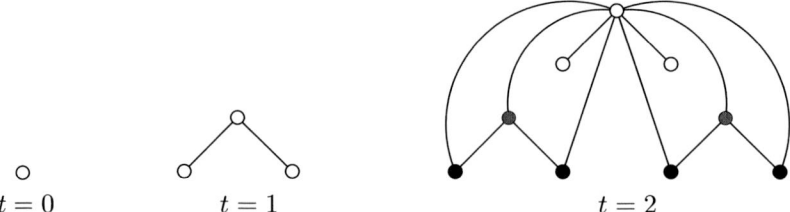

Fig. 6.4 Illustration of the deterministic scale-free graph. At each step, $p = 2$ new copies of the graph are created, and links are added, joining the root vertex in the original to all vertices in the copies. In the graph created at step $t = 2$, nodes in grey are type [112], and nodes in black are type [122].

$(p + 1)^t < m \le (p + 1)^{t+1}$, and zero otherwise. Thus,

$$A(t + 1) = \underbrace{A(t) \oplus A(t) \oplus \cdots \oplus A(t)}_{p+1 \text{ times}} + B(t + 1). \tag{6.41}$$

As in the Barabási-Albert graph process, the edges are naturally ordered, leading from the copies to the root in the original. We already used this ordering in the definition of the matrix $B(t)$. Now we would like to find the degree sequence. We first classify the vertices according to how and when they emerged in the graph. The classes will be formally described by sequences $[s_0 s_1 \ldots s_t]$ characterising the types of the vertices. Initially, there is only one vertex, and its type is denoted by [1]. Next, at time $t = 1$ there is the root with type [11] and p copies of type [1p]. We proceed iteratively further. At time $t + 1$, we make p new copies of the graph. Let us look what happens with vertices which were of type $[s_1 \ldots s_t]$ at time t. At the next time step, $t + 1$, we denote $[s_1 s_2 \ldots s_t 1]$ the type of the same vertices in the original, while $[s_1 s_2 \ldots s_t p]$ is the type of the vertices in any of the p copies. An example can be seen in Fig. 6.4. The reason for using 1's and p's in the type sequence is simple: there are exactly $\prod_{i=0}^{t} s_i$ vertices of type $[s_1 \ldots s_t]$.

To see what the degree is of a vertex with a specified type, we look first at the in-degree. Vertices receive incoming edges as long as they are roots. We can easily read for how long the vertex had been the root from its type sequence, because the first appearance of a p in the sequence means that the vertex appeared in the copy and thus stopped receiving any more edges. If $s_i = 1$ for $i \le l$, but $s_{l+1} = p$ (or $l = t$, if the vertex is still the root at time t), then its in-degree is $(p + 1)^l - 1$, because it is connected to $(p + 1)^l$ vertices existing at time l, except itself. There are $(p+1)^{t-l}$ such vertices, so, denoting $N_t^{\text{in}}(k)$ the number of vertices with in-degree k, we have

$$N_t^{\text{in}}((p + 1)^l - 1) = (p + 1)^{t-l} \tag{6.42}$$

for discrete values $l = 0, 1, \ldots, t$. To compare this result with the degree distribution in Barabási-Albert and other random graphs, we need to calculate the cumulative number $N_t^{\text{in}\ge}(k) = \sum_{k' \ge k} N_t^{\text{in}}(k')$. We get

$$N_t^{\text{in}\ge}(k) = \frac{1}{p} \left[\frac{(p + 1) N_t}{k + 1} - 1 \right], \tag{6.43}$$

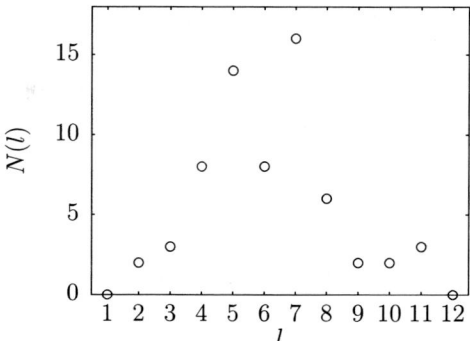

Fig. 6.5 Results of Milgram's experiment demonstrating the small world phenomenon. The plot shows the number $N(l)$ of letters which reached the target person in l steps. Data for this plot are taken from [1239].

and with $P_{\text{in}}(k) \simeq \frac{1}{N_t} \frac{d}{dk} N_t^{\text{in}\geq}(k)$ we conclude that the probability distribution approximately decays as a power law

$$P_{\text{in}}(k) \sim k^{-2}, \tag{6.44}$$

i.e. the degree exponent is $\gamma_{\text{in}} = 2$ independently of p.

The out-degree is also easy to obtain. A vertex sends an edge out each time it is in one of the p copies. So, the out-degree of a vertex is equal to the number of p's in its type. There are $\binom{t}{k}$ different types containing k times p, and there are p^k vertices of each of these types. So, the number of vertices with a given out-degree is given by binomial distribution

$$N_t^{\text{out}}(k) = \binom{t}{k} p^k. \tag{6.45}$$

The number of edges can be obtained easily, $E_t = tp(p+1)^{t-1}$, and recalling the number of vertices, $N_t = (p+1)^t$, we have that the averaged in- and out-degree is

$$\overline{k_{\text{in}}} = \overline{k_{\text{out}}} = \frac{tp}{p+1} = \frac{p}{(p+1)\ln(p+1)} \ln N_t. \tag{6.46}$$

By similar considerations, one can obtain the exact formulae for the clustering coefficient, average distance, and other properties of the graph. In all cases, the procedure relies on iterative construction of the graph. Knowing its properties at time t, we can compute, with not too much effort, the properties at time $t+1$. The same path can also be used, at least in principle, for several other iterative constructions known in literature [1221–1229] or for the so-called Apollonian networks [1230–1232].

6.2 Statistics of real networks

The theory of random graphs started as an abstract discipline, although it was soon realised that it might serve as a mathematical basis for modelling social networks and various other real structures. The boom actually started at the end of the 1990s, when

---

Robots and crawlers   [Box 6.3]

are specialised pieces of software whose purpose is to automatically download web pages, analyse their content, store pertinent information on disk, and then proceed to download other pages. It is relatively easy to write a very simple robot which recognises all URLs on a page and downloads recursively all URLs it comes across. Such activity of the robot is sometimes called a crawl. The business of large internet search engines mainly consists in crawling as large a portion of the Web as possible and copying down the content of the pages. Thus, the owners of the engine have their own copy of (nearly) all relevant files ever published on the WWW and digest the data in various ways to make the response to your search request as quick as possible. Although an individual or a small research group can hardly compete in general, it is possible to write, without an extreme effort, a specialised robot designed to answer a specific question on the Web. See Ref. [1242] for programming hints.

---

the first thorough empirical studies on the structure of the internet and the WWW were published [1188, 1189, 1233–1238].

6.2.1   Milgram's letters

It is fair to point out that the complexity of social networks has been the subject of intense study among sociologists since the 1960s [1239–1241]. Stanley Milgram especially made the subject popular through the term 'six degrees of separation', which is the colloquial summary of his study on letters blindly making their way from one corner of the United States to the other.

Milgram distributed letters among 296 volunteers, 196 of them in Nebraska and the rest in Boston. The instruction was to help to deliver the letter to the target, who was a person living in Sharon, Massachusetts, a suburb of Boston. The condition was that the letter could be sent directly only to a person known to the sender on a personal basis. So, the letter was to proceed toward the target through a chain of personal acquaintances. By measuring the lengths of these chains, Milgram obtained some information on the distances between vertices in the social network of the inhabitants of the USA.

We can see in Fig. 6.5 the result of his experiment. For each of the 64 letters which arrived at the addressee, the sequence of people $(v_0, v_1, \ldots, v_l)$ was registered, where v_0 is the first sender and v_l the target person. The number of letters $N(l)$ going along a path of length l exhibited a two-peak structure, which was traced back to the social group of the people involved. Shorter paths, around $l = 5$, were composed of people related to the target's work, while longer paths, around $l = 7$, reached the target through his hometown. The average length of all completed paths was $\bar{l} = 6.16$. Hence the term 'six degrees of separation'. This is a very low number, considering the hundreds of millions of inhabitants living in the United States. The conclusion is that the graph representing the structure of the social network of North Americans has a very low average vertex distance. We already know that the models of random graphs, beginning with those of Erdős and Rényi, indeed have an average distance which can be considered small, more precisely growing as a logarithm of the size. Therefore, the Erdős-Rényi graphs, as well as other more elaborate models presented in the preceding section, grasp quite well this 'small-world' feature of real networks.

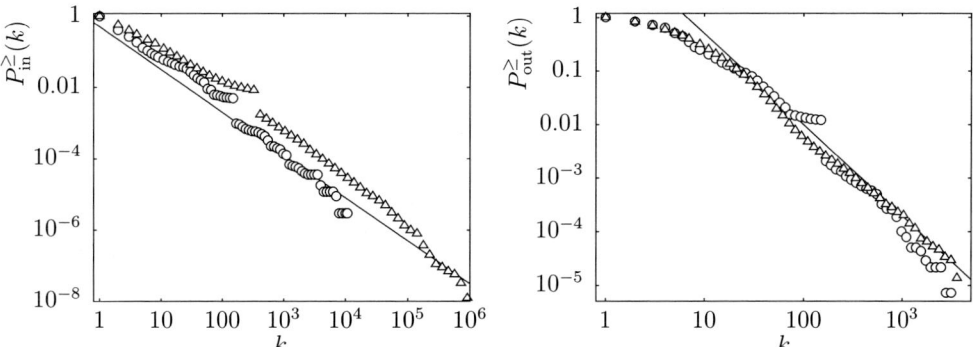

Fig. 6.6 Degree distributions on the WWW. In the left panel, cumulative in-degree distribution. The straight line is the power-law dependence $\propto k^{-1.2}$. In the right panel, the cumulative out-degree distribution, the line indicating the power law $\propto k^{-1.7}$. The symbols distinguish between the two crawls considered. The crawl of nd.edu domain, containing $325,729$ pages, is denoted by $\bigcirc$. The data were downloaded from the page www.nd.edu/~networks/resources.htm. The Alta Vista crawl, comprising 271 million pages [1238], is denoted by $\triangle$. Note the jumps in the data from the smaller sample, which are due to occasional regularities in the structure of the webpages. In the larger sample, most of these accidental features are averaged out. Data from Ref. [1238] replotted with permission of Ravi Kumar.

6.2.2   Degree distributions

However, the average distance is only one of many important features of real networks we are interested in. The next is the degree sequence, and here the Erdős-Rényi graphs completely fail to account for the observed data. The most important finding which aroused the interest of physicists in the field of random networks was the power-law degree distribution in many real-world systems. Such networks are called scale-free, although their self-similar nature is rather subtle [1243].

Technological

Let us look at some typical examples. The first one is the WWW. Fig. 6.6 shows the cumulative distribution of in- and out-degrees, $P_{\mathrm{in}}^{\geq}$ and $P_{\mathrm{out}}^{\geq}(k)$, respectively, i.e. the fraction of nodes with an in- or out-degree larger than or equal to k. Two data sets are shown there. The smaller one, of $\simeq 3 \cdot 10^5$ documents, comes from the crawl (see Box 6.3 if unclear) performed by A.-L. Barabási and coworkers on the nd.edu domain of the University Notre Dame [1189]. The larger set, based on an Alta Vista crawl [1238], contained $\simeq 2 \cdot 10^8$ documents.

We can clearly see the power-law dependence,

$$P_{\mathrm{in,out}}^{\geq}(k) \sim k^{1-\gamma_{\mathrm{in,out}}}$$
$$\gamma_{\mathrm{in}} \simeq 2.2$$
$$\gamma_{\mathrm{out}} \simeq 2.7,$$

(6.47)

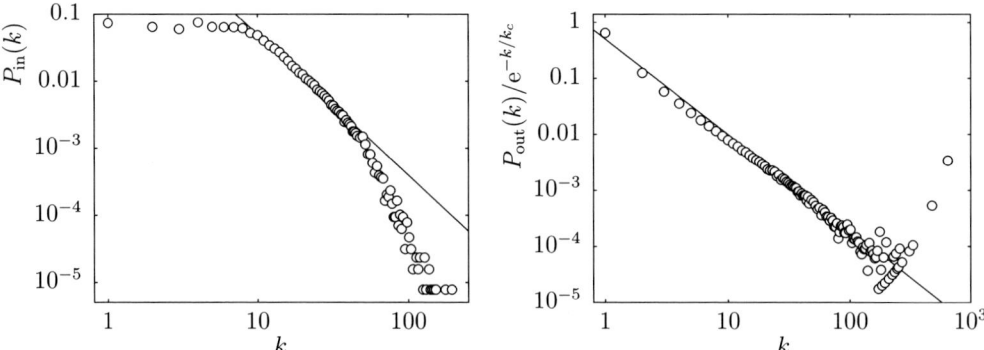

Fig. 6.7 Degree distribution of the bipartite graph of actors-movies network. The edges point from actors to movies. In the left panel, the distribution of in-degrees (actors per movie). The central part should be compared with the line indicating the power law $\propto k^{-2.1}$. The right panel shows the out-degree distribution (movies per actor). The distribution is consistent with the exponentially damped power law, with cutoff value $k_c = 90$. The line is the power law $\propto k^{-1.7}$. The data, published in [1188], were downloaded from the page www.nd.edu/~networks/resources.htm.

extending over more than two orders of magnitude. It is also clear that, besides the accidental effects and noise, the distribution is fairly robust when we increase the size of the sample.

Many other real networks have been analysed in the same way. Table 3.7 in the book [1244] lists 37 studies, most of them giving power-law degree distributions with exponents in the range $2 \lesssim \gamma \lesssim 3$. The most important examples include the physical structure of the internet on the autonomous systems level, with $\gamma \simeq 2.2$ [1237, 1245, 1246]; the citation network of scientific papers [1191, 1247, 1248], where the tail of the distribution was fitted on power laws, with exponent ranging— depending on the method used—from $\gamma \simeq 3$ [1247] to $\gamma \simeq 1.9$ [1248]; the network of e-mail communication, with exponent $\gamma \simeq 1.8$ [1249]; the acquaintance network in the private web club www.wiw.hu was shown to have power-law degree distribution, with exponent $\gamma \simeq 2$ in the tail [1250]; the wiring scheme of a large digital circuit [1251] with exponent $\gamma \simeq 3$. It is very reasonable to model these networks using the Barabási-Albert graph process, but modifications are surely needed to account for details in the structure of the networks. For example, we skipped the very important discussion of the empirically found degree-degree correlations, betweenness and other. All of these features require some specialities in the models used. Instead of going into a deep discussion on them, we refer the reader to dedicated network literature [1177, 1244, 1252–1258] and proceed with a little more empirical data.

Social

Fig. 6.7 shows the degree distribution in the oriented bipartite graph of actor collaborations. The edges lead from actors to movies, according to the cast. So, the in-degree

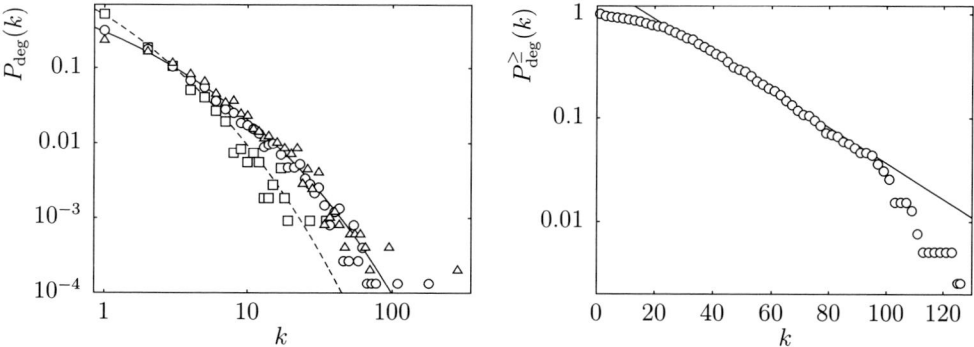

Fig. 6.8 In the left panel, degree distribution of the protein interaction network for the fruit fly *Drosophila melanogaster* (○), baker's yeast *Saccharomyces cerevisiae* (△), and *Homo sapiens* (□). The lines are stretched-exponential fits $\propto e^{-c\,k^{0.3}}$, the solid one with parameter $c = 2.7$, the dashed one with parameter $c = 4$. The data were downloaded from the DIP database at the site dip.doe-mbi.ucla.edu. In the right panel, cumulative degree distribution of the neural network of the worm *Caenorhabditis elegans*. The straight line is the exponential fit $\propto e^{-k/k_c}$ with $k_c = 25$. The data were downloaded from the Wormatlas page, www.wormatlas.org.

is the number of actors who played in a specific movie, and the out-degree is the number of movies the actor in question played in. The in-degree exhibits a power law in the central part of the distribution, with exponent $\gamma_{\rm in} \simeq 2.1$, but the out-degree has a significant exponential cut-off, $P_{\rm out}(k) \sim k^{-\gamma_{\rm out}} e^{-k/k_c}$ with exponent $\gamma_{\rm out} \simeq 1.7$ and a cutoff value of $k_c \simeq 90$. Here it is much less clear if we can call the network scale-free, as there is, in fact, no power-law tail. Nonetheless, the degree distribution is still very broad.

Clearly, it would be overstated to claim that all social networks are scale-free. Other studies on social networks have, for example, shown that the network of sexual contacts in Sweden has a clear power-law degree distribution with an exponent around $\gamma \simeq 3$ [1259], while the acquaintance network in the Mormon community was well fitted on a single-peak Gaussian distribution [1260]. We must be very careful when we decide to 'model the society': for different places, different times and different questions require different graphs to be used.

The network of linguistic relationships exhibits yet other specific network features [1261–1263]. The degree distribution shows a clear distinction between two subsets of the lexicon, the frequent and less-frequent words. The former contribute to the tail of the distribution and are described by a power law with exponent $\gamma \simeq 2.7$, while the latter is apparent for lower degrees, where power law with a different exponent, $\gamma' \simeq 1.5$, holds. A modification of the Barabási-Albert graph process was developed to account for this behaviour [1264].

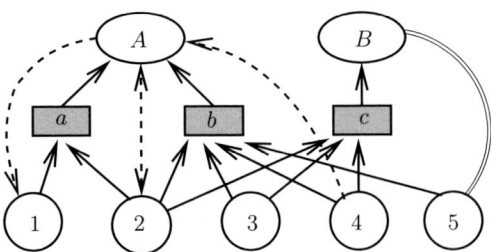

Fig. 6.9 Simplified scheme of the eBay auction site. The bidders 1, 2,... compete for buying the items to be sold a, b,..., which are offered by the sellers A, B,.... The full lines with arrows indicate who places a bid for what item and what item is offered by whom. Besides that, a user can act as both seller and bidder (not for the same item, of course!) so the double line in the figure indicates that 5 and B are in fact the same person. Moreover, the participants can rate each other, writing a report and giving some points to the other side. These second-level relations are indicated by dashed lines with arrows. The reports may also be bilateral, for example both A and 2 in the figure reported on the behaviour of the partner.

Biological

Let us also mention the networks studied by biologists. Two typical situations are shown in Fig. 6.8. The protein-protein interaction network expresses the relationships between various proteins present in the cell. The vertices of the corresponding graph are the proteins themselves, and they are joined by an edge if they come into contact for a chemical reaction. To establish such a network of interactions is a subtle experimental problem [1265–1267] and we can imagine the method used as attaching a lock to one protein, and a key to the other one. If the two interact, they come close to each other; the key fits in the lock, and a signal reaction is triggered, announcing that the interaction has taken place. The data systematically collected in this way have been analysed from the point of view of their graph structure [1268]. The first information is the degree distribution, and we show it in Fig. 6.8 for three organisms: the fruit fly, baker's yeast, and the human. As we can see, in all three cases shown here the distribution has rather fat tails, though it is definitely not a power law. The stretched exponential function

$$P_{\deg}(k) \propto e^{-ck^\alpha} \tag{6.48}$$

fits the data much better. It is interesting to note that, independent of the organism, we find $\alpha \simeq 0.3$, while the parameter c does depend on the species in question. Models trying to account for the observed distributions can be found in Refs. [1269–1271].

Another well-studied case of a biological network is the neural system of a tiny worm *Caenorhabditis elegans* [1272]. When the degree distribution was extracted [1260], it was found that it obeys quite well an exponential law, at least in the middle part of the distribution, as can be seen in Fig. 6.8.

We should also mention the food webs [1273–1276], where it is somewhat disputable as to whether the degree distribution is exponential or power law. The samples are

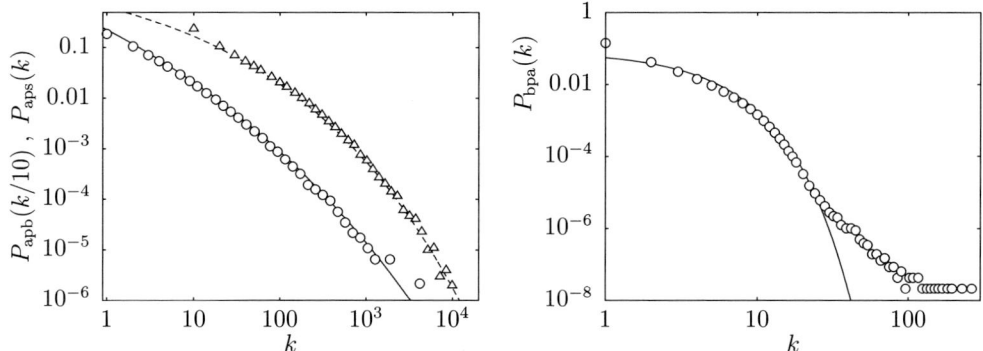

Fig. 6.10 Degree distributions in the tripartite graph of Aukro sellers, auctions, and bidders. In the left panel, we show the distribution of auctions per seller (○) and auctions per bidder (△). The latter distribution was shifted rightwards by the factor 10, for better clarity of the plot. The lines are the functions $\propto \exp(-7.5\,k^{0.12})$ (solid line) and $\propto \exp(-3.5\,(k/10)^{0.21})$ (dashed line). In the right panel, we show distribution of bidders per auction. The line is the function $\propto \exp(-0.38\,x)$. The data were collected from the aukro.cz site during the years 2010 and 2011. In total, the graph consists of about 10^6 bidders, $5 \cdot 10^5$ sellers, and $5 \cdot 10^7$ auctions.

rather small to give a definite answer. However, some other features, e.g. the extremely small average distance about $\overline{l_G} \simeq 2$, are reproduced in several interesting models [749, 1277–1281].

To sum up, a variety of real-world networks can be classified into several groups according to their degree distributions. Most of them are rather broad, with maxima at very small degrees, but allowing much larger degrees with a relatively high probability. None of the networks mentioned here exhibited the Poisson distribution of degrees predicted for the Erdős-Rényi random graphs. This underlies the necessity to use alternative models of random graphs, closer to empirical data. The Barabási-Albert model is a very good candidate for some networks, like the WWW or citation network, as it splendidly reproduces the power-law degree distribution. But many other situations, notably the biological networks, do not obey the power law and alternative models must be looked for. This direction remains to a large extent unexplored yet.

But if we remain within the scope of economic and social phenomena, the power-law distributed, or scale-free, networks seem to prevail, and very often the Barabási-Albert graph is chosen as a 'canonical' model of the social network we want to implement within the model.

6.3 Electronic commerce networks

Let us now look in more detail at a specific group of real-world networks, with specific importance placed on economic activities. With the advent of the WWW, a large portion of trade moved to electronic media [1282]. While the substance of commerce remains largely unchanged, the means used are completely new, and still newer ones

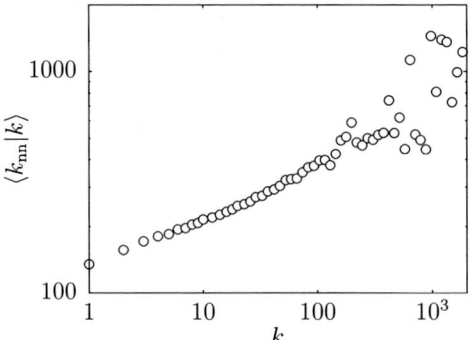

Fig. 6.11 Degree correlations in the network at the Aukro site. We show here average degree of the neighbours of a vertex whose degree is k, in a collapsed bidder graph. The data used for this plot are the same as those of Fig. 6.10.

are being invented. Both producers and customers look happy and, what is most important to us, scientists rejoice as well, because through the web they have access to an immense amount of data on the transaction details, which can be analysed and processed in a plenty of ways. One of the key questions is how the agents in electronic commerce are tied to each other and to their products and services, i.e. what the structure of the underlying commercial network is. We shall discuss two examples where the network is readily accessible to a scientific investigation.

6.3.1 Auctions

One of the biggest WWW sites is eBay, the prominent provider of online auctions. A great many inhabitants of the Earth have used it at least once. The network structure behind that site is sketched in Fig. 6.9. The sellers, A, B,...offer the items a, b, c,...for auctions. The potential buyers 1, 2, 3,...place their bids. When the auction ends, the highest offer wins, and the bidder becomes the buyer of that item. Both sellers and bidders have unique IDs; the participants of each transaction may publish their comments on the reliability of their partner, and each user gains some reputation. Therefore, there is strong feedback encouraging fairness and marginalising those who have committed a fraud. The effect of reputation was studied theoretically [1283], and it was found that it leads to a substantial increase in the overall activity, for users trust the market much more.

Although the bidding process is extremely interesting at all levels of description [1284–1288] and inspires various numerical models as well [1289, 1290], here we are interested only in its network structure. It can be investigated in various ways, since the relationships between nodes in the whole structure of eBay are multiple. We can see in Fig. 6.9 that it is rather intricate, with edges of several kinds. The simplest approach considers the tripartite graph where vertices are bidders on one side, individual auctions are in the middle, and sellers are on the other side. The basic characteristics of this graph are the degree distributions, calculated separately on the bidders' side (auctions per bidder) and on the sellers' side (auctions per seller). The third degree distribution concerns auctions as connected to bidders (bidders per auction). We can see the results in Fig. 6.10. (The data actually plotted there were collected from the auction site aukro.cz, which is a part of the multinational Allegro group [1291]. The functioning is nearly identical to eBay.)

Both the distributions of auctions per seller $P_{\mathrm{aps}}(k)$ and auctions per bidder $P_{\mathrm{apb}}(k)$ seem to be close to a power law, but more careful inspection shows that much better fit to the empirical data is provided by stretched-exponential laws $P_{\mathrm{aps}}(k) \sim \mathrm{e}^{-a\,k^{0.12}}$ and $P_{\mathrm{apb}}(k) \sim \mathrm{e}^{-b\,k^{0.21}}$ with appropriate constants a and b. The third of the degree distributions describes the number of bidders per auction. As we can see in Fig. 6.10, it is exponential for not too large degrees, but develops a fatter tail for large k. It is impossible to determine the precise form of the tail using just the currently available data.

It is also interesting to see what happens with the network when we 'collapse' the tripartite graph in such a way that we retain just one of the three types of vertices, i.e. either sellers, or auctions, or bidders only. For example, we can focus on bidders. Within the collapsed bidder graph, we draw an edge connecting a pair of bidders if both of them participated in the same auction (or auctions).

An important feature to investigate is the correlation between degrees in the collapsed graph. The simplest way to do it is to fix one vertex v with degree k and calculate the average of degrees of all vertices connected to v. Moreover, we average this quantity over all such vertices v having the same degree k. The resulting conditional average for a general graph $G = (\mathcal{V}, \mathcal{E})$ is defined as

$$\langle k_{\mathrm{nn}} | k \rangle = \sum_{\substack{v \in \mathcal{V} \\ d_G(v)=k}} \sum_{\substack{u \in \mathcal{V} \\ (v,u) \in \mathcal{E}}} d_G(u) \Big/ \sum_{\substack{v \in \mathcal{V} \\ d_G(v)=k}} k. \tag{6.49}$$

We can see in Fig. 6.11 that in the collapsed bidder graph this quantity grows with k, indicating that vertices with higher degree have neighbours which have themselves higher degree, and vice versa. Graphs of this type, with positive correlations between degrees of neighbouring vertices, are called assortative; and in the opposite case of negative correlation we speak of disassortative graphs. The empirical investigations [1214, 1292–1295] of various networks have shown that social networks are mostly assortative, while technological (power grid, internet, WWW) or biological (protein interactions, metabolic networks, etc.) are disassortative. As the networks emerging from the online bidding reflect social mechanisms, the assortativity of the collapsed bidders graph is entirely consistent with findings on other social networks.

6.3.2 Reviewers' networks

An essential ingredient for proper functioning of electronic commerce is the immediate feedback from the users. We have already mentioned that in the last paragraph in the context of the eBay auctions, and it is perhaps the site where the reactions of both sellers and buyers have been most developed. To see how it works elsewhere we shall jump now to another prominent site, amazon.com. With Amazon, you typically buy books or music, but effectively all kinds of not too heavy goods are offered, including

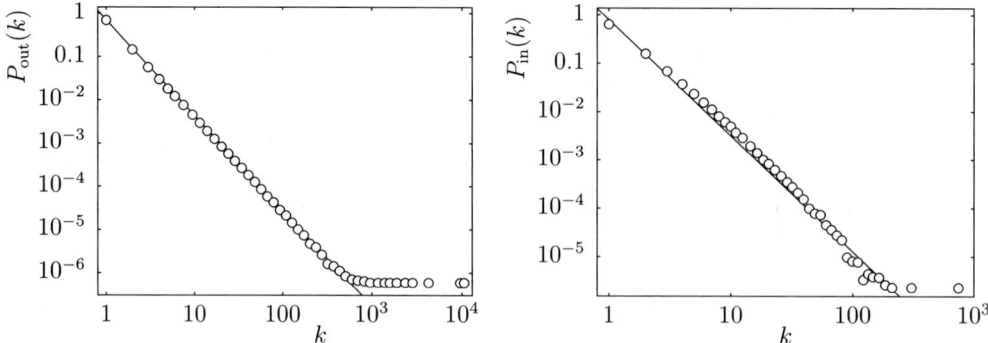

Fig. 6.12 Properties of the reviewer-item network at the amazon.com site. In the left panel, the in-degree and in the right panel, the out-degree distribution is plotted for the bipartite reviewer-item network. The straight lines are powers $\propto k^{-2.2}$ (out-degree) and $\propto k^{-2.4}$ (in-degree).

tyres and hand tools. Quality of every item can be reported by anyone who registers herself as a reviewer. People who read the reviews can vote on their usefulness so that abuse of the refereeing system is discouraged.

The reviewers and the items they write about form a bipartite network, with links going from reviewers to items, much like the bidders-items network in eBay. The reviewer-item network at Amazon is quite large, the number of reviewers exceeding 1.7 million in August 2005. The first information about its structure is provided by the degree distribution, as shown in Fig. 6.12 [1296]. Both in- and out-degree are distributed as power laws with exponents $\gamma_{\text{in}} \simeq 2.4$ and $\gamma_{\text{out}} \simeq 2.2$. This holds for a vast majority of the vertices, but there are some outliers in the tails, with a very large degree. For example the Amazon No. 1 reviewer has written more than 11 thousand reviews.

As with the auctions network, we can also study the collapsed reviewer network. We shall apply a different path here, using spectral analysis, to show how this specialised tool for studying networks [1219, 1228, 1297–1315] reveals new features.

There are several ways to define the collapsed network. One of them, used in the analysis of the eBay (or Aukro) network, assumes a single edge between two reviewers if there is at least one item on which both of them wrote a review. Denoting a_{iu} the adjacency matrix of the bipartite network, with index i denoting a reviewer and u an item, the following matrix results from the collapse

$$M_{ij}^{(1)} = (1 - \delta_{ij})\,\theta\Big(\sum_u a_{iu} a_{ju}\Big). \tag{6.50}$$

The Kronecker δ is here to suppress loops connecting the reviewer with herself. Another way is to take into account how many items the two reviewers have in common. It amounts to placing multiple edges between pairs of vertices. We get the following matrix:

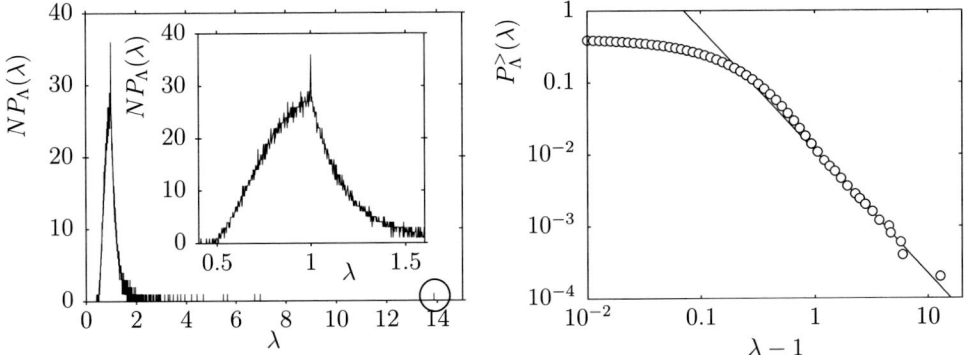

Fig. 6.13 Spectrum of the segment containing the first $N = 5000$ reviewers from the collapsed reviewer network at the amazon.com site. The left panel shows the histogram of eigenvalues of the matrix $M^{(3)}$. There are 5000 equal-sized bins starting at the lowest and ending at the highest eigenvalue. The largest eigenvalue is emphasised by the circle. In the inset, detail of the same histogram is plotted. In the right panel we show the tail of the integrated density of eigenvalues. The straight line is the power law $\propto (\lambda - 1)^{-1.7}$.

$$M_{ij}^{(2)} = (1 - \delta_{ij}) \sum_u a_{iu} a_{ju}. \tag{6.51}$$

However, for the sake of spectral analysis it is useful to normalise the matrix elements with respect to the number of reviews the reviewers wrote, so we arrive at yet another matrix:

$$M_{ij}^{(3)} = \frac{\sum_u a_{iu} a_{ju}}{\sqrt{\left(\sum_u a_{iu} a_{iu}\right)\left(\sum_u a_{ju} a_{ju}\right)}}. \tag{6.52}$$

The entire matrix $M^{(3)}$ is too large to be diagonalised using standard procedures. Indeed, we have already noted that the number of reviewers exceeds one million. Special techniques, like the Lanczos algorithm [1316], should be used instead. However, even a tiny subset of first 5000 reviewers (in the order Amazon itself attributes them a rank) exhibits the essence of the spectral properties; and that is what we show here [412].

In Fig. 6.13 we can see the histogram of eigenvalues. The fraction $P_\Lambda(\lambda)$ of eigenvalues within a bin centred at λ mimics the density of eigenvalues, and we can see that it has a sharp maximum at $\lambda = 1$. It also seems that there is a singularity in the density of eigenvalues at $\lambda = 1$, characterised by different left and right derivatives at this point. Such a 'cusp' was also observed in the spectra of Barabási-Albert networks [1300] and it is believed that it is one of the consequences of power-law degree distribution. An attentive look at the histogram reveals that most of the eigenvalues are concentrated within a relatively narrow interval $(0.1, 1.5)$, but there are some eigenvalues which lie much farther; and the largest eigenvalue, highlighted by a circle in Fig. 6.13, is equal to $\lambda = 13.87\dots$ But there are also large intervals between neighbouring eigenvalues, suggesting that the eigenvalue density may be a slowly-decaying function. Indeed, when we plot in Fig. 6.12 the integrated density of states $P_\Lambda^>(\lambda)$, i.e. the fraction of eigenvalues larger than λ, we can see that the right tail is well described by a power law

$$P_\Lambda^>(\lambda) \sim (\lambda - 1)^{-\eta} , \; \lambda \to \infty \tag{6.53}$$

with exponent $\eta \simeq 1.7$.

The power-law tail of the density of eigenvalues has important implications for the global structure of the network. To see that, consider the adjacency matrix a_{ij} of a certain graph and look at the trace of its powers. The quantity $\operatorname{Tr} a$ is the number of loops, i.e. vertices connected to themselves. Looking at the square of the matrix, $\operatorname{Tr} a^2$ is number of cycles of length 2, which can be composed either of two loops or of a step along an edge connecting two distinct vertices, and back. In a loop-less graph this is twice the number of edges, therefore $\frac{1}{2}(\operatorname{Tr} a^2 - \operatorname{Tr} a)$ is just the number of edges, not counting loops. If we go further to larger powers, we see that $\operatorname{Tr} a^l$ counts the number of cycles of length l. For example, in a graph without loops $\operatorname{Tr} a^3$ is the number of triangles. The traces are easily calculated if we know the spectrum of the matrix. Indeed, the l-th moment of the density of eigenvalues $P_\Lambda(\lambda)$ is exactly the trace of the l-th power of the matrix, so

$$\operatorname{Tr} a^l = \int \lambda^l \, P_\Lambda(\lambda) \, d\lambda. \tag{6.54}$$

For a finite number of vertices N the latter expression is always finite, but when the size of the network grows, $N \to \infty$, it may or may not approach a finite limit, depending on the behaviour of the tail of the density of eigenvalues. In our case the tail follows a power law, $P_\Lambda(\lambda) \sim \lambda^{-1-\eta}$; and hence the moments diverge for $l \geq \eta$. The particular value of the exponent found in the collapsed reviewer network on Amazon already implies divergence for $l = 2$ and more for larger moments. How to understand this result? The second moment is related to the density of edges, and its divergence means that the number of edges grows faster than the number of nodes when the size of the network increases. In other words, the average degree of a vertex becomes larger and larger. Similarly, one could deduce that in a network whose spectrum satisfies $2 < \eta \leq 3$, the average degree remains finite but the density of triangles is divergent for increasing number of vertices.

The interesting point about this finding is that, from a purely static feature as the spectrum is, we can make a prognosis as to the evolution of the network. The hidden assumption which makes this prediction feasible is the invariability of the basic features of the spectrum when the network grows, namely the fact that the tail obeys a power law, and the exponent remains the same, within statistical errors.

6.4   What remains

Random graphs provide an immense field of study, and physical approaches have been proved useful in many areas. We deliberately skipped some of them, and the reader can consult several good books [55, 1177, 1244, 1252, 1253, 1317–1324] or specialised review articles [1254–1257] to complete her education in this area. Also the collection of original papers is now available in book format [1325]. The 'canonic' mathematical monograph on the subject is the previously mentioned book by Béla Bollobás [753].

The topics omitted in this chapter are most notably the questions of resilience and vulnerability to damage. Essentially, it is a percolation problem and was studied thoroughly [1326–1330] with the result that scale-free graphs are very robust to removal of vertices, because they possess zero percolation threshold. On the other hand, if the vertices are removed in decreasing order of their degree, i.e. the most connected nodes of the network are annihilated first, the percolation threshold is low, signalling a high sensitivity to this type of damage.

Another top-importance practical problem is epidemic spreading in random graphs, be it a sexually transmitted disease in a social network or computer-virus infections affecting the whole internet [1331–1334]. It was found that the networks with power-law degree distributions have negligible epidemic threshold, meaning that even the least contagious disease spreads rapidly.

The physics of small-world networks gained immediate popularity when the first article of Watts and Strogatz appeared [1185], but soon the attention of the community shifted to the more dynamic field of scale-free networks [1188]. Indeed, the effect of power-law degree distribution is much deeper than the mere addition of long-range links to regular structures. Nevertheless, many interesting results have been obtained [1260, 1333, 1335–1348], and the work still continues, for example, in the field of synchronisation of coupled nonlinear oscillators [1257, 1349], to cite only one of the applications. Studying the small-world properties of scale-free graphs, it has been found that their diameter is scaled as $\mathrm{diam} \sim \ln \ln N$ with the number of vertices, if the degree exponent lies in the interval $\gamma \in (2, 3)$. This means that they are even 'smaller' than ordinary small-world graphs, and the notion of ultra-small graphs was introduced to describe this phenomenon [1350].

Another very important question is to find communities inside random networks. The community structure determines, for example, interest groups in the WWW, commercial sectors in the ensemble of stocks traded at the stock exchange, etc. From the graph perspective, we can roughly define communities as subgraphs which are more densely connected inside than among each other. The clearest examples of communities are cliques. i.e. complete subgraphs within the graph in question. Most communities are less pronounced, though, and to find them, several smart algorithms are used, including iterative and spectral approaches [342, 386, 412–415, 418, 1216, 1293, 1311, 1315, 1351–1375]. The structure of the networks of electronic commerce influence the information filters and recommendation systems, which are indispensable tools in web-based economy [1283, 1376–1387].

Finally, let us also mention the study of topological phase transitions in random graphs, either static [1388–1393] or dynamic [1394–1396]. For example, when we require that the network provides a service, we also desire that it is somehow optimised to do that [1397–1399]. Therefore, we can identify the quantity to be extremalised with an effective 'Hamiltonian' and try to develop the statistical physics of such a system. Varying the free parameters of the model, we can identify the existence and type of a phase transition, which marks the sudden change of the structure of the graph in question. That is what we call a topological phase transition. For a student of econophysics, there is plenty of room for new discoveries here.

Problems

1. Take the graph ensemble $\mathcal{G}_D$ and find a more accurate formula for the probability that two vertices with degrees k_0 and k_1 are connected by an edge. Show that Eq. (6.19) is valid provided $k_0 k_1 \ll \bar{k} N$.

2. Consider the Barabási-Albert graph process with $m \geq 2$.

   a) For the case of multiple edges allowed, generalise the steps leading to expression (6.37) and thus prove the formula (6.38).

   b) Repeat analogous steps for the case without multiple edges between pairs of vertices and show that for large times the degree distribution is the same, whether we allow multiple edges or not.

3. In the Barabási-Albert graph process, the preferential attachment principle is vital for the power-law tail in the degree distribution. Show that by calculating the degree distribution under assumption that the probability of joining vertex t with vertex $s < t$ is a constant independent of the degree of vertex s.

7
Wealth, poverty, and growth

The perspective for survival in a modern capitalist economy is often reduced to various measures of individual and corporate wealth. Disproportions in wealth distribution across society are also causes, either virtual or actual or both, of social tensions, resulting in incessant dynamics affecting the whole social structure. The distribution of wealth (expressed by diverse indicators) was therefore the first concern in quantitative analyses started in 19th century. Explanation of the empirical facts has remained a serious challenge until now. Very often wealth is confounded with income, although in principle these two quantities are much different. We will mostly be interested in income distributions, but we shall skip to wealth distribution when it will be more convenient. As the actual models work equally well (or poorly) for income and wealth, we shall not make much distinction between the two terms.

This chapter will first review the empirical facts on income distributions, starting with historical discoveries marked by the names of Pareto, Gibrat, and Mandelbrot. Then we turn to numerical and analytical approaches, stressing the importance of the ubiquitous random multiplicative process modified by a small additive term.

7.1 Laws of Pareto and Gibrat

7.1.1 Power-law distribution

Formulation

If there were a gallery of founding fathers of what is now econophysics, surely it would include Vilfredo Pareto. In his book *Cours d'économie politique* [117], published in 1897, he formulated the law for income distribution stating that the number N of individuals having an income greater than v is

$$N = \frac{A}{(v+b)^\alpha} \qquad (7.1)$$

where b is a constant very close to zero and the value of the exponent α lies between 1 and 2 [1400]. Pareto also stressed that the law holds only for incomes higher than a certain threshold, and the distribution of lower incomes escaped the statistics (at his times). In most of the subsequent studies the constant b was set to 0.

So, we may formulate the Pareto law as follows. The probability distribution for individual income v is asymptotically described by a power law

$$P^>(v) \equiv \text{Prob}\{\text{income} > v\} \sim v^{-\alpha}, \quad v \to \infty. \qquad (7.2)$$

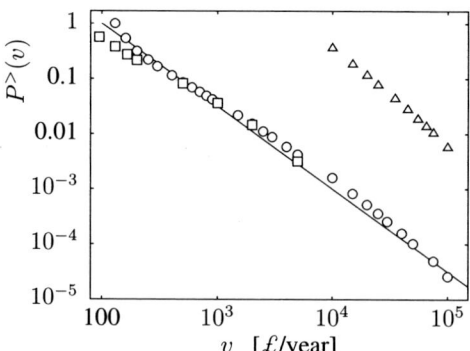

Fig. 7.1 Income distribution in Great Britain compiled from tax revenues. The symbols denote years 1801 ($\square$), 1911–1912 ($\triangle$), and 1918–1919 ($\bigcirc$). The line is the dependence $\propto v^{-1.5}$. Note the surprising fact that the points for years 1801 and 1918–1919 fall on essentially the same curve. Data taken from the tables published in [1401] (1801 and 1911–1912) and [1402] (1918–1919).

Empirical evidence

In spite of its very careful original formulation, many heated debates have been conducted about the validity of the Pareto law. Much data has been collected in support of it and essentially the same numbers have served as demonstrations of its breakdown [1401–1404]. Let us look at some of the historical data on income distribution in Great Britain, shown in Fig. 7.1. For three tax-collecting periods within more than a century, we plot the distribution $P^>(v)$ where the income v is expressed in pounds. Data like those shown are, in principle, easily accessible from taxation records.

First, we note that, in a double logarithmic plot, all three data sets fall approximately on parallel straight lines, indicating power-law dependence $P^>(v) \sim v^{-\alpha}$ which confirms the Pareto law with $\alpha \simeq 1.5$. Moreover, the slope does not seem to change much over time, so all the spectacular developments in economics and society which took place in the course of the 19th century left the exponent α essentially unaffected. This is analogous to the property of universality exhibited by physical systems in a critical state.

A closer look reveals, unfortunately, small but systematic deviations from the power-law dependence, which are more pronounced at low incomes. Eventually, a consensus grew that the universal Pareto law is indeed applicable for a small fraction of society enjoying high incomes, while the rest of society is governed by non-universal laws, i.e. distribution of lower incomes is sensitive to the details of the actual social situation [1405–1415]. In fact, it is not so much the functional form of the Pareto law but its spatial and temporal stability that is intriguing. Indeed, while the value of the exponent α may slightly vary from one society to another, the very fact of the power-law tail in the distribution is valid almost everywhere. Some investigations suggest that the range of validity of the Pareto law may extend as far in the past as to ancient Egypt of the Pharaohs [1416].

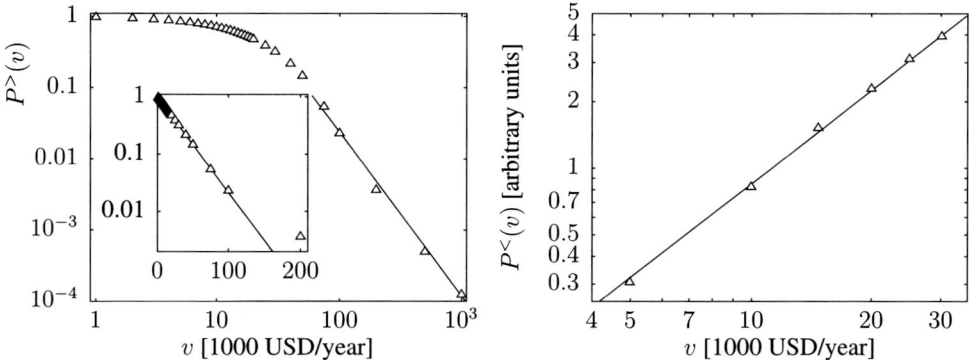

Fig. 7.2 Probability distribution of individual income in the USA in 1997. In the left panel, the complete distribution. The power-law tail is demonstrated by linear dependence in double logarithmic scale. The line is the dependence $\propto v^{-2.3}$. In the inset, the same data in semilogarithmic scale, showing exponential behaviour for medium and lower incomes. The line is the dependence $\propto e^{-v/26}$. Data were extracted from [1419]. In the right panel, the lower end of the household income distribution in the USA in 1997. The line is the dependence $\propto v^{1.4}$. Data were extracted from [1423].

So, the issue of 'wealth' distribution seems to be settled. More recent investigations shed more light into the 'poverty' distribution, observing regular behaviour in the lower income range. The medium part of the distribution pertinent to a large majority of the population was found to obey a simple exponential distribution [1410, 1417–1422]

$$P^{>}(v) \sim e^{-v/v_0} \qquad (7.3)$$

while the probability density for the lowest incomes follow a power law with a positive exponent [1411, 1423]

$$P^{<}(v) \equiv \mathrm{Prob}\{\text{income} < v\} \sim v^{\beta}, \quad v \to 0. \qquad (7.4)$$

These findings are demonstrated in Fig. 7.2. For the year 1997 in the USA the highest incomes are power-law distributed according to (7.2) with $\alpha \simeq 2.3$, medium incomes are governed by an exponential distribution (7.3) with $v_0 \simeq 2.6 \times 10^4$ USD, and the lowest end of the income scale satisfies a power-law dependence (7.4) with $\beta \simeq 1.4$.

More detailed investigations of the Pareto law were carried out for individual and company incomes in several Asian countries [1407–1409, 1412, 1424–1430]. The Pareto index α was found to fluctuate quite dramatically within certain bounds, as seen in Fig. 7.3.

Measures of economic inequality

The Pareto index α measures the width of the distribution, and was used by Pareto himself as a measure of inequality. However, as the Pareto law describes only the tail of the income distribution, other tools have been developed to characterise the

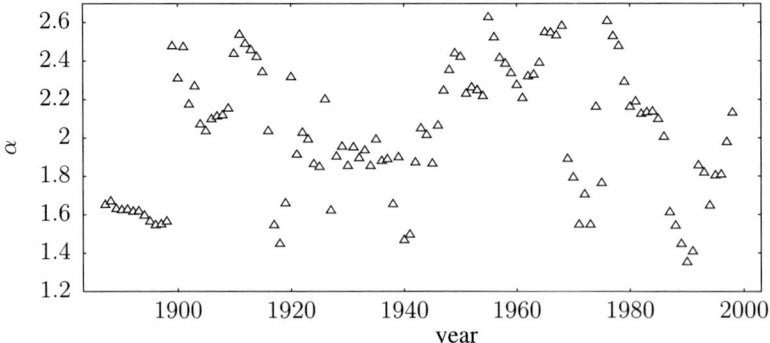

Fig. 7.3 Time dependence of the Pareto index α in Japan during the period from 1887 to 1998. Data were extracted from [1409].

income distribution globally. One of them is the Lorenz curve, which is a particular way of plotting wealth distribution. The horizontal coordinate is the cumulative population $x = \int_0^v P(v')\mathrm{d}v'$, and the vertical one is the fraction of cumulative income, $y = \int_0^v v'P(v')\mathrm{d}v' / \int_0^\infty v'P(v')\mathrm{d}v'$. For exponential income distribution, the Lorenz curve has the form

$$y = x + (1 - x)\ln(1 - x). \tag{7.5}$$

The Gini coefficient, defined as $G = 2\int_0^1 (x - y)\mathrm{d}x$, measures the inequality in terms of the deviation of the Lorenz curve from the diagonal $y = x$. It varies from 0 (full equality) to 1 (maximum inequality). Exponential distribution gives $G = 1/2$. Empirical data for the Lorenz curve and time dependence of the Gini coefficient are shown in Fig. 7.4.

7.1.2 Independent growth rates

After Pareto, the study of wealth distributions and economic inequality in society was marked by the important contribution of Robert Gibrat, a French engineer with a wide range of interests including, among other things, tidal power plants. In his thesis, entitled *Les inégalités économiques* [1431], he formulated a principle which nowadays bears the name Gibrat law, or the law of proportionate effect.

Gibrat law

The Gibrat law states that the rate of income growth is independent of the actual income, and the rates at different times are uncorrelated. But applicability of the Gibrat law is broader than just income statistics. We will proceed with a more general and technical formulation.

Let us consider an economic indicator X_t, which may be income, wealth, size, number of employees, etc. Relative growth of the indicator from time t to t' is $R(t, t') = X_{t'}/X_t$. The Gibrat law requires that two conditions are satisfied

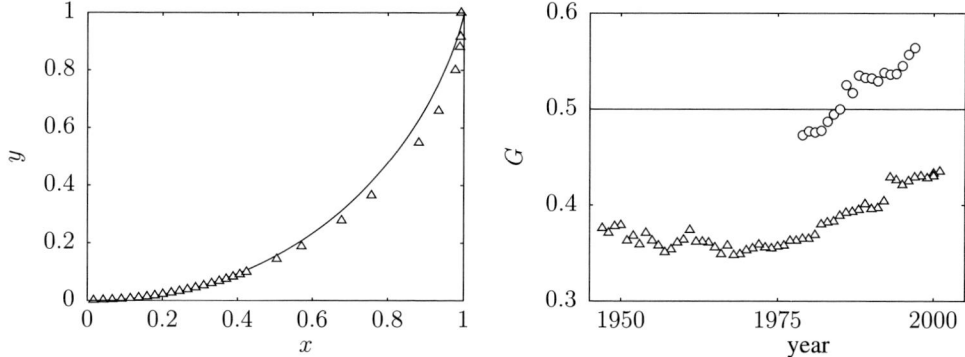

Fig. 7.4 In the left panel, the Lorenz curve for incomes in the USA in 1997. Full line corresponds to exponential distribution according to (7.5). Data were extracted from [1419]. In the right panel, the time evolution of the Gini coefficient in the USA, for individual income (○, data extracted from [1418]) and household income (△, data published by U.S. Census Bureau). The line at $G = 1/2$ corresponds to exponential distribution.

1. The growth rate $R(t, t')$ and the initial value X_t are independent, i.e.
   $\langle R^k(t, t') X_t^l \rangle = \langle R^k(t, t') \rangle \langle X_t^l \rangle$ for any integer k, l.
2. Growth from time t to t' is independent of growth from t' to t'', i.e.
   $\langle R^k(t, t') R^l(t', t'') \rangle = \langle R^k(t, t') \rangle \langle R^l(t', t'') \rangle$ for any integer k, l and $t < t' < t''$.

Multiplicative process

Assuming the Gibrat law holds true, we can model the dynamics of income V_t of a selected individual by a multiplicative stochastic process

$$V_{t+1} = e^{W_t} V_t \tag{7.6}$$

where W_t for all times t are independent and equally distributed random variables with mean $\langle W_t \rangle = C$ and variance $\langle (W_t)^2 \rangle - \langle W_t \rangle^2 = D$. We may map the process onto a simple random walk by substituting $Y_t = \ln V_t$. Starting from a deterministic initial condition $Y_0 = \ln v_0$ the position of the walker at time t is a sum of t independent and identically distributed random variables

$$Y_t = \sum_{t'=1}^{t} W_{t'} + \ln v_0, \tag{7.7}$$

and the central limit theorem tells us that for large times the probability density for Y_t is normally distributed, i.e. it does not depend on the details of the distribution for W_t and assumes a Gaussian form with mean $\ln v_0 + Ct$ and variance Dt. This implies, by substitution, a log-normal distribution for the income at time t

$$P_t(v) = \frac{1}{\sqrt{2\pi Dt}} \frac{1}{v} \exp\left(-\frac{1}{2Dt}\left(\ln\frac{v}{v_0} - Ct\right)^2\right). \tag{7.8}$$

An equivalent result can be obtained in a more elegant fashion if we describe the multiplicative process using a stochastic equation with multiplicative noise

$$\mathrm{d}V_t = V_t \, \mathrm{d}W_t \tag{7.9}$$

where the Stratonovich convention is assumed (see Boxes 2.4 and 2.6 for explanation) and the external noise W_t is determined by $\langle \mathrm{d}W_t \rangle = C \, \mathrm{d}t$ and $\langle (\mathrm{d}W_t)^2 \rangle = D \, \mathrm{d}t$. Then the following Fokker-Planck equation (see Box 2.5 if unclear) can be derived for the probability density of V_t

$$\frac{\partial}{\partial t} P_t(v) = \frac{D}{2} \left(\frac{\partial}{\partial v} v \right)^2 P_t(v) - C \frac{\partial}{\partial v} v P_t(v). \tag{7.10}$$

The solution with initial condition $\lim_{t \to 0+} P_t(v) = \delta(v - v_0)$ is given by the log-normal distribution, Eq. (7.8).

Are Pareto and Gibrat laws compatible?

As we have seen, strict application of the Gibrat law results in log-normal distribution of wealth. The upper tail is very far from the empirically observed power-law dependence. Moreover, the distribution is not stationary, because its width increases in time. It seems that the Gibrat law was falsified by real facts.

However, that would be too quick a judgement. Indeed, the derivation of the log-normal distribution (7.8) relied on various assumptions, which may not always be true. We shall examine them in more depth in the next section. For the moment, we shall only check whether Pareto's power law tails might be compatible with the Gibrat law under fairly general conditions. The argument essentially says that if the the Gibrat law holds true; and if the distribution is stationary, it must be a power law. Let us show it in a slightly more formal way.

To this end, we investigate the joint probability density for the initial value of the income X_t and for the growth $R(t, t') = X_{t'}/X_t$, which is

$$P_{t,t'}(r, x) = \frac{\partial^2}{\partial r \partial x} \text{Prob}\{R(t, t') < r, X_t < x\}. \tag{7.11}$$

The conditional probability density is $P_{t,t'}(r|x) = P_{t,t'}(r, x) / \int P_{t,t'}(r', x) \, \mathrm{d}r'$. The first of the requirements of the Gibrat law implies that this conditional probability in fact does not depend on x and is therefore equal to the probability density for sole R, i.e. $P_{t,t'}(r|x) = P_{R\,t,t'}(r) \equiv \int P_{t,t'}(r, x') \, \mathrm{d}x'$.

Let us also assume that the income distribution is stationary. This can be formulated as time-reversal symmetry of the joint probability density of incomes at times t and t':

$$P_{t,t'}(x, x') = P_{t,t'}(x', x) \tag{7.12}$$

and simultaneously as time independence of the densities for R and X,

$$\int P_{t,t'}(r, x) \mathrm{d}r = P_X(x)$$
$$\int P_{t,t'}(r, x) \mathrm{d}x = P_R(r) \tag{7.13}$$

for all t, t' (for an empirical check of these conditions see [1432]). Under these assumptions we can express the densities through conditional distribution

$$P_{t,t'}(x, x') = \frac{1}{x} P_R\left(\frac{x'}{x}\right) P_X(x) = P_R(r) P_X(x) \frac{1}{x}, \qquad (7.14)$$

and we obtain

$$\frac{P_X(x)}{P_X(x')} = \frac{1}{r} \frac{P_R(\frac{1}{r})}{P_R(r)} \equiv e^{g(r)}. \qquad (7.15)$$

Suppose that the times t and t' are close to each other. Then r is close to 1, and we can expand (7.15) around the point $r = 1$. Using substitutions $x = e^y$, $p(y) = \ln P_X(e^y)$, we arrive at the differential equation for the function $p(y)$

$$\frac{\mathrm{d}p(y)}{\mathrm{d}y} = -g'(1); \qquad (7.16)$$

therefore $p(y) = a - yg'(1)$, and the density for X assumes a power-law form

$$P_X(x) = A\, x^{-g'(1)} \qquad (7.17)$$

in full accordance with the Pareto law [1432].

However, it must be noted that the derivation of (7.17) was based on local (in terms of the values of X_t) properties of the joint probability density. If the time-reversal symmetry, as well as the Gibrat law, holds for x and x' within a certain, narrow enough, interval $\mathcal{I}$, i.e. for a limited range of r, the power-law distribution (7.17) will approximately hold in the interval $\mathcal{I}$. In another short interval $\mathcal{I}'$ we may again find that (7.17) is satisfied, but the exponents may differ. This observation should warn us about vaulting to unfounded conclusions based on spurious power laws observed only within a limited interval of incomes.

7.1.3 Lévy distributions

Although we have just seen that the Gibrat law and the Pareto law are not contradictory in principle, the actual calculations in Sec. 7.1.2 still give a log-normal distribution, independent of the details of the distribution of instantaneous changes in income, prescribed by the stochastic function W_t. This fact follows from the central limit theorem and holds for any W_t provided the variance $\langle (W_t)^2 \rangle - \langle W_t \rangle^2$ is finite.

And here comes Benoît B. Mandelbrot with his breakthrough idea [727]. What if the variance is not finite? What is the distribution of a sum of many independent and equally distributed random variables with diverging variance?

This problem had been encountered previously by mathematicians, and the answer leads us to the family of Lévy-stable distributions. Mandelbrot's crucial contribution was that he brought this abstract subject to the attention of economists and physicists by showing numerous appealing examples.

To see the difference, imagine two pairs of independent and identically distributed random variables. The first pair, G_1 and G_2, will be normally distributed, with Gaussian density $P_G(g) = \exp(-g^2/2D)/\sqrt{2\pi D}$, while the density for second pair, H_1 and H_2, is the Cauchy distribution

---

Formula for Lévy distribution | Box 7.1 |

Except for the Cauchy (or Lorentz) distribution (7.18) with $\mu = 1$, $\beta = 0$, there is no closed expression for the Lévy distributions. The most straightforward representation is through inverse Fourier transform of a stretched exponential. In the symmetric case, $\beta = 0$, we have
$$L_\mu^0(x) = (2\pi)^{-1} \int \exp(-a_\mu|\omega|^\mu - i\omega x)d\omega$$
where a_μ is a certain constant, and the general case is
$$L_\mu^\beta(x) = (2\pi)^{-1} \int \exp\left[-a_\mu|\omega|^\mu(1 - i\beta\,\mathrm{sgn}(\omega)\tan(\tfrac{1}{2}\pi\mu))) - i\omega x\right]d\omega.$$
For more information see [19] or initial sections and appendix in Ref. [194], where many physical examples can be found.

$$P_H(h) = \frac{1}{\pi}\frac{a}{a^2 + h^2} \tag{7.18}$$

(also called the Lorentz distribution by physicists). The probability density of a sum of independent variables is the convolution of their densities; so $G_3 = G_1 + G_2$ is distributed according to $P_{G_3}(g) = \int P_G(g - g')P_G(g)\,dg'$, and similarly for $H_3 = H_1 + H_2$. Performing the integrals we find that $P_{G_3}(g) = \exp(g^2/4D)/\sqrt{4\pi D}$; i.e. the form of the function remained unchanged, only the variance was rescaled, $D \to 2D$. The same conclusion is reached for the other pair. Indeed, $P_{H_3}(h) = 2a/(\pi(4a^2 + h^2))$, and the sum again has Cauchy probability density with parameter a doubled. The essential difference between the Gaussian and Cauchy distributions is that the former has finite variance, while the latter not. Therefore, for a sum of Cauchy-distributed random variables, the central limit theorem cannot hold. Instead, the sum will keep its Lorentzian shape.

Random variables whose distribution preserves its form with respect to additions are said to have stable distributions. While Gaussian is the only stable distribution with a finite variance, there is an entire family of distributions with a diverging second moment, called Lévy-stable distributions. One of them is the Cauchy distribution discussed above. Their common feature is that they decay as a power for large arguments, and that is why Mandelbrot used them to explain the Pareto law.

Generally the Lévy distribution $L_\mu^\beta(x)$ is characterised by two parameters, $\mu \in (0, 2)$ and $\beta \in [0, 1]$ and behaves like

$$L_\mu^\beta(x) \sim \frac{1 \pm \beta}{|x|^{1+\mu}} \text{ for } x \to \pm\infty. \tag{7.19}$$

We can see that the parameter μ corresponds to the Pareto index α. The parameter β measures the asymmetry of the distribution. The Cauchy distribution (7.18) corresponds to $\mu = 1$ and $\beta = 0$. For more information see Box 7.1.

Therefore, the central limit theorem is generalised to variables with infinite variance so that the distribution for their sum converges to a certain Lévy distribution. If $X = \sum_{i=1}^n W_i$ then $\lim_{n\to\infty} P_X(x) \to L_\mu^\beta(x)$ for some μ, β.

While random variables with finite variance give rise to only a single limit distribution, the normal one, here we have a two-parametric set of distributions. Each of them has a certain basin of attraction, i.e. a set of distributions, all of which approach the same Lévy distribution when their corresponding random variables are summed many times.

Coming back to Mandelbrot, his idea was that individual incomes are in fact influenced by large shocks, which may be modelled by random variables with infinite variance. The impact of such shocks on the income is expressed by a sum of these variables. Supposing all individuals started with equal income in some very distant past, the distribution of incomes now follows some of the family of Lévy distributions, characterised by power-law tails. This explains the emergence of the Pareto law and reconciles it with the idea that what we observe in reality is an accumulated effect of many independent random events. Since then, the term Pareto-Lévy distributions has frequently been used for power-law tailed functions.

If we thought that the problem of incomes is now solved, we would be very far from reality. First, it is very reasonable to think that the dynamics of income is essentially a multiplicative process, while Mandelbrot takes us back to additive processes. It is reasonable because it is generally expected that investments will have effects proportional to their sizes; and the same apparently holds for all economic movements. It is also reasonable because it preserves invariance with respect to a change of units in which we measure the income. Finally, additive processes can easily result in negative incomes, while real individual incomes are always positive.

The second reason why the Lévy distributions are not satisfactory is the behaviour at moderate and low incomes. If the empirical distributions consistently differ in a certain range from the Lévy ones, the result certainly cannot be due to summing independent random contributions, the more so because the approach to Lévy distribution is faster in the central part and slower at the tails; so one would expect agreement for smaller incomes and deviation for large ones. As we already know, the opposite is true.

Finally, Mandelbrot does not say anything about the source of the assumed strong random shocks. If we want understand more, we must go deeper. The models presented in the following two sections improve the situation at least partially.

7.2 Individual income

7.2.1 Stochastic modelling

Let us now turn to more modern approaches to income distributions. First of all, we shall make an essential assumption that income (or wealth) distribution is the result of a dynamical process with well-defined laws. Our task will be to find these laws and set up a dynamical model which will reproduce the observed distributions. This apparently obvious assumption excludes, e.g. explanations based on particular historical circumstances as well as static theories considering wealth inequality as imprinted at dark origins by an external force and reproduced since then, generation after generation, by a kind of heredity, be it familial, genetic, social, or maybe some other cause.

As randomness will play major role in the dynamics of income, the process will be a stochastic one. Two types of models can be distinguished from the beginning. First, we can model the dynamics of a representative individual and consider the probability density obtained for that individual as valid for the whole ensemble composing the society. Or, second, we can model the society as a whole, taking into account a strong interaction between individuals. Eventually, we shall see that these two views are not mutually exclusive.

Multiplicative random process with a lower bound

Let us imagine particles in a very tall cylindrical container, kept at a fixed temperature, under the influence of a homogeneous force acting along the axis of the cylinder. An example is a gas in a gravitational field. Without the field the particles would perform a Brownian walk and eventually all escape, but gravitation pushes them to the bottom. What would be the density of the particles in different parts of the cylinder? The answer is given by the well-known barometric formula. Supposing the particles mutually interact only very weakly, the density of particles will be proportional to the probability of finding a chosen particle at a given place.

   The motion of a representative particle is described by the stochastic differential equation

$$\mathrm{d}X_t = \mathrm{d}W_t - f\,\mathrm{d}t \tag{7.20}$$

where X_t is the position of the particle at time t. The movement of the particle is influenced by the constant force f and stochastic perturbations, represented by the differential of the pure Brownian motion $\mathrm{d}W_t$, which has the following property $\langle(\mathrm{d}W_t)^2\rangle = D\mathrm{d}t$, where D is the diffusion constant. The corresponding Fokker-Planck equation is

$$\frac{\partial}{\partial t}P_t(x) = \frac{D}{2}\frac{\partial^2}{\partial x^2}P_t(x) + f\frac{\partial}{\partial x}P_t(x). \tag{7.21}$$

We are interested in the stationary state $P(x) = \lim_{t\to\infty} P_t(x)$. It satisfies a simple equation $\frac{1}{2}D\,P'(x) = -fP(x) + c$; however, it must be complemented by a boundary condition forbidding the particle from escaping from the container. The boundary conditions also fix the constant c. Taking the coordinate of the bottom of the container at $X = 0$, the solution is

$$P(x) = \theta(x)\frac{2f}{D}e^{-\frac{2f}{D}x} \tag{7.22}$$

with, as usual, $\theta(x) = 1$ for $x > 0$, and $\theta(x) = 0$ otherwise. This is the barometric formula, describing dependence of the density of isothermal ideal gas in a gravitational field on its height. Mandelbrot [727] was perhaps the first who tried to use this analogy for income distributions, noting, though, that the exponential distribution disagrees with the empirical findings.

   Imagine now a similar process in discrete time, in which a particle can hop over the set of non-negative integers. If the coordinate X of the particle is positive, in one step it is increased by 1 with probability p and decreased by 1 with probability $1 - p$. The boundary point $X = 0$ is special: with probability 1, the particle jumps to the next point $X = 1$. The probability distribution $P_t(x) = \mathrm{Prob}\{X_t = x\}$ satisfies a master equation

$$\begin{aligned}
P_{t+1}(x) &= p\,P_t(x-1) + (1-p)\,P_t(x+1) \quad \text{for} \quad x > 1\\
P_{t+1}(1) &= P_t(0) + (1-p)\,P_t(2)\\
P_{t+1}(0) &= (1-p)\,P_t(1).
\end{aligned} \tag{7.23}$$

The stationary distribution $P(x) = \lim_{t \to \infty} P_t(x)$ exists for $p < 1/2$ and can easily be found. For $x \geq 1$ we have

$$P(x) = \frac{1 - 2p}{2p(1 - p)} \left(\frac{p}{1 - p}\right)^x. \tag{7.24}$$

The exponential stationary distribution results from two competing mechanisms. The random walk with $p < 1/2$ has a net drift towards lower coordinates, while the fixed boundary at $X = 0$ does not allow the drift to push the particle further. We obtain a discrete analogy of the barometric formula.

Let us see how this model can be adapted to the distribution of incomes [170, 676–678, 681, 687, 1433, 1434]. The idea is to replace the arithmetic sequence of possible positions by a geometric sequence. Suppose a discrete 'income ladder' made of non-negative powers of the number $1 + \epsilon$, where $\epsilon > 0$, applies to the society so the allowed incomes at time t are $V_t \in \{(1 + \epsilon)^n; n = 0, 1, 2, \ldots\}$ [1433]. This corresponds to a society with a built-in mechanism keeping all members above a fixed poverty level, but with no other constraints. Suppose also, following Gibrat, that the income evolves in discrete time according to a multiplicative process, namely

$$V_{t+1} = e^{W_t} V_t \tag{7.25}$$

as in process (7.6), where for simplicity we assume $W_t = \ln(1 + \epsilon)$ with probability p and $W_t = -\ln(1 + \epsilon)$ with probability $1 - p$. There is an exception to these probabilities, though. As nobody is allowed to have income below 1, we have $W_t = \ln(1 + \epsilon)$ with probability 1 if $V_t = 1$. Note that this seemingly marginal modification introduces dependence among the quantities V_t and W_t. The central limit theorem for the sum of the random variables W_t does not apply, because they are not all independent. This is why the stationary distribution for V_t may deviate from the log-normal distribution (7.8).

Now, the process V_t can obviously be mapped onto process X_t by the substitution $X_t = \ln V_t / \ln(1 + \epsilon)$ with the same transition probabilities. Therefore, the stationary distribution of incomes can be expressed as

$$P^>(v) = \lim_{t \to \infty} \text{Prob}\{V_t > v \equiv (1 + \epsilon)^n\}$$
$$= \frac{1}{2(1 - p)} \left(\frac{p}{1 - p}\right)^n. \tag{7.26}$$

We can make the income ladder denser and eventually continuous in the limit $\epsilon \to 0$, $p \to 1/2$, while keeping the combination $\alpha = \left(\ln(1 - p) - \ln p\right)/\ln(1 + \epsilon)$ constant. The result is

$$P^>(v) = \theta(v - 1) v^{-\alpha} \tag{7.27}$$

so that the income distribution follows a power law for all $v > 1$ and is sharply cut off for $v \leq 1$.

In the latter calculation we measured the income in terms of the poverty level v_0 (set here as $v_0 = 1$). In fact, if the Pareto law holds strictly, the ratio of the average income and the lower cutoff uniquely determines the exponent of the power law. Indeed, $\bar{v} = -\int_{v_0}^{\infty} v \, dP^{>}(v) = v_0 \, \alpha/(\alpha - 1)$, so

$$\alpha = \frac{1}{1 - \frac{v_0}{\bar{v}}}. \tag{7.28}$$

This relation provides a straightforward explanation of relative stability of the value of Pareto index α: it is fixed by the social consensus which keeps essentially all individual incomes above a certain fraction of the total income of the society [683].

Multiplicative-additive stochastic process

It seems that a plausible mechanism generating the power-law tailed distributions was successfully identified. Two problems remain open, though. First, the sharp drop of probability density at the lowest income level is completely unrealistic. We must improve the model so that it grasps, at least qualitatively, the empirical findings on the lower income distribution. Second, the hard wall impeding the individuals from falling too low may sound quite arbitrary. Could we find a detailed mechanism implementing such a wall? Or is the existence of the low income limit a mere by-product of some microscopic dynamics?

The simplest model generating the lower bound is the combined multiplicative-additive stochastic process [673]. It can be realised by simply adding a constant term to the multiplicative process (7.9), writing, again in the Stratonovich convention, the equation

$$dV_t = V_t \, dW_t + a \, dt \tag{7.29}$$

with a positive constant a and again $\langle dW_t \rangle = C \, dt$ and $\langle (dW_t)^2 \rangle = D \, dt$. The additive term can be interpreted as a subsidy from a central authority, which helps unfortunate individuals to survive periods when their income falls below a certain level.

Let us first look at the evolution of average income. Changing to the Itô convention, Eq. (7.29) becomes $dV_t = V_t \, dW_t + \left[\frac{D}{2} V_t + a \right] dt$ and the average evolves according to

$$\frac{d}{dt}\langle V_t \rangle = \langle V_t \rangle \left(C + \frac{D}{2} \right) + a. \tag{7.30}$$

In absence of the additive term, the average growth rate would be $\widetilde{C} = C + D/2$. We are interested in the regime $\widetilde{C} < 0$. Without the additive term the average would shrink indefinitely, but positive a ensures that equilibrium value $\lim_{t \to \infty} \langle V_t \rangle = -a/\widetilde{C}$ is approached. This is the first positive sign: non-stationarity of the log-normal distribution (7.8) is cured.

To compute the full probability distribution we need the corresponding Fokker-Planck equation. We obtain

$$\frac{\partial}{\partial t} P_t(v) = \frac{D}{2} \left(\frac{\partial}{\partial v} v \right)^2 P_t(v) - C \frac{\partial}{\partial v} v P_t(v) - a \frac{\partial}{\partial v} P_t. \tag{7.31}$$

As a first step toward a solution we try to find out whether a power-law tail can be expected in the stationary solution $P(v) = \lim_{t \to \infty} P_t(v)$. Inserting as a trial function the Pareto distribution $P(v) \simeq A \, v^{-1-\alpha}$, we get

---

How to solve Eq. (7.31)? $\boxed{\text{Box 7.2}}$

The stationary solution $P(v)$ satisfies the following ordinary differential equation
$$\frac{D}{2}v^2 P''(v) + \left(\frac{3D}{2}v - Cv - a\right)P'(v) + \left(\frac{D}{2} - C\right)P(v) = 0 \ ;$$
and the first step is to take into account the known power-law behaviour for large v. So, we substitute $P(v) = Q(v)\,v^{-1-\alpha}$ and obtain
$$\frac{D}{2}vQ''(v) + \left(\frac{D}{2} + C - a\,v^{-1}\right)Q'(v) + a\left(1 - \frac{2C}{D}\right)v^{-2}Q(v) = 0.$$
Substituting again $v = 1/y$, $R(y) = Q(1/y)$ we have
$$\left[\frac{\mathrm{d}}{\mathrm{d}y} + \frac{1}{y}\left(1 - \frac{2C}{D}\right)\right]\left(\frac{D}{2}R'(y) + a\,R(y)\right) = 0.$$
The latter equation will be satisfied if
$$\frac{D}{2}R'(y) + a\,R(y) = 0$$
which leads to the solution
$$R(y) = A\,\exp\left(-\frac{2a}{D}y\right).$$
Substituting back and fixing the constant A by normalisation $\int P(v)\mathrm{d}v = 1$ eventually leads to (7.34).

$$\frac{D}{2}\alpha^2 + C\alpha + a(1+\alpha)\,v^{-1} \simeq 0, \tag{7.32}$$

and neglecting the last term on the right hand side from $v \to \infty$ we obtain the solution

$$\alpha = -\frac{2C}{D} \tag{7.33}$$

for the Pareto index, apart from the trivial solution $\alpha = 0$ which is to be discarded because it leads to a non-normalisable distribution.

Next, we can obtain the stationary solution of (7.31) explicitly. The details can be found in Box 7.2. The result is

$$P(v) = \frac{1}{\Gamma(\alpha)}\left(\frac{2a}{D}\right)^\alpha v^{-1-\alpha}\exp\left(-\frac{2a}{Dv}\right) \tag{7.34}$$

and obviously behaves in the desired way in both the high-income and low-income regions. Indeed, for $v \to \infty$ it has a power-law tail and for low incomes the probability density is depleted, implying a kind of a soft lower bound for the incomes.

The procedure can be generalised by replacing the constant a with a stochastic additive noise [674, 675, 688, 690, 900, 936, 1426, 1435]. It was found that essentially identical results hold, and only the low-income part of the density $P(v)$ is affected by the specific properties of the additive noise. Generalisation to the broader family of processes described by equation $\mathrm{d}V_t = g(V_t)\,\mathrm{d}W_t + h(V_t)\,\mathrm{d}t$ has also been studied [1436].

As already mentioned, the additive term can be understood as social subsidies. Alternatively, we may think of a redistribution strategy which actually takes place through public expenses covered by the collected tax. It would be interesting to redefine the incomes of the individuals in such a way that they include indirect benefits like elementary education, clean and illuminated streets, environment protection, or even the internal and external security guaranteed by the police and armed forces. Unfortunately, it is hardly possible to carry out such a study quantitatively.

Killed multiplicative process

There is an alternative way of producing distributions with power-law tails from multiplicative random processes. Imagine first a deterministic growth process for income. It is described by an ordinary, non-stochastic, differential equation

$$\frac{\mathrm{d}}{\mathrm{d}t}v(t) = b\,v(t) \tag{7.35}$$

where b is now a fixed constant. Solution of this equation with initial condition $v(0) = v_0$ is straightforward, $v(t) = v_0 \exp(bt)$. Suppose the same process applies to all individuals in a large ensemble. The randomness will be added in the following way. The individuals may enter and exit the ensemble at different times, so at a given instant the duration of the growth T is a random variable, as is the income reached, $V = v(T) = v_0 \exp(bT)$. Assuming an exponential probability density for T, namely $P_T(t) = \gamma \exp(-\gamma t)\,\theta(t)$, we obtain a power-law distribution of incomes above the minimum v_0

$$P_V(v) = \int_0^\infty \delta\left(v - v_0\,\mathrm{e}^{bt}\right) \gamma\,\mathrm{e}^{-\gamma t}\,\mathrm{d}t = \frac{\gamma}{b} v_0^{\gamma/b}\theta(v - v_0)\,v^{-1-\gamma/b} \tag{7.36}$$

which corresponds exactly to the result (7.27) obtained for a multiplicative process with a lower bound.

Such a random process is called a killed multiplicative process [1411, 1423] and can be further modified to comply better with the empirical observations. Assuming the individuals' incomes evolve according to the multiplicative random process (7.9), killed again after a random, exponentially distributed time T, the density for the logarithm of income relative to the initial condition $Y = \ln(V/v_0)$ becomes

$$P_Y(y) = \int_0^\infty \frac{\gamma\,\mathrm{e}^{-\gamma t}}{\sqrt{2\pi Dt}}\,\mathrm{e}^{-\frac{1}{2Dt}(y - Ct)^2}\,\mathrm{d}t$$

$$= \frac{\gamma}{\sqrt{2\gamma D + C^2}}\exp\left(\frac{C}{D}y - \frac{\sqrt{2\gamma D + C^2}}{D}\,|y|\right), \tag{7.37}$$

which means that the income is distributed according to power laws in both lower and higher ends

$$P_V(v) \sim \begin{cases} v^{\beta-1} & \text{for } v < v_0 \\ v^{-1-\alpha} & \text{for } v > v_0 \end{cases} \tag{7.38}$$

where the exponents are given by $\alpha = \frac{1}{D}(\sqrt{2\gamma D + C^2} - C)$ and $\beta = \frac{1}{D}(\sqrt{2\gamma D + C^2} + C)$. The cusp at the level of initial income v_0 is due to an identical initial condition $v(0) = v_0$ for all individuals. Assuming randomness also in the initial conditions, the singularity would be washed out.

We can see that killing the stochastic multiplicative growth at random times explains both the old Pareto law for high incomes and the recently discovered power-law behaviour at the lowest incomes [1423]. The same mechanism was also invoked to explain the power-law distribution of settlement sizes, family name frequency, and other phenomena [1411].

> **Verhulst equation** $\boxed{\text{Box 7.3}}$
>
> It was originally devised in order to amend the model of exponential (geometric) growth of the population, envisioned by Thomas Malthus. For one variable and deterministic growth rate $a > 0$ it reads
> $$\dot{v}(t) = a\,(1 - v(t))v(t),$$
> and from a mathematical point of view it belongs to the (solvable) family of Riccati equations. Its solution is
> $$v(t) = 1/(1 + c\,\exp(-at))$$
> with c being a constant depending on initial conditions. Starting with a very small initial value $0 < v(0) \ll 1$, the general behaviour is characterised by an initial exponential increase, then followed by saturation at time $t \simeq -(\ln v(0))/a$.

7.2.2   Many-particle approaches

The models treated so far were concentrated on the income dynamics of a representative individual. In reality, however, the income varies largely due to mutual interactions between many individuals within a society. In previous sections the stochastic elements were taken from the outside world. It is similar to observing a tagged molecule in a gas where all the rest of the gas is replaced by a stochastic reservoir. We know that this works well for ideal gas or weakly interacting systems, but if the interaction grows stronger, the approach may prove inappropriate.

Such a situation was treated in non-equilibrium statistical mechanics, where the projector techniques have been developed. They provide exact tools to study dynamics of a selected particle (or another small section of the system) strongly interacting with its surroundings. For description of these methods in the context of econophysics see Ref. [29].

Let us now turn to many-particle models of income dynamics. The questions we shall address concern several holes in the stochastic approach used above. First, the noise W_t was so far quite arbitrary. What is the microscopic source of that noise? Can we say anything about its properties beyond its mean and variance? Second, are there features which cannot be reasonably explained by stochastic modelling of a representative individual and require interactions to be considered explicitly? And finally, could we gain more understanding of the remaining empirical facts, namely, the exponential distribution of moderate incomes?

Lotka-Volterra systems

Perhaps the first many-particle (or multi-agent) system introduced in the context of income dynamics was the model based on generalised Lotka-Volterra equations [677]. The Lotka-Volterra equations, originally used in ecology, are themselves generalisations of the Verhulst equation (see Box 7.3). Let us consider a dynamical system composed of N agents, each with income V_{it}. The average income is the random variable $\overline{V}_t = \sum_i V_{it}/N$. The changes in income are influenced by the incomes of other individuals according to the equation

$$\mathrm{d}V_{it} = V_{it}\,\mathrm{d}W_{it} + a\,\overline{V}_t\mathrm{d}t - c\,\overline{V}_t\,V_{it}\mathrm{d}t \tag{7.39}$$

with all $\mathrm{d}W_{it}$ independent of each other. Measuring the incomes relative to the ensemble average, we introduce reduced variables $U_{it} = V_{it}/\overline{V}_t$. It is easy to write a

> **Directed polymers in random media** | Box 7.4
>
> A directed polymer is an elastic spatial object with one preferred direction. All (hyper)planes perpendicular to that direction cut the polymer in at most one point. The energy of the directed polymer is composed of an elastic part due to bending, and a potential part, as some of the spatial positions can have lower potential energy than others. If we consider our wealth-exchange network embedded in space and add an extra coordinate representing time t, then each elementary exchange of wealth can be seen as bending a line tracing the chunk of wealth from one space point to another. Going from i to j costs energy $a_{ij} dt$, while arriving at point i in time t gains energy dW_{it}. As expected, lines tend toward configurations with the lowest energy, which may be very complex due to the randomness, and this tendency is also countered by an entropic effect. W_{it} can be viewed as a sum of Boltzmann factors accumulated along these lines, over all possible realisations of the lines arriving to site i at time t. See Refs. [1442–1448] for more information.

stochastic equation for the collection of U_{it}, but a technical complication arises here. From summation over all individuals, we get, among others, the term

$$\frac{V_{it}}{\overline{V}_t^2} \frac{1}{N} \sum_i V_{it}\, dW_{it}. \tag{7.40}$$

However, this term is a sum of N independent random variables, and we expect that it decreases as $N^{-1/2}$ for large N. Therefore, we neglect it in the thermodynamic limit $N \to \infty$. So, within this limit we find that the coupled system of stochastic differential equations (7.39) leads to decoupled equations for the reduced incomes

$$dU_{it} = U_{it}\, dW_{it} + a(1 - U_{it})\, dt \tag{7.41}$$

which has exactly the form (7.29) investigated in Sec. 7.2.1, and we can use the results derived there without significant change [173, 679, 680, 682–686, 711, 935, 1434, 1437–1441]. Especially it follows that the income distribution is given by Eq. (7.34), with appropriate re-definition of the parameters therein.

It is important, and far from trivial, that a nonlinear set of equations for strongly coupled quantities is reduced to a set of equations describing independent evolution of reduced variables. This fact a posteriori substantiates the use of representative individuals in stochastic models formulated in Sec. 7.2.1.

Wealth-redistribution models and directed polymers

Surprisingly (at first sight), we can arrive at equivalent results from a completely unrelated perspective. In this context it is common to speak about individual wealth instead of income, so we too shall adopt this language in this section. Of course, in reality the difference between total wealth owned by a person and her income is substantial, but formally both quantities can be described using the same dynamical model.

We suppose N individuals have wealth values V_{it} evolving in time by a random multiplicative increase accompanied by an exchange with other individuals. The exchange goes along established links of a social network, quantified by a symmetric matrix a_{ij}. If individuals i and j are disconnected, then $a_{ij} = 0$; if they are connected,

a_{ij} assumes a positive value. In principle we allow for more or less intensely used links, corresponding to a larger or smaller value of a_{ij}. The wealth values evolve according to the set of equations [813, 1449]

$$\mathrm{d}V_{it} = V_{it}\,\mathrm{d}W_{it} + \sum_{j(\neq i)} a_{ij}\,(V_{jt} - V_{it})\,\mathrm{d}t. \tag{7.42}$$

It is interesting that this economic problem can be mapped onto the well-studied problem of directed polymers in random media [1448] (see Box 7.4).

Assuming for simplicity that $a_{ij} = a$ for all $i \neq j$ and repeating the steps leading to (7.41), we arrive again at a set of decoupled equations for the relative wealth values $U_{it} = N\,V_{it}/\sum_j V_{jt}$, in full analogy with Eq. (7.41). Therefore, the distribution (7.34) also applies here.

The model can be further elaborated to investigate the transition to a state where the wealth is 'condensed' in the hands of a single extremely rich individual [407, 922, 1449–1451] and to assess the influence of the non-trivial structure of the economic network. First, it was noticed that on hypercubic networks in lower dimensions (at least lower than 3) the tail of the wealth distribution is no longer power-law, but rather a stretched exponential [813, 1449]. Small-world and scale-free networks were also investigated by numerical simulations [1452–1454], and it seems that the power-law tail is peculiar to networks with enough long-range links; but a final conclusion has not yet been reached.

7.2.3 Agent models

We may proceed further in understanding income or wealth dynamics if we go deeper into microscopic mechanisms underlying wealth production and redistribution. We are led by the analogy with particles in a gas. Indeed, it was Mandelbrot [727] who first suggested that economic exchange can be compared to energy transfer in binary collisions of molecules. However, the well-known Boltzmann distribution $P(e) = \mathrm{e}^{-e/(k_B T)}$ for a particle with energy e contradicted the Pareto law, and the idea was abandoned for forty years.

After the discovery that medium incomes are indeed distributed according to exponential law [1410, 1418, 1419], the analogy between economic activity and binary collisions was resuscitated. Indeed, the exponential form of the Boltzmann distribution is a consequence of a very general feature of inter-particle interaction, which is the conservation of energy. In his pioneering article [1455], J. C. Maxwell introduced the model of ideal gas composed of weakly interacting molecules and showed how conservation of an additive quantity (energy) in each collision, together with assumed statistical independence of particles, expressed by factorisation of the probability density, naturally leads to exponential distribution. With revived interest in modelling economic interactions as a scattering process, the Maxwell model became one of the favourite tools.

Conservative exchanges

Let us first investigate a model where individuals with wealth values V_{it} and V_{jt} meet each other randomly, and at each meeting they exchange a certain proportion of their wealth. The total wealth $V_{it} + V_{jt}$ is conserved and redistributed randomly, so after the 'scattering' we have

$$\begin{pmatrix} V_{i\,t+1} \\ V_{j\,t+1} \end{pmatrix} = \begin{pmatrix} \beta & \beta \\ 1-\beta & 1-\beta \end{pmatrix} \begin{pmatrix} V_{it} \\ V_{jt} \end{pmatrix}. \tag{7.43}$$

If for simplicity we suppose that the random parameter β is distributed uniformly on the interval $(0, 1)$, we obtain a master equation for one-particle probability density. This master equation is a very simplified analogue of the Boltzmann kinetic equation governing scattering in classical (and after appropriate modification, also quantum) gases. The equation is

$$\frac{\partial}{\partial t} P_t(v) + P_t(v)$$
$$= \int_0^1 \int_0^\infty \int_0^\infty \delta((v_1 + v_2)\beta - v)\, P_t(v_1) P_t(v_2)\, \mathrm{d}v_1\, \mathrm{d}v_2\, \mathrm{d}\beta, \tag{7.44}$$

and it can easily be verified that its stationary solution is exponential:

$$\lim_{t \to \infty} P_t(v) = \frac{1}{v_0}\, \mathrm{e}^{-v/v_0}. \tag{7.45}$$

We can conclude that the exponential part of the income distribution corresponds to the regime in which any advantage gained by one individual is compensated by a loss suffered by another individual.

One may wonder if all types of conservative pairwise wealth exchange lead to the same stationary distribution. Surprisingly, it was found that it is not the case [1456, 1457], and especially the behaviour at $v \to 0$ is sensitive to the details of the process. Various modifications of the process (7.43) can be tried, changing the matrix which multiplies the vector of wealth values $\begin{pmatrix} V_{it} \\ V_{jt} \end{pmatrix}$. Therefore, the requirement of wealth conservation is not strong enough to guarantee the Boltzmann-like exponential distribution.

The models in which a certain amount of wealth is saved before the rest is redistributed deserve special attention [1457–1475]. Although the wealth is conserved, numerical simulations have shown that randomness in the fraction of money saved gives rise to distributions with power-law tails. These findings were confirmed analytically in Ref. [1476].

Non-conservative exchanges

The situation becomes more complicated if we allow wealth production in each interaction between the individuals. A similar situation, but with the opposite sign, occurs in shaken systems of granular particles, where energy is dissipated in each collision of the grains. Without energy loss the granular gas, as it is called, would be a macroscopic counterpart of ordinary molecular gasses. But inelastic scattering brings about completely new physical phenomena.

Besides complex effects related to spatial structures which emerge in low (1, 2, 3, ...) Euclidean dimensions [1477], the most important fact is that the velocity distribution does not follow the universal Maxwell-Boltzmann distribution, but generically has fatter tails, either exponential or power-law ones [1478–1482].

As a simple model for granular gasses, the Maxwell model was adapted to inelastic scattering [1479, 1480, 1482–1494]. The rate of dissipation is measured by restitution coefficient $1 + \epsilon$, with $\epsilon < 0$. Power-law tails in energy distribution were observed and within the mean-field approximation, equivalent to infinite spatial dimensionality, an exact solution was found for an arbitrary level of dissipation ϵ.

It is tempting to translate the results known for inelastic gases to the language of wealth exchange. Here, however, instead of dissipating the energy the wealth is produced, corresponding to positive ϵ. Unfortunately the results for $\epsilon < 0$ cannot be used directly in the region $\epsilon > 0$. Production is not a simple inverse of dissipation.

The model of wealth production and exchange in pairwise interactions can be described by a process similar to (7.43) [918]. We shall fix the redistributed fraction β and production rate $\epsilon > 0$ and in each step randomly choose a pair of agents. Requiring that the interaction is symmetric with respect to exchange of agents, $(i, j) \to (j, i)$, we have the following rule

$$\begin{pmatrix} V_{i\,t+1} \\ V_{j\,t+1} \end{pmatrix} = \begin{pmatrix} 1 + \epsilon - \beta & \beta \\ \beta & 1 + \epsilon - \beta \end{pmatrix} \begin{pmatrix} V_{it} \\ V_{jt} \end{pmatrix}. \tag{7.46}$$

Again, the distribution function for the wealth of a single individual obeys a master equation

$$\frac{\partial}{\partial t} P_t(v) + P_t(v)$$
$$= \int P_t(v_1) P_t(v_2)\, \delta((1 - \beta + \varepsilon)v_1 + \beta v_2 - v)\, dv_1\, dv_2. \tag{7.47}$$

The solution of this equation is not straightforward. First, we note that the average wealth $\overline{v}(t) = \int v\, P_t(v)\, dv$ grows indefinitely, $\overline{v}(t) = \overline{v}(0)\, e^{\varepsilon t}$; so strictly speaking there is no stationary solution. This is not a serious obstacle, though, for we encountered a similar situation with Lotka-Volterra and wealth-exchange models. We avoided the problem by introducing variables measured with respect to the mean; and the same will also be done here. We shall look for the solution in the form $P_t(v) = \Phi(v/\overline{v}(t))/\overline{v}(t)$. Moreover, we shall use the Laplace transform $\widehat{\Phi}(z) = \int_0^\infty \Phi(u)\, e^{-zu}\, du$. This way we arrive at a non-local differential equation

$$\epsilon z \widehat{\Phi}'(z) + \widehat{\Phi}(z) = \widehat{\Phi}((1 - \beta + \epsilon)z)\, \widehat{\Phi}(\beta z). \tag{7.48}$$

The tail of the distribution $\Phi(u)$ can be deduced from the behaviour of its Laplace transform at $z \to 0$. Assuming $\widehat{\Phi}(z) = 1 - z + A\, z^\alpha + \ldots$ we have $\Phi(u) \sim u^{-1-\alpha}$ for $u \to \infty$. We have only to check that the assumed behaviour of $\widehat{\Phi}(z)$ is consistent with Eq. (7.48) (it is, indeed) and find the value of the exponent α. This leads to the following equation

$$(1 + \varepsilon - \beta)^\alpha + \beta^\alpha - 1 - \varepsilon\alpha = 0 \tag{7.49}$$

which fixes the exponent of the power-law tail, depending on the parameters β and ϵ.

We have shown that power-law tailed distributions of wealth are generated on condition that wealth is not conserved but produced at each step of the dynamics. Next, it would be desirable to deduce the full form of the distribution $\Phi(u)$. Unfortunately, it is not possible in general, due to the non-local nature of Eq. (7.48). However, we can see that the non-locality is due to finite values of parameters ϵ and β. The smaller the parameters, the closer we are to a well-behaved ordinary differential equation. This suggests making the limit $\epsilon \to 0$ and $\beta \to 0$ with the hope that we get a differential equation amenable to an exact solution. Physically this limit means that the trading is performed in smaller and smaller packets, and we call that the limit of continuous trading.

When doing that, we must carefully choose the way to perform the limit; otherwise all relevant physics would be lost. As the measurable parameter here is the Pareto index α, the proper way is to proceed along the line of constant α. From Eq. (7.49) we find that this line is given by $\beta = \frac{\alpha-1}{2}\epsilon^2 + \ldots$

This is sufficient to derive the equation for the distribution

$$-\frac{1}{2}z\widehat{\Phi}''(z) + \frac{\alpha-1}{2}\left(\widehat{\Phi}'(z) + \widehat{\Phi}(z)\right) = 0, \tag{7.50}$$

which can be explicitly solved using Bessel functions. Inverting the Laplace transform, we arrive at the result we have already seen in Eq. (7.34). Let us repeat it again in the current notation [918]

$$\Phi(u) = \frac{(\alpha-1)^\alpha}{\Gamma(\alpha)}\, u^{-1-\alpha} \exp\left(-\frac{\alpha-1}{u}\right). \tag{7.51}$$

It is indeed worth noting that several apparently unrelated approaches lead to not just qualitatively similar but truly identical results. Surely there is a whole family of models which implement the same basic mechanism in different disguises. Let us list those which have been discussed in this chapter: multiplicative-additive process (Sec. 7.2.1), generalised Lotka-Volterra systems, wealth redistribution models, and non-conservative wealth exchange (studied just now). Looking for the common ingredients among them, we can see there is, first, the multiplicative character of these processes, either scalar or matrix; and second, a stabilising mechanism acting against the geometric increase (or decrease) normally observed in pure multiplicative processes. The stabilising force can be applied either explicitly, as a wall or through an additive term, or implicitly, redistributing the wealth in an essentially uniform fashion among all individuals in the system.

7.3 Corporate growth

7.3.1 Scaling in empirical data

So far, the Gibrat law was regarded as an auxiliary principle with a certain limited validity. It was used to develop the first models for income dynamics, which have subsequently been improved and tuned to accommodate the empirical findings. Now we turn to the very subject touched by the Gibrat law, the statistical properties of growth.

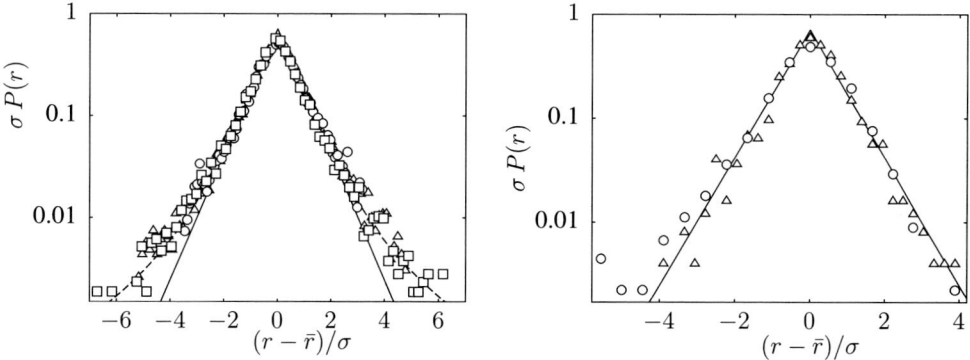

Fig. 7.5 In the left panel, rescaled probability density for annual company growth measured in the period 1974–1993. Different symbols correspond to three bins of initial size: $8^7 < S_0 < 8^8$ ($\bigcirc$), $8^8 < S_0 < 8^9$ ($\triangle$), and $8^9 < S_0 < 8^{10}$ ($\square$). Data were extracted from Ref. [1498]. In the right panel, rescaled distribution of country growth for countries with low GDP ($\bigcirc$) and high GDP ($\triangle$). Data extracted from Ref. [1503]. In both panels, the solid line denotes the exponential distribution (7.52). The dashed line is the power $\propto |r - \bar{r}|^{-3.1}$.

Companies

For historical and practical reasons, most studies on economic growth were performed on companies and other complex entities [1495], while growth of individual income was much less covered [1412]. Let us first present some empirical findings. Like distribution of individual incomes, it is possible to analyse data for firm sizes. A problem arises here as to how to properly define the firm size. Several measures are available to choose from: number of employees, annual sales, total assets, and several others. Unfortunately they do not always give compatible results. While power-law distribution was reported for the number of employees [1496], sales were better reproduced by log-normal distribution [1497–1500] followed by a power-law tail [1432, 1501, 1502]. Interestingly enough, log-normal distribution was also observed for the GDP of entire nations [1503]. We shall see that the analogy between companies and countries goes even farther.

The distribution of annual growth was studied for various times and geographical areas [1412, 1425, 1426, 1432, 1495, 1497–1500, 1504–1507]. Usually the logarithmic growth rate $R = \ln(S_1/S_0)$ is analysed, where S_0 and S_1 are company sizes in two consecutive years, measured in the quantity selected to study (sales, employees, etc.). In terms of growth distribution, little difference is found between the various measures. The first important observation was that, except far in the tail, the variance of growth rates systematically depends on company size; but after subtracting the average growth $\bar{r}$ and rescaling relative to the variance σ, all data for the rescaled quantity $(R - \bar{r})/\sigma$ for various sizes fall onto the same curve (within statistical errors).

Typical data for the distribution $P(r)$ for the quantity R are shown in Fig. 7.5. The striking feature is the cusp at zero growth and the exponential behaviour in the central part. Indeed, it was successfully fitted on the following dependence

$$P(r) = \frac{1}{\sigma\sqrt{2}} \exp\left(-\frac{\sqrt{2}|r - \bar{r}|}{\sigma}\right). \tag{7.52}$$

The tails of the distribution are more difficult to analyse, but they seem to be compatible with the power-law behaviour. Thus, we encounter a situation similar to that prevailing in income distribution: power-law tails accompanied by exponential distribution for lower values of the quantity in question. More thorough investigations also found clear asymmetry of the tails for positive and negative growth [1432]. This asymmetry is even more pronounced in the distribution of individual income growth [1412, 1508].

Countries

A similar study on GDP growth of countries yielded essentially the same results [1503]. We can see the data in Fig. 7.5. If we plotted the data from the left and right panels of Fig. 7.5 in the same graph, they would hardly be distinguishable.

Very interesting results appeared when the variance of the distribution of growth was plotted against initial size (or GDP in the case of countries). While in the tail, where company sizes are power-law distributed, the variance was nearly constant (in accordance with the Gibrat law), in the bulk of the distribution, where log-normal dependence is applicable, the variance decreases with size (contrary to Gibrat) according to

$$\sigma \sim S_0^{-\beta} \tag{7.53}$$

where exponent β varies between 0.15 and 0.2, as can be seen in Fig. 7.6. Such a dependence contradicts the usual scaling of fluctuations in common physical systems like gases, where, for system size N, we have $\sigma \sim N^{-1/2}$. This means that companies as well as states fluctuate significantly more than they would if they were composed of independent units.

7.3.2 Hierarchical model of an organisation

There is an elegant explanation for the anomalous fluctuations in growth rate, based on the hierarchical internal structure of the organisation. Normal scaling of fluctuations depending on their size, described by an inverse square root, holds on the condition that the system is composed of independent units whose number is proportional to the size of the entire system. In fact, it is not necessary that individual particles in the system are nearly free. If that were the case, the inverse square root behaviour would only apply to ideal gases, while in reality it holds for virtually any system not too close to a critical point. We may think of the particles as effectively independent units, which may themselves comprise very many particles. The important point is that each unit behaves as a whole and yet simultaneously the units are mutually quasi-independent. The system is organised in only two levels: the lowest level of particles and the level one step higher, which belongs to the units. There is no more structure above them.

We may generalise the latter idea, introducing many levels in our system [1495, 1509, 1510]. In fact, adding any finite number of levels would not help because the behaviour is then qualitatively the same as with two levels. You can either have only two levels, or you need arbitrarily many of them.

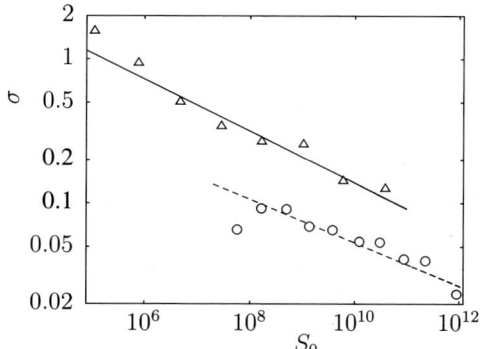

Fig. 7.6 Size dependence of the standard deviation for countries (○) and companies (△). The solid line is the dependence $\sigma \propto S_0^{-0.18}$, the dashed line is the dependence $\sigma \propto S_0^{-0.15}$. Data extracted from Ref. [1503].

Imagine that at each level a manager directs k people at a lower level, each of whom directs again k people at an even lower level, etc. The company is organised as a big tree. Now suppose a manager issues an order to subordinated employees. They will obey that order with probability p and behave independently with probability $1 - p$. It is possible to show that the fluctuations in such a system decrease as $\sigma \sim S^{-\beta}$ with the exponent

$$\beta = \begin{cases} -\frac{\ln p}{\ln k} & \text{for } p > k^{-1/2} \\ \frac{1}{2} & \text{for } p < k^{-1/2} \end{cases} . \tag{7.54}$$

So, hierarchical organisation generically leads to exponents $\beta \le 1/2$, in accordance with empirical findings.

7.4   What remains

Despite numerous models and important progress achieved in understanding income and wealth distributions [43, 1511], some questions are still unanswered. The emergence of the Paretian power-law tail seems to be consistently ascribed to the multiplicative character of income dynamics, countered by a stabilising mechanism. On the other hand, each of the middle and lower incomes has its own explanation, different from the one valid for the tail. Such a state of affairs is certainly unsatisfactory. A theory is desirable which would smoothly interpolate between the three regimes. The theory of killed multiplicative processes seems to us the closest to this goal, but it is not yet clear which microscopic mechanisms may lie behind it. Without deeper insight, the killed multiplicative processes look somewhat arbitrary.

It is also not clear why the various conservative models based on pairwise exchange lead to such different results. We should ask whether it would be possible to devise a plausible model which would give Boltzmann-like exponential distribution for lower values, followed by a power-law tail. The feasibility of such a model is highly probable, recalling the sudden change of behaviour, namely, from exponential for conservative to power-law for non-conservative exchanges. Such an abrupt change may be an artifact of the existing models, and a more realistic one might pass smoothly from one regime to the other. For those interested in possible alternatives, we provide the following references [774, 1512–1539]. In Refs. [1540–1542] the ideas of pure rational thermodynamics are applied.

We must stress again the generic importance of the generalised Lotka-Volterra equations. To see the problem in wider context, the reader is encouraged to consult Ref. [1543]. For various specific applications, see Refs. [561, 818, 1544].

Let us also recall what was mentioned when discussing the wealth-redistribution models in Sec. 7.2.2. The effect of the complex topology of the social network on wealth dynamics has as yet been poorly explored. Finally, a totally untouched question is the source and properties of fluctuations in the Pareto index, which were shown in Fig. 7.3.

The study of growth is relatively less developed with respect to distribution of income and wealth. There have been attempts to explain the tent-shaped exponential distribution (7.52) with the assumption that each company tends towards an intrinsic optimal size despite random fluctuations [1509]. However, this solution is far from being complete. For example, it cannot explain the tails in the growth distribution, which are fatter than those of the exponential one. We should also note that there are explanations based on preferential attachment principle [1545, 1546].

It is probable that the hierarchical model is not the only one that can explain the observed data. For a promising alternative see Refs. [1547–1549]. Indeed, any mechanism which would relate the size of effective units with the size of the whole system would be as good in predicting anomalous scaling of fluctuations. However, it may serve as a transparent phenomenological model even if we are not able to unambiguously determine positions of individuals in our tree. Indeed, for example it may be a daunting task to reliably identify for entire countries who is the one really making the important decisions and who is simply a puppet in front of the TV cameras. It seems that much more data is necessary before we have a reliable model of the internal structures of complex organisations.

Problems

1. Introduce noise in the additive term in Eq. (7.29), so that the dynamics is described by

$$dV_t = V_t \, dW_{1t} + dW_{2t}. \qquad (7.55)$$

Derive the corresponding Fokker-Planck equation and deduce the exponent of the power-law tail.

2. Using the idea of killed multiplicative processes, consider the following model with discrete-time dynamics. The society is composed of a variable number N_t of individuals with incomes V_{it}. At each time step, each of the incomes is multiplied by a random number

$$V_{i\,t+1} = e^{W_{it}} V_{it} \qquad (7.56)$$

where all W_{it} are independent and equally distributed. After this change of income, but still at the same time step, each individual can die with probability p or give birth to an offspring with probability $1 - p$. A newly born individual has an income of 1 while the parent's income is unchanged (but also imagine other possibilities!).

What will the distribution be of the income of a randomly chosen individual in the society? How does this result depend on the parameter p and on the properties of the noise W_{it}?

3. For the one-variable stochastic Lotka-Volterra equation

$$dV_t = V_t \, dW_t - V_t^2 \, dt \qquad (7.57)$$

with noise characterised by $\langle dW_t \rangle = dt$, $\langle (dW_t)^2 \rangle = D \, dt$, $D \ll 1$, find the corresponding Fokker-Planck equation. Can you guess the general features of the solution?

4. We can modify the conservative pair-exchange model (7.43) so that it describes the dynamics of income for families with two earners. Our 'particles' therefore have an internal structure, and their income is a sum of two contributions, $V_{it} = V_{ia\,t} + V_{ib\,t}$. Suppose that in each interaction two such pairs of earners scatter, so that each family member in one pair interacts independently of the other with the corresponding member in the second pair. The resulting process is therefore a combination of two independent instances of the processes described by (7.43). First write the matrix multiplication formula corresponding to (7.43), then find the equation analogous to the master equation (7.44). Are you able to solve it? Hint: use the Laplace transform.

5. How will the equation for the Pareto index (7.49) be modified if we allow ϵ and β to fluctuate randomly? (You may look at Ref. [1550] for inspiration.)

8
Social organisation

If a storm had taken you from your quiet university office, far off your desk full of piles of reprints, preprints, and books; two or three cups with remnants of yesterday's coffee; and abstract ideas dreaming, perhaps, of solving exactly the three-dimensional Ising model... if the storm had tossed you into a factory, amidst workers dissatisfied with their low salary and threatening to go on strike, you would most probably feel rather lost with your wits. But after a while you would start to breathe again, recover your mental equilibrium, Ising model still in mind, and try to figure out what was happening and what would be best to do now.

You will probably notice that the pending strike is certainly a collective phenomenon. Indeed, it would never happen without a large level of coordination among workers. Next, the strike occurs as a sort of singularity in the otherwise smooth operation of the factory. These two features must make a short circuit in your head, producing a spark and illuminating the scene. It is so simple. The phenomenon of a strike is yet another example of a phase transition (so you think). People behave like Ising spins, having two choices: to work (denote it +1) or to strike (−1). And they not only have binary options like Ising spins, but they also interact with each other in a manner familiar from the Ising model, following more or less their close neighbours within the crowd. The extent to which an individual obeys the surrounding opinion is one of crucial parameters determining the fate of the factory. A physicist would call it temperature. The last ingredient is the incentive to work, quantified by the salary. If it is lower than the labour force considers fair, it results in a tendency to strike. If it is higher than the subjectively acceptable level, it motivates people to work.

We can formalise the idea by introducing a dissatisfaction function, summing contributions from the mutual influence between employees and pressure exerted by the management. We can write it as $H = -\sum_{<ij>} J_{ij} S_i S_j - h \sum_i S_i$. The meaning of the symbols is straightforward. $S_i \in \{-1, +1\}$ is the state of the i-th worker, J_{ij} measures the strength of interaction between workers i and j, and h parameterises the incentive to work. It is an increasing function of the salary and equals 0 when the salary is exactly the (subjectively) fair one.

Of course, every physicist instantaneously recognises H as the Hamiltonian (i.e. energy) of the Ising model with couplings' strengths measured by the numbers J_{ij} and immersed into external magnetic field h. Assuming all workers being in touch with all others, and all relations being equally intense, we can set $J_{ij} = J$ for all pairs of individuals, and the model is reduced to a mean-field version of the Ising model. Its solution is well-known and straightforward [1551]. Below a certain critical temperature there are two ordered phases: one with the majority working (+1) and the other with the majority striking (-1). Varying the external stimulus h, the 'magnetisation', or fraction of working people, performs a sudden jump. In other words, there is a first order phase transition with control parameter h. On the other hand, for a temperature above the critical point the magnetisation behaves smoothly when changing h.

What is important for us here is the response of the labourers to the change in salary the manager can offer to the rebellious employees. The results for the mean-field Ising model can be directly translated to the language of the ongoing social conflict. Indeed, if the coordination between labourers is higher, it amounts to a lower temperature and strong cooperation emerges. If the factory is on strike, increasing wages has little effect until it reaches a threshold where the will to go back to work suddenly jumps up. Conversely, the social cohesion prevents the workers from going on strike individually immediately after the payments were lowered; but when the salary drops below a certain unbearable value, the strike begins instantaneously.

An important finding is that the salary at which people jump from work to strike is lower than the salary which forces the striking employees to return to work. This is the effect of hysteresis, common in magnetic materials, where the Ising model rules. Finding the same behaviour in human society was an exciting achievement in 1982, when Serge Galam et al published one of the first articles [1552] in the field of socio-physics [1553, 1554]. We shall devote this chapter to some of the multitude of socio-physical models which emerged since then, or even before [12, 13], maybe much before, as the first person to speak of social physics was Auguste Comte all the way back in the first half of 19th century [1, 2]. Apparently we are in good company.

8.1 Cooperation

People never earn their living alone. From the darkest pre-historical past they stuck together to form gangs, stalking and hunting their prey. Without communication within the band, our ancestors would hardly have been able to catch a mammoth, language would never have evolved, and human brain capacities probably would never have exceeded that of a lemur. In short, cooperation between humans was decisive for shaping the world around us. A tendency toward cooperation is an inherent feature of human nature.

It was not until the middle of 18th century or so that a different view started to spread. People are presented as selfish profit-seekers, and, if any cooperation is observed, it occurs despite the natural tendency to gain the maximum for self, or as a by-product of such a tendency. Under the vague names of neo-liberalism and social Darwinism these and related ideas pervade current thinking on human society.

On the other hand, viewed empirically, the cooperation has much evolved now and assumed more sophisticated and complex forms than in the old times when it was considered self-evident. The situation is rather paradoxical: the more people need each other, the more they proclaim selfishness a norm, or even a desirable behaviour. The evidence from reality, however pungent, is not any more evident in itself but needs external explanation. The facts require proof from a theory. Here, we have no other choice than to proceed the same way. We shall start from the zero hypothesis of no-cooperation and regain cooperation by a non-trivial mechanism we need to discover.

8.1.1   Prisoner's dilemma

Games

There is a very fertile ground to model the starting assumption of selfish individuals, provided by game theory [1555, 1556], introduced in the 1940s by John von Neumann and Oskar Morgenstern. In fact, the agents are supposed to be not only utterly selfish, but also absolutely rational. In a typical setup agents meet in pairs, and each of them chooses one of S possible strategies. If the first agent adopts strategy i and the second agent strategy j, then the first one gains an amount denoted A_{ij}, while the other's gain is A_{ji}. The matrix A is called a payoff matrix and various types of games are distinguished according to its properties. For example, if the gain of one agent equals the loss of the other agent, that is, $A_{ij} + A_{ji} = 0$ for all i and j, we speak of a zero-sum game. If the gain of the wining party is always smaller than the loss of the adversary, the game is a negative-sum one, and we have $A_{ij} + A_{ji} < 0$ for all i and j.

We suppose that the rules of the game, quantified in the matrix A, are known to both players. Therefore, they can build their strategies on rational analysis of the payoff matrix. It may happen that one strategy gives the highest gain irrespective of the action taken by the opponent. Formally, it means that the strategy k of the first player is such that $A_{ij} \leq A_{kj}$ for any strategy i of the first player and any strategy j of the second player. The same may also hold for the second agent. Suppose that l is her best strategy, irrespective of the action of the first player. Obviously then, the first person always plays k while the second always plays l. If either of them changes her strategy unilaterally, she is instantly worse off. Such a situation, if it happens, is called a Nash equilibrium, after John F. Nash, a mathematician who devoted much of his career to applications of game theory in economics [1557, 1558].

It is vital to realise that the notion of the Nash equilibrium differs fundamentally from the usual equilibrium studied in various branches of physics. Whether we think of an equilibrium of solid bodies on a lever, equilibrium configuration of an ensemble of balls tied together by rubber springs, or of thermodynamic equilibrium in the system of steam, water, and ice, we formalise the situation by finding a unique function to be minimised (or maximised, if you like working with entropy). On the contrary, Nash equilibrium means that every player maximises her own function, with the state of all other players fixed, as if the movement of every single molecule in a bottle of wine was governed by an individual Hamilton function, instead of having one Hamiltonian for the whole system [855, 1559].

But if we look around, we do find physical systems in which Nash equilibrium is decisive. For example take a cardboard box and pour a bag of marbles into it. An irregular packing of small spheres will emerge, despite the fact, now proved rigorously [1560], that the most compact configuration is either the face-centred cubic or hexagonal close packed lattice (both of them are equally good, as is any mixture of them). You may try to shake the box, expecting that it would help the beads to get to their optimal arrangement, but you are highly unlikely to enjoy success. The point is that with gentle shaking any of the marbles finds its optimum position with respect to the surrounding marbles, but as a whole the ensemble of the balls is not ordered as well as it could be. Finding the global optimum would require simultaneous rearrangement of many spheres, which is an extremely rare event. The best one can expect is that every ball finds its private optimum, much like the players finding the Nash equilibrium.

Two prisoners

Collaboration was studied in a very simple two-player game with two strategies, called the prisoner's dilemma game. As a classical motivation, imagine that the police caught two accomplices for a suspected burglary, but do not have enough evidence to prove that the two are guilty. The prisoners are kept well separated and the investigator offers a deal to each of them independently. The criminal is given a promise that if he breaks company with the other and confesses the robbery they committed together, he will be rewarded. So, if both of them confess, they will be jailed only for a short period; if one confesses and the other does not, then the first one is released and rewarded and the other gets a severe punishment; and if neither of them confesses, both are released. The point is that each has to choose without knowing the other's choice.

The strategy of collaboration (C) dictates not to confess. In fact, it would be most beneficial to them, if both collaborate. However, individually it is more tempting to defect (D) and confess to the police, as it prevents the situation that the one confesses and the other is punished alone. Because both prisoners reckon in the same way, the result is that both defect and have to suffer some time in prison. Hence the dilemma.

The payoff matrix of this prisoner's dilemma game is characterised by four numbers: the gain if both defect, $A_{DD} = P$, the gain if both cooperate, $A_{CC} = R$, and the gains of the defector, $A_{DC} = T$, and of the collaborator, $A_{CD} = S$, if one of them defects and the other does not. So,

$$A_{PD} = \begin{matrix} & C\ D \\ \begin{pmatrix} R\ S \\ T\ P \end{pmatrix} & \begin{matrix} C \\ D. \end{matrix} \end{matrix} \tag{8.1}$$

In order for the prisoner's dilemma to work as described above, the values must satisfy the inequalities $T > R > P > S$ and $2R > S + T$. The most studied values of the parameters are

$$T = 5,\ R = 3,\ P = 1,\ S = 0. \tag{8.2}$$

It is easy to see that there is a single Nash equilibrium, in which both players defect. Intuitively it was exposed before, and an exact check is even quicker. If the second player collaborates, the rewards of the first are R and T if he collaborates or defects, respectively, and because $T > R$, defection is better. Similarly, if the second player defects, the gains are S and P and as $P > S$, it is again preferable to defect. So, defection is the optimal strategy anyway.

Mixed strategies

The full potential of the game theory is developed introducing the so-called mixed strategies. Intuitively, they mean that the strategies i are not chosen deterministically but are played randomly with prescribed probabilities p_i. If the first player uses a mixed strategy with probabilities p_{1i}, while the second plays according to probabilities p_{2i}, the average gain of the first is $G_1 = \sum_{ij} p_{1i} A_{ij} p_{2j}$ and that of the second is $G_2 = \sum_{ij} p_{2i} A_{ij} p_{1j}$. For example, in the prisoner's dilemma game it is $G_1 = p_{1C} R p_{2C} + p_{1C} S(1 - p_{2C}) + (1 - p_{1C}) T p_{2C} + (1 - p_{1C}) P(1 - p_{2C})$. To be precise, the notion of the Nash equilibrium was formulated for mixed strategies. The point is that the n probabilities p_i are restricted to the interval $[0, 1]$; their sum is always 1, and therefore the set of possible mixed strategies form a $(n-1)$-dimensional polyhedron. If the Nash equilibrium exists, it must lie in one corner of the polyhedron.

In prisoner's dilemma, the analysis of mixed strategies is simple. Differentiating G_1 with respect to p_{1C}, we find that the gain of the first agent always decreases with p_{1C} regardless of the strategy of the second player, quantified by the parameter p_{2C}. Therefore, the optimum lies at $p_{1C} = 0$, which means deterministic strategy of a defector. The same holds for the second player; so we recover the result that defecting is the best choice even if we allow mixed strategies.

The Nash equilibrium in the prisoner's dilemma game is symmetric, i.e. both agents choose the same strategy. Generally it may not be so, even though the game itself is symmetric. The Nash equilibrium may be accompanied by spontaneous symmetry breaking. To see it on a toy example, let us consider the payoff matrix $A = \begin{pmatrix} 0 & 1 \\ 1 & 0 \end{pmatrix}$. It has a simple interpretation: the player gets a point if she manages to play a different strategy than her opponent. The mixed strategies of the two parties are characterised by probabilities p_1 and p_2 that the first strategy is chosen by players 1 and 2, respectively. The average gain $G_1 = p_1(1 - p_2) + (1 - p_1)p_2$ is a decreasing function of p_1 if $p_2 > 1/2$ and increasing for $p_2 < 1/2$; therefore the Nash equilibrium is either $p_1 = 0$ and $p_2 = 1$, or, conversely, $p_1 = 1$ and $p_2 = 0$. The multiplicity of Nash equilibria is a natural consequence of the spontaneous symmetry breaking.

8.1.2 Playing repeatedly

Evolutionarily stable strategies

Game theory proved extremely useful in evolutionary biology. The strategies can be understood as patterns of behaviour the living beings choose, either on the basis of previous experience or according to their genetic code. The framework of game theory is combined with replicator dynamics in the strategy space. The individuals who gain more in the repeated games give birth to more offspring, while those who mostly lose are doomed to extinction. This is the basic setting of evolutionary game theory, elaborated mainly by John Maynard Smith and others [1561–1565].

The basic new concept of this theory is the notion of *evolutionarily stable strategy*. To explain it briefly, imagine a population of individuals, each playing a certain strategy which is encoded in their genes. As the mutations are always present, we should admit that a small change in strategy may occur in one or a few individuals. Now it may happen that the mutated strategy gains more if played against the old strategy. This leads to proliferation of the mutated genome, until the new strategy invades the whole system. The defeated strategy was not evolutionarily stable. If, instead, all mutated strategies perform worse, the original strategy is called evolutionarily stable, because mutations plus selection cannot overhaul it.

In the prisoner's dilemma game, we can think of the parameter p_C characterising the mixed strategy as being encoded in the genes. Elementary mutations cause very small changes in that parameter, which would smoothly move within the interval $[0, 1]$, provided there was no selection pressure. The latter acts as an optimising machine, choosing defection as an evolutionarily stable strategy.

It can happen that there is no evolutionarily stable strategy at all but that the strategies keep replacing each other indefinitely. A nice example of such a situation is the rock-paper-scissors game. Every child knows this game: the rock makes the scissors blunt, the scissors cut the paper, and the paper wraps the rock. The following payoff matrix describes the game entirely:

$$A_{RSP} = \begin{array}{ccc} R & P & S \\ \begin{pmatrix} 0 & -1 & 1 \\ 1 & 0 & -1 \\ -1 & 1 & 0 \end{pmatrix} & \begin{array}{c} R \\ P \\ S. \end{array} \end{array} \tag{8.3}$$

In a population where all individuals play rock, the gain of each is zero. But if a random mutation makes one of them play paper, that individual has an instant advantage, which is translated in abundant offspring, quickly displacing all rock players. But evolution does not stop here. Another random mutation can create a scissors agent, repeating the takeover once more. The scissors population is, in its turn, defeated by 'rockers', and the cycle keeps rotating forever. There is no evolutionarily stable strategy, but a cyclic attractor sets in instead.

At this point it is perhaps appropriate to stress that the rock-paper-scissors game is not a mere pastime. Biology provides examples where rock-paper-scissors fits the observations quite well. One of them consists of three different competing strains of the bacterium *Escherichia coli* [1566]. The first is a common strain, without special abilities. The second strain is able to produce a toxin called colicin, thus killing normal bacteria, and produces an immunity protein for its own protection. The third strain only produces the immunity protein, being resistant against the second strain and at the same time saving resources. This causes its victory over the toxic strain. But now, in the absence of the toxin, the normal strain has an advantage, as it may invest energy into reproduction instead of producing the unnecessary immunity protein. The rock-paper-scissors game is immediately recognised. Another example of cyclic changes was observed in California lizards, which use three different mating strategies, each winning compared to one of the remaining two, but losing when matched to the other. As a result, six-year period oscillations emerge [1567].

Now that we have started our biological digression, we should mention that the prisoner's dilemma game is also realised in nature. There are several examples of fish social behaviour that very closely resemble prisoner's dilemma [1568, 1569], as well as examples from studies on other animals. However, it is very difficult to prove quantitatively that the corresponding payoffs, if they do exist, satisfy the prisoner's dilemma inequalities. A better prospect is provided by the study of viral infections in which two different virus strains attack the same cell. They may cooperate by both manufacturing the products necessary for their replication, or one of them can defect, profiting from the other virus producing most of the chemicals needed. Experiments were carried out on phages $\phi 6$ (cooperating) and $\phi H2$ (defecting) [1570, 1571]. For the sake of the reader's curiosity we show here the experimentally measured payoff matrix,

$$A = \begin{pmatrix} 1 & 0.65 \\ 1.99 & 0.83 \end{pmatrix}.$$

As a final remark concerning the concept of evolutionarily stable strategy, note that it is very close, but not exactly equal, to the Nash equilibrium. While an agent looking for the Nash equilibrium may skip to a distant point in the strategy space when driven by external conditions, the individuals undergoing the evolutionary struggle can change their strategies only smoothly, by gradual accumulation of mutations. It depends on the particular game in question as to whether that difference matters or not.

The evolution of cooperation

After spending some time with biology, let us return to modelling human behaviour. We have explained that the Nash equilibrium excludes cooperation a priori, coinciding with the neo-liberalist view on society. What, then, is the mechanism behind the cooperation seen in reality? If there is a way for emergence of cooperation in prisoner's dilemma, perhaps the reality will follow a similar path.

It was Robert Axelrod who investigated this question in depth [1572–1574]. His approach was certainly inspired by the use of game theory in biology, as explained in the preceding paragraph. It would be a very strange world if every person would get only one unique chance to play a round of the prisoner's dilemma game in her life. Instead, the games are played repeatedly, which opens a much wider space for possible strategies. For example, one can regularly alternate cooperation with defection, or cooperate every third round, etc. If the players are able to recognise each other, they can base their strategies on the observation of the opponent's past actions. As the history is in principle unlimited, the space of all possible strategies is infinite as well. But even if the agents have finite memory, say, M steps, the task to explore all of their strategies is enormous even for moderate values of M.

To test at least a small subset of all strategies Axelrod organised a computer tournament [1572]. Researchers from several disciplines were invited to send their algorithms for intelligent decisions in an iterated prisoner's dilemma game. The tournament was carried out as a round robin, i.e. every algorithm was allowed to play against every other. Moreover, every strategy also played against itself and against the random strategy, which cooperates and defects randomly with equal probabilities. Together with the random strategy, there were 15 participants on the tournament. The scores of all matches played by a strategy were added, and the rank was established according to average score. The winner was the strategy called tit-for-tat (TFT), submitted by Anatol Rapoport. Its idea is simple: cooperate with those willing to cooperate, defect otherwise. It starts with cooperation, and then does what the opponent did in the last step. The cooperation from the other side is rewarded by cooperation, while defection is punished by defection in the next step.

From this first tournament it was still unclear if the victory of the TFT strategy was only a happy coincidence or if it really was the best strategy. To be truthful, TFT was not the absolute best strategy. Certainly not, because other strategies often got a larger score with the same opponent. There was only one opponent against which TFT scored more than any other strategy; and with 5 opponents the best 8 strategies, including TFT, got the same highest score. This means that there were several strategies nearly as good as TFT. Moreover, the organisers of the tournament themselves proposed a strategy in the invitation, which nobody submitted, but when it was tested after the tournament was over, it turned out that it would have been the winner if it had participated! This strategy, called tit-for-two-tats (TFTT), works similarly to TFT, but the agent only defects if the opponent defected in two consecutive steps.

The result being inconclusive, Axelrod organised a second tournament [1573]. This time there were 63 contestants, including TFT and some of the other good strategies from the first tournament. The TFTT strategy was also submitted, as were some other ones which were supposed to win if they had faced only those 15 from the first tournament. Quite surprisingly, under these changed conditions, TFT gained the highest score and won the tournament again.

What is the mystery behind the success of such a simple strategy as TFT? As already shown, and as was seen once more in the second tournament, TFT is rarely the best strategy if the opponent is fixed. There are better specialised strategies tailored to beat a given enemy. However, in a combined and unpredictable environment, TFT performs on average better than any other known strategy. A detailed analysis of the tournament showed that the key to success lies in two features of the TFT strategy. First, it is 'nice', as it is never the first to defect. Two nice strategies always cooperate with each other. Their quality can be distinguished only when they play against non-nice strategies. If the strategy forgives defection too easily, it sooner or later gets exploited by some clever parasitic strategies. That is why the TFTT finally scored significantly lower than it was expected to after the first tournament. On the other hand, the TFT strategy is easily provoked, i.e. defection is its immediate response to the defection of the partner. No forgiveness means no chance to be exploited. As a side effect of the role of ability to be provoked we realise that there are very important strategies, Axelrod calls them kingmakers, which themselves score rather low, but affect crucially the scores of the 'good' strategies. The final order of the best few strategies in the tournament was determined by their performance against the kingmakers. Summarily, the victory of the TFT strategy was due to its niceness and its good average score with the kingmakers as well. Much of its force resides in its universality, as it is capable of competing well in a broad variety of situations.

To complete the evolutionary picture, Axelrod simulated the ecology of strategies. A population of agents evolved so that better-performing strategies got more offspring. It was no surprise that TFT ultimately achieved the highest percentage, but it was also interesting to note that it never dominated entirely, having been closely followed by a few other nearly-as-good strategies.

Competing with noise

Up to now the players were perfect in that they always did what their strategy dictated. But humans are not robots and make errors frequently. This fact brings a certain (possibly rather high) level of noise into the course of the iterated prisoner's dilemma game and makes a big difference in the success of the strategies involved. For example, an occasional defection makes the TFT strategy defect in retaliation. Two agents playing TFT against each other switch into a regime of alternating defection and collaboration, instead of collaborating all the time. A second error may either bring the two to collaboration again, or, with the same probability, to mutual defection. However small the noise is, the TFT strategy loses much of its advantage. There were several attempts to cope with noise in iterated prisoner's dilemma [1575], but the systematic way is provided by numerical simulations of evolving populations of agents, which allow mutations of strategies. Therefore, the strategy space is explored with minimum bias, and certain subsets of possible strategies can be investigated in their entirety [1576, 1577].

The formalism generalises Axelrod's tournaments in an ecological language. The ecosystem consists of N agents, each playing its own strategy; but instead of tracing all the agents individually, we shall follow the population dynamics of the strategies, which can be viewed as various species. If strategies $i = 1, 2, \ldots, S$ are present with concentrations c_i, we first calculate their mutual scores, i.e. the average gain g_{ij} of the agent using strategy i if she plays for an infinitely long time against the strategy j. We suppose that due to the presence of noise, the initial conditions are irrelevant. The score of the strategy i is then

$$u_i = \sum_{j=1}^{S} g_{ij} c_j, \tag{8.4}$$

and the average score in the whole ecosystem is $\overline{u} = \sum_{i=1}^{S} u_i c_i$. The evolution proceeds in continuous time t. The increase of the subpopulation playing strategy i is the larger the more the score of i is above the average, so

$$\frac{\mathrm{d}}{\mathrm{d}t} c_i(t) = (u_i - \overline{u}) \, d \, c_i(t) \tag{8.5}$$

where d is a constant specifying the a priori growth rate. Combining Eqs. (8.4) and (8.5), we get a closed set of evolution equations for the concentrations

$$\frac{\mathrm{d}}{\mathrm{d}t} c_i(t) = \left(\sum_{j=1}^{S} g_{ij} c_j(t) - \sum_{j,k=1}^{S} g_{jk} c_j(t) c_k(t) \right) d \, c_i(t) \tag{8.6}$$

which fully describes the population dynamics once the matrix of the parameters g_{ij} is given.

A good thing about the equations (8.6) is that they manifestly conserve the normalisation condition $\sum_{i=1}^{S} c_i(t) = 1$. We also see that if the concentration of the strategy i vanishes at a certain time t', i. e. $c_i(t') = 0$, it remains zero at any later time $t > t'$. First, it means that in an evolution governed by (8.6) it is dynamically impossible for the concentrations to go outside the allowed interval $0 \le c_i(t) \le 1$. This is an important consistency check. Second, it means that the absorbing states of the dynamics are uniform populations of only one strategy. (A state is called absorbing if there is no way of escape from it. A system which jumps into an absorbing state remains in this state forever.) However, they are a rather theoretical possibility, as they cannot be reached from a generic initial condition by the dynamics (8.6) in finite time.

We have arrived at a theory described by a set of nonlinear differential equations, which is complicated enough by itself, but we must add some more complexity on top of it. Two things are still missing in equations (8.6). The first one is the fact that there are only a finite number N of players in the ecosystem. Therefore, if the concentration of strategy i calculated from (8.6) drops below the level $c_i(t) = 1/N$, it must be artificially set to zero. Doing so, the concentrations of the other strategies must be rescaled in order that the proper normalisation $\sum_i c_i = 1$ be satisfied. This appears to be a marginal change in the evolution rules, and most of the time it is indeed so, but the eventual consequences are profound. The originally perfectly smooth evolution starts exhibiting occasional singularities, marking the complete extinction of a certain strategy. The number of strategies can decrease in time, while in the evolution according to (8.6) it always remains constant.

The second missing ingredient concerns just the opposite process: increasing the number of strategies present in the system. They may appear by mutations of the strategies actually present. To describe the process formally, we first need to encode the strategies in such a way that the mutations make sense. The decisions will be made according to the history of the game, recorded in a bit string. We denote 0 defection and 1 collaboration of an agent. The length of the agent's memory will be M. It means that the agent remembers M past decisions, both by the opponent and by the agent herself. No memory, $M = 0$, means unconditional decisions by the agents: either defection, strategy 0, or collaboration, strategy 1. The shortest non-trivial memory $M = 1$ means that each agent remembers one past step by the opponent. There are $2^M = 2$ past histories, and for each of them the strategy should prescribe the action to be taken. So, the strategy is a 2-bit string. For example, [00] and [11] are the already known strategies of unconditional defection (also called ALLD) and unconditional cooperation (alias ALLC). There are two other possible strategies. The first is [01], meaning action 0 if the history was 0 and action 1 if the history was 1; or, in simple terms, do what the opponent did in the last step. Clearly, we recognise the TFT strategy. The last remaining strategy is [10], which is just the opposite of the TFT strategy; it suggests acting contrary to what the opponent did. This strategy is called anti-tit-for-tat (ATFT); and although it seems counter-intuitive, it is rather successful in our ecology.

For a larger M the notation for strategies proceeds in the same way. The history is an M-bit number (we must be careful when M is odd, as in this case the two players do not remember the same thing: namely the actions of both in the last $(M-1)/2$ steps plus the action of the other party in the $(M+1)/2$-th step in the past), and an action is given for each of them. So, the strategy is a 2^M-bit string, and there are 2^{2^M} different strategies. The reader perhaps remembers the minority game discussed in Chap. 5, where essentially the same notation for strategies was used. For example, the $M = 2$ strategy written as [1001] prescribes cooperation (1) if either both players defected or both cooperated in the last step. If the two took different actions, the suggested action is 0, i.e. defection. We shall see that this particular strategy is rather good, but not the most successful one.

The mutations will occur instantaneously at times t_i, $i = 1, 2, \ldots$ where the waiting times $\Delta t_i = t_{i+1} - t_i > 0$ are independent, exponentially distributed random variables.

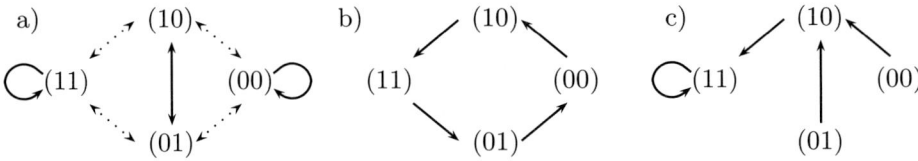

Fig. 8.1 Three examples of transitions between states in the iterated prisoner's dilemma for strategies with memory $M = 1$. Each of the four states is represented by the actions of the two players in the last step. Solid arrows indicate transitions in the absence of noise, and dotted arrows show additional transitions due to weak noise. In panels b) and c) the noise-driven transitions are not shown, as they are identical to those in panel a). The strategies of the opponents are a) both [01], b) [10] and [01], c) [11] and [01].

In fact, the choice of the times to mutate is not crucial as long as they have more or less regular intervals between them. What is crucial, though, is the recipe for the mutations themselves. Three types of mutations will be allowed. The first of them are point mutations, which is the flip of a single bit in the string representing the strategy. An example is the change [1101] → [0101]. The second type is gene duplication, which corresponds to attaching an exact copy of the strategies string at the end of itself, e.g. [10] → [1010]. The meaning of this mutation is simple. The memory length is formally increased by one, but the action of the agent is the same regardless of the information contained in the extra bit. As such, the gene duplication does not make any change in the behaviour of the ecosystem, but it is extremely important because it opens the door for longer and more complex strategies which then evolve through a sequence of point mutations. The third and last type of allowed mutations splits off randomly either the first or the second half of the genome, as if the memory was shortened by one. An example is the replacement [1101] → [01].

In terms of the concentrations, the mutations are implemented as follows. If the strategy i' was created by mutation of strategy i, we replace $c_i(t) \rightarrow c_i(t) - 1/N$ and $c_{i'}(t) \rightarrow c_{i'}(t) + 1/N$, or $c_{i'}(t) = 1/N$ if the strategy i' was absent before.

We have almost all the pieces in place now. The missing information is the matrix of the average gain g_{ij}. When calculating it, we make two assumptions. First, we suppose that the strategies i and j play against each other for an infinitely long time. This is reasonable if the population dynamics is much slower than the dynamics of the game. Second, we introduce a weak noise in the actions of the players. As already explained, this is the ingredient which makes a big difference in the dynamics even if the noise is infinitesimally small. The reason lies in the non-ergodicity of the dynamics in the absence of noise. The state space of our system is the set of possible histories, i.e. sequences of 1s and 0s, with length M if M is even, and $M + 1$ if M is odd. Starting from certain initial conditions, we can explore only a certain part of the state space. The noise reintroduces the ergodicity, as it allows the system to jump from one invariant subspace to another.

Fig. 8.1 explains the idea. In the case of $M = 1$ the state space has 4 elements, corresponding to the 4 combinations of the actions of the two players in the last step.

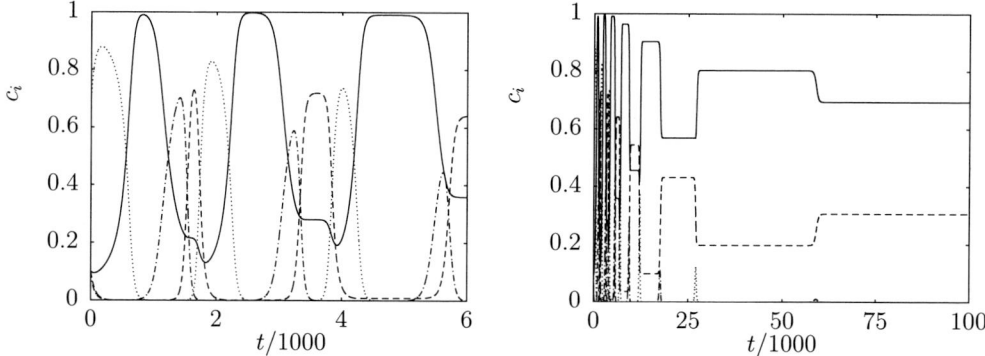

Fig. 8.2 Iterated prisoner's dilemma. Evolution of populations of strategies with memory $M = 1$, found by numerical solution of Eqs. (8.6) with parameters given by (8.7). The strategies considered here are $i = [01]$ (solid line), $[10]$ (dashed line), $[11]$ (dot-dashed line), and $[00]$ (dotted line). Mutations are not allowed. The initial conditions are $c_{[01]} = c_{[10]} = c_{[11]} = 0.1$, and $c_{[00]} = 0.7$. In the right panel we show the long-time behaviour, leading eventually to a stationary mixture of about 70 per cent of the TFT strategy and about 30 per cent of the ATFT strategy. In the left panel we can see the short-time detail of the same data. The initial oscillations gradually disappear.

In the case a), both players use the TFT strategy. Without noise, the dynamics is not ergodic and the state space is split into three invariant subsets: $\{(11)\}$, $\{(10), (01)\}$, and $\{(00)\}$. While in the first one where both agents always cooperate, their gain per step is R; in the second one they alternate cooperation and defection, the gain is on average $(T + S)/2$; and in the third subset they always defect, with gain P. An infinitesimal noise adds transitions as shown by dotted lines in Fig. 8.1. (We ignore much less frequent transitions which involve two errors simultaneously, like $(00) \rightarrow (11)$). The effect of the noise can be equivalently taken into account by averaging the gain of the players over all possible initial conditions. Therefore, still having in mind the TFT strategy, the corresponding element in the matrix g is $g_{[01][01]} = (R + P + T + S)/4$.

The panel b) in Fig. 8.1 shows how the situation is changed when TFT and ATFT play against each other. We can see that the system is ergodic, so the noise plays no role. All four states are visited cyclically, and eventually they are equally probable. The gain is $g_{[01][10]} = (R+P+T+S)/4$, the same as with both agents playing TFT. We can also easily check that two ATFT players gain the same average amount, which leads to a rather surprising result that in a mixed population of TFT and ATFT players nobody has an advantage over any other agent. Such a population would remain stable forever, irrespective of the proportion of the two strategies.

However, the occasional mutations may contaminate the mixture with other strategies, e.g. unconditional cooperation. The panel c) in Fig. 8.1 illustrates the dynamics of the strategy $[11]$ playing against $[01]$. The system is ergodic again, but the attractor of the dynamics is the single state (11), and the gain is $g_{[11][01]} = R$.

A similar analysis can be done for all pairs of strategies. For memory $M = 1$ it is rather simple, and for a larger M we can automate the calculation of the elements of

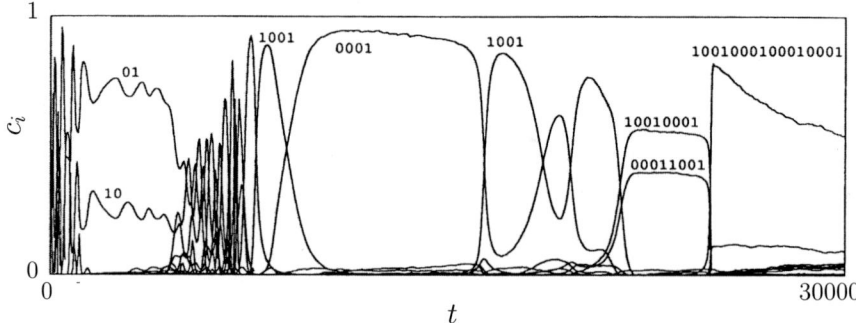

Fig. 8.3 Iterated prisoner's dilemma. Example of the evolution of populations with unlimited memory subject to mutations of the strategies. The strings of 0s and 1s denote the strategies corresponding to the lines shown. Note that the scale of the time used here differs from Fig. 8.2. The figure is reprinted from Ref. [1576], with permission of Kristian Lindgren.

matrix g_{ij} on a computer. The result then serves as an input for the solution of the population equations (8.6). To keep things on the simpler side, we show the result for $M = 1$ and the canonical values of the parameters (8.2)

$$g = \begin{pmatrix} 3 & 3 & 0 & 0 \\ 3 & 9/4 & 1 & 9/4 \\ 5 & 1 & 1 & 5 \\ 5 & 9/4 & 0 & 9/4 \end{pmatrix} \tag{8.7}$$

where the rows and columns are labelled by strategies [11], [01], [00], and [10] (in this order).

We can consider the matrix g the payoff matrix of a new game, which is the prisoner's dilemma played on a higher level. We shall call it the memory-1 iterated prisoner's dilemma game. Note that this game and its generalisations to longer memories (memory-M iterated prisoner's dilemma) are the key pieces of the evolutionary process discussed here, but they must be complemented by the mutations in order to get the full picture. The mutations provide us with the transitions from memory-M iterated prisoner's dilemma to memory-$(M + 1)$ iterated prisoner's dilemma, thus making the space of strategies infinite in principle.

But in the memory-1 iterated prisoner's dilemma level there is already complexity enough. A glance at the matrix (8.7) immediately reveals that there is no pure evolutionary-stable strategy, and the setting resembles the rock-paper-scissors game rather than the original prisoner's dilemma. However, the oscillations are much more complex than in rock-paper-scissors. Let us first investigate them rather verbally. If the TFT strategy (recall that it is [01]) was absent, unconditional defection ([00]) would be the stable strategy, as seen from (8.7). So, it must eventually win the evolutionary race and pervade the system. But if a random mutation does create an agent playing TFT, it does not feel any disadvantage compared to [00], as the latter cannot exploit it. The mutant is not extinct, although it may disappear by reverse mutation, with extremely

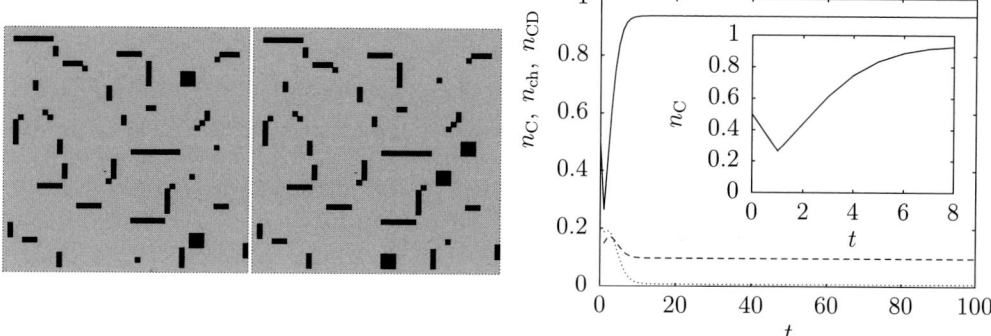

Fig. 8.4 Spatial prisoner's dilemma. In the left two panels, the configuration is drawn in times t_1, t_2, $t_1 < t_2$ on the lattice with $L = 50$. Cooperating agents are shown in grey and the defectors in black. In the right panel, time evolution of the density of cooperators n_C (solid line), density of agents changing their state from one step to the next n_{ch} (dotted line), and density of interfaces n_{CD} (dashed line). System size is $L = 100$, and data are averaged over 100 realisations. The incentive to defect is $b = 1.05$. The initial condition is a random spatial distribution of cooperators with density $n_C = 0.5$. In the inset we show the detail for short times.

small probability. Furthermore, other mutants playing TFT may appear, and, if they are more numerous, they rapidly gain an advantage compared to unconditional defection because TFT playing against TFT gains more, namely 9/4. The TFT strategy multiplies and seems to be destined to win. But in the environment composed of all TFT players, unconditional cooperation performs even better, so that [11] takes over. But this is in turn exploited by the mutant [10] followed by the even more aggressive [00]. The cycle closes. We are back at the beginning.

However, this is only a part of the story. In practice these cycles do not last forever, but eventually lead to a stationary state. We have seen that a mixture of TFT and ATFT is internally stable for any concentration because both strategies have the same gain. The addition of some [00] or [11] players may destabilise the mixture, but there is a concentration where this destabilisation is suppressed. We can easily check, using the matrix (8.7), that for $11/16 < c_{[01]} < 3/4$, $c_{[10]} = 1 - c_{[01]}$, the mixture is stable against intrusion of any other strategy with memory $M = 1$.

The evolution we have just described is illustrated in Fig. 8.2, showing the numerical solution of equations (8.6). We can see both the initial oscillations and the stationary state with concentrations within the bounds predicted theoretically.

The main message from this calculation is that there is no single strategy which can be considered the winner. This finding a posteriori justifies our hesitation to proclaim TFT the best strategy ever found. But the existence of an asymptotic stationary state also bears another lesson. None of the two remaining strategies can survive alone, as each needs the other to beat the intruders. If we consider the binary sequence encoding the strategy as a kind of a 'gene' (and people indeed think in this way [1576, 1577]), then the iterated prisoner's dilemma game does not offer a ground for selfish genes.

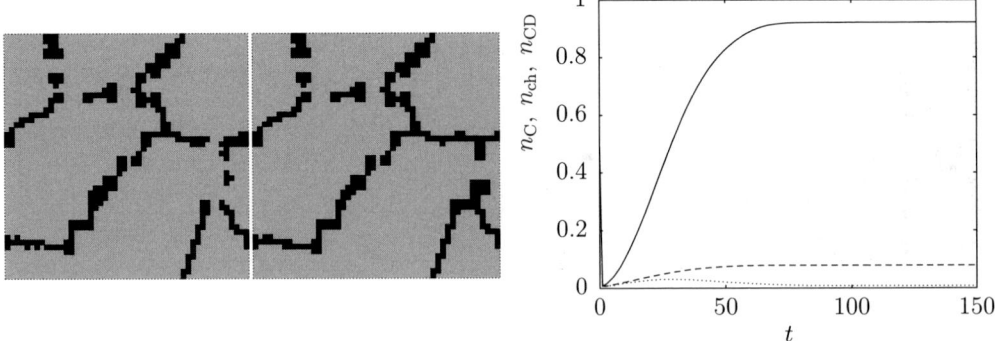

Fig. 8.5 Same as Fig. 8.4, but with $b = 1.5$.

Instead, the best thing the genes [01] and [10] can do is to leave the partner a well-defined share. Unlimited replication of itself does not pay. If you wish, you may call it altruism.

There should be no surprise now that the situation is even less clear-cut when going to general memory-M iterated prisoner's dilemma games with mutations. In numerical simulations it was shown that the evolution never stabilises. An example, taken from the work of Kristian Lindgren [1576], is shown in Fig. 8.3. New, even more complicated strategies emerge all the time, wiping out the predecessors, only to be superseded by some other rivals. We are witness to an open-ended evolution, much like the one that rages in Nature.

To sum up, we have seen that if we allow the players to meet repeatedly and recognise that they inevitably make occasional errors, the simple game of prisoner's dilemma exhibits much of the complexity of human collaborative behaviour. Simple answers should be abandoned. Neither the simple-minded conclusion that defection is always the best choice, nor the slightly less naive belief in the superiority of the TFT strategy survive the test.

8.1.3 Playing in space

We have seen how the trivial prisoner's dilemma game becomes complex if we allow playing again and again with the same opponent. But how it could be that among billions of people you would find the same person you already met? If you were molecules in a well-stirred container, your chance to meet repeatedly would be virtually nil. Some compartmentalisation of the agents must be imposed so that the iterated prisoner's dilemma works as expected. But the influence of compartmentalisation can also be studied independently, preferably in its extreme version, in which the immobile agents are fixed at the vertices of a network, for example, a two-dimensional square lattice. The latter is the structure we shall have in mind throughout this subsection.

Imitate the best

Each agent on the square lattice has eight neighbours, counting also those along the diagonal. The agent plays the usual prisoner's dilemma game with all of them. The

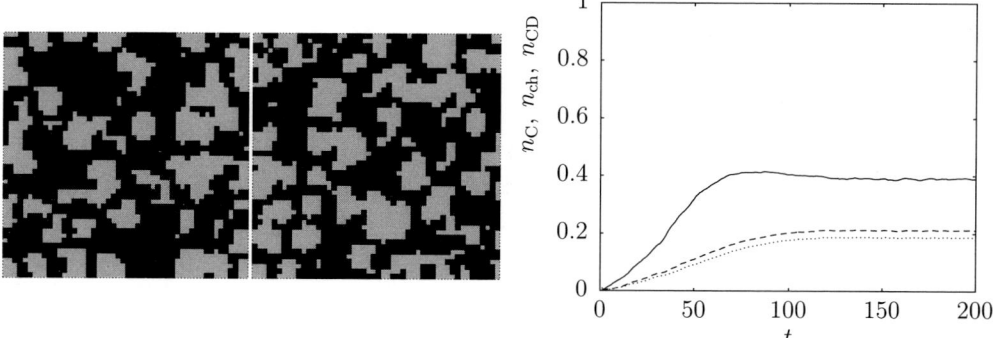

Fig. 8.6 Same as Fig. 8.4, but with $b = 1.6$.

complexity stems from the possibility of changing the strategy from one step to the next [1578]. For simplicity, we allow only the memory-0 strategies, collaboration and defection. In each round, the agent adds together her gain from the eight plays with her neighbours, who do the same in their turn. Every agent then decides about the strategy for the next step. She looks at the gains of her neighbours and compares these numbers with her own gain. If any of the neighbours earns more than herself, the agent adopts the current strategy of the most successful neighbour for the next step. Otherwise the agent preserves her actual strategy for the next time. The evolution of the strategies proceeds by imitation.

We simplify the payoff matrix so that we have only one control parameter, the temptation to defect b; we set $T = b > 1$, $R = 1$, and $P = S = 0$. The spatial prisoner's dilemma game is then simulated, starting from a random configuration, and every agent chooses cooperation or defection with equal probability. Further evolution proceeds by deterministic parallel dynamics, as described above. In fact, the system is a cellular automaton [1579] with specific, relatively complicated, update rules. The resulting configurations are exemplified in Figs. 8.4 to 8.6. We can see that the cooperators survive in significant proportion even if the temptation for defection b is large. However, their spatial arrangement strongly depends on the value of b. Fig. 8.4 shows the situation for the value of $b = 1.05$, only slightly above the lowest limit compatible with prisoner's dilemma inequalities. We observe small islands of defectors within a large sea of cooperators. Most of the defector groups are stable or change cyclically within a short period. When we increase the temptation to $b = 1.5$, isolated islands of defectors grow into strings joined together at some places, as seen in Fig. 8.5. Cooperation still prevails, but only within patches encircled by defectors. The spatial structure exhibits only small variations in time. This feature is changed when we further increase the temptation. In Fig. 8.6, where $b = 1.6$, we already observe more defectors than collaborators, and the arrangement changes chaotically.

Dependence on temptation

It is possible to investigate the dependence on b systematically. In Fig. 8.7 we show the data from simulations, averaged over many realisations of the initial conditions. First,

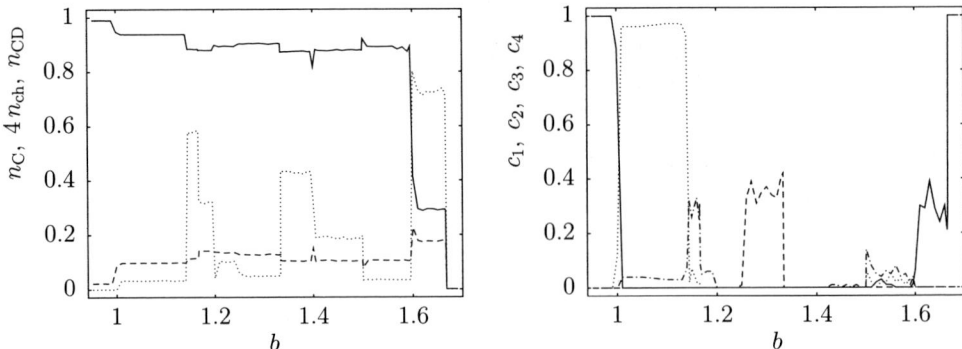

Fig. 8.7 Spatial prisoner's dilemma. Various parameters of the stationary state depending on the incentive to defect b. All data are calculated for lattice with $L = 100$; initial configuration is random spatial distribution of cooperators with density $n_C = 0.5$; data are averaged over 100 realisations. In the left panel, the density of cooperators (solid line), density of sites changing their state from one step to the next (dotted line; note that these data are scaled by factor 4), and density of bonds connecting cooperators with defectors (dashed dine). In the right panel, the fraction of realisations reaching static configuration (solid line) and fractions reaching cyclic attractors with period 2 (dotted line), 3 (dot-dashed line), and 4 (dashed line).

we note that the fraction of cooperators, n_C, remains quite high, above 80 per cent, up to about $b = 1.6$, where it drops suddenly to about 40 per cent; the cooperators vanish only above $b = 1.7$. Clearly, the repeated games induced by the fixed spatial arrangement of the players strongly encourages cooperation. Note that the players do not follow any strategy based on observation of the past behaviour of the agents. The cooperation emerges spontaneously.

Closer inspection of the dependence of n_C on b reveals sudden jumps at specific values of the temptation parameter. The jumps are even more pronounced in some other parameters characterising the stationary state. We observe the concentration n_{ch} of sites changing their state from one step to the next and the concentration of bonds (neighbour pairs) connecting a defector and a cooperator, n_{CD}, which is the measure of the density of interfaces between cooperating and defecting domains. Furthermore, we find that some realisations end in static configurations, while others reach a cyclic attractor with short periods 2, 3, or 4. The fraction of realisations corresponding to these four types are denoted c_1, c_2, c_4, and c_4, respectively. Of course, some initial conditions may also lead to longer stationary periods or to a quasi-chaotic state with barely identifiable periodicity. Interestingly, the quantities c_i depend on b in a very irregular fashion. For example, the period-2 states dominate in an interval from $b = 1$ to about $b = 1.14$, but are rare elsewhere. In contrast, the period-3 states occur almost only in the interval $5/4 < b < 4/3$.

Stability analysis

To understand these features we must perform an analysis of various spatial structures produced in the dynamics [1580]. For example, an isolated defector in the sea

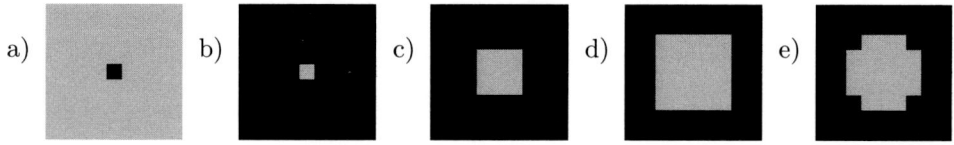

Fig. 8.8 Several configurations of the defectors (black) and cooperators (grey) in the spatial prisoner's dilemma game on a square lattice. Their stability is discussed in the text.

of cooperators, as shown in Fig. 8.8, has 8 cooperating neighbours, so its gain is $g_D = 8p$. The neighbours themselves have 7 cooperating neighbours, resulting in a gain of $g_C = 7$. This means that for $p < 7/8$ the defector becomes a cooperator in the next step, but for $p \geq 7/8$ it survives. Similar analysis shows that an isolated cooperator surrounded by defectors, Fig. 8.8, is never stable. Indeed, the cooperator gains 0, but each of the neighbouring defectors gains p.

We can proceed further to more and more complicated geometries, but we show only some of them here. For example, a 2×2 square of cooperators, Fig. 8.8c, is much more stable than an isolated cooperator. Each of the four gains $g_C = 3$ from its three cooperating neighbours. The defecting neighbours are of two types. At the corners, each of them has one cooperator to exploit, so their gain is p. Each of the defectors adjacent to the edges of the square has two cooperating neighbours, yielding the higher gain $g_C = 2p$. Therefore, the square persists for $p \leq 3/2$; but not only that: the defecting neighbours see that the cooperators gain more, so they become cooperators themselves in the next step. For $p < 3/2$ the initial 2×2 square grows into a 4×4 square of cooperators, Fig. 8.8d, which in turn expands into a 6×6 square and so on. Cooperation spreads despite the relatively large value of the temptation to defect!

We encourage the reader to check the stability and evolution of other configurations, for example, the transition from the square Fig. 8.8d, to the 'cross', Fig. 8.8e. Some configurations may exhibit periodic changes, for example, those offered for your attention in Problem 1. A multitude of possible generalisations can be found in the literature, either concerning the update rules [1581–1585] or the geometry of the links connecting the interacting neighbours [1586–1591]. However, the general features of the spatial prisoner's dilemma game remain in force: namely the fact that repeated plays against the same agents, which are dictated by the geometry of the social network, naturally lead to coexistence of large patches of collaborators alternated with various arrangements of defectors.

Cooperation is viable!

To sum up, spontaneous emergence of cooperation seems to be reproduced in model situations, on the condition that the game is played repeatedly. The overall picture of cooperation is typically very complex, discarding any simplistic ideology-based conclusions.

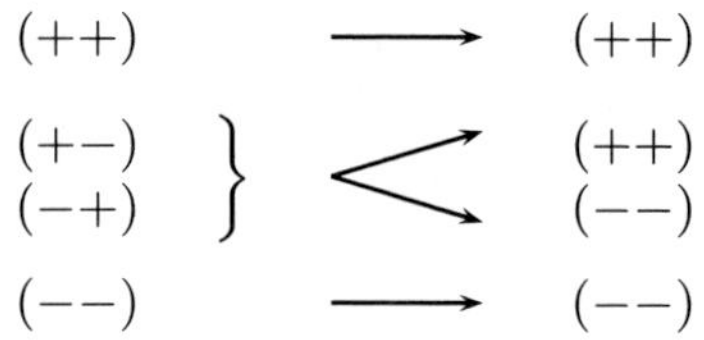

$$
\begin{array}{ccc}
(++) & \longrightarrow & (++) \\[4pt]
\left.\begin{array}{c}(+-) \\ (-+)\end{array}\right\} & < & \begin{array}{c}(++) \\ (--)\end{array} \\[4pt]
(--) & \longrightarrow & (--)
\end{array}
$$

Fig. 8.9 Illustration of the dynamics of opinions in the voter model. The diagram shows how the pairs of neighbouring sites are updated. If the two sites have identical states, they remain unchanged. If the single-site states differ, they can become either both + or both − with equal probability. Conservation of average magnetisation follows directly.

8.2   Consensus

8.2.1   Voter model

People can make up their minds by just looking around and picking the opinion of a randomly chosen neighbour. That is the idea behind the stochastic process introduced in the 1970s [1592, 1593] and called the voter model. It plays a special role among other models of opinion spreading and consensus formation, because it is exactly solvable in any spatial dimension, while showing highly non-trivial dynamics [1594–1603]. Physicists were interested in the voter model, as it can say something about spinodal decomposition [1597] and catalytic reactions [1598, 1599, 1604].

Formulation

In fact, there is a whole family of diverse voter models [1595], and the exactly solvable class consists of the so-called linear voter models. We further limit ourselves to a linear voter model with nearest-neighbour interaction.

We shall mostly work on a d-dimensional hypercubic lattice $\Lambda = \{0, 1, \ldots, L-1\}^d$ with periodic boundary conditions. The coordinates of the point $x \in \Lambda$ will be denoted x_α, $\alpha = 1, 2, \ldots, d$.

On each lattice site we imagine an agent whose state can be either $+1$ or -1. These two choices can represent a person's political preferences in a two-party system; hence the name voter model. The configuration of the entire system is described by a point in the configuration space $\sigma \in \mathcal{S} = \{-1, +1\}^\Lambda$. The state of the site $x \in \Lambda$ is denoted $\sigma(x)$.

The dynamics of the model is very simple. In each step we randomly choose one site x and its neighbour y. Then x adopts the state of y, so $\sigma(x)$ is replaced by $\sigma(y)$, as illustrated in Fig. 8.9. This scheme also demonstrates one important property of the voter model. We can see that the magnetisation defined as $m = \frac{1}{|\Lambda|} \sum_{x \in \Lambda} \sigma(x)$ is conserved when we average over all possible realisations of the process, even though in individual realisations it may fluctuate. It is also evident that the uniform states where all sites are either $+1$ or -1 never change. These two configurations are the absorbing states of the voter model.

In one dimension the dynamics can be easily understood. The configuration is determined by the sequence of 'domain walls', separating regions uniformly populated by $+1$ or -1. The configuration can change just by flipping the state of the sites beside the domain wall. Either the agent on the left from the wall adopts the state of the agent on the right side or vice versa. In the former case the domain wall moves one step leftwards, and in the latter case it jumps rightwards. Both possibilities have the same probability, so the domain wall performs a random walk. Moreover, when two domain walls meet, the region bordered by them disappears, and the domain walls themselves annihilate. So, the one-dimensional voter model is exactly mapped onto a system of annihilating random walkers. Alternatively, we can say that in one dimension the voter model is equivalent to the kinetic Ising model with Glauber dynamics at zero temperature, which is characterised by exactly the same dynamics of domain walls. This model is quite well understood [1605]. Unfortunately, in any dimension larger than one, such equivalence is no longer valid. One of the reasons lies in the difference in surface tension. While the domain walls in the Ising and related models exhibit positive surface tension at all temperatures except the critical point, in the voter model the surface tension is zero [1601]. On the other hand, the voter dynamics is essentially zero-temperature, because there are no spontaneous flips of the state of a site. So, Ising and the voter model can resemble each other only when the critical temperature is $T = 0$, which happens only in the one-dimensional lattice. In fact, the zero-temperature kinetic Ising model in higher dimensions exhibits much more complex dynamics [1606, 1607] than the voter model.

Let us now proceed to a slightly more formal description. The voter model is a continuous-time Markov process (see Box 2.10) denoted σ_t which takes values in the configuration space $\mathcal{S} = \{-1, +1\}^\Lambda$. For any $\sigma \in \mathcal{S}$ denote as σ^x the state which is obtained from σ by flipping the state of site $x \in \Lambda$, so $\sigma^x(y) = (1 - 2\delta_{xy})\,\sigma(y)$. We need to know the set of nearest neighbours of x on the lattice Λ. In the case of a d-dimensional hypercubic lattice there are $2d$ neighbours obtained by shifting the point x by either $+1$ or -1 along the d Cartesian axes. We shall denote the μ-th neighbour of x by x^μ. The transition rates describing a single flip are

$$w(\sigma, \sigma^x) = \frac{1}{2}\left[1 - \sigma(x)\frac{1}{2d}\sum_{\mu=1}^{2d}\sigma(x^\mu)\right] \tag{8.8}$$

while all other transition rates are zero:

$$w(\sigma, \sigma') = 0, \quad \text{if} \ \ |\{x \in \Lambda : \sigma(x) \neq \sigma'(x)\}| \neq 1. \tag{8.9}$$

The dynamics proceeds according to the master equation

$$\frac{\mathrm{d}}{\mathrm{d}t}p_t(\sigma) = \sum_{x \in \Lambda}\left[w(\sigma^x, \sigma)\,p_t(\sigma^x) - w(\sigma, \sigma^x)\,p_t(\sigma)\right]. \tag{8.10}$$

Exact solvability

We can directly write the equation for the average state on a single site $S(x, t) \equiv \langle\sigma_t(x)\rangle = \sum_{\sigma \in \mathcal{S}} p_t(\sigma)\sigma(x)$, starting with the master equation (8.10). The result is

$$\frac{\mathrm{d}}{\mathrm{d}t}S(x, t) = \Delta_x S(x, t) \tag{8.11}$$

where we used the notation $\Delta_x f(x) = -f(x) + \frac{1}{2d}\sum_{\mu=1}^{2d} f(x^\mu)$ for the discrete Laplace operator on the d-dimensional hypercubic lattice. We recognise in (8.11) the discrete diffusion equation, describing the movement of a random walker over the lattice. The mapping of the voter model on Brownian motion assumes a mathematical form.

We are extremely lucky. Usually, in models with interacting particles the equations for one-site averages contain two-site correlation functions. If we attempt to write down the evolution of two-site correlations, we find three- or four-site correlations emerging in the formulae, so we end up with an infinite chain of coupled equations. There is no hope for an exact solution. However, in the voter model, higher correlation functions disappear 'miraculously'. For two-site correlations $R(x - y, t) \equiv \langle \sigma_t(x) \sigma_t(y) \rangle$, we get

$$\frac{\mathrm{d}}{\mathrm{d}t} R(x, t) = 2 \Delta_x R(x, t) \tag{8.12}$$

for $x \neq 0$. If the two points coincide, we have trivially $R(0, t) = 1$ for all times $t \geq 0$. Technically, this is the boundary condition for the solution of the discrete diffusion equation (8.12).

We could continue this way to get closed equations for correlations of all orders. The deep reason why the evolution of correlation functions at order k does not involve the correlation functions at higher orders $k' > k$ is the duality property, valid for all linear voter models [1594]. Without going into detail we can describe the duality as equivalence of the voter model to the evolution of annihilating random walks backward in time. We have already noted that the single-site correlation, i.e. the average single-site state, is mapped exactly onto a random walk. Generally, any k-site correlation function can be mapped to k such walkers, annihilating when they meet, and the actual value of the correlation function at time t should be traced back to the state of the walkers at earlier times. As the number of annihilating walkers can only decrease in the evolution, higher correlation functions never occur, while lower ones do, just at the points where some of the k coordinates coincide [1597].

The equation (8.11) is solved easily by the Fourier transform in space and the Laplace transform in time domain (see Box 2.7). As the initial condition we impose the state $+1$ at the origin, while all other sites are $+1$ or -1 with equal probability, so $S(x, 0) = \delta_{x0}$. We obtain

$$\widehat{\widetilde{S}}(p, z) = \left[z + 1 - \frac{1}{d} \sum_{\alpha=1}^{d} \cos p_\alpha \right]^{-1}, \tag{8.13}$$

and for Re $z > 0$ we can use the trick

$$\widehat{\widetilde{S}}(p, z) = \int_0^\infty \exp \left[-\lambda \left(z + 1 - \frac{1}{d} \sum_{\alpha=1}^{d} \cos p_\alpha \right) \right] \mathrm{d}\lambda. \tag{8.14}$$

When we Fourier-invert the expression (8.14), we can recognise the integral representation of the modified Bessel functions (see Box 8.1). This leads to a compact expression for the solution

$$S(x, t) = \mathrm{e}^{-t} \prod_{\alpha=1}^{d} I_{x_\alpha} \left(\frac{t}{d} \right). \tag{8.15}$$

Using the asymptotic behaviour of the Bessel function, we find that for large times the average state of any site decays to zero as $S(x, t) \sim t^{-d/2}$, $t \to \infty$. Recall that the

---

Modified Bessel functions | Box 8.1

or the so-called Bessel functions of imaginary argument, are defined by (see Ref. [933])

$$I_\nu(z) = \sum_{k=0}^{\infty} \frac{(z/2)^{\nu+2k}}{\Gamma(k+1)\Gamma(k+\nu+1)}.$$

For integer ν, they can be expressed by the very useful integral representation

$$I_\nu(z) = \frac{1}{2\pi} \int_{-\pi}^{\pi} e^{z\cos\phi - i\nu\phi}\, d\phi.$$

Often we need the behaviour of the Bessel function for a large argument. The following asymptotic formula holds in this case

$$I_\nu(z) = e^z (2\pi z)^{-1/2} \left[1 - \frac{4\nu^2 - 1}{4}\frac{1}{2z} + O(|z|^{-2}) \right].$$

---

initial condition was $S(x,0) = \delta_{x0}$. So, the average state of the site at the origin decays to zero monotonously, while the other sites, $x \neq 0$ exhibit first an increase in $S(x,t)$ and start decaying at later times. It can be understood as propagating a diffusive wave of $+1$'s from the origin to the rest of the lattice, eventually vanishing at large times.

More information on the dynamics is contained in the two-site correlation function $R(x,t)$. Apart from the factor 2, it obeys the same equation as $S(x,t)$. However, there is an important difference. Eq. (8.12) holds only for $x \neq 0$; and besides the initial condition $R(x,0) = \delta_{0x}$, the solution must also satisfy the boundary condition $R(0,t) = 1$. Nevertheless, we can proceed again by Fourier- and Laplace-transforming Eq. (8.12). The boundary condition enters through the yet unknown function

$$n_{+-}(t) = \frac{1}{2}\left(1 - \frac{1}{2d}\sum_{\mu=1}^{2d} R(x - x^\mu, t) \right) \tag{8.16}$$

which is just the concentration of interfaces, i.e. the fraction of the bonds connecting sites with unequal state, also called active bonds. This quantity is an important measure of the level of activity in the system, as changes of the configuration can occur only at sites adjacent to the active bonds. For the correlation function we get

$$\widehat{\widehat{R}}(p,z) = \frac{1 + 4\widehat{n}_{+-}(z)}{z + 2 - \frac{2}{d}\sum_{\alpha=1}^{d}\cos p_\alpha}. \tag{8.17}$$

The density of interfaces must be computed self-consistently, and the result should be inserted back in (8.17). After a short algebra we obtain

$$\widehat{n}_{+-}(z) = \frac{1}{4}\left(\left[\frac{z}{2}\,\widehat{S}\left(0, \frac{z}{2}\right)\right]^{-1} - 1 \right) \tag{8.18}$$

and

$$\widehat{R}(x,z) = \frac{1}{z}\frac{\widehat{S}\left(x, \frac{z}{2}\right)}{\widehat{S}\left(0, \frac{z}{2}\right)}. \tag{8.19}$$

Let us now discuss separately the latter results in dimensions $d = 1$, $d = 2$, and $d \geq 3$. In one dimension, the density of interfaces can be expressed in a closed form

$$n_{+-}(t) = \frac{1}{2}e^{-2t}[I_0(2t) + I_1(2t)] \tag{8.20}$$

Elliptic integral | Box 8.2

more precisely said, the complete elliptic integral of the first kind, is defined as [933]
$$K(x) = \int_0^{\pi/2} \left(1 - x^2 \sin^2 \phi\right)^{-1/2} d\phi.$$
Close to the point $x = 1$ it has the following expansion
$$K(1 - z) = \tfrac{1}{2} \ln(8/z) + O(z \ln z).$$

and for the correlation function we obtain

$$\widehat{R}(x, z) = \frac{1}{z} \left(1 + \frac{1}{2}z + \frac{1}{2}\sqrt{(z + 4)\, z}\right)^{-|x|}. \tag{8.21}$$

For large times we have $n_{+-}(t) \simeq (4\pi t)^{-1/2}$ and $\lim_{t \to \infty} R(x, t) = 1$. This means that the activity measured by $n_{+-}(t)$ slowly decays to zero, and eventually all the agents become fully correlated. Both features are signatures of complete ordering, which is the fate of the voter model in one dimension. From Eq. (8.21) it is also possible to estimate how the correlation function approaches its asymptotic value. For small z we can approximate $\widehat{R}(x, z) \simeq \frac{1}{z} e^{-|x|\sqrt{z}}$, and inverting the Laplace transform we get $R(x, t) \simeq \mathrm{erfc}\left(\frac{|x|}{2\sqrt{t}}\right)$. This formula confirms our intuition mentioned earlier, that the coordination between agents in the voter model proceeds as a diffusive wave. Indeed, after time t the system is fully ordered up to distances $\sim \sqrt{t}$.

For $d = 2$ the situation is similar. Again, the density of interfaces decays to zero and the correlation function approaches 1 for large times. However, the evolution is much slower. We have

$$\widehat{n}_{+-}(z) = \frac{1}{4} \left(\left[\frac{z}{\pi} \frac{2}{z + 2} K\left(\frac{2}{z + 2}\right) \right]^{-1} - 1 \right) \tag{8.22}$$

where $K(x)$ is the elliptic integral (see Box 8.2). From the behaviour at $z \to 0$ we can deduce the asymptotic decay of the interface density

$$n_{+-}(t) \simeq \frac{\pi}{2} \frac{1}{\ln(16\, t)}. \tag{8.23}$$

The behaviour changes dramatically for $d \geq 3$. The duality property can tell us why it is so. For example, when we compute the two-site correlation function at distance x, i.e. $R(x, t)$, we effectively investigate what would happen with two annihilating random walkers starting at position x relative to each other, until time t. The correlation function arises from the average over all realisations of the pair of walks. If the walkers meet (and vanish) at time not later than t, the realisation contributes by value 1 to the average; otherwise the contribution is 0. So, the asymptotic value of the correlation function $R(x, \infty)$ is related to the probability that the two walkers meet at some time. It is well known [1608] that this probability is equal to 1 in dimensions $d \leq 2$, but for $d > 2$ there is a finite probability that the walkers never meet. This has the effect that $R(x, \infty) \to 0$ for $|x| \to \infty$, and $n_{+-}(t)$ has a positive limit for $t \to \infty$. In short, the activity never ceases, and the voter model never reaches a totally ordered state.

Let us now support this intuition by more explicit calculations. The stationary density of interfaces becomes

$$n_{+-}(\infty) = \left(2d \int_0^\infty \left[e^{-t} I_0(t) \right]^d dt \right)^{-1}. \tag{8.24}$$

The asymptotic behaviour of the Bessel function guarantees that the integral converges for $d > 2$ and $n_{+-}(\infty) > 0$. Numerical values for some values of d are listed in the table below. We can also find how the density of interfaces approaches its asymptotic value. From Eq. (8.18) we have $\hat{n}_{+-}(z) - n_{+-}(\infty)/z \sim z^{\frac{d}{2}-2}$, which implies $n_{+-}(t) - n_{+-}(\infty) \sim t^{1-\frac{d}{2}}$.

The behaviour of the correlation function in stationary state $R(x, \infty)$ can be easily obtained for large distances, $x \to \infty$. In this regime we can replace the discrete coordinate x by a continuous variable and the correlation function is obtained from the Laplace equation $\Delta R(x, \infty) = 0$ with a unit source at the origin of coordinates, which is an elementary problem from electrostatics [1609]. The solution is $R(x, \infty) \propto |x|^{2-d}$, but we must keep in mind that this is valid only for large $|x|$; in particular at the origin we have the universally valid condition $R(0, \infty) = 1$.

All of the above results for the voter model tacitly assumed an infinitely large lattice. In practice, e.g. in numerical simulations, the system always consists of a finite number $N = L^d$ of sites. This implies that the dynamics eventually leads to one of the two absorbing states, even for $d \geq 3$, where the infinite system never gets ordered. We call the time needed to reach any of the absorbing states in a particular realisation of the process the stopping time τ_{st}. The typical scale for the stopping time, called consensus time $\tau(N)$, diverges in the thermodynamics limit. It is possible to show rigorously [1610] that the consensus time grows as $\tau(N) \sim N^2$ in $d = 1$, $\sim N \ln N$ in $d = 2$ and $\sim N$ for $d \geq 3$. These formulae can be understood on an intuitive level, at least in one dimension, where the dynamics is equivalent to the diffusion of domain walls, annihilating upon encounter. The size of homogeneous areas grows therefore as the mean displacement of a random walk, $\sim t^{-1/2}$. The time needed to cover the whole system of size N is therefore $\sim N^2$. Similar consideration is also feasible in higher dimensions [1598], but we skip it here.

We can summarise the asymptotic behaviour of the voter model on a d-dimensional hypercubic lattice in the following table. Where the behaviour is indicated by '$\sim$', we intend $t \to \infty$, or $x \to \infty$, or $N \to \infty$, according to the context.

d	$n_{+-}(\infty)$	$\dfrac{n_{+-}(t)}{-n_{+-}(\infty)}$	$R(x, \infty)$	$\tau(N)$	d	$n_{+-}(\infty)$
1	0	$\sim t^{-1/2}$	1	$\sim N^2$	3	$0.329731\ldots$
2	0	$\sim (\ln t)^{-1}$	1	$\sim N \ln N$	4	$0.403399\ldots$
≥ 3	> 0	$\sim t^{1-\frac{d}{2}}$	$\sim |x|^{2-d}$	$\sim N$	5	$0.432410\ldots$
					∞	$1/2$

On a complete graph

Let us now look at the voter model on a complete graph, which can be considered a kind of mean-field approximation [1611, 1612]. This is not just an approximation to a

---

Linear operator   Box 8.3

is a prescription which takes a function f from a set $\mathcal{A}$ and makes from f a new function g. For a linear operator $\mathcal{L}$ we use the notation $\mathcal{L} : f \to g$, or $g = \mathcal{L}f$. In order that an operator $\mathcal{L}$ was called linear it must satisfy the property
$$\mathcal{L}(af + bg) = a\mathcal{L}f + b\mathcal{L}g$$
for all numbers a and b and all functions $f, g \in \mathcal{A}$. As an example, we can take as $\mathcal{A}$ the set of functions which can be differentiated arbitrarily many times and as $\mathcal{L}$ the operator of differentiation, so $\mathcal{L}f(x) = \frac{\mathrm{d}}{\mathrm{d}x}f(x)$.

There may be special functions which are called eigenvectors or eigenfunctions of the linear operator $\mathcal{L}$. The action of a linear operator on its eigenvector is the same as multiplication by a number. Thus, if
$$\mathcal{L}f = \lambda f$$
then the function f is called an eigenvector and the number λ the corresponding eigenvalue of the operator $\mathcal{L}$. The eigenvectors are very useful in solving linear partial differential equations, such as the Fokker-Planck equation.

---

finite-dimensional model, but it bears a relevance in itself. The reader may remember Kirman's ant model [790] discussed in Sec. 3.2.1. The insects imitated each other's choice just as the agents in the voter model do, with the only difference being that the ants were allowed to change their option spontaneously. Moreover, the mechanism of copying strategies of the other agents is the main ingredient of the Lux-Marchesi model of stock-market fluctuations, as we have seen in Sec. 3.2.3. The voter model on a complete graph is therefore a good starting point for understanding a broader range of econophysical models.

The completeness of the graph implies that every agent is a neighbour of every other one, and the evolution amounts to randomly choosing two agents and copying the state of the second agent as the new state of the first one. As all geometric structure is gone, the state of the system with N agents at time t is completely described by the number of 'pluses', which we denote N_{+t}. In the thermodynamic limit $N \to \infty$ the most convenient variable to work with is the magnetisation $M_t = 2N_{+t}/N - 1$. When one site is flipped, the magnetisation changes by $\pm\frac{2}{N}$, and the rate for such change is proportional to $1 - M_t^2$. Hence we guess that the magnetisation wanders diffusively within the allowed interval $M_t \in [-1, 1]$, but the diffusion constant depends on the position. Indeed, we can easily obtain the Fokker-Planck equation for the time-dependent probability density of the magnetisation

$$N\frac{\partial}{\partial t}P_{Mt}(m) = \frac{\partial^2}{\partial m^2}\left[(1 - m^2)\,P_{Mt}(m)\right]. \qquad (8.25)$$

This equation is quite well understood [1596, 1613–1616]. It describes the so-called Fisher-Wright process and finds applications in areas as distant as catalysis [1604] and genetics [1613].

The solution can we be written as an expansion in eigenvectors of the linear operator $\mathcal{L} : f(x) \to [(x^2 - 1)f(x)]''$, which appears on the right-hand side of Eq. (8.25). (For explanations of linear operators and their eigenvectors, see Box 8.3) Denoting by $\Phi_c(x)$ the eigenvector of $\mathcal{L}$ corresponding to the eigenvalue c, we have the following equation

$$(1 - x^2)\,\Phi_c''(x) - 4x\,\Phi_c'(x) + (c - 2)\,\Phi_c(x) = 0. \qquad (8.26)$$

The full solution of (8.25) can then be expanded as

$$P_{Mt}(m) = \sum_c A_c e^{-ct/N} \, \Phi_c(m) \tag{8.27}$$

with coefficients A_c determined from the initial condition. For large times only the lowest eigenvalues in the expansion (8.27) are relevant, because they contribute to the sum (8.27) by the most slowly decaying exponentials. The lowest eigenvalue is always $c = 0$, due to the conservation of probability that every Fokker-Planck equation must satisfy. The stationary regime corresponds to this 'ground state' $\Phi_0(m)$.

It turns out that the eigenvectors are composed of two δ-functions located at $+1$ and -1 and a continuous part on support $(-1, 1)$; thus

$$\Phi_c(x) = \phi_{c+}\, \delta(x - 1) + \phi_{c-}\, \delta(x + 1) + \phi_c(x)\, \theta(x - 1)\, \theta(x + 1). \tag{8.28}$$

(We denote by $\theta(x)$ the Heaviside function, equal to 1 for $x > 0$ and 0 otherwise.) For the lowest eigenvalue $c = 0$ we have $\phi_0(x) = 0$, and the weights $\phi_{0\pm}$ of the δ-functions are arbitrary, i.e. the ground state is twice degenerate.

For $c > 0$ we must solve the equation for $\phi_c(x)$, which has a form identical to Eq. (8.26), but the solution is looked for within the set of double-differentiable functions on $[-1, 1]$. We find that the eigenvalues form a sequence $c_l = (l + 1)(l + 2)$, $l = 0, 1, 2, \ldots$ and the functions $\phi_{c_l}(x)$ are proportional to the so-called Gegenbauer polynomials [644, 933]

$$C_l^{3/2}(x) = \sum_{k=0}^{\lfloor l/2 \rfloor} (-1)^k \, 2^{-l-1} \, \frac{(2(l - k + 1))!}{k!\,(l - k + 1)!\,(l - 2k)!} \, x^{l - 2k}. \tag{8.29}$$

Having the continuous part of the eigenvector, we complete the solution by determining the coefficients $\phi_{c\pm}$ in Eq. (8.28) from the conditions

$$\lim_{x \to \pm 1} \phi_c(x) = -\frac{c}{2}\, \phi_{c\pm}. \tag{8.30}$$

Let us now interpret the results obtained. The stationary solution of Eq. (8.25) is the mixture of δ-functions located at $m = \pm 1$. Each of the δ-functions represents one of the two absorbing states, and the weights $\phi_{0\pm}$ are just the probabilities of reaching the corresponding absorbing state. Since the magnetisation averaged over realisations of the process is conserved, we easily deduce that the probability $P_+ = \phi_{0+}$ of hitting the absorbing state with $m = +1$, starting from the configuration with magnetisation $M_0 = m_0$, is $P_+ = (m_0 + 1)/2$, i.e. the initial concentration of agents in state $+1$. In fact, this property also holds on the hypercubic lattice of any dimension because it relies only on the conservation of the average magnetisation.

Often we need to know how long we shall wait before we reach the absorbing state. The probability distribution of stopping times τ_{st} can be deduced from the solution (8.27). Indeed, the probability $P_{\mathrm{st}}^>(\tau)$ that the stopping time is larger than τ is equal to the weight of the continuous component of the distribution $P_{M\tau}(m)$. This is in turn the complement to the weight of the δ-functions, so

$$P_{\mathrm{st}}^>(\tau) = \sum_{c>0} \frac{2}{c} \left(\phi_c(-1) + \phi_c(1) \right) A_c \, e^{-c\tau/N}. \tag{8.31}$$

We can see that the distribution has an exponential tail $P_{st}^>(\tau) \sim e^{-2\tau/N}$, $\tau \to \infty$. The full distribution (8.31) depends, of course, on the initial condition through the coefficients A_c. If the process starts with fixed magnetisation $M_0 = m_0$ for all realisations, we can compute the average stopping time using a general technique of backward Fokker-Planck equation [1596, 1604, 1617], which in our case implies the differential equation for the average stopping time as the function of m_0

$$\frac{1}{N} \left(1 - m_0^2\right) \frac{d^2}{dm_0^2} \langle \tau_{st} \rangle (m_0) = -1 \qquad (8.32)$$

which has a straightforward solution:

$$\frac{1}{N} \langle \tau_{st} \rangle (m_0) = -\frac{m_0}{2} \ln \frac{1+m_0}{1-m_0} - \frac{1}{2} \ln \frac{1-m_0^2}{4}. \qquad (8.33)$$

Having dealt with the paths leading into absorbing states, we turn now to the subset of such realisations of the process which have not hit any absorbing state until time t. The continuous component on the probability density $P_{Mt}(m)$ carries the desired information, i.e.

$$P_{\text{not absorbed}}(m, t) = \frac{\sum_{c>0} A_c e^{-ct/N} \phi_c(m)}{\sum_{c>0} 2 A_c e^{-ct/N} \left(\phi_c(+1) + \phi_c(-1)\right)/c}. \qquad (8.34)$$

For large times the lowest non-zero eigenvalues are dominant, and we shall keep only two terms with eigenvalues $c_0 = 2$ and $c_1 = 6$. From Eq. (8.29) we can see that the lowest Gegenbauer polynomial is constant and the next one is proportional to x; so we can choose $\phi_2(x) = 1$ and $\phi_6(x) = x$. Then,

$$P_{\text{not absorbed}}(m, t) \simeq \frac{1}{2} + \frac{A_6}{2A_2} m e^{-4t/N}, \quad t \to \infty. \qquad (8.35)$$

Interestingly, for large times the probability density for non-absorbed realisations is uniform; and if we average the magnetisation over only these non-absorbed cases, we obtain 0 independently of the initial condition! How can we understand this result? If we look at the evolution of the density of interfaces, which is here $n_{+-}(t) = (1 - \langle M_t^2 \rangle)/2$, we can compute its average over the set of non-absorbed realisations using the distribution (8.35). This provides the limit $n_{+-\text{not absorbed}}(\infty) = 1/3$. However, if we calculate from (8.35) not just the average, but the entire distribution of asymptotic values of the density of interfaces, we find that the most probable value is $1/2$, as opposed to $1/3$ for the average.

It means that in a typical realisation, after time $\sim N/4$, the density of interfaces either has already dropped to zero, reaching an absorbing state, or fluctuates around the plateau value $1/2$. If we instead take the average of the interface density over all realisations, we get exponential decay $n_{+-}(t) \sim e^{-2t/N}$ without an apparent plateau. Such behaviour should be compared with what we have seen on the hypercubic lattice. Here the non-zero asymptotic value of the density of interfaces is also a sign of incessant activity and lack of ordering. The ultimate hit to the absorbing state is rather a finite-size effect. Note, however, a subtle difference between the complete graph and the hypercubic lattice in high dimensions. If we make the limit $d \to \infty$ in the formula (8.24) for asymptotic density of interfaces on a d-dimensional hypercubic lattice, we get the value $1/2$, at odds with the average value $1/3$, but coinciding with the most probable value obtained in this paragraph. What is more important, the time needed to reach the plateau value on the complete graph is itself proportional to N, while on the hypercubic lattice it stays finite when $N \to \infty$. That is why the plateau cannot be seen in the quantity $n_{+-}(t)$.

Network effects

To think of people sitting quietly on the nodes of a hypercubic lattice requires a certain level of perversity. Having everybody interacting with everybody else seems slightly more sound but still is not very realistic. Obviously, social contacts are tangled in a much more complex network. So, it is natural to put the voter model on some of the many sorts of random graphs we have learnt about in Chap. 6.

A principal difference between either a hypercubic lattice or complete graph and a heterogeneous network, i.e. such that the degree of all nodes is not equal, is that the voter model on the latter geometries may not conserve the average magnetisation [1618]. Let us now report some specific results.

As a first step beyond the complete graph we can choose a complete bipartite graph $K_{a,b}$ which consists of two disjoint sets $\mathcal{N}_a$ and $\mathcal{N}_b$ of nodes, and every node from $\mathcal{N}_a$ is connected to every node from $\mathcal{N}_b$. We denote by N_a, N_b the numbers of nodes, by m_a, m_b the magnetisation values in the respective sets, and by $N = N_a + N_b$ the total system size. We can proceed a manner similar to that applied in a complete graph [1619]. The first result is that the evolution quickly (characteristic time stays finite when $N \to \infty$) reaches the equilibrium magnetisation in the two sets, namely, their common value m. For the mean stopping time we obtain an equation similar to Eq. (8.32), namely [1619]

$$\left(\frac{1}{N_a} + \frac{1}{N_b} \right)(1 - m^2)\frac{\mathrm{d}^2}{\mathrm{d}m^2}\langle \tau_{\mathrm{st}} \rangle(m) = -1. \tag{8.36}$$

Thus, the solution has the same form as Eq. (8.33), except the number N of nodes is replaced with $N_a N_b/(N_a + N_b)$, and instead of the initial magnetisation we have the common magnetisation m of the two sets. Note that on the star graph $N_a = 1$ and $N_b = N - 1$, the mean stopping time is thus of order $O(1)$, while if both N_a and N_b are of order $\sim N$, then the average stopping time also grows as $\langle \tau_{\mathrm{st}} \rangle \sim N$. The analysis can be broadened to bipartite graphs with power-law degree distribution $P_{\mathrm{deg}}(k) \sim k^{-\gamma}$. The behaviour of the stopping time depends on the exponent γ. Specifically, $\langle \tau_{\mathrm{st}} \rangle \sim N$ for $\gamma > 3$, $\sim N/\ln N$ for $\gamma = 3$ and $\sim N^{2(\gamma-2)/(\gamma-1)}$ for $2 < \gamma < 3$ [1619].

Analytical results are also available for the Watts-Strogatz small-world network in annealed approximation, i.e. the long-range links are rewired randomly at each step of the evolution of the voter model [1620]. In this case the system does not reach the ordered state asymptotically, but the correlation length, i.e. the typical size of the ordered domains, diverges as $p^{-1/2}$ when the Watts-Strogatz parameter p goes to zero.

Numerical simulations on small-world, Barabási-Albert, and several other types of networks have been performed [1621–1624], with the generic conclusion that the random character of these networks, including long-range bonds, prevents complete ordering, and the overall behaviour resembles those of the hypercubic lattices in dimensions $d \geq 3$. The details depend on the parameters of the network, e.g. the asymptotic value of the interface density grows with the average node degree.

8.2.2   Galam, Sznajd, and related models

Hierarchical majority voting

In democracy, consensus on an issue is rarely achieved by simply waiting until one of the options pervades the whole system through pairwise contact of individuals. Instead, there are various hierarchical levels of decision making, each of which comes to a conclusion based on the principle of majority. This is supposed to lead to a state in which most people are satisfied, for the opinion of the majority of those who participate in the decision process is always declared a law. If there was only a single hierarchical level, namely a popular referendum (Switzerland may serve as a model example), one could be quite sure that the outcome really represents the majority opinion in the society. However, as soon as there are more levels and decisions made on a lower level are passed on to the level above, there is no obvious guarantee that the opinions are not distorted or even reversed. Serge Galam devised a simple model demonstrating that the distortions may not be an exception but rather a rule [1625–1630].

The reason for breaking a society into several hierarchical levels is that the cost of communication in a too-large collective is prohibitive. It is interesting that this is also the main technical obstacle hindering the introduction of secure electronic elections. One can imagine voting for your presidential candidate from your home computer over the Internet, but to make the protocol reliable to ensure proper secrecy and resistance to any attempt of fraud on the scale of a whole nation seems to be beyond the currently available technical capabilities [1631].

So, we take for granted that society is organised hierarchically, and on each level the decision is made within a small group, say, of 3 people. Within each group, the majority rule tells us what opinion shall be held by the representative of the group when sent to make decision on the upper level. Now, suppose there are only two possible choices, A and B, and the fraction of people with opinion A on level l is $n_A(l)$. On the basic level, $n_A(0)$ represents the concentration of A in the entire population.

Supposing that people are not correlated in any way, the concentrations $n_A(l)$ provide full information on the system. Essentially we make a kind of mean-field approximation, neglecting any social structure or network within one hierarchical level. Anticipating the results, we can say that the model of majority decisions on a complete graph, introduced later, [1612, 1632] is to a large extent equivalent to the Galam model.

The dynamics of the model is given by determining the fraction of A at level $n+1$ on the basis of the concentration on level n. For groups of size 3 it leads to the recurrence relation

$$n_A(l+1) = n_A^3(l) + 3\left(1 - n_A(l)\right)n_A^2(l). \tag{8.37}$$

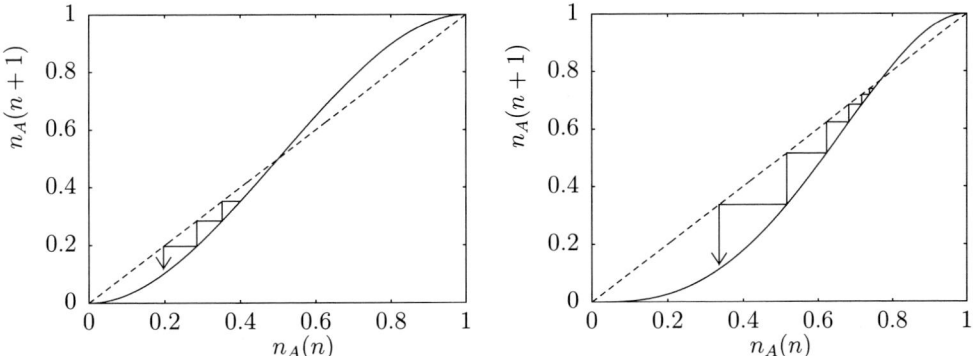

Fig. 8.10 Galam model. In the left panel, graph of the recurrence relation for groups of size 3. The 'stairs' indicate how the iterations of the relation drive the system away from the unstable fixed point $n_{A\,\mathrm{fix}} = 1/2$. In the right panel, an analogous scheme for groups of size 4 with infinitesimal bias in favour of option B. Note the shift of the unstable fixed point to value $n_{A\,\mathrm{fix}} = (1 + \sqrt{13})/6$.

The evolution according to this rule is depicted in Fig. 8.10. We can see that there are three fixed points at values $n_{A\,\mathrm{fix}} = 0$, 1, and $1/2$, which can be easily checked by insertion into Eq. (8.37). The two extremal values are stable, while the central point is an unstable fixed point. This reminds us of physical systems with a phase transition, treated by the renormalisation group technique [1633]. In such formalism, the unstable fixed point corresponds to the critical value of a control parameter and the stable fixed points represent the various possible phases. In the Galam model there are two phases, populated uniformly by either all A or all B opinions, marking total consensus on the issue.

The transition occurs at the symmetric point $n_{A\,\mathrm{fix}} = 1/2$, where exactly half of the people have opinion A. In this sense the decision making is 'fair' because it leads to consensus according to the initial majority. However, a surprise is waiting for us if we make a seemingly minor modification. Imagine that the groups formed at each level are composed of an even number of persons, e.g. 4. We must decide what to do if exactly half of the group pushes for decision A but the rest pushes for B. No majority emerges and we must resolve the tie. If there is an arbitrary small bias towards one of the choices, say, B, the recurrence relation for the concentrations of opinion A is

$$n_A(l + 1) = n_A^4(l) + 4 \left(1 - n_A(l)\right) n_A^3(l). \tag{8.38}$$

In Fig. 8.10 we can see what happens. The unstable fixed point is shifted significantly to higher values of n_A, namely to

$$n_{A\,\mathrm{fix}} = \frac{1 + \sqrt{13}}{6} = 0.76759\ldots \tag{8.39}$$

which means that the opinion B may win even if it was initially in a minority! Note that similar phenomenon of single 'zealots' was studied within the voter model [1634, 1635].

The situation can be even more complicated if some of the people systematically vote against the majority instead of following the crowd. We call such agents contrarians [573, 1636, 1637]. For simplicity, again take groups of size 3. First, we find which is the majority option within the group in question. Then, minority option is chosen with probability p_c, which is to be interpreted as the concentration of contrarians in the population. The recursion relation is modified to

$$n_A(l+1) = (1 - p_c)\left[n_A^3(l) + 3\left(1 - n_A(l)\right)n_A^2(l)\right]$$
$$+ p_c\left[(1 - n_A(l))^3 + 3\left(1 - n_A(l)\right)^2 n_A(l)\right] \tag{8.40}$$

and results in a shift of the stable fixed points away from the endpoints of the interval $[0, 1]$. This means that a total consensus is never reached; the contrarians always introduce a certain level of dissidence, but generally the final decision respects the initial majority opinion in the society. For $p_c < 1/6$ this picture remains valid, as there are three fixed points, with the unstable one keeping its position at $n_{A\,\text{fix}} = 1/2$. However, if the concentration of contrarians rises above the value $p_c = 1/6$, the three fixed points coalesce into a single stable fixed point at $n_{A\,\text{fix}} = 1/2$. No consensus is ever reached, and the distribution of opinions tends toward a precisely equilibrated state of equally represented opinions A and B. Several recent cases of popular referenda or presidential elections ending in extremely narrow victories surface in our memories, showing that the presence of contrarians may have palpable consequences for our lives [1638].

A natural question arises as to how fast we approach the consensus when we climb higher and higher on the ladder of hierarchies. It is evident that if we start closer to the unstable fixed point, it takes more time to approach one of the stable ones. To get a quantitative estimate, we turn to the simplest case described by Eq. (8.37) and replace the discrete level index l by a continuous variable l which can be interpreted as time elapsed during the formation of the consensus. Thus, we get the differential equation

$$\frac{\mathrm{d}}{\mathrm{d}l}n_A(l) = -n_A(l)\left(n_A(l) - 1\right)\left(2n_A(l) - 1\right) \tag{8.41}$$

which can be solved relatively easily:

$$n_A(l) = \frac{1}{2}\left(1 \pm \frac{1}{\sqrt{1 + b\,\mathrm{e}^{-l}}}\right) \tag{8.42}$$

where the parameter b and the choice of sign depends on the initial condition.

The solution for several initial conditions is shown in Fig. 8.11. Of course, starting from any point $n_A(0)$ inside the interval $(0, 1)$, the time to reach either of the stable fixed points 0 or 1 is infinite. However, in reality we do not need to come infinitely close to it, because the population consists of a finite number N of people and the full consensus is reached if we stop at the distance $1/N$ from the fixed point. The time to achieve that situation diverges for increasing N as $\sim \ln N$, as we shall see from the explicit calculation.

To be more precise, the stopping time t_{st} as a function of the initial concentration $n_0 \equiv n_A(0) < 1/2$ will be defined by the formulae $n_A(0) = n_0$ and $n_A(t_{\text{st}}(n_0)) = 1/N$

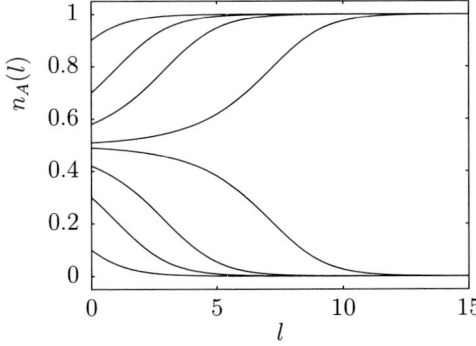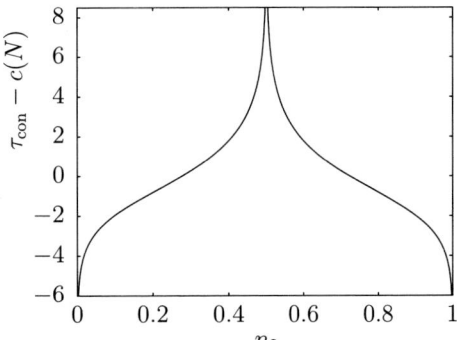

Fig. 8.11 Galam model with continuous levels, for group size 3. The evolution according to Eq. (8.41) is shown in the left panel, with initial concentration of the opinion A $n_0 = 0.9$, 0.7, 0.58, 0.51, 0.49, 0.42, 0.3, and 0.1 (from top to bottom). In the right panel, the time to reach consensus in the society of N people. In this graph, we denote $c(N) = \ln \frac{(N-2)^2}{N-1}$.

expressing the initial and final conditions. The case $n_0 > 1/2$ differs only in the final condition, which is $n_A(t_{\rm st}(n_0)) = 1 - 1/N$. Knowing the general solution (8.42) we get

$$t_{\rm st}(n_0) = \ln \frac{(1 - n_0)n_0}{(2n_0 - 1)^2} + \ln \frac{(N - 2)^2}{N - 1}. \tag{8.43}$$

Hence, the already announced logarithmic divergence of the stopping time for $N \to \infty$ occurs. We can see the N-independent part of the stopping time in Fig. 8.11. Note the divergence for $n_0 \to 1/2$, which can be regarded as a sign of a certain kind of dynamic phase transition.

The Galam model has potential for extensions in various directions, and indeed it was further generalised [1639–1649]. The book [1554] is largely devoted to it.

For us, the most physically relevant question is how the dynamics changes if we distribute the people on a fixed lattice or network [1650–1652]. After all, the Galam model has all the attributes of a mean-field version of some more complicated model, even though it may not be obvious which one [1653]. The next section will make the point clearer.

Local majority rule

Let us return to the set of agents on a hypercubic lattice, with possible opinions $+1$ or -1, as in the voter model. The configuration of the system changes according to local majorities. In each step, we first randomly choose a group of neighbouring sites, and then all members adopt the opinion which prevails among the agents in that group. This defines the dynamics of the majority-rule model [1632, 1654–1657]. Contrary to the voter model, for an agent to flip her opinion it is necessary that at least two neighbouring agents are in the same state. This means that the agents must be sufficiently correlated at a short distance before they can significantly influence others. Recall that the voter model does not require any such correlation because any single agent can influence its neighbour.

$$(+ + +)$$
$$(+ + -)(+ - +)(- + +)\Big\} \quad \longrightarrow \quad (+ + +)$$

$$(- - +)(- + -)(+ - -)\Big\}$$
$$(- - -) \quad\quad\quad\quad\quad \Big\} \quad \longrightarrow \quad (- - -)$$

Fig. 8.12 Illustration of the dynamics of opinions in the majority-rule model. The minority spin in the triple is flipped.

The simplest case is a one-dimensional chain. Assume groups of size 3 and choose a site x and its two neighbours $x \pm 1$. The triple $(\sigma(x - 1), \sigma(x), \sigma(x + 1))$ is then updated according to the local majority, as illustrated in Fig. 8.12.

The mathematical description of the majority-rule model closely follows the concepts introduced in the voter model. The configuration evolves according to a Markov process with transition rates for flipping the state of site x as

$$
\begin{aligned}
w(&\sigma, \sigma^x) \\
&= \frac{1}{12}\Big[\sigma(x - 2)\sigma(x - 1) + \sigma(x - 1)\sigma(x + 1) + \sigma(x + 1)\sigma(x + 2) \\
&\quad - \sigma(x)\Big(\sigma(x - 2) + 2\sigma(x - 1) + 2\sigma(x + 1) + \sigma(x + 2)\Big) + 3\Big].
\end{aligned}
\tag{8.44}
$$

In principle, the strategy for solution of this model seems simple. First, we write down the master equation and then deduce from it the equations for the evolution of the average state of a single site $\langle \sigma_t(x) \rangle$ and the correlation function $\langle \sigma_t(x)\sigma_t(y) \rangle$. However, we immediately realise that the happy coincidence leading to an exact solution for the voter model is absent here. The equations do not close, and in order to find the single-site average we need the two-site correlation function, which in turn requires knowledge of the four-point correlations, etc. Why is this so?

The basic answer is simple. What we called the voter model in the preceding section was an example of linear voter models, a class which is characterised by linear dependence of the transition rates for a single flip at site x on the state of neighbours of x. (There are also some more subtle additional requirements, but they are of little interest here.) A short look at Eq. (8.44) immediately reveals that in the majority-rule model the dependence on neighbours' states is quadratic; thus there is no hope that the equations for the correlations would close. The majority-rule model belongs to a wider class of nonlinear voter models, about which very little is known, with the exception of the so-called threshold voter models [1595]. They can be briefly described by saying that the site flips its state if the number of sites in its fixed neighbourhood which are in the opposite state exceed a certain fixed threshold. Unfortunately, this is not the case with the majority rule model. Eventually we have to resort to simulations and some approximations, which are quite good, after all.

We suppose that on average the state of the system is invariant with respect to translations. Then all correlation functions depend only on the differences of the coordinates, and in particular the one-site average is just the magnetisation

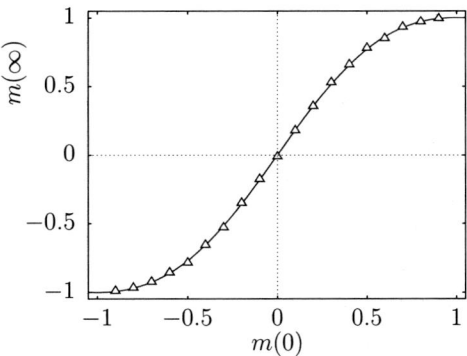

Fig. 8.13 One-dimensional majority-rule model. Average magnetisation in the final state as a function of the initial condition. The points are obtained by numerical simulation, the line is the analytical result (8.48) following from the Kirkwood approximation.

$m(t) = \langle \sigma_t(x) \rangle$. We also need the two-, three- and four-site correlations $R(x - y, t) = \langle \sigma_t(y)\sigma_t(x) \rangle$, $R(x - z, y - z, t) = \langle \sigma_t(z)\sigma_t(y)\sigma_t(x) \rangle$, and $R(x - u, y - u, z - u, t) = \langle \sigma_t(u)\sigma_t(z)\sigma_t(y)\sigma_t(x) \rangle$. Using the transition rates (8.44) we get the exact evolution equations

$$\frac{\mathrm{d}}{\mathrm{d}t}m(t) = \frac{1}{2}\big[m(t) - R(1, 2, t)\big]$$
$$\frac{\mathrm{d}}{\mathrm{d}t}R(1, t) = \frac{1}{3}\big[2 + R(2, t) + R(3, t) - 3R(1, t) - R(1, 2, 3, t)\big]. \tag{8.45}$$

Now comes the promised approximation. First we assume that the two-point correlations decay slowly in space, so $R(1, t) \simeq R(2, t) \simeq R(3, t)$. This is surely wrong shortly after the start of the dynamics from a random uncorrelated initial condition, but we assume that the correlations over a few lattice steps quickly develop close to their stationary values, which only weakly depend on the distance.

The next approximation provides a closure of the evolution equations. It decouples the set of sites at which the correlation is calculated into two subsets which are considered approximately independent. The higher correlation functions are then approximated as products of two lower correlation functions. This approach is called the Kirkwood approximation and often provides very good results, although the procedure is in principle ill-defined and uncontrollable.

In our case, the Kirkwood approximation assumes that $R(1, 2, t) \simeq m(t)R(1, t)$ and $R(1, 2, 3, t) \simeq \big(R(1, t)\big)^2$. If you are about to ask why we use just this decoupling and not something else, you are on a good track to understanding why the approximation is deemed uncontrollable. As it often happens, the choice is justified only a posteriori when it proves successful.

Then, the evolution of the magnetisation and the nearest-neighbour correlation function is described by a closed set of two nonlinear differential equations

$$\frac{\mathrm{d}}{\mathrm{d}t}m(t) = \frac{1}{2}m(t)\big(1 - R(1,t)\big)$$
$$\frac{\mathrm{d}}{\mathrm{d}t}R(1,t) = \frac{1}{3}\big(1 - R(1,t)\big)\big(R(1,t) + 2\big)$$
(8.46)

with the only fixed point determining the asymptotic value of the correlation function, $\lim_{t\to\infty} R(1,t) = 1$. Unfortunately, the knowledge of the fixed point is not enough to calculate the long-time limit of the magnetisation. To this end, we must explicitly solve the equations (8.46). We find

$$R(1,t) = \frac{\rho\,e^t - 2}{\rho\,e^t + 1}$$
$$m(t) = m(0)\left[\frac{\rho + 1}{\rho + e^{-t}}\right]^{\frac{3}{2}}$$
(8.47)

where we introduced the parameter $\rho = \big(R(1,0) + 2\big)/\big(1 - R(1,0)\big)$ depending on the initial condition for the correlation function. Usually we assume that the initial state has prescribed magnetisation but otherwise it is completely random, so $R(1,0) = m^2(0)$. From the solution (8.47) we read off the asymptotic value the magnetisation approaches for large times

$$m(\infty) = m(0)\left[\frac{3}{2 + m^2(0)}\right]^{\frac{3}{2}}.$$
(8.48)

In Fig. 8.13 we can compare the formula (8.48) with the results of numerical simulations; and we dare to conclude that the agreement is excellent despite the crude approximations we made.

As in the one-dimensional voter model, the final state of the dynamics is one of the two absorbing states, i.e. the configuration with either all sites $+1$ or all -1. The result (8.48) should be interpreted as giving the probability of reaching the ultimate state of all $+1$, given the concentration of $+1$ in the initial configuration. The result differs from the voter model in the sense that the dependence is not a linear function, but shares the property that it is continuous everywhere, namely at the symmetric point $m(0) = 0$. This means that there is no dynamical phase transition like that observed in the Galam model.

It turns out that this is a peculiarity of the one-dimensional case. In any higher dimension the final magnetisation $m(\infty)$ as a function of $m(0)$ jumps from the value -1 to $+1$ at $m(0) = 0$. Thus, the dynamic phase transition is present in dimensions $d \geq 2$.

Influencing neighbours

Clearly, a group of people puts stronger pressure on an individual than all members of the group separately. Humans are designed so that they follow the crowd. If you see two or more people sharing an opinion on a certain issue, you are tempted to join. This is the basic idea behind the model invented by Katarzyna Sznajd-Weron and Józef Sznajd (daughter and father) [1658]. In its first version, the model was defined on a

$$\text{model of Ref. [1658]} \begin{cases} (\,\cdot +\, +\, \cdot\,) & \longrightarrow & (+\, +\, +\, +) \\ (\,\cdot -\, -\, \cdot\,) & \longrightarrow & (-\, -\, -\, -) \\ (\,\cdot -\, +\, \cdot\,) & \longrightarrow & (-\, +\, -\, +) \\ (\,\cdot +\, -\, \cdot\,) & \longrightarrow & (-\, +\, -\, +) \end{cases} \Big\} \text{ Sznajd model}$$

Fig. 8.14 Illustration of the dynamics of opinions in the model of Sznajd-Weron and Sznajd, as it appears in the first paper [1658] and as it was subsequently modified and named the 'Sznajd model'. The dots replace any state of the site.

one-dimensional lattice of length N. Each site is inhabited by an agent which can be in one of the two states, denoted $+1$ and -1, as in the voter or majority-rule models. We may think of people choosing between two dominant brands of a certain product in the market or voting in a two-party political system. In each step of the dynamics, a pair of neighbours is randomly chosen. If they are in the same state, say, $+1$, then the two sites adjacent to the pair adopt the same opinion $+1$, propagating the consensus outwards. Conversely, if they differ in opinion, they propagate the dissent. We can see the rule schematically in Fig. 8.14.

Formally, suppose the chosen pair is in the state $(\sigma(x), \sigma(x + 1))$, at time t. If $\sigma(x) = \sigma(x+1)$, the neighbours of the pair are updated as $\sigma_{t+1/N}(x-1) = \sigma_{t+1/N}(x + 2) = \sigma(x)$, while for $\sigma(x) \neq \sigma(x + 1)$ the update rule is $\sigma_{t+1/N}(x - 1) = \sigma(x + 1)$, $\sigma_{t+1/N}(x + 2) = \sigma(x)$. We can see that, unlike the voter or majority-rule model, there are three absorbing states. Besides the obvious 'ferromagnetic' states of all $+1$ or all -1, there is the 'antiferromagnetic' state where the sites in states $+1$ and -1 alternate regularly. Strictly speaking, there are two such states, one characterised by $+1$ at sites with odd coordinates and the other by $+1$ at even sites.

Looking at the rules more thoroughly we can see that the linear chain can be divided into two sublattices, one of them containing all odd coordinates x, the other all even sites. The states of the agents in one sublattice are never influenced by the agents in the other one. Moreover, the dynamics within one sublattice is a trivial modification of the voter model. The only difference is that in the voter model one site induces change of the state of one of its neighbours, while here the agent makes both of its neighbours adopt its state. Indeed, the update rule can be written as $\sigma_{t+1/N}(x - 1) = \sigma(x + 1)$, $\sigma_{t+1/N}(x + 2) = \sigma(x)$, irrespectively of the relation between $\sigma(x)$ and $\sigma(x + 1)$ [1659]. This observation trivialises the model to a large extent. Indeed, it is equivalent to two copies of the voter models evolving in parallel, and the only coupling between them comes from the fact that the sites to be updated are neighbours on the original lattice. We can say that the two voter models evolve so that the update occurs at the same place in both copies. However, the initial state of both sublattices is chosen randomly and independently, and many properties, for example the concentrations of $+1$ in either of the sublattices, remain uncorrelated forever.

Restricting the study to only one sublattice, we can follow the way we used successfully for the voter model. We can write the flipping rates analogous to (8.8), only to see that they differ in factor 2, originating from the fact that a site induces a change in state of its two neighbours. This factor only means rescaling of time, so we can safely neglect it. The result known for the voter model can be translated directly to our case.

As a first application, let us look at the asymptotic states. The probability to reach the state of all $+1$ is just the concentration of the $+1$ opinions in the initial state. This holds for both sublattices independently. From here, we easily deduce the probabilities P_+, P_-, and P_{AF} such that the system ends in the absorbing states with all $+1$, all -1, and the antiferromagnetic state, respectively, as a function of the initial concentration of $+1$ opinions. Indeed, the state with all $+1$ means that both sublattices reached the uniform $+1$ state (and similarly for -1) while the antiferromagnetic state is obtained if one of the sublattices ended in $+1$ state and the other in -1 state. Hence,

$$
\begin{aligned}
P_+(n_+) &= n_+^2 \\
P_-(n_+) &= (1 - n_+)^2 \\
P_{AF}(n_+) &= 2(1 - n_+)\, n_+.
\end{aligned}
\tag{8.49}
$$

Now we focus on a single agent. During the evolution it may change its state several (including zero) times. The time elapsed between two subsequent changes of state is called decision time t_{dec}, and it is natural to ask what the probability distribution is of decision times of all agents. The answer relies on mapping the evolution of the voter model to diffusive motion of domain walls. As explained in Sec. 8.2.1, the domain walls separating the regions of $+1$ and -1 opinions evolve in time like annihilating random walkers. Selecting an agent at site x, its state is flipped if and only if the domain wall crosses the point x. The decision time is nothing else than the time between two successive visits of the random walker at the same position x, or, using language familiar to experts in random walks, the decision time is the time of first return of the random walk to the origin. Calculation of this quantity is a standard exercise of probability textbooks [575], and the reader may remember that we did it already in Sec. 3.1.1. The decision time distribution behaves asymptotically as a power law

$$
P_{\mathrm{dec}}^{>}(t) \equiv \mathrm{Prob}\{t_{\mathrm{dec}} > t\} \sim t^{-1/2},\ t \to \infty,
\tag{8.50}
$$

and this is exactly what was found numerically in the founding work of the Sznajds.

Everything having been solved by mapping in the voter model, the story seems to be over. But just the opposite is true. Things start to be exciting again when we 'simplify' the dynamic rules introduced above by allowing only neighbour pairs with equal states to influence their neighbourhood. If the two agents in the pair do not agree in their opinions, nothing happens. In the scheme in Fig. 8.14 it corresponds to taking only the first two rows as allowed updates. It is this modification that has been widely studied afterwards, and common consensus assigned it the name Sznajd model (although calling it 'Sznajds' would perhaps do more justice to the authors).

As a first step in the analysis of the Sznajd model, we note that there are again three absorbing states, all $+1$, all -1, and the antiferromagnetic one. But now the antiferromagnetic state is unstable, because randomly flipping a single site results in a nucleus of three sites in the same state, which irresistibly invades the whole system. But, although unstable, the existence of such an absorbing state leaves important traces in the dynamics, as will be clearer later.

Following the customary route, we write the transition rates for the underlying Markov process,

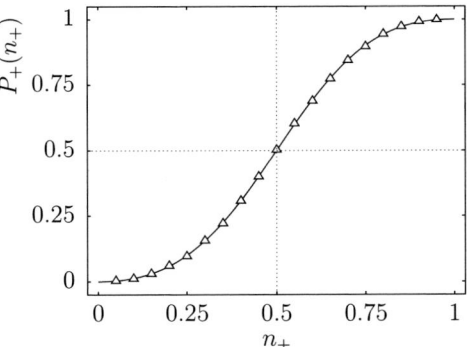

Fig. 8.15 One-dimensional Sznajd model. Probability of reaching the final configuration of all sites in state +1, depending on the initial concentration n_+ of +1 sites. The points are numerical simulation data; the line is the analytical result (8.52) obtained using the Kirkwood approximation.

$$w(\sigma, \sigma^x)$$
$$= \frac{1}{8}\Big[\sigma(x-2)\sigma(x-1) + \sigma(x+1)\sigma(x+2) \tag{8.51}$$
$$- \sigma(x)\Big(\sigma(x-2) + \sigma(x-1) + \sigma(x+1) + \sigma(x+2)\Big) + 2\Big].$$

The similarity with Eq. (8.44) describing the majority-rule dynamics is striking. The Sznajd model is yet another member of the family of nonlinear voter models, and the difference between the transition rates (8.51) and (8.44) stems from the flip induced to a site if the left and the right neighbours agree, which is possible in the majority-rule, but not in the Sznajd model.

We can again use the Kirkwood approximation and calculate the probability for ending in the state of all +1

$$P_+(n_+) = \frac{n_+^2}{(1-n_+)^2 + n_+^2}. \tag{8.52}$$

Fig. 8.15 compares this result with numerical simulations, again showing very good agreement [923, 1660], as in the case of the majority-rule model. We leave to the reader the pain of showing how the result (8.52) is obtained (Problem 5).

Two more questions were asked about the one-dimensional Sznajd model. The first is the decision time, as we introduced it a short while ago. It was found numerically [1659] that it follows the same power law (8.50) as in the previous case, where it resulted from the underlying linear voter dynamics. For the Sznajd model we do not have any proof at hand showing that this is the exact result, but we can still understand this result on an intuitive level. Although the transition rates (8.51) correspond to a nonlinear rather than a linear voter model, the evolution of domain walls, responsible for the behaviour of the decision time, remains very similar. If the domain walls are far from each other, they again perform a random walk. The nonlinearity comes into play only when the domain walls come close together; more precisely, when they are just one lattice-spacing apart. Then they cannot move freely anymore; they can only make one step towards the other domain wall, thus annihilating each other. The dynamics is again a diffusion-annihilation process, but the domain walls are not independent; in addition to annihilation, they interact at short distance.

The interaction is in fact a trace of the unstable antiferromagnetic absorbing state. Two walls at a distance of one lattice unit form a small nucleus of the antiferromagnetic state, thus slowing down the dynamics with respect to free annihilating random walkers. If by chance several domain walls come so close that the distance between neighbouring walls is 1, they form an antiferromagnetic ordered cluster, where the walls squeezed between two other walls from both left and right sides cannot move at all and the dynamics is hindered even more.

However, for large times the density of domain walls is low, and the influence of such antiferromagnetic islands can be neglected. Therefore, it is not so big a surprise that the power-law dependence of the decision time remains unaffected, at least for large times.

The second interesting question we ask is, how many agents have never changed their opinion up to a certain time t? In this context, we speak of persistence in the dynamics of the Sznajd model. In the voter model in one dimension, and equivalently the Ising model with Glauber zero-temperature dynamics, there is an exact analytical solution [1605] showing that the fraction of sites which retain their initial opinion at least to time t decays as $\sim t^{-3/8}$. It may seem quite surprising that the same behaviour has been found numerically in the one-dimensional Sznajd model [1661, 1662]. However, applying the same arguments as above and showing that the long-time dynamics is essentially dominated by annihilating random walks of the domain walls, we come to the conclusion that it is quite plausible that the long-time behaviour of the persistence is governed by the same exponent 3/8 both in the voter and the Sznajd models in one dimension.

Having understood the agents lined up in a chain, we may ask what gets changed if we arrange them in a two-dimensional mesh, like people scattered on the surface of the Earth (neglecting skyscrapers). For example, on a square lattice, we can generalise the one-dimensional dynamics in such a way that a pair of neighbouring agents is chosen; and if the two are in the same state, they make all their 6 nearest neighbours share their state [1663, 1664]. Slightly more formally, let x be a random coordinate on the square lattice with $N = L^2$ sites and $y = x^\nu$ one of the 4 neighbours of x. If $\sigma_t(x) = \sigma_t(y)$, then we update the configuration as $\sigma_{t+1/N}(x^\mu) = \sigma_{t+1/N}(y^\mu) = \sigma_t(x)$ for all $\mu \in \{1, 2, 3, 4\}$. Numerical simulations show an interesting difference from the one-dimensional case. The probability $P_+(n_+)$ has discontinuity at $n_+ = 1/2$, indicating a dynamic phase transition, while, as we remember, the dependence is smooth in one dimension. More strongly, the data suggest that $P_+(n_+) = 0$ for $n_+ < 1/2$ and $P_+(n_+) > 0$ in the opposite case $n_+ > 1/2$. This result is equally valid in any dimension larger than 1. In the limit of infinite dimension the behaviour should coincide with the mean-field approximation, which implies putting the Sznajd model on a complete graph [1612]. Such modification is analytically solvable, as we shall show now.

As in the case of the voter model on a complete graph with N vertices, the configuration of the system is fully described by the number N_{+t} of agents in state $+1$ or by the magnetisation $M_t = 2N_{+t}/N - 1$. The evolution of the latter is governed by transition rates which in the limit $N \to \infty$ yield the Fokker-Planck equation

$$\frac{\partial}{\partial t} P_{Mt}(m) = -\frac{\partial}{\partial m} \left[\frac{1}{2}(1 - m^2)\, m\, P_{Mt}(m) \right] \tag{8.53}$$

for the probability density $P_{Mt}(m)$ of the magnetisation at time t. Before proceeding further we note that the average magnetisation $\langle M_t \rangle = \int m P_{Mt}(m)\, \mathrm{d}m$ satisfies the equation

$$\frac{\mathrm{d}}{\mathrm{d}t} \langle M_t \rangle = \frac{1}{2} \left(\langle M_t \rangle - \langle M_t^3 \rangle \right) \tag{8.54}$$

which has a similar structure as the equation (8.41) describing the dynamics of the Galam model. Indeed, if we equate $M_t = 2n_A - 1$, identify the level index with time, $l = t$, and take the approximation $\langle M_t^3 \rangle \simeq \langle M_t \rangle^3$, thus neglecting the fluctuations, the equations (8.41) and (8.54) coincide. This is not the last coincidence we find within this section. If we try to solve the majority vote model on a complete graph, we get a virtually identical Fokker-Planck equation as (8.53), the only difference consisting in rescaling the time variable. So, we are solving Galam, majority-rule, and Sznajd models in one shot.

It can be easily verified that the general solution of Eq. (8.53) has the form

$$P_{Mt}(m) = \frac{1}{(1 - m^2)\, m}\, f\left(\mathrm{e}^{-t/2} \frac{m}{\sqrt{1 - m^2}} \right) \tag{8.55}$$

for an arbitrary function $f(y)$. The latter has to be determined from the initial condition, and assuming that in the beginning the magnetisation was m_0 for all realisations of the process, we have $f(m/\sqrt{1 - m^2}) = (1 - m^2)m\, \delta(m - m_0)$. Clearly, in this case the probability density consists of a single δ-function which moves, as time passes, towards one of the ends of the allowed interval for the magnetisation, $m \in [-1, 1]$. The dynamics is a pure deterministic drift, and no diffusion ever blurs the evolving probability packet, as can immediately be seen from the absence of a second-derivative term in the equation (8.53). Clearly, this is due to the limit of infinite system size, and for finite N there is an additional diffusive term in the Fokker-Planck equation, proportional to N^{-1}.

The deterministic nature of the evolution of opinions has profound consequences. First, as we have seen, the initial sharp δ-function distribution of magnetisation remains sharp until the end, thus justifying the neglect of fluctuations and the replacement $\langle M_t^3 \rangle = \langle M_t \rangle^3$. The fluctuations only set in at times which diverge when $N \to \infty$, and the equivalence to the Galam model is exact in the thermodynamic limit. In a finite system of size N we can estimate the typical time t_{st} needed to reach the absorbing state, starting from magnetisation m_0, by the requirement that the drift brings the magnetisation to the distance of order $\sim 1/N$ from either of the extremal points $m = \pm 1$. Expressed in terms of the initial concentration of the $+1$ opinion, $n_+ = (m_0 + 1)/2$, it is

$$t_{\mathrm{st}} \simeq \ln \left(\frac{(1 - n_+)n_+ N}{(2n_+ - 1)^2} \right). \tag{8.56}$$

Now, when we understand the close relation to the Galam model, we are not surprised that the equations (8.56) and (8.43) coincide for large N.

For any positive initial magnetisation, the final state is always the uniform configuration of all agents in state $+1$ and vice versa. The probability of ending in the all $+1$ state is the step function $P_+ = \theta(n_+ - \frac{1}{2})$. As we have seen, this is consistent with simulations of the Sznajd model in dimensions larger than 1 and confirms the existence of dynamic phase transition at $n_+ = 1/2$. It also shows that in a society where everybody interacts with everybody else but no person changes her opinion unless she occasionally meets at least two other people who make her want to do so, the initial majority, however narrow it is, always takes all. No chance is left to minorities. Fortunately enough, reality is more complex. In the next paragraph we shall see how it is possible to implement in a model one common 'complication' so that the number of choices is not limited to two but can be large, or even very large.

Can we predict election results?

The rule of people in a democracy becomes explicit on periodic occasions: in elections. In our days they have become large festivals where cheerful people meet on squares listening to popular singers rather than to candidates' speeches. Newspapers are filled with piquancies more than serious proposals of how to solve the crises of the medical system. The campaign is run by the same experts and in exactly the same way as marketing of soap. The very mechanism of competition has two effects. The open and fair elections prevent an evidently malicious individual from getting into power and at the same time they equalise the programmes of the candidates to such an extent that they become hardly distinguishable. The same is true for soap or cars. Due to competition, a really bad product has zero chance of being sold, and the qualities are so close to each other that your choice is based mostly on irrational or random factors.

Such a state of affairs may not please our desire for an ideal world, but from the scientific point of view it largely simplifies the description. People who are thinking too much can hardly be modelled as inanimate particles, but the stochastic nature of choice in the political marketplace promises at least partial scientific success.

Imagine now that N agents choose from among q alternatives, for example parties, candidates or commercial products. Suppose that everybody interacts with everybody else, so the underlying social structure is our familiar complete graph. Besides the usual assumption that N is large, we also consider a large number q of options. The dynamics will be that of the voter model and the underlying social network will again be the complete graph, i.e. in each step a randomly picked agent adopts the state of another randomly chosen agent. The configuration of the system is fully described by the numbers N_σ of agents choosing option $\sigma \in \{1, 2, \ldots, q\}$. We define the distribution of votes

$$D(n) = \frac{1}{q} \sum_{\sigma=1}^{q} \delta(n - N_\sigma/N) \tag{8.57}$$

where in this formula we denote $\delta(x) = 1$ for $x = 0$ and $\delta(x) = 0$ elsewhere. It would be rather difficult to solve the evolution equation for the quantity $D(n)$ (among others, it is not Markovian), so we make an approximation, replacing the distribution of votes

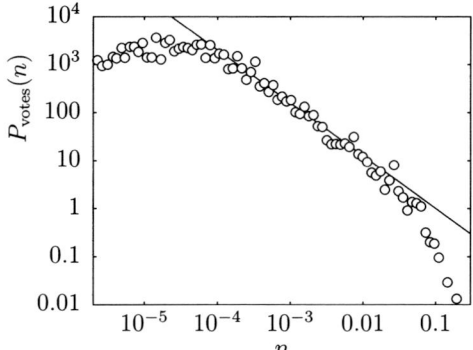

Fig. 8.16 Election results for the federal deputies in Brazil in 2002. In this plot, n is the percentage of votes a candidate obtained in her or his district. The distribution $P_{\text{votes}}(n)$ is calculated as the numerical derivative of the cumulative distribution $N^{\geq}(n)/N$, where N is the total number of candidates and $N^{\geq}(n)$ the number of candidates who gained at least the fraction n in their districts. The straight line is the power $\sim n^{-1.1}$ The data were downloaded from the web page of the Tribunal Superior Eleitoral, www.tse.gov.br.

with its average $P_n(n) = \langle D(n) \rangle$. In the limit $N \to \infty$ and $q \to \infty$ we arrive at the equation

$$N \frac{\partial}{\partial t} P_n(n, t) = \frac{\partial^2}{\partial n^2} \left[(1 - n)n\, P_n(n, t) \right] \qquad (8.58)$$

which we know very well from the analysis of the voter model on complete graph, Eq. (8.25). We found yet another occasion where the Fisher-Wright process is at work! However, we must note that the similarity is rather formal and the interpretation of the solution is different here. There is no need to solve Eq. (8.58) if we only want to know the stationary state. There are q absorbing states, in which all agents choose the same option. The distribution of votes is the same for any of the q possibilities, $D(n) = \left(1 - \frac{1}{q}\right)\delta(n) + \frac{1}{q}\delta(n - N)$.

In real life it would mean that all people vote for the same party. The society reaches consensus on a one-party system, and further evolution of the political system is inhibited. But this picture not only contradicts the everyday experience of a citizen in a democratic country but it is also internally inconsistent with the model itself. If we try to estimate the time needed to reach one of the absorbing states, we find that it is scaled proportionally to the number of agents N, and the time is measured with respect to the typical rate of opinion change of an agent who is exposed to an opinion different from her own. Guessing that people do not change their views more often than once per day, we conclude that a medium-sized country with 60 million inhabitants would need about 10^5 years to reach the uniform state if the opinions evolved according to our model. Clearly it means that the absorbing states are irrelevant for description of the reality. What we see are rather some long-lived quasi-stationary states, and we should try to identify them among the solutions of the equation (8.58).

It can be easily seen that the time-independent distributions $P_{n,0}(n) = 1/n$ and $P_{n,1}(n) = 1/(1-n)$ satisfy Eq. (8.58). They cannot constitute the stationary solution of the original problem, as they cannot be normalised and therefore cannot be reached from any normalised initial state. But they can be interpreted as stationary states of an open variant of the model. We can calculate the probability current $j = -\frac{\partial}{\partial n}\left[(1-n)nP_n(n)\right]$ which is $+1$ and -1 for the solutions $P_{n,0}(n)$ and $P_{n,1}(n)$, respectively. For us, the relevant distribution is the former one,

$$P_{n,0}(n) = \frac{1}{n} \tag{8.59}$$

exhibiting a uniform flux towards a larger number of votes n. It means that there is a source in the system, located at $n = 0$, constantly adding parties with a very small number of votes. Indeed, new parties join politics all the time, but their initial support is negligible and mostly remains negligible all the time. So, we conclude that formula (8.59) describes a snapshot of the distribution of votes in an open political system.

It is fine to see that the empirical data are in quite a good agreement with the formula (8.59). It would be difficult to check it in a country with two-party system, as it relies on the assumption $q \gg 1$, i.e. existence of very large number of competing options. On the other hand, for example in Brazil, the number of candidates is large and reasonable statistics can be gathered [1665–1668]. We show some results in Fig. 8.16, demonstrating that the distribution of votes obeys the power law

$$P_{\text{votes}}(n) \propto n^{-1.1} \tag{8.60}$$

close to the analytical result, at least in the middle part of the distribution, i.e. except for the very popular and very unpopular candidates.

To answer the question in the title of this subsection: no, of course we cannot predict who will win the elections, but nevertheless we can identify the regularities in the distribution of support people give to the politicians, whatever the names and labels attached to them might be.

8.2.3 Axelrod model

Culture is not a disease

Sometimes the spreading of opinions is compared to waves of infection propagating across a society. Both viruses and views cannot live without their human bearers and both phenomena rely on contacts between individuals, be it touch, speech, newspaper or e-mail. But while for any disease there exists a more or less clearly defined healthy situation to which an individual struggles to return, it is hardly possible to find such an indisputable reference state when we deal with cultural tastes. In formal parlance, we expect that the disease spreading models would be characterised by a single absorbing state, all people being healthy and no viruses circulating around. On the contrary, opinion-spreading models have typically at least two absorbing states, and dissemination of culture should take into account the multitude of possible stable configurations; and an extensive number of possible absorbing states should be no surprise.

But there is another, even deeper difference from the mere spread of epidemics. When we look at the ways culture is shared and propagated around the globe as well as inside tight Caucasian valleys, we cannot miss the observation that similar cultures are much more prone to mutual convergence, while incompatible lifestyles often coexist side by side without visibly influencing each other. Robert Axelrod introduced a model [1669] nicely describing such a situation.

Limited confidence

In the Axelrod model, contrary to the voter or Sznajd model, the character of each of the agents is given by more than one feature. One can think of tastes regarding food, sports, music, etc. These categories represent the features. For each feature the taste can assume various values, e.g. somebody likes eating raw vegetables, spending whole days in a fitness centre and listening to Mozart in the evenings, while somebody else feeds on French fries, watches football on TV and adores the pop star of the season. If two neighbours do not agree on any of the features, they do not find any common theme for conversation. No conversation implies no spread of ideas. The two are so much different that they do not influence each other. Conversely, if they find at least one feature where they share the same preference, they talk to each other. One day one of them perhaps looks up a second feature in which they do differ and changes the preference on that second feature so that it agrees with the preference of the neighbour. We call the fact that the agents do not always interact but only if they have something in common limited confidence.

More precisely, there are $N = L^d$ agents placed again on the d-dimensional hypercubic lattice $\Lambda = \{0, 1, \ldots, L-1\}^d$ with periodic boundary conditions. Each agent is characterised by $F > 1$ features, which are represented by integer numbers. Therefore, the state of the agent at site $x \in \Lambda$ is described by a vector with coordinates $\sigma(x, i) \in \mathbb{Z}$, $i = 1, 2, \ldots, F$. The configuration space of the model is $\mathbb{Z}^{\Lambda \times \{1, \ldots, F\}}$ and the evolution of the configuration σ_t is a Markov process determined by transition rates

$$w(\sigma, \sigma^{x,f,a}) = \frac{1}{2d} \sum_{\mu=1}^{2d} \theta\Big(A(x, x^\mu)\Big) \Big[F - A(x, x^\mu)\Big]^{-1} \delta_{a, \sigma(x^\mu, f)} \qquad (8.61)$$

where the summation goes over the set of $2d$ neighbours x^μ of the site x. We denoted by $\sigma^{x,f,a}$ the configuration which differs from σ only in feature f of the agent at site x, so $\sigma^{x,f,a}(x, f) = a$ and $\sigma(x, f) \neq a$. The function $A(x, y) = \sum_{g=1}^{F} \delta_{\sigma(x,g), \sigma(y,g)}$ counts the number of features on which the agents at positions x and y agree, and $\theta(x)$ is the Heaviside function, $\theta(x) = 1$ for $x > 0$ and zero otherwise. The first factor after the sum in (8.61) accounts for the condition that the neighbours must agree in at least one feature, and the second factor is here due to the fact that in each update each agent chooses randomly from among $F - A\big(\sigma(x), \sigma(x^\mu)\big)$ features in which she disagrees with her neighbour. Finally, the last factor with the Kronecker delta function assures that the new value of the feature f of the agent at site x is the same as the value of the same feature f of the neighbour at site x^μ.

Fig. 8.17 Typical evolution of the Axelrod model on a square lattice with periodic boundary conditions. In each row, we show snapshots of the configuration in four times $t_1 < t_2 < t_3 < t_4$, from left to right, where the last configuration is the absorbing state. The active bonds are drawn in grey, the bonds with full consensus in black, the bonds with absolute disagreement are left white. The initial condition is drawn from $\mathbb{Z}_q$ with the uniform probability. The parameters are, from top to bottom, $F = 3$ and $q = 2$; $F = 3$ and $q = 14$; $F = 2$ and $q = 2$. The frame in light grey around each configuration is there only for visual convenience.

Simulations

If the two factors after the sum in Eq. (8.61) were absent, the transition rates would depend linearly on the states of the neighbours, and we would recover a certain generalisation of the linear voter model, with all its beautiful solvability. However, the nonlinearity is there and makes the model non-trivial. Lacking suitable analytical tools, we resort to numerical simulations.

For a computer-minded reader we repeat once again the rules of the dynamics of the Axelrod model, as they are implemented in simulations. On a lattice of size $N = L^d$ the time t is discrete and proceeds in chunks of $1/N$, so from time t to $t+1$ we make as many elementary updates as there are sites. One update step goes as follows:

1. Choose a site x at random; choose randomly one of its neighbours x^μ. Count number A of features in which the agent at x agrees with the chosen neighbour.
2. If $A = 0$ or $A = F$, do nothing. If $0 < A < F$, choose randomly a feature f from among those in which the agent and the neighbour differ. Then set equal the value of the feature f, i.e. $\sigma_{t+1/N}(x, f) = \sigma_t(x^\mu, f)$.

One of the principal quantities of interest will be the density of active bonds, which is a generalisation of the density of interfaces we investigated in the voter model. A

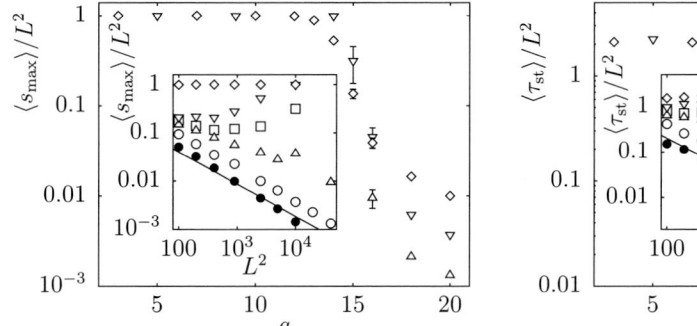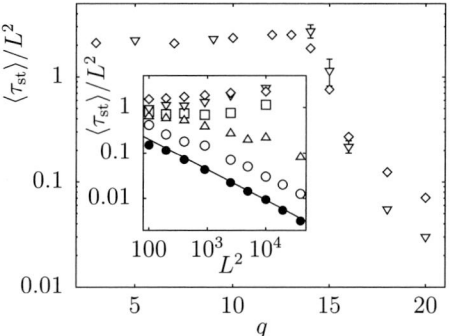

Fig. 8.18 Phase transition in the Axelrod model on a square lattice of size $L \times L$, for $F = 3$, when changing the localisation parameter q of the initial state. In the left panel, relative size of the largest cluster; in the right panel, the average stopping time relative to the system size. Different symbols correspond to sizes $L = 50$ ($\Diamond$), $L = 100$ ($\triangledown$), and $L = 200$ ($\triangle$). Where not shown, the error bars are smaller than the symbol size. In the insets, dependence of the corresponding quantities on the system size L^2. Symbols correspond to parameters $q = 5$ ($\Diamond$), 14 ($\triangledown$), 15 ($\square$), 16 ($\triangle$), 20 ($\bigcirc$), and 30 ($\bullet$). The straight line is the power $\propto L^{-4/3}$.

bond connecting agents at sites x and y is called active if the agents differ in at least one feature and also agree in at least one feature, i.e. $0 < A(\sigma(x), \sigma(y)) < F$. Their fraction relative to the total number of bonds in the lattice will be denoted $n_A(t)$.

When the number of active bonds drops to zero (the time at which it occurs is called stopping time and denoted τ_{st}), the evolution freezes and the system reaches one of its absorbing states. It means that all neighbours are either identical or absolutely incompatible, and we can identify clusters of agents who share the same values of all features, while the borders of such clusters are marked by bonds with no shared value. Even for the simplest case of two features $F = 2$ and the set of allowed values for these features constrained to only two elements, there are infinitely many possible absorbing states characterised by various cluster configurations. (For a finite lattice this number is of course finite, but grows very fast with the system size.) This reminds us of what we have said at the beginning of this subsection: speaking of culture, there might be a good many stable configurations, and we must not be too discriminative and say which is the best one.

The full characterisation of the set of absorbing states would be rather difficult. We limit ourselves to a rough measure, which is the number of agents in the largest cluster s_{max} compared to the total number of agents N. When it is close to the total system size, it means that the absorbing state is uniform, or very close to it. If the size of the largest cluster is small, or even remains finite when $N \to \infty$, the absorbing state is 'multicultural' or fragmented into many isolated islands without mutual interaction, much like the hundreds of villages in Papua New Guinea, each of them having its own particular language.

We shall see that the evolution and the shapes of the eventual absorbing states depend critically on the initial condition. First, note that if all the values of the features belonged to certain subset $V \subset \mathbb{Z}$ of integers at the beginning, all features would remain within the set V forever. The most important thing is how large the set V is. But even if V is infinite, the values may be mostly concentrated on a finite subset of V, very rarely going beyond it. We would need a simple but reliable measure of the effective size of the set V, taking into account how often each member is actually present. Suppose that in the initial state the probability for a chosen feature of a randomly picked agent to have i-th value from the set V is p_i. The simplest measure of the localisation of values within V is the inverse participation ratio

$$q^{-1} = \sum_{i=1}^{|V|} p_i^2. \tag{8.62}$$

It is the same quantity we already used for measuring the localisation of eigenvectors of correlation matrices in Sec. 1.4.1. The definition (8.62) is motivated by a simple consideration. If the set V consists of q lowest non-negative integers, i.e. $V = \mathbb{Z}_q \equiv \{0, 1, \ldots, q-1\}$, and all members have the same probability $p_i = 1/q$, the inverse participation ratio is exactly q^{-1}. The integer q is just the number of elements which are participating; hence the name. Of course, the same result holds for any set of q equally probable elements. Even the fact that they are integers is unessential. We shall soon see that q is the most important parameter of the initial condition, and it would be desirable to be able to tune it continuously. Therefore, besides the choice of $\mathbb{Z}_q$ with uniform probability for the starting configuration, the initial state is often drawn from non-negative integers with Poisson distribution, $i \in \{0, 1, \ldots\}$, $p_i = e^{-\lambda} \lambda^i / i!$. The calculation of the corresponding inverse participation ratio is an exercise from the theory of Bessel functions. The result is

$$q^{-1} = e^{-2\lambda} I_0(2\lambda) \simeq \frac{1}{\sqrt{4\pi\lambda}} \quad \text{for} \quad \lambda \to \infty, \tag{8.63}$$

and the reader can easily recognise it in Box 8.1. The simulations show [1670] that all relevant properties of the Axelrod model depend on the quantity q only, for initial condition distributed either uniformly in $\mathbb{Z}_q$ or according to Poisson distribution.

Phase transition

Let us look at how the Axelrod model behaves on a two-dimensional square lattice. Recall that the linear size of the lattice is L, so there are $N = L^2$ agents. The initial condition will set the values of all features for all agents independently from the same uniform distribution on $\mathbb{Z}_q$. In Fig. 8.17 we can see how the configurations typically evolve. For larger F and q the absorbing state is very fragmented, but when q decreases it becomes totally uniform, and a single cluster covers all the lattice. On the contrary, for small F the absorbing state contains a few clusters of moderate size.

These vague observations are put on a quantitative basis in Figs. 8.18 and 8.19. Let us first look at the behaviour of the average size of the largest cluster, $\langle s_{\max} \rangle$, when we change the parameter q. The most important finding is that at a certain value $q = q_c$ there is a phase transition separating the regime $q < q_c$ in which the maximum cluster makes up a finite fraction of the whole lattice, from the phase with $q > q_c$ where

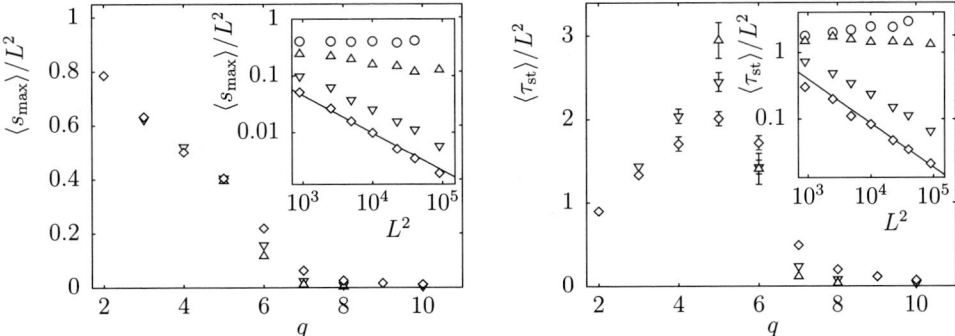

Fig. 8.19 Phase transition in the Axelrod model on a square lattice of size $L \times L$, for $F = 2$, when changing the localisation parameter q of the initial state. In the left panel, relative size of the largest cluster; in the right panel, the average stopping time relative to the system size. In both panels, results for the uniform initial distribution are plotted for sizes $L = 50$ ($\Diamond$), 100 ($\triangledown$), and 200 ($\triangle$). Where not shown, the error bars are smaller than the symbol size. In the insets, dependence of the corresponding quantities on the system size, for $q = 5$ ($\bigcirc$), 6 ($\triangle$), 7 ($\triangledown$), and 8 ($\Diamond$). The straight lines are power laws $\propto L^{-4/3}$.

the fraction of sites within the largest cluster goes to zero for $N \to \infty$. The quantity $\langle s_{\max}\rangle/N$ can serve as an order parameter. If you wonder how the clusters on the lattice can be quickly found and their size measured, look at the book on percolation theory [751], where the beautiful Hoshen-Kopelman algorithm [1671] is explained.

Another quantity which indicates the phase transition is the average stopping time, i.e. the time to reach the absorbing state. Generically, it grows with the system size as $\langle \tau_{\mathrm{st}}\rangle \sim N^{\eta(q)}$; but below the transition it is proportional to the system size, so $\eta(q) = 1$, while for $q > q_c$ it grows more slowly, with the exponent $\eta(q) \simeq 1/3$. Interestingly, comparing the stopping time with the size of the largest cluster, we conclude that for $F = 3$ the two quantities are roughly proportional, $\langle \tau_{\mathrm{st}}\rangle \propto \langle s_{\max}\rangle$, both of them exhibiting a jump at the transition, suggesting that the clusters grow linearly in time until they reach the absorbing state. On the contrary, we do not observe such proportionality for $F = 2$. Instead, the average maximum cluster decreases continuously to zero when we approach q_c from below, and at the same time the stopping time seems to diverge.

With discrete q it is impossible to determine the position of the phase transition with high precision, but in practice we can make a reasonable estimate based on the dependence of $\langle s_{\max}\rangle/N$ on N, as shown in the insets. For $F = 3$ the critical value of the parameter q lies somewhere close to $q_c \simeq 15$, while for $F = 2$ the estimated value is $q_c \simeq 5$.

A more important difference between the cases of $F = 2$ and $F = 3$ than the numerical value of the transition point q_c is the very nature of the phase transition. We have already seen that the stopping time normalised to the system size blows up at the transition only for $F = 2$, remaining finite for $F = 3$. We can say with fair trustworthiness that for $F = 2$ the transition is continuous, i.e. second-order, while for $F = 3$ the transition is clearly a first-order one. Although it may be dangerous to rely only on the numerical data for the order parameter, the conclusion on the nature of the phase transition has also been supported by the results for the distribution of cluster sizes close to the transition. We can see in Fig 8.20 that the probability of finding in the absorbing state a cluster of a size larger than s, decays as a power, i.e. $P^>_{\text{clust}}(s) \sim s^{-\alpha}$, with an obvious cutoff at large s due to the finiteness of the lattice. The value of the exponent α plays a critical role. For $F = 2$ we estimate $\alpha_{F=2} \simeq 0.6$ and for $F = 3$ we obtain $\alpha_{F=3} \simeq 1.65$. It has been shown that the exponent does not depend on the number of features as long as $F \geq 3$ [1670]. How does the distribution of cluster sizes relate to the type of phase transition? Denoting $N_{\text{clust}}(N)$ the total number of clusters and $P_{\text{clust}}(s, N) = P^>_{\text{clust}}(s) - P^>_{\text{clust}}(s + 1)$ the fraction of clusters of size s on a lattice with N sites, we must have

$$N = N_{\text{clust}}(N) \sum_{s=1}^{N} s P_{\text{clust}}(s, N). \tag{8.64}$$

If $P_{\text{clust}}(s, N) \sim s^{-1-\alpha}$ and $\alpha \leq 1$, the sum on the right hand side diverges when $N \to \infty$. But the sum is just the average cluster size, in other words, the quantity which plays the role of correlation length. The behaviour resembles the percolation transition, and the diverging correlation length is an unmistakable sign of a second-order phase transition.

On the other hand, if $P_{\text{clust}}(s, N) \sim s^{-1-\alpha}$ with $\alpha > 1$, the sum converges; and to keep the equality in (8.64) for $N \to \infty$, we need that either the number of clusters is proportional to the system size, or, in addition to the power-law component, there is an additional term in the distribution of cluster sizes, accounting for the largest cluster of size $s_{\max} \simeq N$, i.e.

$$P_{\text{clust}}(s, N) \simeq a s^{-1-\alpha} + b \, \delta_{s, s_{\max}}. \tag{8.65}$$

The former possibility occurs for $q > q_c$ and the latter for $q < q_c$. Indeed, the simulation data in Fig. 8.20 show the presence of δ-function part in (8.65), with positive weight $b > 0$. Below the transition, the largest cluster spans essentially the whole system, and at q_c its size drops discontinuously to a value negligible with respect to N. This is typical of first-order transitions.

These considerations are also compatible with the behaviour of the stopping time as function of q. For $F = 2$ the data indicate that $\langle \tau_{\text{st}} \rangle / N$ diverges as $q \to q_c$, while for $F = 3$ it remains constant. Interpreting the stopping time as the correlation time of the dynamics, we have exactly the behaviour commonly observed in the second- and first-order phase transitions, respectively.

To conclude, the phase transition in the Axelrod model is driven by the parameter q, measuring the localisation of values in the initial condition, and belongs to the class of first-order transitions if the number of features is at least 3, while for only $F = 2$ features the transition is second-order. Interestingly, the same dependence in terms of number of components is known in the Potts model. However, one should bear in mind that the transition in the Axelrod model has a purely dynamical origin. There is no equilibrium besides the absorbing states, so the analogy with phase transitions

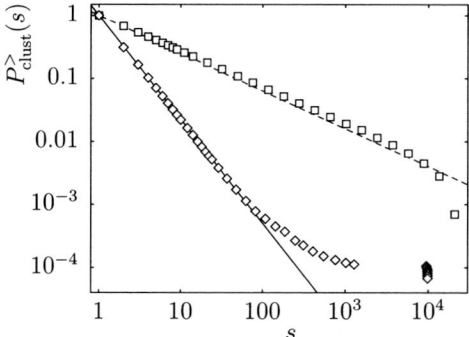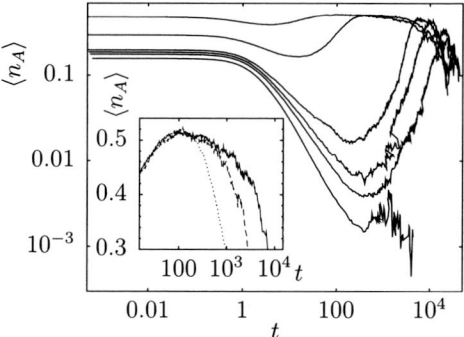

Fig. 8.20 Axelrod model on a square lattice. In the left panel, distribution of cluster sizes close to the transition $q \simeq q_c$, averaged over 100 realisations. The parameters are $F = 2$, $q = 5$, $L = 200$ ($\square$), and $F = 3$, $q = 15$, $L = 100$ ($\diamond$). The straight lines are the power laws $\propto s^{-0.6}$ (dashed) and $\propto s^{-1.65}$ (solid). In the data corresponding to $F = 3$ (symbol $\diamond$), note the isolated points around $s \simeq 10^4 = L^2$ indicating that the distribution is composed of a power-law part plus a δ-function located close to L^2. In the right panel, time dependence of the number of active bonds, averaged over realisations of the process. The lattice size is $L = 100$, number of features $F = 3$, and different lines correspond to $q = 5, 9, 14, 15, 16$, and 18, from top to bottom. In the inset, the dynamics is shown for $F = 3$, $q = 5$, and three lattice sizes, $L = 30$ (dotted line), 50 (dashed line), and 100 (solid line).

in equilibrium statistical physics must be taken cautiously and with a certain reserve. You may recall that a similar warning was also made in Sec. 5.1.3 when we dealt with the dynamic phase transition in the minority game model.

Although we spend most of the time analysing the absorbing states, it is instructive to also see how the system approaches them. In Fig. 8.20 we can see the evolution of the average number of active bonds. The averages are taken only over realisations which have not yet reached the absorbing state (recall a similar point of view when we analysed the voter model in Sec. 8.2.1). Around the time $t \simeq 1$, i.e. after as many updates on the computer as there are lattice sites, the density of interfaces decreases drastically. For $q > q_c$ it quickly brings the system to the absorbing state, while in the opposite case $q < q_c$ the activity rises again and n_A increases to a value close to $n_A \simeq 0.5$. Then, a very slow evolution follows. Mostly the evolution ends due to an occasional hit into an absorbing state, but this is rather a finite size effect, as demonstrated in the inset in Fig. 8.20. When the system size increases, the slow evolution is more and more prolonged. It is very difficult to see the behaviour at these very long times on a computer, but it seems that the density of active bonds decays very slowly to zero, resembling the logarithmic decay in the density of interfaces in the two-dimensional voter model.

One-dimensional case

The two-dimensional square lattice is just one of the possible geometries the agents can sit on. The one-dimensional case is also very interesting, [1672, 1673] as the dynamics has the special property that we can define a Lyapunov function, i.e. the quantity which never increases during the evolution [1674, 1675].

Indeed, summing the number of equal features $A(x, y)$ over all pairs of neighbouring agents, and changing the sign, we get the number

$$H = - \sum_x A(x, x^1) \tag{8.66}$$

where x^1 is the right neighbour of the site x, in accordance with the general notation x^μ we introduced for the neighbours of x on any lattice. The proof that H is the Lyapunov function relies on the fact that each site has exactly 2 neighbours. Thus, it is not the question of dimensionality but connectivity. The same would be true on any lattice where all sites have at most two neighbours, for example on a regular graph with all vertices having order 2, but not on a one-dimensional lattice with interactions between next-nearest neighbours.

Let us see how H changes in one update. Choose a site x and one of its neighbours, which we denote y. When comparing the features of agents at x and y the number of agreements either remains unchanged or increases by 1. The latter case occurs if one of the features of x has changed its value. If x does not have any other neighbour besides y, the change in H is accordingly either 0 or -1. If x does have another neighbour, say z, the value of $A(x, z)$ may be affected too. But as only one feature is changed, $A(x, z)$ cannot increase more than by 1. The change in H is the negative of the sum of changes in $A(x, y)$ and $A(x, z)$, but as we have seen, it cannot exceed 0. Therefore H is the Lyapunov function of the Axelrod model dynamics.

The quantity $(1 - H)/N \in [0, 1]$ vanishes in the totally uniform state, and in several studies it was used as an alternative order parameter [1674, 1675]. Interestingly, the simulation shows that the transition can be discontinuous in the order parameter $\langle s_{\max} \rangle / N$ while $(1 - H)/L$ exhibits a continuous transition [1674]. The discrepancy can be attributed to the dynamic nature of the phase transition, where the usual distinction between first- and second-order transitions may not be fully appropriate, as we have already pointed out.

Topological and thermal disorder

As with the voter and Sznajd model, we naturally ask what happens if the agents are placed on a more realistic social network than on a two-dimensional lattice or a linear chain [1676]. The two choices which are first at hand are the Watts-Strogatz small-world networks; the second class are the scale-free networks created via the Barabási-Albert process. In the former case, the main finding is that the phase transition occurs at a value q_c which grows with the probability of rewiring, p_{WS}. In the latter case, it was found that the transition shifts to larger and larger values of q when the size of the network grows. More precisely, on the Barabási-Albert network of N nodes, with degree exponent $\gamma = 3$, the dependence is well fitted on a power law $q_c \sim N^{0.39}$. As a third network topology, the so-called structured scale-free networks were used, which have effectively a one-dimensional topology [1676]. Contrary to the Barabási-Albert network, the transition occurs at a value independent of the network size.

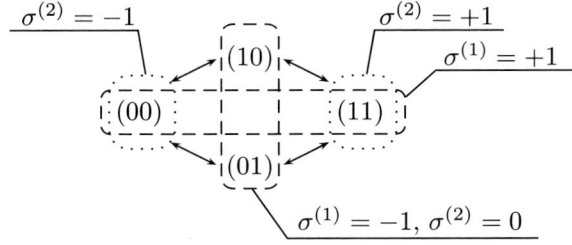

Fig. 8.21 Illustration of the dynamics of opinions in the Axelrod model with $F = 2$ and $q = 2$. Allowed transitions between states of an agent are indicated by arrows. The dashed and dotted ovals denote groupings of the 4 states according to the values of $\sigma^{(1)}$ and $\sigma^{(2)}$.

The evolution of the Axelrod model exposed here does not allow spontaneous changes of the state of individual agents due to random perturbations. This corresponds to a zero-temperature dynamics. One may ask what new effects can arise if we 'heat up' the system a little.

There are several ways of implementing the perturbations. Let us look what happens if they are infinitesimally small. We let the Axelrod model evolve until an absorbing state is reached, then change the value of one feature in one agent and let the evolution proceed. When the dynamics stops at another absorbing state, we perturb it again and repeat these perturbation cycles again and again. The interesting conclusion from simulations of this process is [1674, 1677] that the average size of the maximum cluster grows until it reaches the uniform configuration with $\langle s_{\max} \rangle = N$. The fragmented states achieved for $q > q_c$ in one run, starting from a random initial configuration, are rather metastable 'traps', and the system escapes from them if it gets a chance to do so.

We can also introduce a noise of finite strength. With the rate $r > 0$, a random change in a single feature is made. We naturally expect that the phase transition at $q = q_c$ is blurred, and this was actually confirmed in simulations [1674, 1677, 1678]. Moreover, on a two-dimensional lattice, a scaling form $\langle s_{\max} \rangle / N = g((1 - q^{-1}) r \, N \ln N)$ with a universal, smoothly decreasing function $g(x)$ was found.

A (partially) solvable case

The case $F = 2$ and $q = 2$ is special as it yields several exact results [1679–1681]. There are four possible states for each agent, and the transitions between the states can go along the arrows indicated in Fig. 8.21.

The structure of the diagram in Fig. 8.21 allows us to group the agents in state (11) together with agents in state (00) into a larger group, with the common parameter $+1$. Similarly agents in states (01) and (10) form a group labelled by -1. Formally, the agent at site x, who has $\sigma(x, f)$ as the value of f-th feature, is given a new scalar variable $\sigma^{(1)}(x)$ calculated as

$$\sigma^{(1)}(x) = 1 - 2|\sigma(x, 1) - \sigma(x, 2)|. \tag{8.67}$$

Now it is easy to see that the new state variable $\sigma^{(1)}(x)$ evolves according to the standard rules of the voter model. Indeed, any of the agents in the state (01) or (10) can force her neighbour in state (00) or (11) to adopt her state, and vice versa. This way the Axelrod model is simplified to what we already know. For example, the active bonds are just those which connect an agent from group $+1$ with a neighbour from the group -1. The active bonds are equivalent to interfaces in the voter model, and their time evolution, starting from a totally random initial configuration, decays as $n_A \sim t^{-1/2}$ in one dimension, as $n_A \sim (\ln t)^{-1}$ in two dimensions and $n_A \sim t^{1-d/2}$ in $d \geq 3$ dimensions. As the voter model eventually reaches an ordered state in dimensions $d = 1$ and 2, we conclude that the Axelrod model in these dimensions also evolves into a state in which all agents are either from group -1 or from group $+1$. What we do not know, though, is the dynamics within the groups, for example the density of inactive bonds connecting states (00) and (11). But we can get some information on it by refining our division into groups.

In fact, because of the symmetry, it is enough to distinguish between the states in one of the two groups $+1$, or -1. Exchanging values $0 \leftrightarrow 1$ in the second feature we get information on the internal dynamics inside the other group, which must be statistically identical. So, we define yet another state variable

$$\sigma^{(2)}(x) = \sigma(x,1) + \sigma(x,2) - 1 \tag{8.68}$$

with possible values ± 1 and 0. We may think of them as representing three opinions on an issue, namely two extremes and a centrist position. The rules of the Axelrod model imply that the neighbouring agents interact only if the difference in their opinions is at most 1, i.e. the extremists do not interact among themselves but only with the centrists.

Such mapping can be done in any spatial dimension, but in dimension one the dynamics may be conveniently represented in terms of interacting interfaces, or domain walls. The bonds separating an opinion $+1$ or -1 from 0 are the usual active bonds of the original Axelrod model, or the interfaces in the voter model. Let us denote these two types of walls by M_+ and M_-, respectively. They evolve as the usual annihilating random walkers, and their density decays in the way already mentioned previously, namely $\sim t^{-1/2}$. On the contrary, the domain walls separating the extremists, $+1$ and -1, cannot move, as the agents do not have influence over each other. We denote these stationary walls as S. Their spatial configuration can change only via their interaction with the moving walls of the two other types. Schematically, we can denote the interaction of the three species of walls as $M_\pm + M_\pm \to \emptyset$, $M_\pm + M_\mp \to S$, and $M_\pm + S \to M_\mp$.

Contrary to the moving walls, the density of the stationary walls $n_S(t)$ decays in time in a non-universal way. Numerical simulations [1679] show that

$$n_S(t) \sim t^{-\psi}, \text{ for } t \to \infty \tag{8.69}$$

where the exponent ψ depends on the initial concentration n_0 of centrists, and for the small and high concentrations it approaches $\psi(n_0) \to 0$ for $n_0 \to 0$ and $\psi(n_0) \to 0.5$ for $n_0 \to 1$. An approximate mapping on kinetic q-state Potts model with effective number of components $q = 1/(1 - n_0)$ was proposed, resulting in the following formula [1605, 1679]

$$\psi(n_0) = -\frac{1}{8} + \frac{2}{\pi^2} \arccos^2 \left(\frac{1 - 2\,n_0}{\sqrt{2}} \right) \qquad (8.70)$$

which agrees well with the simulations in the range $n_0 \lesssim 0.4$.

More exact results are available in the mean-field approximation, where the agents are placed on a complete graph [1680]. The calculation goes essentially along the lines presented for the voter model on a complete graph (see Sec. 8.2.1), but the analysis is a lot more complicated because there are infinitely many absorbing states. Indeed, in addition to the three obvious uniform states when all agents have either $+1$ or -1 or 0 opinion, also all mixtures of the extremists, i.e. configurations with no centrists but with arbitrary concentration of the $+1$ and -1 opinions, are absorbing states. In essence, we have again a diffusion problem, analogous to Eq. (8.25), but instead of being bound to an interval with absorbing endpoints, we are limited to a triangle with absorbing corners and one edge.

Due to the length of the calculations we present here only the results [1680]. Denote n_+ and n_- the concentration of the opinions $+1$ and -1 in the initial state. The easiest quantity to calculate is the probability $P_0(n_+, n_-)$ that the final absorbing state consists of all centrists. It is

$$P_0(n_+, n_-) = 1 - n_+ - n_- \qquad (8.71)$$

which can be seen directly from mapping on the voter model. The probabilities $P_\pm$ of reaching the homogeneous ± 1 states can be obtained in closed form as an infinite series, but we show here the result only in the simpler case of symmetric initial condition $n_+ = n_-$. Then

$$P_+(n_+, n_-) = P_-(n_+, n_-) = n_+ - \frac{1}{2} + \frac{1 - 4n_+^2}{2\sqrt{1 + 4n_+^2}}. \qquad (8.72)$$

On the same condition, the probability of reaching any frozen state with mixed $+1$ and -1 opinions is

$$P_{+-}(n_+, n_-) = 1 - \frac{1 - 4n_+^2}{\sqrt{1 + 4n_+^2}}. \qquad (8.73)$$

It would be tempting to extend this approach to a general case of Axelrod model with an arbitrary number of components F and for any q. We encourage the reader to try her luck.

8.2.4   Bounded confidence with continuous opinions

There may be a different view on consensus formation, supposing that the opinions on a certain issue can vary continuously, but a discussion on that subject and possible convergence of the opinions cannot take place unless the actual opinions are close enough. It the people differ too much, they may even avoid mutual contact completely or decide to use 'different means', an often used euphemism for killing the opponent. If, on the contrary, the difference in their opinions, measured by a continuous variable, does not exceed certain threshold, then the difference may be further diminished by discussions and peaceful persuasion. This idea is called bounded confidence and we already encountered a kind of it when we discussed the Axelrod model.

Linear dynamics

Modelling the consensus formation using continuous variables has been around for quite a long time. Let us denote by F_i the variable describing the opinion of the individual i. The simplest way to implement the convergence of opinions, introduced by DeGroot [1682], is to assume that in the next period the opinions are linear combinations of the current opinions of all individuals

$$F_i(t+1) = \sum A_{ij} F_j(t) \tag{8.74}$$

where all of the elements of the matrix A are non-negative, $A_{ij} \geq 0$, and the sum of each of its row's entries is one, $\sum_j A_{ij} = 1$. We call it a confidence matrix. We shall also suppose that the diagonal elements are strictly positive, $A_{ii} > 0$, which means that the individuals have at least some non-vanishing belief in their own opinions. The new values of F_i are weighted averages of the previous opinions, the weights being stored in the matrix A. As such, the dynamics is very simple, and it is equivalent to a Markov chain with transition probabilities A_{ij}.

The bounded confidence is missing in this model, but it is worth studying, as it can teach us an important lesson about when a consensus can be reached and when it is not possible [1682, 1683]. The latter question is answered by the so-called stabilisation theorems, which rely on the notion of zero pattern of the confidence matrix. The zero pattern of the matrix A is another matrix A_0 of the same dimensionality whose elements are zero at the same places where A has zero elements, and one elsewhere. Thus

$$A_{0\,ij} = \begin{cases} 1 \text{ if } A_{ij} \neq 0 \\ 0 \text{ if } A_{ij} = 0. \end{cases} \tag{8.75}$$

The zero pattern can be interpreted as an adjacency matrix for a graph representing direct interactions of the individuals.

Starting with the vector of opinions $F(0)$, after t steps of the dynamics the opinions will be given by t-th power of the confidence matrix, $F(t) = A^t F(0)$. For large times, the opinions should converge to an eigenvector of the matrix A, corresponding to the eigenvalue 1. The fact that A is a stochastic matrix guarantees existence of at least one such eigenvector. If it is unique, it is also uniform, as can easily be checked. Indeed, if $F_i = f$ for all i, then $\sum_j A_{ij} F_j = f \sum_j A_{ij} = f = F_i$, so F_i is an eigenvector with eigenvalue 1. The interpretation of such a state is straightforward. Since all individuals have the same value of F_i, they all share the same opinion, and a complete consensus is reached.

So far it has seemed that full consensus is an inevitable outcome, but this is not the case. The problem is that there may be several linearly independent eigenvectors for the same eigenvalue 1, and all of them may be the asymptotic states, together with any linear combination of them. They form a vector subspace of dimensionality at least two. We can choose the basis vectors in this subspace so that they do not have any non-zero elements in common. Let us denote them $F_{\alpha i}$, where $\alpha = 1, \ldots, r$ numbers the r eigenvectors. Their non-zero elements must again be uniform, and we may freely set them equal to 1. This way we have identified r communicating subgroups of individuals, which are the sets $G_\alpha = \{i : F_{\alpha i} \neq 0\}$. Now, the consensus is guaranteed inside each communicating subgroup, but not necessarily across them.

It is fairly easy to determine whether there are more than one communicating subgroup and eventually to find all of them. The point is that the zero pattern of the matrix A^t is equal to the zero pattern of t-th power of the zero pattern of A, i.e. $(A^t)_0 = ((A_0)^t)_0$. But this means that the non-zero elements in A^t correspond to such pairs of individuals for which there is a path of length t connecting them in the graph of direct interactions. The requirement of positive diagonal elements introduces loops on each vertex of the graph, so if there is a path of length t between two individuals, we can construct also a path of any larger length $t' > t$.

What does it mean for the evolution of our model? As time passes, the zeros in the matrix A^t can only disappear; they never occur at those matrix elements that were non-zero before. If diam_G is the diameter of the graph of direct interaction, for times $t > \mathrm{diam}_G$ the zero pattern of the matrix A^t will not change anymore. (See Sec. 6.1.1 for the definition of graph diameter.)

If the graph is connected, for sufficiently large t the zero-pattern of A^t contains no zeros, and the eigenvector corresponding to eigenvalue 1 is unique. In the opposite case, each connected component of the graph is just composed of the individuals from the corresponding communicating subgroup G_α.

The question of reaching a consensus or not in DeGroot's model is reduced to the study of connected components of the graph of direct interactions. When the graph is composed of several connected components, each of them reaches its own consensus, independent of the others. This is quite simple. Now we shall see how the bounded confidence complicates the thing.

Hegselmann and Krause

The new opinion of the individual i takes into account the others' opinions with weights A_{ij}. What if these weights are not constant, but depend on the current opinions themselves? Although some attempts in this direction had been made earlier [1684], the principle of bounded confidence was first applied in the models developed by the groups of Deffuant et al [1685] and Krause and Hegselmann [1686, 1687]. Let us discuss the latter model first.

The Hegselmann-Krause model [1687] considers opinion formation as a fully deterministic process, as in DeGroot's model investigated in the last section. The new opinion of an individual is the average of current opinions of those individuals (including the individual herself), whose opinions lie within a fixed confidence bound $\epsilon > 0$ from her own opinion.

For those pairs of individuals who do not differ more than ϵ in their opinions the weights A_{ij} are uniform, so

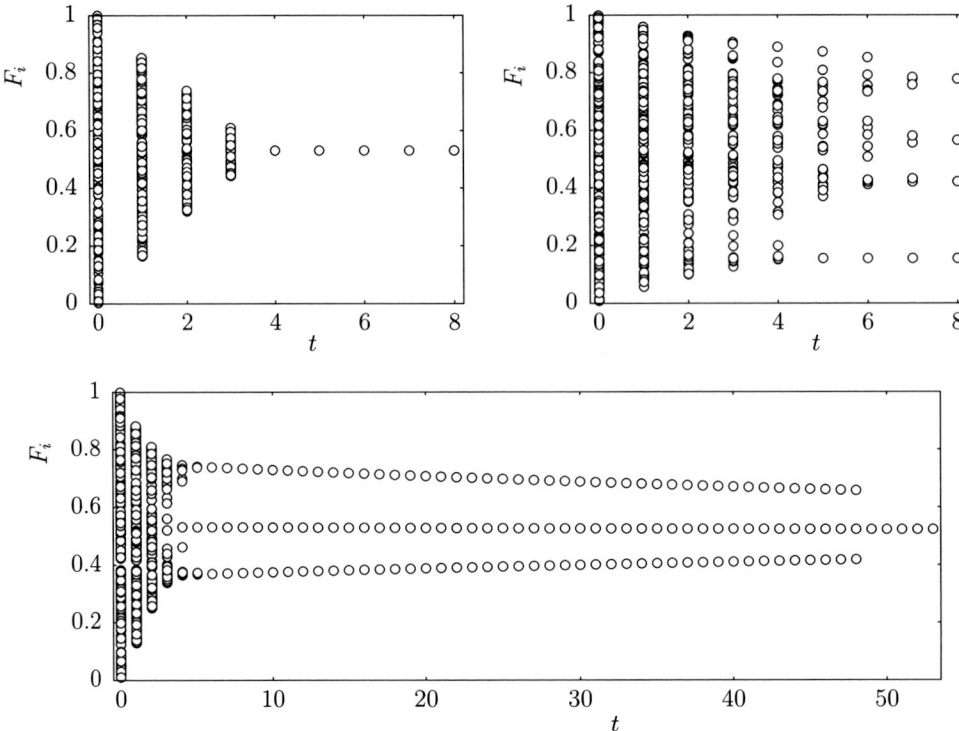

Fig. 8.22 Evolution of the Hegselmann-Krause model. Each point represents one or more agents with a specified value F_i. The total number of agents is $N = 200$. In the upper left panel, the confidence bound is $\epsilon = 0.3$, while in the upper right panel it is $\epsilon = 0.1$. In the lower panel we show how the approach to consensus is slowed down when the confidence bound is close to its critical value. Here we have $\epsilon = 0.24$.

$$A_{ij} = \begin{cases} 0 & \text{for } |F_i - F_j| > \epsilon \\ \dfrac{1}{N_{i\,\epsilon}} & \text{for } |F_i - F_j| \le \epsilon. \end{cases} \tag{8.76}$$

Obviously, the normalisation constant is the number of individuals within the confidence bound $N_{i\,\epsilon} = |\{j : |F_i - F_j| \le \epsilon\}|$. Note that the dependence of the confidence matrix on the current opinions is very strongly nonlinear. From the formal point of view the most important thing is that also the zero-pattern of the confidence matrix can change in time. We may interpret it as changing the structure of the graph of directly communicating individuals. To see if we are going to reach the consensus, it is important to watch if the graph remains connected during evolution. Clearly, once it splits into disconnected parts, they will never get joined together again, and the consensus will not be achieved.

To get an impression of how the evolution of the Hegselmann-Krause model proceeds, look at Fig. 8.22. For large enough confidence bound ϵ the system approaches full consensus, while lower values of ϵ induce several stable communicating subgroups, which do not interact with each other, and the system splits into several clusters with different opinions. The number of such clusters grows as $1/\epsilon$ when the confidence bound shrinks. The numerical simulations show that the critical value ϵ_c at which the full consensus breaks down approaches the value $\epsilon_c \simeq 0.2$ when the number of individuals increases. Interestingly, when the confidence bound comes close to the critical value ϵ_c, the number of steps needed to reach consensus increases, suggesting that we indeed come across a kind of a dynamic phase transition [1688]. This behaviour is rather robust, as it persists in various modifications of the Hegselmann-Krause model, most notably if we allow the individuals to communicate only through a regular lattice or a random network [1689–1692]. It was confirmed both by simulations and by solution of a corresponding partial integro-differential equation [1689–1700]. For this purpose, a smart way of discretisation of the integro-differential equation, called the interactive Markov chain, was developed [1698–1702].

The detailed analysis of the transition [1688] reveals the decisive role of a handful of agents who can be called mediators [1688, 1702]. Imagine that the agents separate into two big groups of about the same size plus a single agent located at about the middle between the big groups. Suppose the groups are at a distance larger than ϵ, so that they do not directly influence each other, but the single agent in the middle has distance smaller than ϵ from both the first and the second group. This is the mediator, as that agent transmits the influence from the first big group to the other one. Because the groups are of comparable size, the position of the mediator does not change substantially. However, the mediator very slowly attracts both of the groups towards the middle. An example can be seen in the lower panel of Fig. 8.22. Such very slow evolution lasts until the groups come close enough to overcome the confidence bound ϵ. Then, all agents collapse into one single group, and consensus is reached. One single mediator decided the fate of the entire system. Certainly there is a lesson we can translate into real life.

Deffuant et al

Apart from the initial condition, the model of Hegselmann and Krause is fully deterministic and the opinions are updated in parallel. A similar model, but with random sequential update was introduced by G. Deffuant et al [1685].

We have again N individuals with opinions F_i. In each update step, we randomly choose two of them, say, i and j, and see if their opinions differ less than (or equal to) the confidence bound ϵ. In a positive case, their opinions are slightly shifted towards each other

$$\left.\begin{array}{l} F_i(t + 1/N) = (1 - \mu)F_i(t) + \mu F_j(t) \\ F_j(t + 1/N) = \mu F_i(t) + (1 - \mu)F_j(t) \end{array}\right\} \text{ for } |F_i(t) - F_j(t)| \le \epsilon \qquad (8.77)$$

where μ is a parameter fixing the rate of convergence.

For a very large number of individuals, $N \to \infty$, the dynamics can be expressed in terms of the continuous distribution of opinions $P_{Ft}(f)$, which can be written formally as $\int_0^f P_{Ft}(f')\mathrm{d}f' = \lim_{N\to\infty} \frac{1}{N}\sum_j \theta(f - F_j(t))$. We get the following innocent-looking rate equation [1703]

$$\frac{\partial}{\partial t} P_{Ft}(f) = \int_{|f_1 - f_2| \leq \epsilon} P_{Ft}(f_1) \, P_{Ft}(f_2)$$
$$\times \left[\delta((1-\mu)f_1 + \mu f_2 - f) - \delta(f_1 - f) \right] \mathrm{d}f_1 \mathrm{d}f_2. \tag{8.78}$$

which exhibits a fairly complex behaviour. Starting from the uniform initial condition $P_{F0}(f) = 1$ for $f \in [0,1]$ the configuration evolves into a stationary state composed of one or several δ-functions, each of them corresponding to one cluster of individuals sharing the same opinion.

To see it on a trivial example, consider the case $\epsilon > 1$. All individuals interact with each other, so their opinions converge to a common limit $\lim_{t \to \infty} F_i(t) = 1/2$. We can see it from Eq. (8.78) when we investigate the time evolution of the moments of the distribution. Obviously $\frac{\mathrm{d}}{\mathrm{d}t}\langle F \rangle = 0$, so the average opinion is independent of time, $\langle F \rangle = 1/2$. The dispersion from the mean $\langle F^2 \rangle_c = \langle F^2 \rangle - \langle F \rangle^2$ obeys

$$\frac{\mathrm{d}}{\mathrm{d}t} \langle F^2 \rangle_c = -2\mu(1-\mu)\langle F^2 \rangle_c \tag{8.79}$$

which means that the dispersion decays exponentially to zero with a rate of $2\mu(1-\mu)$. This confirms our intuition noted earlier that μ determines the speed of the evolution towards the stationary state.

The behaviour in the complementary regime of very small ϵ can be guessed if we suppose that $P_{Ft}(f)$ does not vary too wildly at scales comparable to ϵ or shorter. This also means that f is assumed to be farther than ϵ from the extremal values 0 and 1. In this case we expand $P_{Ft}(f_1, t)$ and $P_{Ft}(f_2, t)$ in the Taylor series and perform the integral on the right-hand side of Eq. (8.78) explicitly. Finally we get partial differential equation

$$\frac{\partial}{\partial t} P_{Ft}(f) = -\frac{\epsilon^3}{3} \mu(1-\mu) \frac{\partial^2}{\partial f^2} P_{Ft}^2(f). \tag{8.80}$$

There is a trivial homogeneous solution $P_{Ft}(f) = C$, independent of f and t, which is however unstable at all length scales. To see it, we can linearise (8.80) close to the uniform solution, $P_{Ft}(f) = C + D(f, t)$, where $|D(f, t)| \ll C$, and express the result in terms of the Fourier transform $\tilde{D}(k, t)$ of the small perturbation. We obtain $\frac{\partial}{\partial t} \tilde{D}(k, t) = \frac{2}{3} k^2 \epsilon^3 C \mu(1-\mu) \tilde{D}(k, t)$, indicating that perturbations are amplified more the larger is the wavevector k, i. e. the shorter is their wavelength. On the other hand, if a stationary state is reached, the structures created either collapse into a single point or are at distance larger than ϵ. This feature is lost in the derivation of Eq. (8.80), as we supposed that the probability is smooth on scale comparable with ϵ. Therefore, Eq. (8.80) describes well the initial stage of the breakdown of the uniform solution, but does not apply to the ultimate state of the evolution. Nevertheless we can conclude that for small ϵ the stationary state will be composed of more or less regularly spaced δ-functions with the period proportional to ϵ.

To see explicitly how the stationary state is approached for a general value of the confidence bound ϵ, the solution of Eq. (8.78) should be found numerically. Examples of such evolution are shown in Fig. 8.23. For large enough ϵ, all opinions converge to the common value 1/2, as in the case $\epsilon > 1$. However, low ϵ produces two or more distinct peaks. Detailed analysis [1703, 1704] reveals that a series of bifurcations

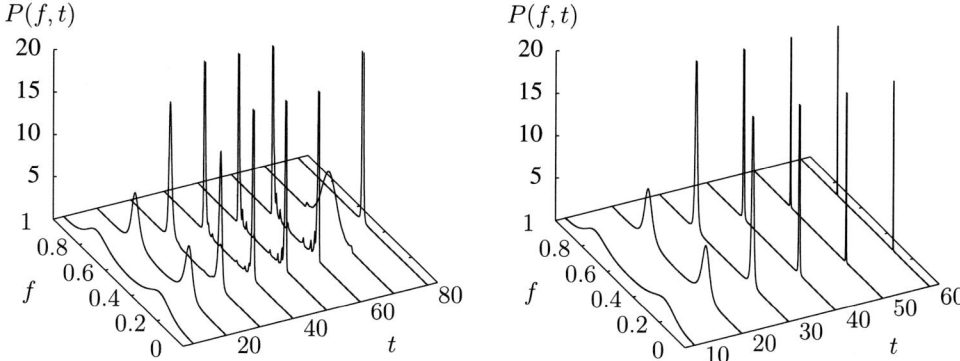

Fig. 8.23 Evolution of the Deffuant model with $\mu = 0.5$. In the left panel, approach to consensus for confidence bound $\epsilon = 0.27$, slightly larger than the critical value ϵ_{c1}. In the right panel, two peaks are formed in the distribution of opinions, for $\epsilon = 0.25$.

occurs in the system when we decrease the confidence bound. The consensus breaks into separated groups suddenly at a critical value of ϵ_{c1}. The groups themselves break up into more groups at $\epsilon_{c2} < \epsilon_{c1}$, and so forth. The numerically found value of the first critical point is $\epsilon_{c1} = 0.269\ldots$. But how do the peaks emerge from the originally homogeneous distribution? In Fig. 8.23 we can observe that they are triggered by the original inhomogeneity at the edges of the interval $[0, 1]$. A structure of gradually sharpening peaks propagates towards the centre of the interval, where the nonlinear waves from the opposite endpoints meet and form the resulting pattern.

We can translate this observation into common language by saying that the ultimate fate of the opinions in the society is largely in the hands of the extremists. Indeed, the individuals close to the edges of the scale of possible opinions are from the very beginning drawn towards the middle of the interval $[0, 1]$, i.e. close to the centrist position, while those who are farther from the edges than ϵ keep their opinions unchanged for some finite time. The crucial question is: how far from the extreme will the former extremists move before they settle? It may happen that they will drift so close to the midpoint that the centrists will attract them all. Full consensus follows. But the extremists themselves can attract enough centrists, and instead of creating one peak at the centre, two (or even more) peaks occur. The resulting society lives with polarised (or diversified) opinions. The parameter which decides about the outcome is the confidence bound ϵ. Higher confidence leads to an overwhelming consensus among people, while lower confidence causes the society split into several non-communicating communities.

Stabilisation theorem

As you may have noticed, the rigorous analysis of the models of Hegselmann and Krause and Deffuant et al would be very difficult, but there is a very interesting, although partial, result [1705–1708].

The Hegselmann-Krause and Deffuant models differ from the linear dynamics of DeGroot in two ways. First, the confidence matrix A_{ij} depends on the current configuration of the opinions F_i; and second, it may depend explicitly on time. In the Deffuant model, the explicit time dependence is due to the stochastic choice of interacting pairs of individuals. For each realisation of the dynamics we get a sequence of confidence matrices $A_{ij}(t)$, where under the argument t we hide both the direct explicit time dependence and the indirect dependence through the current opinion configuration. For the stabilisation theorem, there is no need to distinguish between these two sources of the time dependence.

What does matter, on the other hand, are three properties the confidence matrix must fulfil at all times. First, we need that all of its diagonal elements are positive, $A_{ii}(t) > 0$. It means that each individual has some self-confidence. Second, the zero-pattern is a symmetric matrix, so if $A_{ij}(t) > 0$, then also $A_{ji}(t) > 0$. This requirement is interpreted so that the confidence is mutual, even though it may not be as strong in one direction as it is in the other. Third, there is a fixed lower bound for positive elements of the confidence matrix, i.e. there is a constant $c > 0$ such that either $A_{ij}(t) = 0$ or $A_{ij}(t) > c$ for all times t. This is a necessary requirement to avoid certain pathological situations.

A theorem can be proved provided the latter three properties are satisfied. In their formulation we shall use the notation $A(t_2, t_1) = A(t_2 - 1) \ldots A(t_1 + 1)A(t_1)$ for the product of confidence matrices. We shall also call the square matrix K with all non-negative elements a consensus matrix if all its rows are equal and the sum of elements in every column is 1. Leaving aside the proof, here is the theorem.

Theorem. For any realisation of the sequence of matrices $A(t)$ there is a finite time t_0 such that $A(t, 0)$ converges to the product of $A(t_0, 0)$ and a block-diagonal matrix, so

$$\lim_{t \to \infty} A(t, 0) = \begin{pmatrix} K_1 & & & 0 \\ & K_2 & & \\ & & \ddots & \\ 0 & & & K_m \end{pmatrix} A(t_0, 0) \tag{8.81}$$

where $K_1, K_2, \ldots, K_m$ are consensus matrices.

The meaning of the theorem is that after a certain finite time t_0 the process definitely tends towards a consensus within each of the m groups on which the matrices K_i are acting, but each of the subgroups reaches a different opinion. To be sure, the consensus is not reached at time t_0, nor at any finite time whatsoever, but at time t_0 it is already decided what individuals shall form the m consensus groups. In this sense, t_0 can be considered a consensus time in Hegselmann-Krause and Deffuant et al models, even though for the full convergence of opinions within the consensus groups we must wait longer, and in the model of Deffuant et al with $\mu \leq 1/2$ the consensus is not achieved before infinity.

8.3   Stratification

People are not equal. They not only differ in the colour of their eyes and in their capabilities to play chess or to run a marathon race, but even individuals with very similar talents may find themselves at very different social statuses. All human attempts to bring more justice into such evident disequilibrium have ended in desperate or even catastrophic failure. Perhaps the best one can do is to make the diaphragms separating the social levels as permeable as possible, so that no one is a priori disqualified. The ubiquity of social stratification in animal as well as human collectives is certainly a phenomenon which calls for an explanation, and the fact that rabbits, dogs, apes, and *Homo sapiens* all exhibit similar behaviour patterns suggests some common mechanisms which may not be too complicated after all, although they produce highly complex outcomes. We shall show two very simple but appealing models, at the same time bearing in mind that this means merely groping on a very uncertain ground.

8.3.1   Emerging hierarchies

Bonabeau model

A newcomer in an animal group always has to undergo some fighting before its position in the social ladder is commonly accepted. If, on the other hand, an individual leaves the group for a prolonged period and then returns, it has to fight again, as the previously established level has faded away. These two observations motivated Bonabeau et al [1709] to introduce a model of self-organised hierarchies [1709–1715].

To be clear, we do not speak of hierarchies in the sense of trees with a king or a marshal on the top and lesser ranks below. We rather have in mind an ordering, in which each individual bears a single number we shall call strength, indicating its position among others. When two of them meet, the stronger one has a higher probability to be strengthened, while the weaker is most likely pushed down even more. The strengths of the agents who do not meet at this time relax towards zero by a fixed fraction. More formally, the configuration of the system at time t is described by a collection of strengths $F_i(t)$ of the agents $i = 1, 2, \ldots, N$. In each step, a pair of agents is chosen to fight. The strength of the winner is increased by 1, while the strength of the loser is decreased by 1. Moreover, all strengths relax to the reference (zero) level deterministically. Thus

$$F_i\left(t + \frac{1}{N}\right) = \left(1 - \frac{\mu}{N}\right)F_i(t) + \Delta_{ij}$$
$$F_j\left(t + \frac{1}{N}\right) = \left(1 - \frac{\mu}{N}\right)F_j(t) - \Delta_{ij} \tag{8.82}$$
$$F_k\left(t + \frac{1}{N}\right) = \left(1 - \frac{\mu}{N}\right)F_k(t), \ k \neq i, k \neq j$$

with

$$\text{Prob}\{\Delta_{ij} = \pm 1\} = \frac{1}{1 + e^{\mp \eta(F_i(t) - F_j(t))}} \tag{8.83}$$

expressing the probability that i is the winner and j the looser. The parameter η tunes the level of randomness in the dynamics, where $\eta \to \infty$ corresponds to purely deterministic outcomes, the stronger agent always beating the weaker, while $\eta = 0$ means that the strengths are increased and decreased by mere chance.

So far we have not touched upon the question of which pairs of agents do interact and when. The most natural choice is to place the individuals on a network, or simply on a square lattice, leaving some sites empty, and allow the agents to diffuse along the edges of the network. When two agents happen to meet at the same site, they fight. More than two agents on a site are not allowed. They behave like hard hemispheres, each site being able to accommodate one full sphere. Numerical simulations of such system show [1709–1713] that, if we change the density of agents, or, equivalently, the relaxation rate μ, there is a phase transition from an uniform state with all agents' strength close to 0 to a hierarchical state where the strengths are highly non-homogeneous. As an order parameter we choose the dispersion of the number of fights won by the agents, i.e. $\sigma = \sqrt{\frac{1}{N} \sum_{i=1}^{N} (w_i - \frac{1}{2})^2}$, where $w_i = n_i^+ / (n_i^+ + n_i^-)$ and n_i^+, n_i^- are number of encounters won and lost by the agent i, respectively. In the homogeneous phase $\sigma = 0$ while in the hierarchical phase it gets a finite value.

Mean-field solution

Analytical study is possible in mean-field approximation. Indeed, suppose that the diffusion is fast enough to ensure many encounters with various agents during the typical time given by the speed of relaxation of the strengths towards zero. Then, the spatial structure of the lattice on which the diffusion takes place becomes irrelevant. In other words, at each time step we choose two agents at random and let them fight. In this case the dynamics is much simpler, and for a large system we obtain deterministic evolution equations for the strengths

$$\frac{\mathrm{d}F_i}{\mathrm{d}t} = -\mu F_i + \frac{1}{N} \sum_{j=1}^{N} \frac{\sinh \eta (F_i - F_j)}{1 + \cosh \eta (F_i - F_j)}. \tag{8.84}$$

We neglected here the stochastic term which decreases as $1/N$ for large N. It can easily be seen that the average strength $\overline{F} = \frac{1}{N} \sum_i F_i$ relaxes exponentially to 0 according to $\frac{\mathrm{d}\overline{F}}{\mathrm{d}t} = -\mu \overline{F}$. So, it is sufficient to consider only stationary states with zero mean, $\overline{F} = 0$. The simplest of them, satisfying Eq. (8.84), is the trivial uniform state $F_i = 0$. It can be unstable, though, and we shall investigate its linear stability now.

Using the notation $\frac{\mathrm{d}F_i}{\mathrm{d}t} = R_i(F_1, \ldots, F_N)$ for the expression occurring in Eq. (8.84), we need to investigate the eigenvalues of the matrix $H_{ij} = \frac{\partial}{\partial F_j} R_i(0, \ldots, 0)$. We have

$$H_{ij} = \left(-\mu + \frac{\eta}{2} \right) \delta_{ij} - \frac{\eta}{2N}. \tag{8.85}$$

One of the eigenvectors is uniform, $x_i = 1$, corresponding to eigenvalue $-\mu$, which is always negative. The remaining $N - 1$ eigenvectors have the forms $x_i = 1 - N\delta_{ik}$ for some k, and they all belong to the same eigenvalue $\frac{\eta}{2} - \mu$. If the latter is positive, the uniform solution of Eq. (8.84) is unstable. This happens for

$$2\mu < \eta, \tag{8.86}$$

and we are naturally led to a question: what are the stationary configurations beyond this critical point? If we go just a short distance beyond the critical point, the linear stability analysis can offer a useful hint, but it needs to be complemented by the influence of the lowest nonlinear terms. Therefore, we expand the right-hand side of Eq. (8.84) up to the third order in the differences $F_i - F_j$, and for the stationary state we obtain

$$(\eta - 2\mu)F_i = \frac{\eta^3}{12N} \sum_j (F_i - F_j)^3. \tag{8.87}$$

We look for a solution in the form

$$F_i = a\delta_{ik} - b \tag{8.88}$$

suggested by the eigenvectors of H_{ij} corresponding to unstable modes. Inserting a trial solution (8.88) into Eq. (8.87) we obtain a set of equations for the parameters a and b which can be solved easily. Finally we have

$$F_i = \begin{cases} 0 & \text{for } 2\mu \geq \eta \\ \pm\sqrt{\dfrac{12(\eta - 2\mu)}{\eta^3}}(N\delta_{ik} - 1) & \text{for } 2\mu < \eta. \end{cases} \tag{8.89}$$

Hence we obtain the order parameter σ. We should note that in the stationary state the fraction w_i of fights won by the agent i should be balanced by the relaxation of her strength. Indeed, $2w_i - 1$ is the average increase of the strength of agent i in one step, which should be equal to μF_i. So, $\sigma = \mu\sqrt{\sum_i F_i^2/N}\,/2$, and inserting the result (8.89) we have

$$\sigma = \begin{cases} 0 & \text{for } 2\mu \geq \eta \\ \dfrac{\mu}{\eta^{3/2}}\sqrt{\dfrac{3}{N}\left(1 - \dfrac{1}{N}\right)(\eta - 2\mu)} & \text{for } 2\mu < \eta. \end{cases} \tag{8.90}$$

How can we interpret this solution? One can compare it to a situation in a society with one master and many servants, a single emperor while all the rest are servants, equal to each other in their subordination. This looks nice but the enthusiasm fades away when we realise from the formulae (8.89) that this 'realistic' solution always comes together with a mirror image of itself, a society of a single servant subject to many equal masters. This is an absurd situation, revealing something artificial in the Bonabeau model itself. Indeed, the dynamics is invariant with respect to inversion of all strengths, $F_i \to -F_i$, because we never introduced explicitly any a priori advantage of being stronger. The only thing we supposed was that strong grow stronger and weak become weaker.

Let us now turn back to mathematical aspects of the model. Recall that the results (8.89) and (8.90) were obtained assuming that we are not far from the critical point, i.e. for small value of the parameter $\eta - 2\mu$. It can be shown that a solution of the type (8.88) exists for any value of μ and η, with the only complication that the equations for a and b become transcendental, and thus analytically insolvable (Problem 7). This would be too simple, though. In fact, we can easily find a stable solution in the limit $\eta \to \infty$, with μ kept finite, and see that it differs completely from (8.88). The expression within the sum in (8.84) becomes the sign function of the difference $F_i - F_j$, and the set of equations for the stationary state becomes

$$\mu F_i = \frac{1}{N} \sum_{j=0}^{N} \text{sign}(F_i - F_j). \tag{8.91}$$

If we reorder the agents so that their strengths make an increasing sequence $F_1 < F_2 < \cdots < F_N$, we get

$$F_i = \frac{1}{\mu N}(2i - N - 1), \tag{8.92}$$

a society organised as a regular ladder of ranks. The order parameter corresponding to such a state,

$$\sigma = \frac{1}{2}\sqrt{\frac{1}{3}\left(1 - \frac{1}{N^2}\right)} \tag{8.93}$$

approaches a non-zero limit for $N \to \infty$, contrary to the configuration (8.88), which is appropriate only in a close vicinity of the critical point and whose order parameter decreases as $N^{-1/2}$, which can be seen from (8.90).

Despite the obvious criticism that the essential mirror symmetry $F_i \to -F_i$ is unrealistic and leads to unacceptable solutions, the Bonabeau model quite well manifests the basic idea of how the various social classes emerge. Within this framework, the hierarchies are due to dynamic instabilities of the uniform state, and we believe that the same general mechanism is (at least partially) also responsible for the stratification of human society. Any departure from the mean, be it positive or negative, is amplified. Any accidental misfortune sends you almost invariably even deeper. That is the whole mystery.

8.3.2   Dynamics of social classes

It is also possible to break explicitly the non-realistic mirror symmetry of the Bonabeau model and allow only positive integer strengths, with a reflecting wall at $F_i = 0$ [1716, 1717]. The non-trivial features of the model stem from the interplay between advancement in pairwise competitions and decline due to inactivity. Thus, two processes are at work in the ensemble of N agents. With probability $1/(1 + r)$ we choose a pair of agents i and j at random and let them compete. If $F_i(t) \geq F_j(t)$, then

$$\begin{aligned} F_i(t + 1/N) &= F_i(t) + 1 \\ F_j(t + 1/N) &= F_j(t) \end{aligned} \tag{8.94}$$

and vice versa. This means that the stronger agent is strengthened, while the weaker one remains at its current position. With a complementary probability $r/(1 + r)$ we randomly choose an agent i and decrease its strength by one,

$$F_i(t + 1/N) = F_i(t) - 1, \tag{8.95}$$

provided that it is positive. If $F_i(t) = 0$, nothing happens. This way we implement a reflecting wall, prohibiting negative strengths.

Let us denote by $N_k = \sum_{i=1}^{N} \delta_{k,F_i}$ the number of agents with strength k and by $n_k = N_k/N$ their fraction among all the agents. In the limit of large N we can make the time continuous and write a master equation for the cumulative fraction $g_k = \sum_{l=0}^{k} n_l$,

$$\frac{\mathrm{d}}{\mathrm{d}t} g_k(t) = r\big(g_{k+1}(t) - g_k(t)\big) + \big(g_{k-1}(t) - g_k(t)\big) g_k(t). \tag{8.96}$$

The presence of the reflecting wall is implemented as a boundary condition $g_{-1}(t) = 0$, and an initial condition is chosen so that all agents have zero strength, so that $g_k(0) = 1$ for all $k \geq 0$.

We can safely suppose that the values of strength will not typically be small numbers, so we can make a continuum limit in Eq. (8.96) and obtain a partial differential equation

$$\frac{\partial g(k,t)}{\partial t} = (r - g(k,t)) \frac{\partial g(k,t)}{\partial k}. \tag{8.97}$$

For large times we must distinguish between two cases. The strengths of some agents remain finite when $t \to \infty$, while strengths of the others can grow linearly with t. We shall call the former 'lower class' while the latter will be referred to as 'middle class'. We can restrict the treatment to the lower class by requiring that k is finite in the stationary solution of (8.97). It is only possible if $g(k,\infty) = r$ or $\frac{\partial g(k,\infty)}{\partial k} = 0$. Because $g(k,t) \leq 1$ by definition, we conclude that the solution is $g(k,\infty) = \min(1,r)$. This means that the lower class comprises all agents for $r \geq 1$, while for $r < 1$ only the fraction r belongs to the lower class, and the rest consists of agents whose strengths diverge with increasing time. To grasp the behaviour of these middle-class agents, we assume that Eq. (8.97) has a scaling solution in the form $g(k,t) = h(k/t)$. The equation for the scaling function $h(x)$ is obtained by substitution into Eq. (8.97), which results in a very simple equation

$$x h'(x) = (h(x) - r) h'(x). \tag{8.98}$$

This is satisfied if either $h(x) = x + r$ or $h'(x) = 0$, hence

$$h(x) = \min(x + r, 1). \tag{8.99}$$

This is a very interesting result. The finite jump at $x = 0$ justifies the distinction between lower and middle classes. Indeed, the height of the jump is equal to the fraction of the lower class, while the rest of the population has strengths linearly increasing with time. As we have already seen, for $r < 1$ both the lower and middle classes comprise finite fractions, and we can speak of a hierarchical society. Besides the poor, who do not perceive any change in their status, there is an entire hierarchy of different linearly growing strengths in the middle class. Interestingly, the occupation of each hierarchical level is the same, as can be immediately seen from the linear dependence of the scaling function on x. On the other hand, we observe a lower-class society for $r > 1$. There is no growth, and all population stagnates. The crucial parameter which determines the fate of the people is the decay rate of the social status r. If every victory is very soon neutralised by unfavourable overall conditions, the middle class never emerges. This mimics the phenomenon of pauperisation.

8.4   What remains

Some readers will probably miss certain important older approaches which we decided to skip—not because we do not recognise their value, but because we feel they follow a slightly different way of thinking. Most importantly, they include many sociophysical applications based on synergetics [15]. There are also innumerable variants of the Ising model applied to the human society [1718–1720]. The analogy is so appealing, and it is so easy to explain the idea even to a layman, that now (but not in the early times of sociophysics) the use of the Ising model in social modelling is perhaps too widespread.

We do not have enough space to speak of modelling via active Brownian particles [1721, 1722], but enthusiasts will certainly be satisfied by an excellent monograph written by Frank Schweitzer [35].

The prisoner's dilemma game attracts ever-growing attention, but there are also other games of similar type, showing yet different dilemmas. One of them is the snowdrift game [1723–1725]. The emergence of cooperation was also studied within different frameworks [1726], so it would be short-sighted to limit our attention to only the prisoner's dilemma game.

Much of the physics of models investigated in this chapter is dominated by absorbing states. More general treatment of absorbing states can be found in [1727–1730].

The voter model is not the only possible way to treat the mechanisms of elections. There are alternative approaches, backed also by empirical data [1731–1734]. Many extensions of the Sznajd model are possible. For example, we may put the agents on various kinds of networks [1735–1744] or introduce long-range influence [1745–1747]. Network effects were also studied in the majority-rule model [1748–1753]. There is also a so-called Ochrombel simplification of the Sznajd model [1754] which brings it back to the linear voter model regime. Sznajd model has also been tried as a model for stock-market dynamics [1755, 1756]; people have asked what happens if we include advertising effects in it [1757–1759]; and on a more formal level, implications of parallel updating was studied [1760, 1761], as well as various other modifications [1762–1785]. An interesting ramification was a model of a network with evolving topology according to Sznajd-like rules [1786, 1787]. The implications of the absence of detailed balance was investigated in Ref. [1788].

Network effects have also been thoroughly investigated [1675] in the Axelrod model, as well as the role of finite temperature (i.e. noise) [1674, 1677, 1678], influence of mass media [1789, 1790], and other variations [1791–1796]. Other exact results on Axelrod model can be found in [1797].

As for the bounded confidence models, various modifications of Deffuant et al and Hegselmann-Krause models were investigated. For example, a model which interpolates between Deffuant et al and Hegselmann-Krause was introduced [1798]. Heterogeneous confidence thresholds [1699, 1799], influence of extremists [1800, 1801], and presence of a 'true truth' [1697, 1802] were studied. Introduction of multi-dimensional opinion space [1696, 1701, 1704, 1803] is also a natural generalisation. Interestingly, introduction of noise into the dynamics alters the behaviour profoundly [1804]. This might be interpreted so that Hegselmann-Krause and Deffuant et al models follow a strictly zero-temperature dynamics, which is unstable with respect to noise. For further information we refer the reader to papers [1704, 1800, 1805–1821]. For a comparison of the models of Hegselmann-Krause and Deffuant et al with the Sznajd model, see [1694, 1695, 1822].

Instead of adding more and more items into the list of achievements of sociophysics, let us conclude our treatise with a few words about a work that has made a really wide impact. This work is Garrett Hardin's classical essay *The Tragedy of the Commons* [1823].

The article can be (and has been) read from many perspectives, but roughly it can be said that Hardin questions the fundamentals of the prevailing neo-liberal doctrine rooted in Adam Smith's idea of an 'invisible hand' of the market. He is mainly concerned with the overpopulation of the planet and the consequences of free access of too many people to resources which belong to 'nobody', as were open pastures in a not very distant past. Letting people self-organise in the trivial meaning of the word—argues Hardin—has disastrous consequences: pollution in the third world, extinction of whales, etc. All these examples show that the invisible hand has failed. To quote: 'Freedom in a commons brings ruin to all.' This is the tragedy of the commons. At this point, humankind must consider what the word 'freedom' really means. If it brings ruin to all, should we discard it? Or should we rather make its meaning clearer? For example: does it imply the freedom to pick somebody's pocket? Or does it mean the freedom to rationally maximise profit by cutting down tropical rainforests? Or does it include the freedom to start a family and to bring up children? Or to have free access to water? The answers are not as simple as one might naively think.

Faced with new problems, humankind must agree on creating new structures, previously unknown. In order to promote the public interest, it is necessary to institute certain social arrangements, such as issuing a law against robbers, who act as if a bank was a commons, and organising an armed squadron to persecute the perpetrators. The robber is not caught by an invisible hand of the market, but by a fairly visible police officer. It does not mean that self-organisation is futile; it means it must be accomplished on a more sophisticated level.

The overall moral of the tragedy of the commons is that a society grows increasingly complex as it faces increasingly difficult problems. Many of them can be described in terms of a commons, i.e. a place void of any social structure. For the survival of humankind it is absolutely necessary to fill the void by a functional *dispositif*, as the French would say. It is the mission of scientists, sociophysicists included, to provide considered professional judgement on the available technical solutions and warn us when the technical solution is impossible for profound, rationally justified reasons. Science must provide the knowledge, and at the same time say where the knowledge has its principal limits. Science must also resist the eternal temptation to confuse knowledge with power. The rest is politics. And in politics all people have an equal right to speak and to take part in decisions, whether they are scientists or not.

We think that this idea may be the summary not only of this chapter, but of the entire book.

Problems

1. In the spatial prisoner's dilemma game, establish the range of the parameter b in which the following cyclic attractors are stable.

   a) The 'blinker', periodically exchanging a single defector surrounded by all cooperators, with 3×3 square of defectors surrounded by cooperators. Look right for a pictorial representation.

   b) The cycle of period 3, going as shown in the sequence of pictures on the right.

2. When we define the voter model on heterogeneous graphs, we distinguish between site-initiated and bond-initiated dynamics. In the former case we first choose agent 1 and then choose agent 2 among neighbours of 1. In the latter version we randomly choose a bond and then decide, again randomly, which of the agents attached at the end of the bond is 1 and which is 2. In both cases we complete the update by taking the state of 2 as a new state of 1, as usual. Show that the average magnetisation is conserved in the bond-initiated dynamics, while in the site-initiated one the conservation breaks down.

3. Analyse more general versions of the Galam model.
   a) Derive the formula for concentration of opinion A at level l, supposing that the size of the groups is a general number which may even depend on l. What if the society is divided into groups of many different sizes? See [1640] if obscure.
   b) Assume the agents choose from among three possible choices A, B, and C [1649]. With groups of size 3 we resolve the situation in which all three members have different opinions introducing probabilities p_A, p_B, $p_C = 1 - p_A - p_B$ that the group adopts conclusion A, B, or C, respectively. What is the final state in different situations?

4. The majority rule can be combined with the minority rule into one model [1654], where each triple as a whole adopts the majority opinion with a probability of p and the minority opinion with a probability of $1 - p$. For which value of p do the evolution equations for average magnetisation and two-site correlation function not contain higher-order correlations, thus enabling exact analytical solution? With the experience from the voter model, try to find as explicit a solution as you can.

5. For the one-dimensional Sznajd model, perform explicitly the calculations leading from the transition rates (8.51) through the Kirkwood approximation to the formula (8.52).

6. Consider a variant of the Axelrod model. In each step, choose a pair of neighbouring agents at sites x and y, and a pair of features f and g (we admit $f = g$). If the agents agree in the first feature, then the first agent copies the value of the second feature from the second agent, i.e. if $\sigma_t(x, f) = \sigma_t(y, f)$, then $\sigma_{t+1/N}(x, g) = \sigma_t(y, g)$. Find the transition rates analogous to (8.61) and write

down the equation for the correlation function $R_{f,a}(x-y,t) = \langle \delta_{\sigma_t(x,f),a} \, \delta_{\sigma_t(y,f),a} \rangle$. Try to solve it using the Kirkwood approximation.

7. Show that for the Bonabeau model in the mean-field approximation, a stationary solution always exists in the form (8.88). Discuss the stability of this solution.

8. Examine a model of people aggregating in political parties [1824]. There are N agents, each of them with an opinion quantified by an integer number. Denote by $N_k(t)$ the number of agents with opinion k at time t. In each step of the dynamics, perform one of the following two alternatives. With a probability of $r \in [0,1]$ choose two agents at random, and if the difference in their opinions is exactly 2, they reach consensus at their average opinion, i.e. $(k+1, k-1) \to (k,k)$. If the difference is not 2, nothing happens. With a complementary probability of $1-r$, choose instead one agent and change her opinion by $+1$ or -1, i.e. $k \to k \pm 1$.

Write the equation for the evolution of the average number of agents in each opinion group $\langle N_k(t) \rangle$. Show that it allows for a uniform stationary solution and, using the linear stability analysis, show at which value of r the uniform solution breaks into separate peaks (if you like, these peaks can be interpreted as political parties).

References

[1] A. Comte, *Plan des travaux scientifiques nécessaires pour réorganiser la société* (1822).
[2] A. Comte, *Cours de philosophie positive*, tome IV, 46e leçon (Bachelier, Paris, 1839).
[3] A. Quételet, *Du système social et des lois qui le régissent* (Guillaumin, Paris, 1848).
[4] F. A. Hayek, *The Counter-revolution of Science: Studies on the Abuse of Reason* (Free Press, Glencoe, 1952).
[5] P. Coveney and R. Highfield, *Frontiers of Complexity* (Fawcett Columbine, New York, 1995).
[6] R. M. Robertson, Journal of Political Economy **57**, 523 (1949).
[7] D. Štys and M. Vlačihová, arXiv:1007.0472.
[8] G. von Buquoy, *Die Theorie der Nationalwirthschaft nach einem neuen Plane und nach mehrern eigenen Ansichten dargestellt* (Breitkopf und Härtel, Leipzig, 1815).
[9] E. Majorana, Scientia **36**, 58 (1942).
[10] R. N. Mantegna, Quant. Finance **5**, 133 (2005).
[11] B. B. Mandelbrot, *The Fractal Geometry of Nature* (W. H. Freeman, New York, 1983).
[12] N. A. Chigier and E. A. Stern (Eds.), *Collective Phenomena and the Applications of Physics to Other Fields of Science* (Brain Research Publications, Fayetteville, 1975).
[13] E. Callen and D. Shapero, Phys. Today **27**, 23 (July 1974).
[14] H. Haken, *Synergetics: an Introduction* (Springer, Berlin, 1977).
[15] W. Weidlich, Physics Reports **204**, 1 (1991).
[16] P. W. Anderson, K. J. Arrow, and D. Pines (Eds.), *The Economy as an Evolving Complex System* (Addison Wesley, Reading, 1988).
[17] R. N. Mantegna, Physica A **179**, 232 (1991).
[18] J.-P. Bouchaud and M. Potters, *Théorie des risques financiers* (Aléa, Saclay, 1997).
[19] J.-P. Bouchaud and M. Potters, *Theory of Financial Risk and Derivative Pricing* (Cambridge University Press, Cambridge, 2003).
[20] F. Slanina, in: *Encyclopedia of Complexity and Systems Science*, p. 8379 (Springer, New York, 2009).
[21] F. Slanina, in: *Order, disorder, and criticality, vol. 3*, ed. Y. Holovatch, p. 201 (World Scientific, Singapore, 2012).
[22] N. F. Johnson, P. Jefferies, and P. M. Hui, *Financial Market Complexity* (Oxford University Press, Oxford, 2003).
[23] M. Levy, H. Levy, and S. Solomon, *Microscopic Simulation of Financial Markets* (Academic Press, San Diego, 2000).
[24] R. N. Mantegna and H. E. Stanley, *Introduction to Econophysics: Correlations and Complexity in Finance* (Cambridge University Press, Cambridge, 2000).
[25] B. M. Roehner, *Patterns of Speculation: A Study in Observational Econophysics* (Cambridge University Press, Cambridge, 2002).
[26] B. M. Roehner, *Driving Forces in Physical, Biological and Socio-economic Phenomena: A Network Science Investigation of Social Bonds and Interactions* (Cambridge University Press, Cambridge, 2007).
[27] B. M. Roehner, *Hidden Collective Factors in Speculative Trading: A Study in Analytical Economics* (Springer, Berlin, 2009).
[28] B. M. Roehner and T. Syme, *Pattern and Repertoire in History* (Harvard University Press, Cambridge, 2002).
[29] M. Schulz, *Statistical Physics and Economics* (Springer-Verlag, New York, 2003).
[30] S. Sinha, A. Chatterjee, A. Chakraborti, and B. K. Chakrabarti, *Econophysics: An Introduction* (Wiley-VCH, Weinheim, 2011).
[31] D. Sornette, *Why Stock Markets Crash: Critical Events in Complex Financial Systems* (Princeton University Press, Princeton, 2003).
[32] J. Voit, *The Statistical Mechanics of Financial Markets* (Springer, Berlin, 2003).
[33] D. Challet, M. Marsili, and Y.-C. Zhang, *Minority Games* (Oxford University Press, Oxford, 2005).

[34] A. C. C. Coolen, *The Mathematical Theory of Minority Games* (Oxford University Press, Oxford, 2005).

[35] F. Schweitzer, *Brownian Agents and Active Particles* (Springer, Berlin, 2003).

[36] J. Šesták, *Science of Heat and Thermophysical Studies: A Generalized Approach to Thermal Analysis* (Elsevier, Amsterdam, 2005).

[37] L. T. Wille (Ed.), *New Directions in Statistical Physics. Econophysics, Bioinformatics, and Pattern Recognition* (Springer, Berlin, 2004).

[38] M. F. M. Osborne, *The Stock Market and Finance from a Physicist's Viewpoint* (Published by the author, Temple Hills, 1977).

[39] J. Kertész and I. Kondor (Eds.), *Econophysics: An Emergent Science. Proceedings of the 1st Workshop on Econophysics, Budapest, 1997*, http://newton.phy.bme.hu/~kullmann/Egyetem/konyv.html (Budapest, 2002).

[40] H. Takayasu (Ed.), *Empirical Science of Financial Fluctuations: The Advent of Econophysics* (Springer, Berlin, 2002).

[41] H. Takayasu (Ed.), *Application of Econophysics. Proceedings of the Second Nikkei Econophysics Symposium* (Springer, Berlin, 2004).

[42] H. Takayasu (Ed.), *Practical Fruits of Econophysics. Proceedings of The Third Nikkei Econophysics Symposium* (Springer, Berlin, 2006).

[43] A. Chatterjee, S. Yarlagadda, and B. K. Chakrabarti (Eds.), *Econophysics of Wealth Distributions. Proceedings of the Econophys-Kolkata I* (Springer, Milan, 2005).

[44] A. Chatterjee and B. K. Chakrabarti (Eds.), *Econophysics of Stock and other Markets. Proceedings of the Econophys-Kolkata II* (Springer, Milan, 2006).

[45] A. Chatterjee and B. K. Chakrabarti (Eds.), *Econophysics of Markets and Business Networks. Proceedings of the Econophys-Kolkata III* (Springer, Milan, 2007).

[46] B. Basu, B. K. Chakrabarti, S. R. Chakravarty, and K. Gangopadhyay, *Econophysics and Economics of Games, Social Choices and Quantitative Techniques* (Springer, Milan, 2010).

[47] F. Abergel, B. K. Chakrabarti, A. Chakraborti, and M. Mitra (Eds.), *Econophysics of Order-driven Markets* (Springer, Milan, 2011).

[48] W. B. Arthur, S. N. Durlauf, and D. A. Lane (Eds.), *The Economy as an Evolving Complex System II* (Perseus, Reading, 1997).

[49] L. Blume and S. Durlauf (Eds.), *The Economy as an Evolving Complex System III* (Oxford University Press, Oxford, 2005).

[50] A. Kirman and J.-B. Zimmermann (Eds.), *Economics with Heterogeneous Interacting Agents* (Springer, Berlin, 2001).

[51] M. Gallegati, A. P. Kirman, and M. Marsili (Eds.), *The Complex Dynamics of Economic Interaction* (Springer, Berlin, 2004).

[52] M. Salzano and A. Kirman (Eds.), *Economics: Complex Windows* (Springer, Milan, 2005).

[53] T. Lux, S. Reitz, and E. Samanidou (Eds.), *Nonlinear Dynamics and Heterogeneous Interacting Agents* (Springer, Berlin, 2005).

[54] B. K. Chakrabarti, A. Chakraborti, and A. Chatterjee (Eds.), *Econophysics and Sociophysics: Trends and Perspectives* (Wiley-VCH, Weinheim, 2006).

[55] A. Namatame, T. Kaizoji, and Y. Aruka (Eds.), *The Complex Networks of Economic Interaction* (Springer, Berlin, 2006).

[56] M. Faggini and T. Lux (Eds.), *Coping with the Complexity of Economics* (Springer, Milan, 2009).

[57] M. Takayasu, T. Watanabe, and H. Takayasu (Eds.), *Econophysics Approaches to Large-Scale Business Data and Financial Crisis. Proceedings of the Tokyo Tech-Hitotsubashi Interdisciplinary Conference + APFA7* (Springer, Tokyo, 2010).

[58] J.-P. Bouchaud, M. Marsili, B. M. Roehner, and F. Slanina (Eds.), Physica A **299**, Issues 1–2 (2001).

[59] F. Schweitzer, S. Battiston, and C. J. Tessone (Eds.), Eur. Phys. J. B **71**, Issue 4 (2009).

[60] T. Di Matteo and T. Aste (Eds.), Eur. Phys. J. B **55**, Issue 2 (2007).

[61] J. A. Hołyst and M. A. Nowak (Eds.), Physica A **344**, Issues 1–2 (2004).

[62] A. Carbone, G. Kaniadakis, and A. M. Scarfone (Eds.), Physica A **382**, Issue 1 (2007).

[63] T. Kaizoji, A. Namatame, and E. Scalas (Eds.), Physica A **383**, Issue 1 (2007).

[64] D. A. Mendes, O. Gomes, and R. Menezes (Eds.), Physica A **387**, Issue 15 (2008).

[65] J. D. Farmer and T. Lux (Eds.), Journal of Economic Dynamics and Control **32**, Issue 1 (2008).

[66] Acta Physica Polonica B **36**, Issue 8 (2005).

[67] M. Takayasu, T. Watanabe, Y. Ikeda, and H. Takayasu (Eds.), J. Phys.: Conf. Ser. **221**, Number 1 (2010).

[68] S. Alfarano, T. Lux, and M. Milaković (Eds.), Eur. Phys. J. B **73**, Issue 1 (2010).

[69] T. Preis, Eur. Phys. J. Special Topics **194**, (2011).

[70] M. Ausloos, in: *Econophysics and Sociophysics: Trends and Perspectives*, ed. B. K. Chakrabarti, A. Chakraborti, and A. Chatterjee, p. 249 (Wiley-VCH, Weinheim, 2006).

[71] A. Chakraborti, I. Muni Toke, M. Patriarca, and F. Abergel, arXiv:0909.1974.

[72] A. Chakraborti, I. Muni Toke, M. Patriarca, and F. Abergel, Quant. Finance **11**, 991 (2011).

[73] A. Chakraborti, I. Muni Toke, M. Patriarca, and F. Abergel, Quant. Finance **11**, 1013 (2011).

[74] A. Chakraborti, M. Patriarca, and M. S. Santhanam, arXiv:0704.1738.

[75] Y.-C. Zhang, Europhysics News **29**, 51 (1998).

[76] M. Ausloos, Europhysics News **29**, 70 (1998).

[77] S. Solomon and M. Levy, Quant. Finance **3**, C12 (2003).

[78] M. Gallegati, S. Keen, T. Lux, and P. Ormerod, Physica A **370**, 1 (2006).

[79] J. L. McCauley, Physica A **371**, 601 (2006).

[80] B. M. Roehner, arXiv:1004.3229.

[81] C. Schinckus, Physica A **388**, 4415 (2009).

[82] C. Schinckus, Physica A **389**, 3814 (2010).

[83] C. Schinckus, Am. J. Phys. **78**, 325 (2010).

[84] R. Cont, IEEE Signal Processing Magazine **28**, 16 (2011).

[85] M. Brooks, New Scientist p. 12 (18 October 1997).

[86] M. Stix, Sci. Am. p. 70 (May 1998).

[87] J. D. Farmer, Computing in Science and Engineering **1**, 26 (1999).

[88] M. Buchanan, New Scientist, p. 22 (19 August 2000).

[89] P. Ball, Nature **415**, 371 (2002).

[90] J. D. Farmer, M. Shubik, and E. Smith, Phys. Today **58**, 37 (2005).

[91] P. Ball, Nature **441**, 686 (2006).

[92] J. D. Farmer and D. Foley, Nature **460**, 685 (2009).

[93] L. Neal, *The Rise of Financial Capitalism* (Cambridge University Press, Cambridge, 1990).

[94] L. Neal, in: *Business and Economic History, Second Series, Vol. 11*, ed. J. Atack, p. 81 (University of Illinois, 1982).

[95] E. Stringham, Journal of Private Enterprise **17**, 1 (2002).

[96] D. Defoe, *The Anatomy of Exchange Alley* (Smith, London, 1719).

[97] R. Dale, *The First Crash* (Princeton University Press, Princeton, 2004).

[98] C. MacKay, *Memoirs of Extraordinary Popular Delusions and the Madness of Crowds* (Bentley, London, 1841).

[99] J. Carswell, *The South Sea Bubble* (Stanford University Press, Stanford, 1960).

[100] C. M. Reinhart and K. Rogoff, *This Time Is Different: Eight Centuries of Financial Folly* (Princeton University Press, Princeton, 2009).

[101] E. F. Fama, Journal of Finance **25**, 383 (1970).

[102] L. Bachelier, Annales Scientifiques de l'École Normale Supérieure, 3e série **17**, 21 (1900).

[103] J. Regnault, *Calcul des chances et philosophie de la bourse* (Mallet-Bachelier et Castel, Paris, 1863).

[104] F. Jovanovic, in: *Pioneers of Financial Economics, Vol. 1*, ed. G. Poitras, p. 191 (Edward Elgar, Cheltenham, 2006).

[105] E. F. Fama, Journal of Business **38**, 34 (1965).

[106] A. Cowles, Econometrica **1**, 309 (1933).

[107] D. Applebaum, *Lévy Processes and Stochastic Calculus*, (Cambridge University Press, Cambridge, 2004).

[108] P. A. Samuelson, Industrial Management Review **6**, 41 (1965).

[109] J. L. McCauley, K. E. Bassler, and G. H. Gunaratne, Physica A **387**, 202 (2008).

[110] R. C. Merton, *Continuous-time Finance* (Blackwell Publishing, Cambridge, 1990).

[111] M. F. M. Osborne, Operations Research **7**, 145 (1959).

[112] M. G. Kendall, J. Roy. Stat. Soc. A **116**, 11 (1953).

[113] J. Laherrère and D. Sornette, Eur. Phys. J. B **2**, 525 (1998).

[114] H. E. Stanley, L. A. N. Amaral, S. V. Buldyrev, P. Gopikrishnan, V. Plerou, and M. A. Salinger, Proc. Natl. Acad. Sci. USA **99**, 2561 (2002).

[115] A. Clauset, C. R. Shalizi, and M. E. J. Newman, SIAM Rev. **51**, 661 (2009).

[116] B. Mandelbrot, Journal of Business **36**, 394 (1963).

[117] V. Pareto, *Cours d'économie politique* (Lausanne, F. Rouge, 1897).

[118] G. K. Zipf, *Human Behavior and the Principle of Least Effort*, (Addison-Wesley, Cambridge, 1949).

[119] J. Korčák, Bulletin de l'Institut International de Statistique **3**, 295 (1938).

[120] E. F. Fama, Journal of Business **36**, 420 (1963).

[121] B. B. Mandelbrot, Physica A **263**, 477 (1999).

[122] A. W. Lo, Econometrica **59**, 1279 (1991).

[123] J. Y. Campbell, A. W. Lo, and A. C. MacKinlay, *The Econometrics of Financial Markets* (Princeton University Press, Princeton, 1996).

[124] A. W. Lo and A. C. MacKinlay, *A Non-random Walk down Wall Street* (Princeton University Press, Princeton, 2001).

[125] S. J. Grossman and J. E. Stiglitz, The American Economic Review **70**, 393 (1980).

[126] G. J. Stigler, Journal of Political Economy **69**, 213 (1961).

[127] G. A. Akerlof, Quarterly Journal of Economics **84**, 488 (1970).

[128] T. E. Copeland and D. Friedman, Journal of Finance **46**, 265 (1991).

[129] Y.-C. Zhang, Physica A **269**, 30 (1999).

[130] J. L. McCauley, in: *Complexity from Microscopic to Macroscopic Scales: Coherence and Large Deviations*, ed. A. T. Skjeltorp and T. Vicsek, p. 181 (Kluwer, Dordrecht, 2002).

[131] R. N. Mantegna and H. E. Stanley, Nature **376**, 46 (1995).

[132] P. Gopikrishnan, M. Meyer, L. A. N. Amaral, and H. E. Stanley, Eur. Phys. J. B **3**, 139 (1998).

[133] P. Gopikrishnan, V. Plerou, L. A. N. Amaral, M. Meyer, and H. E. Stanley, Phys. Rev. E **60**, 5305 (1999).

[134] V. Plerou, P. Gopikrishnan, L. A. N. Amaral, M. Meyer, and H. E. Stanley, Phys. Rev. E **60**, 6519 (1999).

[135] J. Rotyis and G. Vattay, in: *Econophysics: An Emergent Science. Proceedings of the 1st Workshop on Econophysics, Budapest, 1997*, ed. J. Kertész and I. Kondor, http://newton.phy.bme.hu/~kullmann/Egyetem/konyv.html, p. 65 (Budapest, 2002).

[136] S. Galluccio, G. Caldarelli, M. Marsili, and Y.-C. Zhang, Physica A **245**, 423 (1997).

[137] N. Vandewalle, M. Ausloos, and P. Boveroux, in: *Econophysics: An Emergent Science. Proceedings of the 1st Workshop on Econophysics, Budapest, 1997*, ed. J. Kertész and I. Kondor, http://newton.phy.bme.hu/~kullmann/Egyetem/konyv.html, p. 36 (Budapest, 2002).

[138] R. Cont, in: *Econophysics: An Emergent Science. Proceedings of the 1st Workshop on Econophysics, Budapest, 1997*, ed. J. Kertész and I. Kondor, http://newton.phy.bme.hu/~kullmann/Egyetem/konyv.html, p. 1 (Budapest, 2002).

[139] L. Kullmann, J. Töyli, J. Kertész, A. Kanto, and K. Kaski, Physica A **269**, 98 (1999).

[140] P. Harrison, Journal of Business **71**, 55 (1998).

[141] R. Cont, arXiv:cond-mat/9705075.

[142] Z. Palágyi and R. N. Mantegna, Physica A **269**, 132 (1999).

[143] I. M. Jánosi, B. Janecskó, and I. Kondor, Physica A **269**, 111 (1999).

[144] Z.-F. Huang, Physica A **287**, 405 (2000).

[145] S. V. Muniandy, S. C. Lim, and R. Murugan, Physica A **301**, 407 (2001).

[146] A. Ganchuk, V. Derbentsev, and V. Soloviev, arXiv:physics/0608009.

[147] P. Norouzzadeh and G. R. Jafari, Physica A **356**, 609 (2005).

[148] K. Kim, S.-M. Yoon, S. Kim, K.-H. Chang, Y. Kim, and S. H. Kang, Physica A **376**, 525 (2007).

[149] J.-S. Yang, S. Chae, W.-S. Jung, and H.-T. Moon, Physica A **363**, 377 (2006).

[150] K. Kim, S.-M. Yoon, J.-S. Choi, and H. Takayasu, J. Korean Phys. Soc. **44**, 647 (2004).

[151] N. Sazuka, T. Ohira, K. Marumo, T. Shimizu, M. Takayasu, and H. Takayasu, Physica A **324**, 366 (2003).

[152] T. Mizuno, S. Kurihara, M. Takayasu, and H. Takayasu, Physica A **324**, 296 (2003).

[153] J. J.-L. Ting, Physica A **324**, 285 (2003).

[154] K. E. Lee and J. W. Lee, J. Korean Phys. Soc. **44**, 668 (2004).

[155] H. Situngkir and Y. Surya, arXiv:cond-mat/0403465.

[156] H. F. Coronel-Brizio and A. R. Hernandez-Montoya, Revista Mexicana de Fisica **51**, 27 (2005).

[157] M. Kozłowska and R. Kutner, Acta Physica Polonica B **37**, 3027 (2006).

[158] R. K. Pan and S. Sinha, Europhys. Lett. **77**, 58004 (2007).

[159] R. K. Pan and S. Sinha, Physica A **387**, 2055 (2008).

[160] A.Sharkasi, H. J. Ruskin, and M. Crane, arXiv:physics/0607182.

[161] I. M. Dremin and A. V. Leonidov, arXiv:physics/0612170.

[162] G. R. Jafari, M. S. Movahed, P. Noroozzadeh, A. Bahraminasab, M. Sahimi, F. Ghasemi, and M. R. R. Tabar, Int. J. Mod. Phys. C **18**, 1689 (2007).
[163] A. Rasoolizadeh and R. Solgi, arXiv:cond-mat/0410289.
[164] T. Di Matteo, T. Aste, and M. M. Dacorogna, Physica A **324**, 183 (2003).
[165] W. K. Bertram, Physica A **341**, 533 (2004).
[166] R. L. Costa and G. L. Vasconcelos, Physica A **329**, 231 (2003).
[167] Y. Malevergne, V. Pisarenko, and D. Sornette, arXiv:physics/0305089.
[168] R. N. Mantegna and H. E. Stanley, Phys. Rev. E **73**, 2946 (1994).
[169] R. N. Mantegna and H. E. Stanley, J. Stat. Phys. **89**, 469 (1997).
[170] D. Sornette and R. Cont, J. Phys I France **7**, 431 (1997).
[171] A. Matacz, arXiv:cond-mat/9710197.
[172] R. N. Mantegna and H. E. Stanley, Physica A **254**, 77 (1998).
[173] S. Solomon, in: *Decision Technologies for Computational Finance* ed. A.-P. Refenes, A. N. Burgess, and J. E. Moody (Kluwer Academic Publishers, 1998).
[174] B. Podobnik, P. C. Ivanov, Y. Lee, and H. E. Stanley, Europhys. Lett. **52**, 491 (2000).
[175] P. C. Ivanov, B. Podobnik, Y. Lee, and H. E. Stanley, Physica A **299**, 154 (2001).
[176] A. C. Silva, R. E. Prange, and V. M. Yakovenko, Physica A **344**, 227 (2004).
[177] H. Kleinert and X. J. Chen, arXiv:physics/0609209.
[178] Y. Liu, P. Gopikrishnan, P. Cizeau, M. Meyer, C.-K. Peng, and H. E. Stanley, Phys. Rev. E **60**, 1390 (1999).
[179] Y. Liu, P. Cizeau, M. Meyer, C.-K. Peng, and H. E. Stanley, Physica A **245**, 437 (1997).
[180] P. Cizeau, Y. Liu, M. Meyer, C.-K. Peng, and H. E. Stanley, Physica A **245**, 441 (1997).
[181] Y. Liu, P. Cizeau, P. Gopikrishnan, M. Meyer, C.-K. Peng, and H. E. Stanley, in: *Econophysics: An Emergent Science. Proceedings of the 1st Workshop on Econophysics, Budapest, 1997*, ed. J. Kertész and I. Kondor, http://newton.phy.bme.hu/~kullmann/Egyetem/konyv.html, p. 50 (Budapest, 2002).
[182] S. Micciché, G. Bonanno, F. Lillo, and R. N. Mantegna, Physica A **314**, 756 (2002).
[183] F. Wang, K. Yamasaki, S. Havlin, and H. E. Stanley, Phys. Rev. E **73**, 026117 (2006).
[184] F. Wang, P. Weber, K. Yamasaki, S. Havlin, and H. E. Stanley, Eur. Phys. J. B **55**, 123 (2007).
[185] I. Vodenska-Chitkushev, F. Z. Wang, P. Weber, K. Yamasaki, S. Havlin, and H. E. Stanley, Eur. Phys. J. B **61**, 217 (2008).
[186] F. Lillo and R. N. Mantegna, Phys. Rev. E **62**, 6126 (2000).
[187] F. Lillo and R. N. Mantegna, Eur. Phys. J. B **15**, 603 (2000).
[188] F. Lillo and R. N. Mantegna, Physica A **299**, 161 (2001).
[189] F. Lillo, R. N. Mantegna, J.-P. Bouchaud, and M. Potters, arXiv:cond-mat/0107208.
[190] F. Lillo and R. N. Mantegna, Eur. Phys. J. B **20**, 503 (2001).
[191] F. Lillo, G. Bonanno, and R. N. Mantegna, in: *Empirical Science of Financial Fluctuations: The Advent of Econophysics*, ed. H. Takayasu, p. 77 (Springer, Berlin, 2002).
[192] V. Plerou, P. Gopikrishnan, L. A. N. Amaral, X. Gabaix, and H. E. Stanley, Phys. Rev. E **62**, R3023 (2000).
[193] P. C. Ivanov, A. Yuen, B. Podobnik, and Y. Lee, Phys. Rev. E **69**, 056107 (2004).
[194] J.-P. Bouchaud and A. Georges, Phys. Rep. **195**, 127 (1990).
[195] P. Weber, Phys. Rev. E **75**, 016105 (2007).
[196] M. Raberto, E. Scalas, R. Gorenflo, and F. Mainardi, arXiv:cond-mat/0012497.
[197] F. Mainardi, M. Raberto, R. Gorenflo, and E. Scalas, Physica A **287**, 468 (2000).
[198] M. Raberto, E. Scalas, and F. Mainardi, Physica A **314**, 749 (2002).
[199] L. Sabatelli, S. Keating, J. Dudley, and P. Richmond, Eur. Phys. J. B **27**, 273 (2002).
[200] J. Masoliver, M. Montero, and G. H. Weiss, Phys. Rev. E **67**, 021112 (2003).
[201] P. Repetowicz and P. Richmond, Physica A **343**, 677 (2004).
[202] R. Bartiromo, Phys. Rev. E **69**, 067108 (2004).
[203] E. Scalas, R.Gorenflo, H. Luckock, F. Mainardi, M. Mantelli, and M. Raberto, Quant. Finance **4**, 695 (2004).
[204] N. Sazuka, Physica A **376**, 500 (2007).
[205] N. Sazuka and J. Inoue, Physica A **383**, 49 (2007).
[206] N. Sazuka and J. Inoue, arXiv:physics/0702003.
[207] N. Sazuka, J. Inoue, and E. Scalas, Physica A **388**, 2839 (2009).
[208] J. Inoue and N. Sazuka, Quant. Finance **10**, 121 (2010).
[209] K. Matia, Y. Ashkenazy, and H. E. Stanley, Europhys. Lett **61**, 422 (2003).
[210] N. Vandewalle and M. Ausloos, Physica A **246**, 454 (1997).

[211] N. Vandewalle and M. Ausloos, Eur. Phys. J. B **4**, 257 (1998).
[212] K. Ivanova and M. Ausloos, Eur. Phys. J. B **8**, 665 (1999).
[213] M. Ausloos, N. Vandewalle, P. Boveroux, A. Minguet, and K. Ivanova, Physica A **274**, 229 (1999).
[214] M. Ausloos, N. Vandewalle, and K. Ivanova, in: *Noise, Oscillators and Algebraic Randomness*, ed. M. Planat, p. 156 (Springer Berlin, 2000).
[215] M. Ausloos, Physica A **285**, 48 (2000).
[216] J. W. Kantelhardt, S. A. Zschiegner, E. Koscielny-Bunde, S. Havlin, A. Bunde, and H. E. Stanley, Physica A **316**, 87 (2002).
[217] J.-P. Bouchaud, M. Potters, and M. Meyer, Eur. Phys. J. B **13**, 595 (2000).
[218] Z.-Q. Jiang and W.-X. Zhou, Physica A **387**, 3605 (2008).
[219] J. de Souza and S. M. D. Queirós, arXiv:0711.2550.
[220] L. G. Moyano, J. de Souza, and S. M. D. Queirós, Physica A **371**, 118 (2006).
[221] M. Ausloos and K. Ivanova, Computer Physics Communications **147**, 582 (2002).
[222] M.-E. Brachet, E. Taflin, and M. Tcheou, Chaos, Solitons and Fractals **11**, 2343 (2000).
[223] L. Calvet and A. Fisher, Review of Economics and Statistics **84**, 381 (2002).
[224] H. Katsuragi, Physica A **278**, 275 (2000).
[225] K. Kim, S.-M. Yoon, and J.-S. Choi, J. Korean Phys. Soc. **44**, 643 (2004).
[226] P. Oświęcimka, J. Kwapień, S. Drożdż, and R. Rak, Acta Physica Polonica B **36**, 2447 (2005).
[227] P. Oświęcimka, J. Kwapień, and S. Drożdż, Physica A **347**, 626 (2005).
[228] J. Kwapień, P. Oświęcimka, and S. Drożdż, Physica A **350**, 466 (2005).
[229] P. Oświęcimka, J. Kwapień, S. Drożdż, A.Z. Górski, and R. Rak, Acta Physica Polonica B **37**, 3083 (2006).
[230] M. Pasquini and M. Serva, Economics Letters **65**, 275 (1999).
[231] M. Pasquini and M. Serva, Physica A **269**, 140 (1999).
[232] M. Pasquini and M. Serva, Eur. Phys. J. B **16**, 195 (2000).
[233] R. Baviera, M. Pasquini, M. Serva, D. Vergni, and A. Vulpiani, Physica A **300**, 551 (2001).
[234] R. Baviera, M. Pasquini, M. Serva, D. Vergni, and A. Vulpiani, Eur. Phys. J. B **20**, 473 (2001).
[235] U. L. Fulco, M. L. Lyra, F. Petroni, M. Serva, and G. M. Viswanathan, Int. J. Mod. Phys. B **18**, 681 (2004).
[236] M. Serva, U. L. Fulco, I. M. Gléria, M. L. Lyra, F. Petroni, and G. M. Viswanathan, Physica A **363**, 393 (2006).
[237] Z. Xu and R. Gençay, Physica A **323**, 578 (2003).
[238] A. Bershadskii, J. Phys. A: Math. Gen. **34**, L127 (2001).
[239] A. Turiel and C. J. Pérez-Vicente, Physica A **322**, 629 (2003).
[240] A. Turiel and C. J. Pérez-Vicente, Physica A **355**, 475 (2005).
[241] X. Sun, H. Chen, Y. Yuan, and Z. Wu, Physica A **301**, 473 (2001).
[242] X. Sun, H. Chen, Z. Wu, and Y. Yuan, Physica A **291**, 553 (2001).
[243] A. Z. Górski, S. Drożdż, and J. Speth, Physica A **316**, 496 (2002).
[244] Y. Wei and D. Huang, Physica A **355**, 497 (2005).
[245] Z.-Q. Jiang and W.-X. Zhou, Physica A **381**, 343 (2007).
[246] M. Constantin and S. Das Sarma, Phys. Rev. E **72**, 051106 (2005).
[247] Z. Eisler and J. Kertész, Europhys. Lett. **77**, 28001 (2007).
[248] R. N. Mantegna and H. E. Stanley, Nature **383**, 587 (1996).
[249] R. N. Mantegna and H. E. Stanley, Physica A **239**, 255 (1997).
[250] W. Breymann and S. Ghashghaie, in: *Econophysics: An Emergent Science. Proceedings of the 1st Workshop on Econophysics, Budapest, 1997*, ed. J. Kertész and I. Kondor, http://newton.phy.bme.hu/~kullmann/Egyetem/konyv.html, p. 92 (Budapest, 2002).
[251] J.-P. Bouchaud, A. Matacz, and M. Potters, arXiv:cond-mat/0101120.
[252] Z. Ding, C. W. J. Granger, and R. F. Engle, Journal of Empirical Finance **1**, 83 (1993).
[253] A. Arneodo, J. F. Muzy, and D. Sornette, Eur. Phys. J. B **2**, 277 (1998).
[254] J.-F. Muzy, J. Delour, and E. Bacry, Eur. Phys. J. B **17**, 537 (2000).
[255] J.-F. Muzy, D. Sornette, J. Delour, and A. Arneodo, Quant. Finance **1**, 131 (2001).
[256] J.-P. Bouchaud, A. Matacz, and M. Potters, Phys. Rev. Lett. **87**, 228701 (2001).
[257] J.-P. Bouchaud and M. Potters, Physica A **299**, 60 (2001).
[258] J. Perelló, J. Masoliver, and J.-P. Bouchaud, Applied Mathematical Finance **11**, 27 (2004).
[259] J. Perelló and J. Masoliver, Phys. Rev. E **67**, 037102 (2003).
[260] J. Perelló, J. Masoliver, and N. Anento, Physica A **344**, 134 (2004).
[261] J.-P. Bouchaud, M. Mézard, and M. Potters, Quant. Finance **2**, 251 (2002).

[262] D. Challet and R. Stinchcombe, Physica A **300**, 285 (2001).
[263] F. Lillo, Eur. Phys. J. B **55**, 453 (2007).
[264] S. Maslov and M. Mills, Physica A **299**, 234 (2001).
[265] M. Potters and J.-P. Bouchaud, Physica A **324**, 133 (2003).
[266] I. Zovko and J. D. Farmer, Quant. Finance **2**, 387 (2002).
[267] D. Challet and R. Stinchcombe, Physica A **324**, 141 (2003).
[268] R. Stinchcombe, in: *Econophysics and Sociophysics: Trends and Perspectives*, ed. B. K. Chakrabarti, A. Chakraborti, and A. Chatterjee, p. 35 (Wiley-VCH, Weinheim, 2006).
[269] A. W. Lo, A. C. MacKinlay, and J. Zhang, Journal of Financial Economics **65**, 31 (2002).
[270] E. Scalas, T. Kaizoji, M. Kirchler, J. Huber, and A. Tedeschi, Physica A **366**, 463 (2006).
[271] P. Weber and B. Rosenow, Quant. Finance **5**, 357 (2005).
[272] J. Maskawa, Physica A **383**, 90 (2007).
[273] J. Hasbrouck, Journal of Finance **46**, 179 (1991).
[274] J. A. Hausman, A. W. Lo, and A. C. MacKinlay, Journal of Financial Economics **31**, 319 (1992).
[275] D. B. Keim and A. Madhavan, Review of Financial Studies **9**, 1 (1996).
[276] A. Kempf and O. Korn, Journal of Financial Markets **2**, 29 (1999).
[277] V. Plerou, P. Gopikrishnan, X. Gabaix, and H. E. Stanley, Phys. Rev. E **66**, 027104 (2002).
[278] X. Gabaix, P. Gopikrishnan, V. Plerou, and H. E. Stanley, Nature **423**, 267 (2003).
[279] V. Plerou, P. Gopikrishnan, X. Gabaix, and H. E. Stanley, Quant. Finance **4**, C11 (2004).
[280] X. Gabaix, P. Gopikrishnan, V. Plerou, and H. E. Stanley, Physica A **324**, 1 (2003).
[281] X. Gabaix, P. Gopikrishnan, V. Plerou, and H. E. Stanley, Physica A **382**, 81 (2007).
[282] H. E. Stanley, X. Gabaix, P. Gopikrishnan, and V. Plerou, Physica A **382**, 286 (2007).
[283] J. D. Farmer and F. Lillo, Quant. Finance **4**, C7 (2004).
[284] J. D. Farmer, P. Patelli, and I. I. Zovko, Proc. Natl. Acad. Sci. USA **102**, 2254 (2005).
[285] J.-P. Bouchaud, J. D. Farmer, and F. Lillo, in: *Handbook of Financial Markets: Dynamics and Evolution*, ed. T. Hens and K. R. Schenk-Hoppé, p. 57 (Elsevier, Amsterdam, 2009).
[286] F. Lillo, J. D. Farmer, and R. N. Mantegna, arXiv:cond-mat/0207428.
[287] F. Lillo, J. D. Farmer, and R. N. Mantegna, Nature **421**, 129 (2003).
[288] W.-X. Zho, arXiv:0708.3198.
[289] B. Tóth, Y. Lempérière, C. Deremble, J. de Lataillade, J. Kockelkoren, and J.-P. Bouchaud, Phys. Rev. X **1**, 021006 (2011).
[290] J. D. Farmer, L. Gillemot, F. Lillo, S. Mike, and A. Sen, Quant. Finance **4**, 383 (2004).
[291] P. Weber and B. Rosenow, Quant. Finance **6**, 7 (2006).
[292] J.-P. Bouchaud, Y. Gefen, M. Potters, and M. Wyart, Quant. Finance **4**, 176 (2004).
[293] F. Lillo and J. D. Farmer, Studies in Nonlinear Dynamics and Econometrics **8**, No. 3 Article 1 (2004).
[294] F. Lillo, S. Mike, and J. D. Farmer, Phys. Rev. E **71**, 066122 (2005).
[295] E. Moro, J. Vicente, L. G. Moyano, A. Gerig, J. D. Farmer, G. Vaglica, F. Lillo, and R. N. Mantegna, Phys. Rev. E **80**, 066102 (2009).
[296] V. Plerou, P. Gopikrishnan, and H. E. Stanley, Phys. Rev. E **71**, 046131 (2005).
[297] M. Wyart, J.-P. Bouchaud, J. Kockelkoren, M. Potters, and M. Vettorazzo, Quant. Finance **8**, 41 (2008).
[298] A. Ponzi, F. Lillo, and R. N. Mantegna, arXiv:physics/0608032.
[299] A. Ponzi, F. Lillo, and R. N. Mantegna, Phys. Rev. E **80**, 016112 (2009).
[300] B. Tóth, J. Kertész, and J. D. Farmer, Eur. Phys. J. B **71**, 499 (2009).
[301] B. Biais, P. Hillion, and C. Spatt, Journal of Finance **50**, 1655 (1995).
[302] D. B. Keim and A. Madhavan, Journal of Financial Economics **37**, 371 (1995).
[303] J. D. Farmer and S. Joshi, Journal of Economic Behavior and Organization **49**, 149 (2002).
[304] T. Chordia, R. Roll, and A. Subrahmanyam, Journal of Finance **56**, 501 (2001).
[305] J. Hasbrouck and D. J. Seppi, Journal of Financial Economics **59**, 383 (2001).
[306] B. Rosenow, Int. J. Mod. Phys. C **13**, 419 (2002).
[307] J.-P. Bouchaud, J. Kockelkoren, and M. Potters, Quant. Finance **6**, 115 (2006).
[308] A. G. Zawadowski, J. Kertész, and G. Andor, Physica A **344**, 221 (2004).
[309] J. D. Farmer, A. Gerig, F. Lillo, and S. Mike, Quant. Finance **6**, 107 (2006).
[310] J. D. Farmer and N. Zamani, Eur. Phys. J. B **55**, 189 (2007).
[311] Z. Eisler, J. Kertész, and F. Lillo, Proc. SPIE **6601**, 66010G (2007).
[312] Z. Eisler, J. Kertész, F. Lillo, and R. N. Mantegna, Quant. Finance **9**, 547 (2009).
[313] M. Bartolozzi, C. Mellen, F. Chan, D. Oliver, T. Di Matteo, and T. Aste, Proc. SPIE **6802**, 680203 (2007).

[314] G. La Spada, J. D. Farmer, and F. Lillo, Eur. Phys. J. B **64**, 607 (2008).
[315] D. Challet, Physica A **387**, 3831 (2008).
[316] J.-P. Bouchaud, in: *Encyclopaedia of Quantitative Finance*, ed. R. Cont, **vol. III**, p. 1402 (Wiley, 2010).
[317] Z. Eisler, J.-P. Bouchaud, and J. Kockelkoren, arXiv:0904.0900.
[318] G.-F. Gu, W. Chen, and W.-X. Zhou, Physica A **387**, 3173 (2008).
[319] G.-F. Gu, W. Chen, and W.-X. Zhou, Physica A **387**, 5182 (2008).
[320] G.-F. Gu, F. Ren, X.-H. Ni, W. Chen, and W.-X. Zhou, Physica A **389**, 278 (2010).
[321] D. Challet, Physica A **382**, 29 (2007).
[322] D. Challet, arXiv:physics/0702210.
[323] I. Muni Toke, arXiv:1003.3796.
[324] P. Malo and T. Pennanen, Quant. Finance **12**, 1025 (2012).
[325] R. Cont, A. Kukanov, and S. Stoikov, arXiv:1011.6402.
[326] A. Carollo, G. Vaglica, F. Lillo, and R. N. Mantegna, Quant. Finance **12**, 517 (2012).
[327] B. Tóth, Z. Eisler, F. Lillo, J. Kockelkoren, J.-P. Bouchaud, and J. D. Farmer, arXiv:1104.0587.
[328] F. Ren and L.-X. Zhong, Physica A **391**, 2667 (2012).
[329] G.-F. Gu, F. Ren, and W.-X. Zhou, arXiv:1112.6085.
[330] L. E. Harris and V. Panchapagesan, Journal of Financial Markets **8**, 25 (2005).
[331] B. Hollifield, R. A. Miller, and P. Sandås, Review of Economic Studies **71**, 1027 (2004).
[332] B. Zheng, E. Moulines, and F. Abergel, arXiv:1204.1381.
[333] R. N. Mantegna, Eur. Phys. J. B **11**, 193 (1999).
[334] R. N. Mantegna, in: *Econophysics: An Emergent Science. Proceedings of the 1st Workshop on Econophysics, Budapest, 1997*, ed. J. Kertész and I. Kondor, http://newton.phy.bme.hu/~kullmann/Egyetem/konyv.html, p. 27 (Budapest, 2002).
[335] G. Bonanno, N. Vandewalle, and R. N. Mantegna, Phys. Rev. E **62**, 7615 (2000).
[336] G. Bonanno, F. Lillo, and R. N. Mantegna, Physica A **299**, 16 (2001).
[337] G. Bonanno, F. Lillo, and R. N. Mantegna, Quant. Finance **1**, 96 (2001).
[338] J.-P. Onnela, A. Chakraborti, K. Kaski, J. Kertész, and A. Kanto, Phys. Rev. E **68**, 056110 (2003).
[339] J.-P. Onnela, A. Chakraborti, K. Kaski, and J. Kertész, Physica A **324**, 247 (2003).
[340] S. Micciché, G. Bonanno, F. Lillo, and R. N. Mantegna, Physica A **324**, 66 (2003).
[341] G. Bonanno, G. Caldarelli, F. Lillo, and R. N. Mantegna, Phys. Rev. E **68**, 046130 (2003).
[342] J.-P. Onnela, K. Kaski, and J. Kertész, Eur. Phys. J. B **38**, 353 (2004).
[343] G. Bonanno, G. Caldarelli, F. Lillo, S. Micciché, N. Vandewalle, and R. N. Mantegna, Eur. Phys. J. B **38**, 363 (2004).
[344] T. Di Matteo, T. Aste, and R. N. Mantegna, Physica A **339**, 181 (2004).
[345] V. Tola, F. Lillo, M. Gallegati, and R. N. Mantegna, arXiv:physics/0507006.
[346] C. Coronnello, M. Tumminello, F. Lillo, S. Micciché, and R. N. Mantegna, Acta Physica Polonica B **36**, 2653 (2005).
[347] C. Coronnello, M. Tumminello, F. Lillo, S. Micciché, and R. N. Mantegna, arXiv:physics/0609036.
[348] M. Tumminello, C. Coronnello, F. Lillo, S. Micciché, and R. N. Mantegna, Int. J. Bifurcation Chaos **17**, 2319 (2007).
[349] R. Coelho, S. Hutzler, P. Repetowicz, and P. Richmond, Physica A **373**, 615 (2007).
[350] R. Coelho, C. G. Gilmore, B. Lucey, P. Richmond, and S. Hutzler, Physica A **376**, 455 (2007).
[351] M. Tumminello, T. Di Matteo, T. Aste, and R. N. Mantegna, Eur. Phys. J. B **55**, 209 (2007).
[352] M. Tumminello, F. Lillo, and R. N. Mantegna, arXiv:0809.4615.
[353] F. Hüffner, C. Komusiewicz, H. Moser, and R. Niedermeier, in: *Combinatorial Optimization and Applications*, ed. B. Yang, D.-Z. Du, and C. A. Wang, Lecture Notes in Computer Science Vol. 5165, p. 405 (Springer, Berlin, 2008).
[354] T. Di Matteo, F. Pozzi, and T. Aste, Eur. Phys. J. B **73**, 3 (2010).
[355] A. Chakraborti, arXiv:physics/0605246.
[356] K. Urbanowicz, P. Richmond, and J. A. Hołyst, Physica A **384**, 468 (2007).
[357] N. Basalto, R. Bellotti, F. De Carlo, P. Facchi, E. Pantaleo, and S. Pascazio, Physica A **379**, 635 (2007).
[358] M. McDonald, O. Suleman, S. Williams, S. Howison, and N. F. Johnson, Phys. Rev. E **72**, 046106 (2005).
[359] N. F. Johnson, M. McDonald, O. Suleman, S. Williams, and S. Howison, Proc. SPIE **5848**, 86 (2005).

[360] W.-S. Jung, S. Chae, J.-S. Yang, and H.-T. Moon, Physica A **361**, 263 (2006).
[361] N. Basalto, R. Bellotti, F. De Carlo, P. Facchi, and S. Pascazio, Physica A **345**, 196 (2005).
[362] M. Potters, J.-P. Bouchaud, and L. Laloux, Acta Physica Polonica B **36**, 2767 (2005).
[363] R. Rammal, G. Toulouse, and M. A. Virasoro, Rev. Mod. Phys. **58**, 765 (1986).
[364] L. Kullmann, J. Kertész, and K. Kaski, Phys. Rev. E **66**, 026125 (2002).
[365] J. Kertész, L. Kullmann, A. G. Zawadowski, R. Karádi, and K. Kaski, Physica A **324**, 74 (2003).
[366] M. A. Serrano, M. Boguñá, and A. Vespignani, Journal of Economic Interaction and Coordination **2**, 111 (2007).
[367] V. Plerou, P. Gopikrishnan, B. Rosenow, L. A. N. Amaral, and H. E. Stanley, Phys. Rev. Lett. **83**, 1471 (1999).
[368] S. Galluccio, J.-P. Bouchaud, and M. Potters, Physica A **259**, 449 (1998).
[369] L. Laloux, P. Cizeau, J.-P. Bouchaud, and M. Potters, Phys. Rev. Lett. **83**, 1467 (1999).
[370] V. Plerou, P. Gopikrishnan, B. Rosenow, L. A. N. Amaral, and H. E. Stanley, Physica A **287**, 374 (2000).
[371] H. E. Stanley, P. Gopikrishnan, V. Plerou, and L. A. N. Amaral, Physica A **287**, 339 (2000).
[372] H. E. Stanley, L. A. N. Amaral, X. Gabaix, P. Gopikrishnan, and V. Plerou, Physica A **302**, 126 (2001).
[373] V. Plerou, P. Gopikrishnan, B. Rosenow, L. A. N. Amaral, and H. E. Stanley, Physica A **299**, 175 (2001).
[374] P. Gopikrishnan, B. Rosenow, V. Plerou, and H. E. Stanley, Phys. Rev. E **64**, 035106 (2001).
[375] V. Plerou, P. Gopikrishnan, B. Rosenow, L. A. N. Amaral, T. Guhr, and H. E. Stanley, Phys. Rev. E **65**, 066126 (2002).
[376] B. Rosenow, P. Gopikrishnan, V. Plerou, and H. E. Stanley, Physica A **324**, 241 (2003).
[377] P. Repetowicz and P. Richmond, arXiv:cond-mat/0403177.
[378] A. Utsugi, K. Ino, and M. Oshikawa, Phys. Rev. E **70**, 026110 (2004).
[379] D. Wilcox and T. Gebbie, Physica A **344**, 294 (2004).
[380] D. Wilcox and T. Gebbie, Int. J. Theor. Appl. Finance **11**, 739 (2008).
[381] D.-H. Kim and H. Jeong, Phys. Rev E **72**, 046133 (2005).
[382] G. Tibély, J.-P. Onnela, J. Saramäki, K. Kaski, and J. Kertész, Physica A **370**, 145 (2006).
[383] C. Cattuto, C. Schmitz, A. Baldassarri, V. D. P. Servedio, V. Loreto, A. Hotho, M. Grahl, and G. Stumme, AI Communications **20**, 245 (2007).
[384] M. Gligor and M. Ausloos, Eur. Phys. J. B **57**, 139 (2007).
[385] M. Gligor and M. Ausloos, Eur. Phys. J. B **63**, 533 (2008).
[386] T. Heimo, G. Tibély, J. Saramäki, K. Kaski, and J. Kertész, Physica A **387**, 5930 (2008).
[387] J.-P. Bouchaud and M. Potters, in: *The Oxford Handbook on Random Matrix Theory*, ed. G. Akemann, J. Baik, and P. Di Francesco, p. 824 (Oxford University Press, Oxford, 2011).
[388] F. O. Redelico, A. N. Proto, and M. Ausloos, Physica A **388**, 3527 (2009).
[389] F. Kyriakopoulos, S. Thurner, C. Puhr, and S. W. Schmitz, Eur. Phys. J. B **71**, 523 (2009).
[390] J. Kwapień, S. Drożdż, F. Grümmer, F. Ruf, and J. Speth, Physica A **309**, 171 (2002).
[391] R. Rak, S. Drożdż, J. Kwapień, and P. Oświęcimka, Acta Physica Polonica B **37**, 3123 (2006).
[392] S. Drożdż, J. Kwapień, F. Grümmer, F. Ruf, and J. Speth, Physica A **299**, 144 (2001).
[393] L. Giada and M. Marsili, Phys. Rev. E **63**, 061101 (2001).
[394] M. Marsili, Quant. Finance **2**, 297 (2002).
[395] J. D. Noh, Phys. Rev. E **61**, 5981 (2000).
[396] H.-J. Kim, Y. Lee, B. Kahng, and I.-M. Kim, J. Phys. Soc. Japan **71**, 2133 (2002).
[397] S. Maslov, Physica A **301**, 397 (2001).
[398] P. Ormerod and C. Mounfield, Physica A **280**, 497 (2000).
[399] Z. Burda and J. Jurkiewicz, Physica A **344**, 67 (2004).
[400] V. Kulkarni and N. Deo, Eur. Phys. J. B **60**, 101 (2007).
[401] M. L. Mehta, *Random Matrices* (Academic Press, San Diego, 1991).
[402] E. Hückel, Z. Physik **70**, 204 (1931).
[403] F. R. K. Chung, *Spectral Graph Theory* (American Mathematical Society, 1997).
[404] V. A. Marčenko and L. A. Pastur, Math. USSR–Sbornik **1**, 457 (1967).
[405] F. Slanina, Phys. Rev. E **83**, 011118 (2011).
[406] P. Cizeau and J.-P. Bouchaud, Phys. Rev. E **50**, 1810 (1994).
[407] Z. Burda, J. Jurkiewicz, M. A. Nowak, G. Papp, and I. Zahed, Acta Physica Polonica B **34**, 4747 (2003).
[408] Z. Burda, J. Jurkiewicz, M. A. Nowak, G. Papp, and I. Zahed, Physica A **343**, 694 (2004).
[409] Z. Burda, A. T. Görlich, and B. Wacław, Phys. Rev. E **74**, 041129 (2006).

[410] G. Biroli, J.-P. Bouchaud, and M. Potters, Europhys. Lett. **78**, 10001 (2007).
[411] G. Biroli, J.-P. Bouchaud, and M. Potters, Acta Physica Polonica B **38**, 4009 (2007).
[412] F. Slanina and Z. Konopásek, Adv. Compl. Syst. **13**, 699 (2010).
[413] S. Fortunato, Phys. Rep. **486**, 75 (2010).
[414] M. Ausloos and R. Lambiotte, Physica A **382**, 16 (2007).
[415] L. Kullmann, J. Kertész, and R. N. Mantegna, Physica A **287**, 412 (2000).
[416] J. M. Kumpula, J. Saramäki, K. Kaski, and J. Kertész, Eur. Phys. J. B **56**, 41 (2007).
[417] M. Blatt, S. Wiseman, and E. Domany, Phys. Rev. Lett. **76**, 3251 (1996).
[418] E. Domany, J. Stat. Phys. **110**, 1117 (2003).
[419] R. N. Mantegna, Z. Palágyi, and H. E. Stanley, Physica A **274**, 216 (1999).
[420] P. Gopikrishnan, V. Plerou, Y. Liu, L. A. N. Amaral, X. Gabaix, and H. E. Stanley, Physica A **287**, 362 (2000).
[421] K. Matia, L. A. N. Amaral, S. P. Goodwin, and H. E. Stanley, Phys. Rev. E **66**, 045103 (2002).
[422] B. Podobnik, P. C. Ivanov, I. Grosse, K. Matia, and H. E. Stanley, Physica A **344**, 216 (2004).
[423] K. Matia, M. Pal, H. Salunkay, and H. E. Stanley, Europhys. Lett. **66**, 909 (2004).
[424] P. Weber, F. Wang, I. Vodenska-Chitkushev, S. Havlin, and H. E. Stanley, Phys. Rev. E **76**, 016109 (2007).
[425] H. E. Stanley, V. Plerou, and X. Gabaix, Physica A **387**, 3967 (2008).
[426] X. Gabaix, P. Gopikrishnan, V. Plerou, and H. E. Stanley, Journal of Economic Dynamics and Control **32**, 303 (2008).
[427] V. Plerou and H. E. Stanley, Phys. Rev. E **77**, 037101 (2008).
[428] V. Plerou and H. E. Stanley, Phys. Rev. E **76**, 046109 (2007).
[429] E. Rácz, Z. Eisler, and J. Kertész, Phys. Rev. E **79**, 068101 (2009).
[430] V. Plerou and H. E. Stanley, Phys. Rev. E **79**, 068102 (2009).
[431] V. Plerou, P. Gopikrishnan, and H. E. Stanley, arXiv:cond-mat/0111349.
[432] V. Plerou, P. Gopikrishnan, and H. E. Stanley, Nature **421**, 130 (2003).
[433] M. Potters and J.-P. Bouchaud, arXiv:cond-mat/0304514.
[434] T. Lux, Empirical Economics **25**, 641 (2000).
[435] M. Boguñá and J. Masoliver, Eur. Phys. J. B **40**, 347 (2004).
[436] S. M. D. Queirós, Physica A **344**, 279 (2004).
[437] J. de Souza, L. G. Moyano, and S. M. D. Queirós, Eur. Phys. J. B **50**, 165 (2006).
[438] S. M. D. Queirós, Eur. Phys. J. B **60**, 265 (2007).
[439] S. M. D. Queirós and L. G. Moyano, Physica A **383**, 10 (2007).
[440] R. Weron, K. Weron, and A. Weron, Physica A **264**, 551 (1999).
[441] A. Weron, S. Mercik, and R. Weron, Physica A **264**, 562 (1999).
[442] S. Mercik and R. Weron, Physica A **267**, 239 (1999).
[443] R. Weron and B. Przybyłowicz, Physica A **283**, 462 (2000).
[444] E. Broszkiewicz-Suwaj, A. Makagon, R. Weron, and A. Wyłomańska, Physica A **336**, 196 (2004).
[445] S. Drożdż, M. Forczek, J. Kwapień, P. Oświęcimka, and R. Rak, Physica A **383**, 59 (2007).
[446] J. Kwapień, S. Drożdż, and J. Speth, Physica A **337**, 231 (2004).
[447] S. Drożdż, F. Grümmer, F. Ruf, and J. Speth, Physica A **294**, 226 (2001).
[448] K. Karpio, M. A. Załuska–Kotur, and A. Orłowski, Physica A **375**, 599 (2007).
[449] M. Montero, J. Perelló, J. Masoliver, F. Lillo, S. Miccichè, and R. N. Mantegna, Phys. Rev. E **72**, 056101 (2005).
[450] G. Vaglica, F. Lillo, E. Moro, and R. N. Mantegna, Phys. Rev. E **77**, 036110 (2008).
[451] G. Bonanno, F. Lillo, and R. N. Mantegna, Physica A **280**, 136 (2000).
[452] K. Ivanova and L. T. Wille, Physica A **313**, 625 (2002).
[453] M. Ausloos and K. Ivanova, Eur. Phys. J. B **27**, 177 (2002).
[454] M. Pasquini and M. Serva, Physica A **277**, 228 (2000).
[455] R. Baviera, D. Vergni, and A. Vulpiani, Physica A **280**, 566 (2000).
[456] R. Baviera, M. Pasquini, M. Serva, D. Vergni, and A. Vulpiani, Physica A **312**, 565 (2002).
[457] G. M. Viswanathan, U. L. Fulco, M. L. Lyra, and M. Serva, Physica A **329**, 273 (2003).
[458] F. Petroni and M. Serva, Eur. Phys. J. B **34**, 495 (2003).
[459] F. Petroni and M. Serva, Physica A **344**, 194 (2004).
[460] L. Berardi and M. Serva, Physica A **353**, 403 (2005).
[461] T. Kaizoji, Physica A **343**, 662 (2004).
[462] L. Pichl, T. Kaizoji, and T. Yamano, Physica A **382**, 219 (2007).
[463] T. Kaizoji, Physica A **326**, 256 (2003).

[464] T. Kaizoji and M. Kaizoji, Physica A **344**, 240 (2004).
[465] T. Kaizoji and M. Kaizoji, Physica A **344**, 138 (2004).
[466] T. Kaizoji, Eur. Phys. J. B **50**, 123 (2006).
[467] A.-H. Sato, Eur. Phys. J. B **50**, 137 (2006).
[468] M.-C. Wu, M.-C. Huang, H.-C. Yu, and T. C. Chiang, Phys. Rev. E **73**, 016118 (2006).
[469] G. Oh, S. Kim, and C.-J. Um, J. Korean Phys. Soc. **48**, S197 (2006).
[470] G. Oh, S. Kim, and C. Eom, Physica A **387**, 1247 (2008).
[471] G. Oh, S. Kim, and C. Eom, Physica A **382**, 209 (2007).
[472] J.-S. Yang, W. Kwak, T. Kaizoji, and I. Kim, Eur. Phys. J. B **61**, 241 (2008).
[473] A.-H. Sato, Physica A **382**, 258 (2007).
[474] Z.-Q. Jiang, L. Guo, and W.-X. Zhou, Eur. Phys. J. B **57**, 347 (2007).
[475] E. Scalas, Physica A **253**, 394 (1998).
[476] G. Cuniberti, M. Raberto, and E. Scalas, Physica A **269**, 90 (1999).
[477] J. Voit, Physica A **321**, 286 (2003).
[478] G. Cuniberti, M. Porto, and H. E. Roman, Physica A **299**, 262 (2001).
[479] R. Kitt and J. Kalda, Physica A **345**, 622 (2005).
[480] S. Bianco and R. Renò, arXiv:physics/0610023.
[481] A. Leonidov, V. Trainin, A. Zaitsev, and S. Zaitsev, Physica A **386**, 240 (2007).
[482] R. Vicente, C. de B. Pereira, V. B. P. Leite, and N. Caticha, Physica A **380**, 317 (2007).
[483] I. Simonsen, P. T. H. Ahlgren, M. H. Jensen, R. Donangelo, and K. Sneppen, Eur. Phys. J. B **57**, 153 (2007).
[484] G. R. Jafari, M. S. Movahed, S. M. Fazeli, M. R. R. Tabar, and S. F. Masoudi, J. Stat. Mech. P06008 (2006).
[485] N. B. Ferreira, R. Menezes, and D. A. Mendes, Physica A **382**, 73 (2007).
[486] A. Dionisio, R. Menezes, and D. A. Mendes, Physica A **382**, 58 (2007).
[487] M. Raberto, E. Scalas, G. Cuniberti, and M. Riani, Physica A **269**, 148 (1999).
[488] E. Cisana, L. Fermi, G. Montagna, and O. Nicrosini, arXiv:0709.0810.
[489] I. M. Jánosi, Eur. Phys. J. B **17**, 333 (2000).
[490] L. Gillemot, J. Töyli, J. Kertész, and K. Kaski, Physica A **282**, 304 (2000).
[491] S. Pafka and I. Kondor, Eur. Phys. J. B **27**, 277 (2002).
[492] J. Kertész and Z. Eisler, in: *Practical Fruits of Econophysics. Proceedings of The Third Nikkei Econophysics Symposium*, ed. H. Takayasu, p. 19 (Springer, Berlin, 2006).
[493] Z. Eisler and J. Kertész, Eur. Phys. J. B **51**, 145 (2006).
[494] Z. Eisler and J. Kertész, Phys. Rev. E **73**, 046109 (2006).
[495] Z. Eisler and J. Kertész, Physica A **382**, 66 (2007).
[496] P. Gopikrishnan, V. Plerou, X. Gabaix, and H. E. Stanley, Phys. Rev. E **62**, R4493 (2000).
[497] V. Plerou, P. Gopikrishnan, X. Gabaix, L. A. N. Amaral, and H. E. Stanley, Quant. Finance **1**, 262 (2001).
[498] L. Gillemot, J. D. Farmer, and F. Lillo, Quant. Finance **6**, 371 (2006).
[499] K. Ivanova, M. Ausloos, Physica A **265**, 279 (1999).
[500] M. Bartolozzi, D. B. Leinweber, and A. W. Thomas, Physica A **350**, 451 (2005).
[501] S. Gluzman and V. I. Yukalov, arXiv:cond-mat/9803059.
[502] S. Gluzman and V. I. Yukalov, Mod. Phys. Lett. B **12**, 61 (1998).
[503] S. Gluzman and V. I. Yukalov, Mod. Phys. Lett. B **12**, 75 (1998).
[504] S. Gluzman and V. I. Yukalov, Mod. Phys. Lett. B **12**, 575 (1998).
[505] V. I. Yukalov and S. Gluzman, Int. J. Mod. Phys. B **13**, 1463 (1999).
[506] J. Kertész and Z. Eisler, arXiv:physics/0512193.
[507] M. Wyart and J.-P. Bouchaud, Journal of Economic Behavior and Organization **63**, 1 (2007).
[508] O. V. Precup and G. Iori, Physica A **344**, 252 (2004).
[509] B. Tóth and J. Kertész, Physica A **383**, 54 (2007).
[510] B. Tóth and J. Kertész, Physica A **360**, 505 (2006).
[511] B. Rosenow, P. Gopikrishnan, V. Plerou, and H. E. Stanley, Physica A **314**, 762 (2002).
[512] J. Miśkiewicz and M. Ausloos, Int. J. Mod. Phys. C **17**, 317 (2006).
[513] J.-P. Bouchaud, N. Sagna, R. Cont, N. El-Karoui, M. Potters, arXiv:cond-mat/9712164.
[514] W.-X. Zhou and D. Sornette, Physica A **387**, 243 (2008).
[515] P. Richmond, Physica A **375**, 281 (2007).
[516] T. Kaizoji and M. Kaizoji, Physica A **336**, 563 (2004).
[517] M. Ausloos, J. Miśkiewicz, and M. Sanglier, Physica A **339**, 548 (2004).
[518] F. O. Redelico, A. N. Proto, and M. Ausloos, Physica A **387**, 6330 (2008).
[519] P. Ormerod and C. Mounfield, Physica A **293**, 573 (2001).

[520] E. Gaffeo, M. Gallegati, G. Giulioni, and A. Palestrini, Physica A **324**, 408 (2003).
[521] P. Ormerod, Physica A **341**, 556 (2004).
[522] C. Di Guilmi, E. Gaffeo, and M. Gallegati, Physica A **333**, 325 (2004).
[523] P. Ormerod and A. Heineike, Journal of Economic Interaction and Coordination **4**, 15 (2009).
[524] I. Wright, Physica A **345**, 608 (2005).
[525] C. Di Guilmi, E. Gaffeo, M. Gallegati, and A. Palestrini, International Journal of Applied Econometrics and Quantitative Studies **2**, 5 (2005).
[526] I. Simonsen, M. H. Jensen, and A. Johansen, Eur. Phys. J. B **27**, 583 (2002).
[527] M. H. Jensen, A. Johansen, and I. Simonsen, Physica A **324**, 338 (2003).
[528] A. Johansen, Physica A **324**, 157 (2003).
[529] M. H. Jensen, A. Johansen, F. Petroni, and I. Simonsen, Physica A **340**, 678 (2004).
[530] W.-X. Zhou and W.-K. Yuan, Physica A **353**, 433 (2005).
[531] H. Ebadi, M. Bolgorian, and G. R. Jafari, Physica A **389**, 5439 (2010).
[532] F. Lillo and R. N. Mantegna, Physica A **338**, 125 (2004).
[533] F. Lillo and R. N. Mantegna, Phys. Rev. E **68**, 016119 (2003).
[534] A. Watanabe, N. Uchida, and N. Kikuchi, arXiv:physics/0611130.
[535] N. Vandewalle and M. Ausloos, Physica A **268**, 240 (1999).
[536] B. Podobnik and H. E. Stanley, Phys. Rev. Lett. **100**, 084102 (2008).
[537] M. Ausloos and K. Ivanova, Phys. Rev. E **68**, 046122 (2003).
[538] A. A. G. Cortines and R. Riera, Physica A **377**, 181 (2007).
[539] B. M. Roehner, International Journal of Systems Science **15**, 917 (1984).
[540] B. M. Roehner, Eur. Phys. J. B **8**, 151 (1999).
[541] B. M. Roehner, Eur. Phys. J B **13**, 175 (2000).
[542] B. M. Roehner, Eur. Phys. J B **13**, 189 (2000).
[543] B. M. Roehner, Eur. Phys. J. B **17**, 341 (2000).
[544] B. M. Roehner, Eur. Phys. J. B **14**, 395 (2000).
[545] B. M. Roehner, Int. J. Mod. Phys. C **11**, 91 (2000).
[546] B. M. Roehner and C. H. Shiue, Int. J. Mod. Phys. C **11**, 1383 (2000).
[547] B. M. Roehner, Int. J. Mod. Phys. C **12**, 43 (2001).
[548] B. M. Roehner, Physica A **299**, 71 (2001).
[549] S. Maslov and B. M. Roehner, Int. J. Mod. Phys. C **14**, 1439 (2003).
[550] S. Maslov and B. M. Roehner, Physica A **335**, 164 (2004).
[551] B. M. Roehner, Physica A **347**, 613 (2005).
[552] B. M. Roehner, arXiv:physics/0502045.
[553] B. M. Roehner, Evolutionary and Institutional Economic Review **2**, 167 (2006).
[554] K. Watanabe, H. Takayasu, and M. Takayasu, Physica A **382**, 336 (2007).
[555] B. M. Roehner and D. Sornette, Eur. Phys. J. B **4**, 387 (1998).
[556] B. M. Roehner and D. Sornette, Eur. Phys. J. B **16**, 729 (2000).
[557] J. V. Andersen, S. Gluzman, and D. Sornette, Eur. Phys. J. B **14**, 579 (2000).
[558] D. Sornette and W.-X. Zhou, Quant. Finance **2**, 468 (2002).
[559] B. M. Roehner and D. Sornette, Int. J. Mod. Phys. C **10**, 1099 (1999).
[560] B. M. Roehner, D. Sornette, and J. V. Andersen, Int. J. Mod. Phys. C **15**, 809 (2004).
[561] D. Challet, S. Solomon, and G. Yaari, Economics **3**, 36 (2009).
[562] D. Sornette, A. Johansen, and J.-P. Bouchaud, J. Phys. I France **6**, 167 (1996).
[563] D. Sornette and A. Johansen, Physica A **245**, 411 (1997).
[564] N. Vandewalle, M. Ausloos, P. Boveroux, and A. Minguet, Eur. Phys. J. B **4**, 139 (1998).
[565] N. Vandewalle, P. Boveroux, A. Minguet, and M. Ausloos, Physica A **255**, 201 (1998).
[566] A. Johansen and D. Sornette, Eur. Phys. J. B **1**, 141 (1998).
[567] D. Sornette and A. Johansen, Physica A **261**, 581 (1998).
[568] A. Johansen and D. Sornette, arXiv:cond-mat/9901035.
[569] A. Johansen and D. Sornette, arXiv:cond-mat/9901268.
[570] A. Johansen and D. Sornette, Eur. Phys. J. B **9**, 167 (1999).
[571] N. Vandewalle, M. Ausloos, P. Boveroux, and A. Minguet, Eur. Phys. J. B **9**, 355 (1999).
[572] D. Sornette and J. V. Andersen, Int. J. Mod. Phys. C **13**, 171 (2002).
[573] A. Corcos, J.-P. Eckmann, A. Malaspinas, Y. Malevergne, and D. Sornette, Quant. Finance **2**, 264 (2002).
[574] W.-X. Zhou and D. Sornette, Physica A **337**, 243 (2004).
[575] G. Grimmett and D. Stirzaker, *Probability and Random Processes* (Oxford University Press, Oxford, 2001).
[576] B. Øksendal, *Stochastic Differential Equations* (Springer, Berlin, 1998).

[577] E. W. Montroll and G. H. Weiss, J. Math. Phys. **6**, 167 (1965).

[578] E. Scalas, R. Gorenflo, and F. Mainardi, Physica A **284**, 376 (2000).

[579] J. Masoliver, M. Montero, J. Perelló, and G. H. Weiss, Physica A **379**, 151 (2007).

[580] R. Kutner and F. Świtała, Quant. Finance **3**, 201 (2003).

[581] P. Repetowicz and P. Richmond, Physica A **344**, 108 (2004).

[582] E. Scalas, in: *The Complex Network of Economic Interactions*, ed. A. Namatame, T. Kaizoji, and Y. Aruka, p. 3 (Springer, Berlin, 2006).

[583] J. Masoliver, M. Montero, and J. Perelló, Phys. Rev. E **71**, 056130 (2005).

[584] R. F. Engle, Econometrica **50**, 987 (1982).

[585] T. Bollerslev, Journal of Econometrics **31**, 307 (1986).

[586] T. Bollerslev, R. Y. Chou, and K. F. Kroner, Journal of Econometrics **52**, 5 (1992).

[587] F. X. Diebold and J. A. Lopez, Federal Reserve Bank of New York, Research Paper No. 9522 (1995).

[588] A. Pagan, Journal of Empirical Finance **3**, 15 (1996).

[589] T. Bollerslev, Journal of Econometrics **100**, 41 (2001).

[590] T. Bollerslev, Review of Economics and Statistics **69**, 542 (1987).

[591] Z. Ding and C. W. J. Granger, Journal of Econometrics **3**, 185 (1996).

[592] T. Bollerslev and H. O. Mikkelsen, Journal of Econometrics **73**, 151 (1996).

[593] R. T. Baillie, T. Bollerslev, and H. O. Mikkelsen, Journal of Econometrics, **4**, 3 (1996).

[594] S. Roux and A. Hansen, in: *Disorder and Fracture*, ed. J. C. Charmet, S. Roux, and E. Guyon, p. 17 (Plenum Press, New York, 1990).

[595] W. D. McComb, Rep. Prog. Phys. **58**, 1117 (1995).

[596] L. D. Landau and E. M. Lifshitz, *Fluid Mechanics* (Butterworth-Heinemann, Oxford, 1987).

[597] A. N. Kolmogorov, Doklady Akademii Nauk SSSR **30**, 299 (1941).

[598] U. Frisch, P.-L. Sulem, and M. Nelkin, J. Fluid Mech. **87**, 719 (1978).

[599] U. Frisch and G. Parisi, in: *Turbulence and Predictability in Geophysical Fluid Dynamics and Climate Dynamics*, ed. M. Ghil, R. Benzi, and G. Parisi, p. 71 (North-Holland, Amsterdam, 1985).

[600] R. Benzi, G. Paladin, G. Parisi, and A. Vulpiani, J. Phys. A: Math. Gen **17**, 3521 (1984).

[601] R. Benzi, L. Biferale, and G. Parisi, Physica D **65**, 163 (1993).

[602] R. Benzi, L. Biferale, G. Paladin, A. Vulpiani, and M. Vergassola, Phys. Rev. Lett. **67**, 2299 (1991).

[603] A. Arneodo et al., Europhys. Lett. **34**, 411 (1996).

[604] S. Ghashghaie, W. Breymann, J. Peinke, P. Talkner, and Y. Dodge, Nature **381**, 767 (1996).

[605] B. Holdom, arXiv:cond-mat/9709141.

[606] R. N. Mantegna and H. E. Stanley, arXiv:cond-mat/9609290.

[607] A. Arneodo, J.-P. Bouchaud, R. Cont, J.-F. Muzy, M. Potters, and D. Sornette, arXiv:cond-mat/9607120.

[608] B. Mandelbrot, A. Fisher, and L. Calvet, Cowles Foundation Discussion Paper 1164 (1997).

[609] L. Calvet, A. Fisher, and B. Mandelbrot, Cowles Foundation Discussion Paper 1165 (1997).

[610] A. Fisher, L. Calvet, and B. Mandelbrot, Cowles Foundation Discussion Paper 1166 (1997).

[611] L. Borland, J.-P. Bouchaud, J.-F. Muzy, and G. Zumbach, arXiv:cond-mat/0501292.

[612] L. Calvet and A. Fisher, Journal of Econometrics **105**, 27 (2001).

[613] L. E. Calvet and A. J. Fisher, Journal of Financial Econometrics **2**, 49 (2004).

[614] T. Lux, University of Kiel Economics working paper 2004-11 (2004).

[615] T. Lux, Journal of Business and Economic Statistics **26**, 194 (2008).

[616] L. E. Calvet, A. J. Fisher, and S. B. Thompson, Journal of Econometrics **131**, 179 (2006).

[617] R. Liu, T. Di Matteo, and T. Lux, arXiv:0704.1338.

[618] L. E. Calvet and A. J. Fisher, Journal of Financial Economics **86**, 178 (2007).

[619] L. E. Calvet and A. J. Fisher, Journal of Mathematical Economics **44**, 207 (2008).

[620] L. E. Calvet and A. J. Fisher, *Multifractal Volatility* (Academic Press, Amsterdam, 2008).

[621] T. Lux, Journal of Business and Economic Statistics **26**, 194 (2008).

[622] T. Lux, University of Kiel Economics working paper 2003-13 (2003).

[623] Z. Eisler and J. Kertész, Physica A **343**, 603 (2004).

[624] O. Vasicek, Journal of Financial Economics **5**, 177 (1977).

[625] J. C. Cox, J. E. Ingersoll Jr., and S. A. Ross, Econometrica **53**, 385 (1985).

[626] W. Feller, Ann. Math. **54**, 173 (1951).

[627] J. Hull and A. White, Review of Financial Studies **3**, 573 (1990).

[628] A. Einstein, Annalen der Physik **17**, 549 (1905).

[629] L. S. Ornstein, Proc. Acad. Amsterdam **21**, 96 (1919).

[630] G. E. Uhlenbeck and L. S. Ornstein, Phys. Rev. **36**, 823 (1930).
[631] L. O. Scott, Journal of Financial and Quantitative Analysis **22**, 419 (1987).
[632] J. Hull and A. White, Journal of Finance **42**, 281 (1987).
[633] E. M. Stein and J. C. Stein, Review of Financial Studies **4**, 727 (1991).
[634] C. A. Ball and A. Roma, Journal of Financial and Quantitative Analysis **29**, 589 (1994).
[635] R. Schöbel and J. Zhu, European Finance Review **3**, 23 (1999).
[636] S. L. Heston, Review of Financial Studies **6**, 327 (1993).
[637] A. A. Drăgulescu and V. M. Yakovenko, Quant. Finance **2**, 443 (2002).
[638] A. C. Silva and V. M. Yakovenko, Physica A **324**, 303 (2003).
[639] R. Vicente, C. M. de Toledo, V. B. P. Leite, and N. Caticha, Physica A **361**, 272 (2006).
[640] A. C. Silva and V. M. Yakovenko, Physica A **382**, 278 (2007).
[641] T. S. Biró and R. Rosenfeld, Physica A **387**, 1603 (2008).
[642] P. Jizba, H. Kleinert, and P. Haener, Physica A **388**, 3503 (2009).
[643] J. Perelló, R. Sircar, and J. Masoliver, J. Stat. Mech. P06010 (2008).
[644] I. S. Gradshteyn and I. M. Ryzhik, *Table of Integrals, Series, and Products (5th edition)* (Academic Press, San Diego, 1994).
[645] J. W. Dash, *Functional path integrals and the average bond price*, AT&T Bell Laboratories preprint, unpublished (1986).
[646] J. W. Dash, *Path integrals and options - I*, Merrill Lynch Capital Markets preprint, unpublished (1988).
[647] J. W. Dash, *Path integrals and options- II: one-factor term-structure models*, Merrill Lynch Capital Markets preprint, unpublished (1989).
[648] B. E. Baaquie, J. Phys. I France **7**, 1733 (1997).
[649] V. Linetsky, Computational Economics **11**, 129 (1998).
[650] C. Chiarella and N. El-Hassan, Journal of Financial Engineering **6**, 121 (1997).
[651] C. Chiarella, N. El-Hassan, and A. Kucera, Journal of Economic Dynamics and Control **23**, 1387 (1999).
[652] M. Otto, arXiv:cond-mat/9812318.
[653] E. Bennati, M. Rosa-Clot, and S. Taddei, Int. J. Theor. Appl. Finance **2**, 381 (1999).
[654] L. Ingber, Physica A **283**, 529 (2000).
[655] A. Matacz, Journal of Computational Finance **6**, No. 2 (2002).
[656] G. Montagna, O. Nicrosini, and N. Moreni, Physica A **310**, 450 (2002).
[657] H. Kleinert, Physica A **311**, 536 (2002).
[658] H. Kleinert, Physica A **312**, 217 (2002).
[659] H. Kleinert, Physica A **338**, 151 (2004).
[660] G. Bormetti, G. Montagna, N. Moreni, and O. Nicrosini, Quant. Finance **6**, 55 (2006).
[661] M. Decamps, A. De Schepper, and M. Goovaerts, Physica A **363**, 404 (2006).
[662] P. Jizba and H. Kleinert, arXiv:0712.0083.
[663] P. Jizba and H. Kleinert, Phys. Rev. E **78**, 031122 (2008).
[664] B. E. Baaquie, *Quantum Finance* (Cambridge University Press, Cambridge, 2004).
[665] J. W. Dash, *Quantitative Finance and Risk Management: A Physicist's Approach* (World Scientific, Singapore, 2004).
[666] H. Kleinert, *Path Integrals in Quantum Mechanics, Statistics, Polymer Physics, and Financial Markets* (World Scientific, Singapore, 2006).
[667] E. Scalas, arXiv:physics/0608217.
[668] M. M. Meerschaert and E. Scalas, Physica A **370**, 114 (2006).
[669] V. Gontis and B. Kaulakys, Physica A **343**, 505 (2004).
[670] V. Gontis and B. Kaulakys, Physica A **344**, 128 (2004).
[671] V. Gontis and B. Kaulakys, arXiv:physics/0606115.
[672] J. Masoliver, M. Montero, and J. M. Porrà, Physica A **283**, 559 (2000).
[673] H. Kesten, Acta Mathematica **131**, 207 (1973).
[674] J. M. Deutsch, Physica A **208**, 433 (1994).
[675] I. T. Drummond, J. Phys. A: Math. Gen. **25**, 2273 (1992).
[676] M. Levy and S. Solomon, Int. J. Mod. Phys. C **7**, 595 (1996).
[677] S. Solomon and M. Levy, Int. J. Mod. Phys. C **7**, 745 (1996).
[678] M. Levy and S. Solomon, Int. J. Mod. Phys. C **7**, 65 (1996).
[679] O. Biham, O. Malcai, M. Levy, and S. Solomon, Phys. Rev. E **58**, 1352 (1998).
[680] S. Solomon, in: *Application of Simulation to Social Sciences*, ed. G. Ballot and G. Weisbuch, (Hermes Science Publications, Oxford, 2000).
[681] O. Malcai, O. Biham, and S. Solomon, Phys. Rev. E **60**, 1299 (1999).

[682] A. Blank and S. Solomon, Physica A **287**, 279 (2000).
[683] S. Solomon and P. Richmond, Physica A **299**, 188 (2001).
[684] P. Richmond and S. Solomon, Int. J. Mod. Phys. C **12**, 333 (2001).
[685] S. Solomon and P. Richmond, Eur. Phys. J. B **27**, 257 (2002).
[686] Z.-F. Huang and S. Solomon, Physica A **306**, 412 (2002).
[687] O. Biham, Z.-F. Huang, O. Malcai, and S. Solomon, Phys. Rev. E **64**, 026101 (2001).
[688] H. Takayasu, A.-H. Sato, and M. Takayasu, Phys. Rev. Lett. **79**, 966 (1997).
[689] U. Frisch and D. Sornette, J. Phys. I France **7**, 1155 (1997).
[690] D. Sornette, arXiv:cond-mat/9708231.
[691] D. Sornette, D. Stauffer, and H. Takayasu, in: *The Science of Disasters*, ed. A. Bundle, J. Kropp, and H. J. Schellnhuber, p. 411 (Springer, Berlin 2002).
[692] A. Krawiecki, J. A. Hołyst, and D. Helbing, Phys. Rev. Lett. **89**, 158701 (2002).
[693] C. Tsallis, J. Stat. Phys. **52**, 479 (1988).
[694] C. Tsallis, S. V. F. Levy, A. M. C. Souza, and R. Maynard, Phys. Rev. Lett. **75**, 3589 (1995).
[695] C. Tsallis, Physica A **324**, 89 (2003).
[696] S. M. D. Queirós, Europhys. Lett. **71**, 339 (2005).
[697] S. M. D. Queirós, C. Anteneodo, and C. Tsallis, Proc. SPIE **5848**, 151 (2005).
[698] S. M. D. Queirós, L. G. Moyano, J. de Souza, and C. Tsallis, Eur. Phys. J. B **55**, 161 (2007).
[699] S. M. D. Queirós and C. Tsallis, Eur. Phys. J. B **48**, 139 (2005).
[700] B. Podobnik, D. Horvatic, A. Lam Ng, H. E. Stanley, and P. C. Ivanov, arXiv:0709.0838.
[701] A.-H. Sato and H. Takayasu, arXiv:cond-mat/0109139.
[702] L. Borland and J.-P. Bouchaud, arXiv:physics/0507073.
[703] S. R. Bentes, R. Menezes, and D. A. Mendes, arXiv:0709.2178.
[704] M. I. Krivoruchenko, Phys. Rev. E **70**, 036102 (2004).
[705] M. Paluš, Phys. Rev. Lett. **101**, 134101 (2008).
[706] A. Arneodo, J.-F. Muzy, and D. Sornette, arXiv:cond-mat/9708012.
[707] A. Arneodo, E. Bacry, and J.-F. Muzy, J. Math. Phys. **39**, 4142 (1998).
[708] E. Bacry, J. Delour, and J.-F. Muzy, Phys. Rev. E **64**, 026103 (2001).
[709] A. Saichev and D. Sornette, Phys. Rev. E **74**, 011111 (2006).
[710] L. Palatella, J. Perelló, M. Montero, and J. Masoliver, Eur. Phys. J. B **38**, 671 (2004).
[711] P. Richmond and L. Sabatelli, Physica A **336**, 27 (2004).
[712] C. Anteneodo and R. Riera, Phys. Rev. E **72**, 026106 (2005).
[713] P. Henry-Labordère, Quant. Finance **7**, 525 (2007).
[714] P. Henry-Labordère, arXiv:cond-mat/0504317.
[715] G. Bonanno, D. Valenti, and B. Spagnolo, Phys. Rev. E **75**, 016106 (2007).
[716] G. Bonanno, D. Valenti, and B. Spagnolo, Eur. Phys. J. B **53**, 405 (2006).
[717] D. Valenti, B. Spagnolo, and G. Bonanno, Physica A **382**, 311 (2007).
[718] J. Villarroel, Physica A **382**, 321 (2007).
[719] J. Masoliver and J. Perelló, Phys. Rev. E **75**, 046110 (2007).
[720] J. Masoliver and J. Perelló, Quant. Finance **6**, 423 (2006).
[721] Z. Eisler, J. Perelló, and J. Masoliver, Phys. Rev. E **76**, 056105 (2007).
[722] G. Bormetti, V. Cazzola, G. Montagna, and O. Nicrosini, J. Stat. Mech. P11013 (2008).
[723] J. Perelló and J. Masoliver, Int. J. Theor. Appl. Finance **5**, 541 (2002).
[724] D. Sornette, Eur. Phys. J. B **3**, 125 (1998).
[725] P. Santa-Clara and D. Sornette, Review of Financial Studies **14**, 149 (2001).
[726] A. L. Lewis, *Option Valuation under Stochastic Volatility* (Finance Press, Newport Beach, 2000).
[727] B. Mandelbrot, International Economic Review **1**, 79 (1960).
[728] P. Bak, C. Tang, and K. Wiesenfeld, Phys. Rev. Lett. **59**, 381 (1987).
[729] P. Bak and C. Tang, J. Geophys. Res. **94**, 15635 (1989).
[730] Z. Olami, H.J.S. Feder, and K. Christensen, Phys. Rev. Lett. **68**, 1244 (1992).
[731] P. Bak and K. Sneppen, Phys. Rev. Lett. **71**, 4083 (1993).
[732] S. I. Zaitsev, Physica A **189**, 411 (1992).
[733] F. Slanina, Phys. Rev. E **60**, 1940 (1999).
[734] K. Nagel and M. Schreckenberg, J. Phys. I France **2**, 2221 (1992).
[735] K. Nagel and M. Paczuski, Phys. Rev. E **51**, 2909 (1995).
[736] F. Slanina, Phys. Rev. E **59**, 3947 (1999).
[737] P. Bak, K. Chen, J. Scheinkman, and M. Woodford, Ricerche Economiche **47**, 3 (1993).
[738] J. Scheinkman and M. Woodford, The American Economic Review **84**, 417 (1994).
[739] D. Dhar and R. Ramaswamy, Phys. Rev. Lett. **63**, 1659 (1989).

[740] P. Bak, C. Tang, and K. Wiesenfeld, Phys. Rev. A **38**, 364 (1988).
[741] A. Chessa, H.E. Stanley, A. Vespignani, and S. Zapperi, Phys. Rev. E **59**, R12 (1999).
[742] C. Tang and P. Bak, J. Stat. Phys. **51**, 797 (1988).
[743] P. Alstrøm, Phys. Rev. A **38**, 4905 (1988).
[744] S. Zapperi, K. B. Lauritsen, and H. E. Stanley, Phys. Rev. Lett. **75**, 4071 (1995).
[745] K. B. Lauritsen, S. Zapperi, and H. E. Stanley, Phys. Rev. E **54**, 2483 (1996).
[746] E. V. Ivashkevich, Phys. Rev. Lett. **76**, 3368 (1996).
[747] A. Vespignani and S. Zapperi, Phys. Rev. E **57**, 6345 (1998).
[748] F. Slanina, Eur. Phys. J. B **25**, 209 (2002).
[749] F. Slanina and M. Kotrla, Phys. Rev. E **62**, 6170 (2000).
[750] R. Cont and J.-P. Bouchaud, Macroeconomic Dynamics **4**, 170 (2000).
[751] D. Stauffer and A. Aharony, *Introduction to Percolation Theory* (Taylor & Francis, London, 1994).
[752] J. W. Essam, Rep. Prog. Phys. **43**, 833 (1980).
[753] B. Bollobás, *Random Graphs* (Academic Press, London, 1985).
[754] D. Stauffer and D. Sornette, Physica A **271**, 496 (1999).
[755] D. Stauffer, P. M. C. de Oliveira, and A. T. Bernardes, Int. J. Theor. Appl. Finance **2**, 83 (1999).
[756] D. Stauffer and T. J. P. Penna, Physica A **256**, 284 (1998).
[757] I. Chang and D. Stauffer, Physica A **264**, 294 (1999).
[758] I. Chang and D. Stauffer, arXiv:cond-mat/0105573.
[759] I. Chang, D. Stauffer, and R. B. Pandey, arXiv:cond-mat/0108345.
[760] D. Chowdhury and D. Stauffer, Eur. Phys. J. B **8**, 477 (1999).
[761] L. R. da Silva and D. Stauffer, Physica A **294**, 235 (2001).
[762] F. Castiglione and D. Stauffer, Physica A **300**, 531 (2001).
[763] F. Castiglione, R. B. Pandey, and D. Stauffer, Physica A **289**, 223 (2001).
[764] G. Ehrenstein, arXiv:cond-mat/0205320.
[765] G. Ehrenstein, F. Westerhoff, and D. Stauffer, arXiv:cond-mat/0311581.
[766] D. Stauffer and N. Jan, Physica A **277**, 215 (2000).
[767] A. Chakraborti, arXiv:cond-mat/0109280.
[768] C. Schulze, Int. J. Mod. Phys. C **13**, 207 (2002).
[769] D. Makowiec, P. Gnaciski, and W. Miklaszewski, Physica A **331**, 269 (2004).
[770] J. Wang, C.-X. Yang, P.-L. Zhou, Y.-D. Jin, T. Zhou, and B.-H. Wang, arXiv:cond-mat/0412097.
[771] V. M. Eguíluz and M. G. Zimmermann, Phys. Rev. Lett. **85**, 5659 (2000).
[772] R. D'Hulst and G. J. Rodgers, Int. J. Theor. Appl. Finance **3**, 609 (2000).
[773] Y. Xie, B.-H. Wang, H. Quan, W. Yang, and P. M. Hui, Phys. Rev. E **65**, 046130 (2002).
[774] Y.-B. Xie, B.-H. Wang, B. Hu, and T. Zhou, Phys. Rev. E **71**, 046135 (2005).
[775] W. Feller, *An Introduction to Probability Theory and Its Applications, 3rd ed.* (Willey, New York, 1970).
[776] R. D'Hulst and G. J. Rodgers, Eur. Phys. J. B **20**, 619 (2001).
[777] R. D'Hulst and G. J. Rodgers, Physica A **280**, 554 (2000).
[778] D. F. Zheng, P. M. Hui, K. F. Yip, and N. F. Johnson, Eur. Phys. J. B **27**, 213 (2002).
[779] D. Zheng, P. M. Hui, and N. F. Johnson, arXiv:cond-mat/0105474.
[780] D. Zheng, G. J. Rodgers, P. M. Hui, and R. D'Hulst, Physica A **303**, 176 (2002).
[781] D. Zheng, G. J. Rodgers, and P. M. Hui, Physica A **310**, 480 (2002).
[782] F. Ren and B. Zheng, Phys. Lett. A **313**, 312 (2003).
[783] B. Zheng, F. Ren, S. Trimper, and D. F. Zheng, Physica A **343**, 653 (2004).
[784] R. D'Hulst and G. J. Rodgers, Eur. Phys. J. B **21**, 447 (2001).
[785] R. D'Hulst and G. J. Rodgers, Physica A **299**, 328 (2001).
[786] G. J. Rodgers and D. Zhang, Physica A **308**, 375 (2002).
[787] G. J. Rodgers and Y. J. Yap, Eur. Phys. J. B **28**, 129 (2002).
[788] S. Rawal and G. J. Rodgers, Physica A **344**, 50 (2004).
[789] G. J. Stigler, Journal of Business **37**, 117 (1964).
[790] A. Kirman, Quarterly Journal of Economics **108**, 137 (1993).
[791] J. Arifovic, Journal of Political Economy **104**, 510 (1996).
[792] J. Kareken and N. Wallace, Quarterly Journal of Economics **96**, 207 (1981).
[793] J. Arifovic and R. Gençay, Journal of Economic Dynamics and Control **24**, 981 (2000).
[794] J. Arifovic, Journal of Economic Dynamics and Control **25**, 395 (2001).
[795] T. Lux and S. Schornstein, Journal of Mathematical Economics **41**, 169 (2005).

[796] V. I. Arnold, *Teoriya katastrof* (MGU, Moskva, 1983).

[797] E. C. Zeeman, Journal of Mathematical Economics **1**, 39 (1974).

[798] T. Rheinlaender and M. Steinkamp, Studies in Nonlinear Dynamics and Econometrics **8**, No. 4, Article 4 (2004).

[799] M. Youssefmir, B. A. Huberman, and T. Hogg, arXiv:adap-org/9409001.

[800] M. Youssefmir and B. A. Huberman, Journal of Economic Behavior and Organization **32**, 101 (1997).

[801] W. A. Brock and C. H. Hommes, Journal of Economic Dynamics and Control **22**, 1235 (1998).

[802] J. A. Frankel and K. A. Froot, The Dollar as a Speculative Bubble: A Tale of Fundamentalists and Chartists, National Bureau of Economic Research, working paper No. 1854, http://papers.nber.org/papers/w1854 (1987).

[803] R. H. Day and W. Huang, Journal of Economic Behavior and Organization **14**, 299 (1990).

[804] S.-H. Chen and C.-H. Yeh, Journal of Economic Dynamics and Control **21**, 1043 (1997).

[805] S.-H. Chen and C.-H. Yeh, Journal of Economic Dynamics and Control **25**, 363 (2001).

[806] S.-H. Chen and C.-H. Yeh, Journal of Economic Behavior and Organization **49**, 217 (2002).

[807] J. H. Holland, *Adaptation in Natural and Artificial Sysyems: An Introductory Analysis with Applications to Biology, Control, and Artificial Intelligence* (University of Michigan Press, Ann Arbor, 1975).

[808] J. R. Koza, *Genetic Programming: On the Programming of Computers by Means of Natural Selection* (MIT Press, Cambridge, 1992).

[809] R. M. W. J. Beetsma and P. C. Schotman, Economic Journal **111**, 821 (2001).

[810] J. L. Kelly, Bell System Technical Journal **35**, 917 (1956).

[811] E. O. Thorp, in: *Stochastic Optimization Models in Finance*, ed. W. T. Ziemba and R. G. Vickson, (reprinted from: *American Statistical Association, Business and Economics Statistics section, Proceedings, 1971, pp. 215-224*), p. 599 (World Scientific, Singapore, 2006).

[812] S. Maslov and Y.-C. Zhang, Int. J. Theor. Appl. Finance **1**, 377 (1998).

[813] M. Marsili, S. Maslov, and Y.-C. Zhang, Physica A **253**, 403 (1998).

[814] D. Sornette, Physica A **256**, 251 (1998).

[815] R. Baviera, M. Pasquini, M. Serva, and A. Vulpiani, Int. J. Theor. Appl. Finance **1**, 473 (1998).

[816] S. Maslov and Y.-C. Zhang, Physica A **262**, 232 (1999).

[817] E. W. Piotrowski and M. Schroeder, arXiv:physics/0607166.

[818] G. Yaari and S. Solomon, Eur. Phys. J. B **73**, 625 (2010).

[819] F. Slanina, Physica A **269**, 554 (1999).

[820] M. Levy, H. Levy, and S. Solomon, Economics Letters **45**, 103 (1994).

[821] M. Levy, H. Levy, and S. Solomon, J. Phys. I France **5**, 1087 (1995).

[822] M. Levy, N. Persky, and S. Solomon, Int. J. High Speed Computing **8**, 93 (1996).

[823] E. Zschischang and T. Lux, Physica A **291**, 563 (2001).

[824] M. Shatner, L. Muchnik, M. Leshno, and S. Solomon, arXiv:cond-mat/0005430.

[825] S. Solomon and M. Levy, in: *New Directions in Statistical Physics. Econophysics, Bioinformatics, and Pattern Recognition*, ed. L. T. Wille p. 69 (Springer, Berlin, 2004).

[826] T. Lux, Economic Journal **105**, 881 (1995).

[827] T. Lux, Journal of Economic Behavior and Organization **33**, 143 (1998).

[828] T. Lux and M. Marchesi, Nature **397**, 498 (1999).

[829] T. Lux and M. Marchesi, Int. J. Theor. Appl. Finance **3**, 675 (2000).

[830] S.-H. Chen, T. Lux, and M. Marchesi, Journal of Economic Behavior and Organization **46**, 327 (2001).

[831] E. Egenter, T. Lux, and D. Stauffer, Physica A **268**, 250 (1999).

[832] E. Samanidou, E. Zschischang, D. Stauffer, and T. Lux, arXiv:cond-mat/0110354.

[833] A. G. Zawadowski, R. Karádi, and J. Kertész, Physica A **316**, 403 (2002).

[834] S. Alfarano and T. Lux, in: *Long Memory in Economics*, ed. G. Teyssière and A. P. Kirman, p. 345 (Springer, Berlin, 2007).

[835] S. Alfarano, F. Wagner, and T. Lux, Computational Economics **26**, 19 (2005).

[836] S. Alfarano and T. Lux, Macroeconomic Dynamics **11**, 80 (2007).

[837] S. Alfarano, T. Lux, and F. Wagner, Eur. Phys. J. B **55**, 183 (2007).

[838] S. Solomon, G. Weisbuch, L. de Arcangelis, N. Jan, and D. Stauffer, Physica A **277**, 239 (2000).

[839] J. Goldenberg, B. Libai, S. Solomon, N. Jan, and D. Stauffer, Physica A **284**, 335 (2000).

[840] R. G. Palmer, W. B. Arthur, J. H. Holland, B. LeBaron, and P. Tayler, Physica D **75**, 264 (1994).

[841] W. B. Arthur, J. H. Holland, B. LeBaron, R. Palmer, and P. Tayler, in: *The Economy as an Evolving Complex System II*, ed. W. B. Arthur, S. N. Durlauf, and D. A. Lane, p. 15 (Perseus, Reading, 1997).

[842] B. LeBaron, W. B. Arthur, and R. Palmer, Journal of Economic Dynamics and Control **23**, 1487 (1999).

[843] B. LeBaron, Journal of Economic Dynamics and Control **24**, 679 (2000).

[844] N. S. P. Tay and S. C. Linn, Journal of Economic Dynamics and Control **25**, 321 (2001).

[845] N. Ehrentreich, *Agent-Based Modeling: The Santa Fe Institute Artificial Stock Market Model Revisited* (Springer, Berlin, 2008).

[846] G. Caldarelli, M. Marsili, and Y.-C. Zhang, Europhys. Lett. **40**, 479 (1997).

[847] F. Slanina and Y.-C. Zhang, Physica A **272**, 257 (1999).

[848] R. Donangelo and K. Sneppen, Physica A **276**, 572 (2000).

[849] D. Stauffer and J. P. Radomski, Physica A **291**, 583 (2001).

[850] G. Iori, Int. J. Mod. Phys. C **10**, 1149 (1999).

[851] G. Iori, Int. J. Theor. Appl. Finance **3**, 467 (2000).

[852] G. Iori, Journal of Economic Behavior and Organization **49**, 269 (2002).

[853] G. Iori, AIP Conf. Proc. **553**, 297 (2001).

[854] S. Salop and J. Stiglitz, Review of Economic Studies **44**, 493 (1977).

[855] M. Marsili and Y.-C. Zhang, Physica A **245**, 181 (1997).

[856] J.-P. Bouchaud and R. Cont, Eur. Phys. J. B **6**, 543 (1998).

[857] S. Bornholdt, arXiv:cond-mat/0105224.

[858] L. Kullmann and J. Kertész, Physica A **299**, 234 (2001).

[859] A. Ponzi and Y. Aizawa, Physica A **287**, 507 (2000).

[860] A. Ponzi, Eur. Phys. J. B **20**, 565 (2001).

[861] R. Mansilla, Physica A **284**, 478 (2000).

[862] G. Cuniberti, A. Valleriani, and J. L. Vega, Quant. Finance **1**, 332 (2001).

[863] R. Rothenstein and K. Pawelzik, Physica A **326**, 534 (2003).

[864] M. Potters and J.-P. Bouchaud, arXiv:physics/0508104.

[865] F. Ghoulmié, R. Cont, and J.-P. Nadal, J. Phys.: Condens. Matter **17**, S1259 (2005).

[866] F. Ghoulmié, M. Bartolozzi, C. P. Mellen, and T. Di Matteo, Proc. SPIE **6802**, 68020D (2007).

[867] G. De Masi, G. Iori, and G. Caldarelli, Phys. Rev. E **74**, 066112 (2006).

[868] G. Raffaelli and M. Marsili, J. Stat. Mech. L08001 (2006).

[869] T. Kaizoji, in: *Nonlinear Dynamics and Heterogeneous Interacting Agents*, ed. T. Lux, S. Reitz, and E. Samanidou, p. 237 (Springer, Berlin, 2005).

[870] P. Ormerod, Physica A **370**, 60 (2006).

[871] S. Sinha, arXiv:physics/0606078.

[872] S. Sinha and S. Raghavendra, arXiv:physics/0607071.

[873] D. Horváth, M. Gmitra, and Z. Kuscsik, Physica A **361**, 589 (2006).

[874] Z. Kuscsik and D. Horváth, arXiv:0710.0459.

[875] Z. Kuscsik, D. Horváth, and M. Gmitra, Physica A **379**, 199 (2007).

[876] E. Samanidou, E. Zschischang, D. Stauffer, and T. Lux, Rep. Prog. Phys. **70**, 409 (2007).

[877] A. De Martino, M. Marsili, and I. Prez Castillo, Macroeconomic Dynamics **11 Supplement S1**, 34 (2007).

[878] V. Alfi, M. Cristelli, L. Pietronero, and A. Zaccaria, Eur. Phys. J. B **67**, 385 (2009).

[879] V. Alfi, M. Cristelli, L. Pietronero, and A. Zaccaria, Eur. Phys. J. B **67**, 399 (2009).

[880] S. Cantono and S. Solomon, New J. Phys. **12**, 075038 (2010).

[881] T. Tokár, D. Horváth, and M. Hnatich, arXiv:1110.2603.

[882] J. Janová, Eur. J. Phys. **32**, 1443 (2011).

[883] M. F. M. Osborne, Econometrica **33**, 88 (1965).

[884] L. R. Glosten, Journal of Finance **49**, 1127 (1994).

[885] D. K. Gode and S. Sunder, Journal of Political Economy **101**, 119 (1993).

[886] D. K. Gode and S. Sunder, Quarterly Journal of Economics **112**, 603 (1997).

[887] B. Tóth, E. Scalas, J. Huber, and M. Kirchler, Eur. Phys. J. B **55**, 115 (2007).

[888] J. D. Murray, *Mathematical Biology*, 3rd edition (Springer, Berlin, 2003).

[889] D. C. Mattis and M. L. Glasser, Rev. Mod. Phys. **70**, 979 (1998).

[890] F. Slanina, Eur. Phys. J. B **61**, 225 (2008).

[891] M. Marchesi, S. Cincotti, S. Focardi, and M. Raberto, preprint (2000).

[892] M. Raberto, S. Cincitti, S. M. Focardi, and M. Marchesi, Physica A **299**, 319 (2001).

[893] S. Cincotti, S. M. Focardi, M. Marchesi, and M. Raberto, Physica A **324**, 227 (2003).

[894]   M. Raberto and S. Cincotti, Physica A **355**, 34 (2005).

[895]   M. Raberto, S. Cincotti, S. M. Focardi, and M. Marchesi, Computational Economics **22**, 255 (2003).

[896]   M. Raberto, S. Cincotti, C. Dose, S. M. Focardi, and M. Marchesi, in: *Nonlinear Dynamics and Heterogeneous Interacting Agents*, ed. T. Lux, S. Reitz, and E. Samanidou, p. 305 (Springer, Berlin, 2005).

[897]   S. Cincotti, L. Ponta, and M. Raberto, preprint, presented at WEHIA 2005 (2005).

[898]   S. Cincotti, S. M. Focardi, L. Ponta, M. Raberto, and E. Scalas, in: *The Complex Network of Economic Interactions*, ed. A. Namatame, T. Kaizoji, and Y. Aruka, p. 239 (Springer, Berlin, 2006).

[899]   D. Sornette, Phys. Rev. Lett. **72**, 2306 (1994).

[900]   D. Sornette, Phys. Rev. E **57**, 4811 (1998).

[901]   P. Bak, M. Paczuski, and M. Shubik, Physica A **246**, 430 (1997).

[902]   D. Eliezer and I. I. Kogan, arXiv:cond-mat/9808240.

[903]   G. T. Barkema, M. J. Howard, and J. L. Cardy, Phys. Rev. E **53**, 2017 (1996).

[904]   P. L. Krapivsky, Phys. Rev. E **51**, 4774 (1995).

[905]   M. A. Rodriguez and H. S. Wio, Phys. Rev. E **56**, 1724 (1997).

[906]   F. Leyvraz and S. Redner, Phys. Rev. Lett. **66**, 2168 (1991).

[907]   S. Cornell and M. Droz, Phys. Rev. Lett. **70**, 3824 (993).

[908]   L.-H. Tang and G.-S. Tian, Physica A **264**, 543 (1999).

[909]   S. Maslov, Physica A **278**, 571 (2000).

[910]   E. Smith, J. D. Farmer, L. Gillemot, and S. Krishnamurthy, Quant. Finance **3**, 481 (2003).

[911]   M. G. Daniels, J. D. Farmer, G. Iori, E. Smith, arXiv:cond-mat/0112422.

[912]   M. G. Daniels, J. D. Farmer, L. Gillemot, G. Iori, and E. Smith, Phys. Rev. Lett. **90**, 108102 (2003).

[913]   G. Iori, M. G. Daniels, J. D. Farmer, L. Gillemot, S. Krishnamurthy, and E. Smith, Physica A **324**, 146 (2003).

[914]   J. D. Farmer, L. Gillemot, G. Iori, S. Krishnamurthy, D. E. Smith, and M. G. Daniels, in: *The Economy as an Evolving Complex System III*, ed. L. Blume and S. Durlauf, p. 133 (Oxford University Press, 2005).

[915]   S. Mike and J. D. Farmer, Journal of Economic Dynamics and Control **32**, 200 (2008).

[916]   L. Muchnik, F. Slanina, and S. Solomon, Physica A **330**, 232 (2003).

[917]   A. Svorenčík and F. Slanina, Eur. Phys. J. B **57**, 453 (2007).

[918]   F. Slanina, Phys. Rev. E **69**, 046102 (2004).

[919]   R. D. Vigil, R. M. Ziff, and B. Lu, Phys. Rev. B **38**, 942 (1988).

[920]   P. L. Krapivsky and S. Redner, Phys. Rev. E **54**, 3553 (1996).

[921]   S. N. Majumdar, S. Krishnamurthy, and M. Barma, J. Stat. Phys. **99**, 1 (2000).

[922]   Z. Burda, D. Johnston, J. Jurkiewicz, M. Kamiński, M. A. Nowak, G. Papp, and I. Zahed, Phys. Rev. E **65**, 026102 (2002).

[923]   F. Slanina, K. Sznajd-Weron, and P. Przybyła, Europhys. Lett. **82**, 18006 (2008).

[924]   R. Rajesh and S. N. Majumdar, Phys. Rev. E **63**, 036114 (2001).

[925]   S. N. Coppersmith, C.-H. Liu, S. N. Majumdar, O. Narayan, and T. A. Witten, Phys. Rev. E **53**, 4673 (1996).

[926]   M. von Smoluchowski, Phys. Z. **17**, 557 (1916).

[927]   H. Takayasu, I. Nishikawa, and H. Tasaki, Phys. Rev. A **37**, 3110 (1988).

[928]   H. Takayasu, Phys. Rev. Lett. **63**, 2563 (1989).

[929]   S. N. Majumdar and C. Sire, Phys. Rev. Lett. **71**, 3729 (1993).

[930]   D. ben-Avraham, M. A. Burschka, and C. R. Doering, J. Stat. Phys. **60**, 695 (1990).

[931]   S. N. Majumdar and D. A. Huse, Phys. Rev. E **52**, 270 (1995).

[932]   S. N. Majumdar, S. Krishnamurthy, and M. Barma, Phys. Rev. Lett. **81**, 3691 (1998).

[933]   N. N. Lebedev, *Special Functions and Their Applications* (Dover, New York, 1972).

[934]   F. Slanina, Phys. Rev. E **64**, 056136 (2001).

[935]   Z.-F. Huang and S. Solomon, Eur. Phys. J. B **20**, 601 (2001).

[936]   D. Sornette, Physica A **250**, 295 (1998).

[937]   A. Crisanti, G. Paladin, and A. Vulpiani, *Products of Random Matrices in Statistical Physics* (Springer-Verlag, Berlin, 1993).

[938]   M. Weigt, J. Phys. A: Math. Gen. **31**, 951 (1998).

[939]   A. L. Fetter and J. D. Walecka, *Quantum Theory of Many-Particle Systems* (McGraw-Hill, New York, 1971).

[940]   M. B. Garman, Journal of Financial Economics **3**, 257 (1976).

[941] H. Mendelson, Econometrica **50**, 1505 (1982).
[942] A. S. Kyle, Econometrica **53**, 1315 (1985).
[943] L. R. Glosten and P. R. Milgrom, Journal of Financial Economics **14**, 71 (1985).
[944] K. Back and S. Baruch, Econometrica **72**, 433 (2004).
[945] A. S. Kyle, Review of Economic Studies **56**, 317 (1989).
[946] S. Chakravarty and C. W. Holden, Journal of Financial Intermediation **4**, 213 (1995).
[947] T. Foucault, Journal of Financial Markets **2**, 99 (1999).
[948] C. A. Parlour, Rev. Financ. Stud. **11**, 789 (1998).
[949] B. Hollifield, R. A. Miller, P. Sandås, and J. Slive, Journal of Finance **6**, 2753 (2006).
[950] I. Roşu, Rev. Financ. Stud. **22**, 4601 (2009).
[951] R. L. Goettler, C. A. Parlour, and U. Rajan, Journal of Finance **60**, 2149 (2005).
[952] T. Foucault, O. Kadan, and E. Kandel, Rev. Financ. Stud. **18**, 1171 (2005).
[953] P. Lerner, arXiv:1204.1410.
[954] K. J. Cohen, S. F. Maier, R. A. Schwartz, and D. K. Whitcomb, Journal of Political Economy **89**, 287 (1981).
[955] A. Langnau and Y. Punchev, arXiv:1105.4789.
[956] A. Alfonsi, A. Fruth, and A. Schied, Quant. Finance **10**, 143 (2010).
[957] A. Alfonsi and A. Schied, SIAM J. Financial Math. **1**, 490 (2010).
[958] A. Alfonsi, A. Schied, and A. Slynko, preprint: *Order book resilience, price manipulation, and the positive portfolio problem* (2009).
[959] A. Alfonsi and J. I. Acevedo, arXiv:1204.2736.
[960] A. Weiss, arXiv:0904.4131.
[961] S. Predoiu, G. Shaikhet, and S. Shreve, SIAM J. Finan. Math. **2**, 183 (2011).
[962] A. Fruth, T. Schöneborn, and M. Urusov, arXiv:1109.263.
[963] O. Guéant, C.-A. Lehalle, and J. F. Tapia, arXiv:1106.3279.
[964] F. Guilbaud and H. Pham, arXiv:1106.5040.
[965] E. Bayraktar and M. Ludkovski, arXiv:1105.0247.
[966] R. Cont and A. de Larrard, arXiv:1104.4596.
[967] R. Cont and A. de Larrard, arXiv:1202.6412.
[968] R. Cont, S. Stoikov, and R. Talreja, Operations Research **58**, 549 (2010).
[969] I. Muni Toke, in: *Econophysics of Order-driven Markets*, ed. F. Abergel, B. K. Chakrabarti, A. Chakraborti, and M. Mitra, p. 49 (Springer, Milan, 2011).
[970] I. Domowitz and J. Wang, Journal of Economic Dynamics and Control **18**, 29 (1994).
[971] T. Bollerslev, I. Domowitz, and J. Wang, Journal of Economic Dynamics and Control **21**, 1471 (1997).
[972] M. Šmíd, Kybernetika **48**, 50 (2012).
[973] F. Abergel and A. Jedidi, in: *Econophysics of Order-driven Markets*, ed. F. Abergel, B. K. Chakrabarti, A. Chakraborti, and M. Mitra, p. 93 (Springer, Milan, 2011).
[974] B. Derrida, Phys. Rep. **301**, 65 (1998).
[975] R. D. Willmann, G. M. Schütz, and D. Challet, Physica A **316**, 430 (2002).
[976] T. Preis, S. Golke, W. Paul, and J. J. Schneider, Europhys. Lett. **75**, 510 (2006).
[977] T. Preis, S. Golke, W. Paul, and J. J. Schneider, Phys. Rev. E **76**, 016108 (2007).
[978] T. Preis, J. Phys.: Conf. Ser. **221**, 012019 (2010).
[979] S. Mike and J. D. Farmer, arXiv:physics/0509194.
[980] G.-F. Gu and W.-X. Zhou, Eur. Phys. J. B **67**, 585 (2009).
[981] A. Zaccaria, M. Cristelli, V. Alfi, F. Ciulla, and L. Pietronero, Phys. Rev. E **81**, 066101 (2010).
[982] M. Cristelli, V. Alfi, L. Pietronero, and A. Zaccaria, Eur. Phys. J. B **73**, 41 (2010).
[983] D. Challet and R. Stinchcombe, arXiv:cond-mat/0208025.
[984] C. Chiarella and G. Iori, Quant. Finance **2**, 346 (2002).
[985] C. Chiarella, G. Iori, and J. Perelló, Journal of Economic Dynamics and Control **33**, 525 (2009).
[986] B. LeBaron and R. Yamamoto, Physica A **383**, 85 (2007).
[987] G. Tedeschi, G. Iori, and M. Gallegati, Eur. Phys. J. B **71**, 489 (2009).
[988] Z. Eisler, J.-P. Bouchaud, and J. Kockelkoren, in: *Market Microstructure: Confronting Many Viewpoints*, ed. F. Abergel, J.-P. Bouchaud, T. Foucault, and C.-A. Lehalle, p. 115 (Willey, Chichester, 2012).
[989] T. Hirabayashi, H. Takayasu, H. Miura, and K. Hamada, Fractals **1**, 29 (1993).
[990] H. Takayasu, H. Miura, T. Hirabayashi, and K. Hamada, Physica A **184**, 127 (1992).
[991] A.-H. Sato and H. Takayasu, Physica A **250**, 231 (1998).

[992] J. Maskawa, Physica A **382**, 172 (2007).

[993] M. Bartolozzi, Eur. Phys. J. B **78**, 265 (2010).

[994] J.-J. Tseng, C.-H. Lin, C.-T. Lin, S.-C. Wang, and S.-P. Li, Physica A **389**, 1699 (2010).

[995] F. Franci and L. Matassini, arXiv:cond-mat/0008466.

[996] L. Matassini and F. Franci, arXiv:cond-mat/0103106.

[997] L. Matassini and F. Franci, Physica A **289**, 526 (2001).

[998] F. Franci, R. Marschinski, and L. Matassini, Physica A **294**, 213 (2001).

[999] A. Bovier, J. Černý, and O. Hryniv, Int. J. Theor. Appl. Finance **9**, 91 (2006).

[1000] N. Vvedenskaya, Y. Suhov, and V. Belitsky, arXiv:1102.1104.

[1001] S. N. Cohen and L. Szpruch, arXiv:1110.4811.

[1002] C. A. Parlour and D. J. Seppi, in: *Handbook of Financial Intermediation and Banking*, ed. A. V. Thakor and A. W. A. Boot, p. 63 (Elsevier, Amsterdam, 2008).

[1003] M. D. Gould, M. A. Porter, S. Williams, M. McDonald, D. J. Fenn, and S. D. Howison, arXiv:1012.0349.

[1004] W. B. Arthur, Amer. Econ. Review (Papers and Proceedings) **84**, 406 (1994).

[1005] D. Challet and Y.-C. Zhang, Physica A **246**, 407 (1997).

[1006] D. Challet and Y.-C. Zhang, Physica A **256**, 514 (1998).

[1007] R. Savit, R. Manuca, and R. Riolo, arXiv:adap-org/9712006.

[1008] R. Savit, R. Manuca, and R. Riolo, Phys. Rev. Lett. **82**, 2203 (1999).

[1009] D. Challet and M. Marsili, Phys. Rev. E **60**, R6271 (1999).

[1010] M. Mézard, G. Parisi, and M. A. Virasoro, *Spin Glass Theory and Beyond* (World Scientific, Singapore 1987).

[1011] P. C. Martin, E. D. Siggia, and H. A. Rose, Phys. Rev. A **8**, 423 (1973).

[1012] C. De Dominicis, Phys. Rev. B **18**, 4913 (1978).

[1013] A. Cavagna, Phys. Rev. E **59**, R3783 (1999).

[1014] D. Challet and M. Marsili, Phys. Rev. E **62**, 1862 (2000).

[1015] D. Sherrington and S. Kirkpatrick, Phys. Rev. Lett. **35**, 1792 (1975).

[1016] J. J. Hopfield, Proc. Natl. Acad. Sci. USA **79**, 2554 (1982).

[1017] D. J. Amit, *Modeling Brain Function* (Cambridge University Press, Cambridge, 1989).

[1018] N. Brunel and R. Zecchina, Phys. Rev. E **49**, 1823 (1994).

[1019] A. Cavagna, J. P. Garrahan, I. Giardina, and D. Sherrington, Phys. Rev. Lett. **83**, 4429 (1999).

[1020] D. Challet, M. Marsili, and R. Zecchina, Phys. Rev. Lett. **85**, 5008 (2000).

[1021] A. Cavagna, J. P. Garrahan, I. Giardina and D. Sherrington, Phys. Rev. Lett. **85**, 5009 (2000).

[1022] J. P. Garrahan, E. Moro, and D. Sherrington, Phys. Rev. E **62**, R9 (2000).

[1023] D. Sherrington, J. P. Garrahan, and E. Moro, arXiv:cond-mat/0010455.

[1024] J. P. Garrahan, E. Moro, and D. Sherrington, Quant. Finance **1**, 246 (2001).

[1025] D. Sherrington and T. Galla, Physica A **324**, 25 (2003).

[1026] D. Sherrington, Physica A **370**, 7 (2006).

[1027] D. Challet, M. Marsili, and R. Zecchina, Phys. Rev. Lett. **84**, 1824 (2000).

[1028] D. Challet, M. Marsili, and R. Zecchina, arXiv:cond-mat/9904392.

[1029] M. Marsili, D. Challet, and R. Zecchina, Physica A **280**, 522 (2000).

[1030] D. Challet, M. Marsili, and R. Zecchina, Int. J. Theor. Appl. Finance **3**, 451 (2000).

[1031] M. Marsili and D. Challet, Phys. Rev. E **64**, 056138 (2001).

[1032] A. De Martino and M. Marsili, J. Phys. A: Math. Gen. **34**, 2525 (2001).

[1033] D. Challet, M. Marsili, and Y.-C. Zhang, Physica A **276**, 284 (2000).

[1034] D. Challet, M. Marsili, and Y.-C. Zhang, Physica A **294**, 514 (2001).

[1035] D. Challet, A. Chessa, M. Marsili, and Y.-C. Zhang, Quant. Finance **1**, 168 (2001).

[1036] L. D. Landau and E. M. Lifshitz, *Statistical Physics* (Butterworth-Heinemann, Oxford, 1980).

[1037] D. Challet and M. Marsili, Phys. Rev. E **68**, 036132 (2003).

[1038] I. Giardina, J.-P. Bouchaud, and M. Mézard, Physica A **299**, 28 (2001).

[1039] I. Giardina and J.-P. Bouchaud, Eur. Phys. J. B **31**, 421 (2003).

[1040] I. Giardina and J.-P. Bouchaud, Physica A **324**, 6 (2003).

[1041] M. Marsili, Physica A **324**, 17 (2003).

[1042] D. Challet, M. Marsili, and Y.-C. Zhang, Physica A **299**, 228 (2001).

[1043] A. De Martino and M. Marsili, J. Phys. A: Math. Gen. **39**, R465 (2006).

[1044] T. Galla, G. Mosetti, and Y.-C. Zhang, arXiv:physics/0608091.

[1045] J. A. F. Heimel and A. C. C. Coolen, Phys. Rev. E **63**, 056121 (2001).

[1046] A. C. C. Coolen, J. A. F. Heimel, and D. Sherrington, Phys. Rev. E **65**, 016126 (2002).

[1047] A. C. C. Coolen and J. A. F. Heimel, J. Phys. A: Math. Gen. **34**, 10783 (2001).
[1048] T. Galla, A. C. C. Coolen, and D. Sherrington, J. Phys. A: Math. Gen. **36**, 11159 (2003).
[1049] A. C. C. Coolen, J. Phys. A: Math. Gen. **38**, 2311 (2005).
[1050] A. C. C. Coolen, arXiv:cond-mat/0205262.
[1051] J. A. F. Heimel and A. De Martino, J. Phys. A: Math. Gen. **34**, L539 (2001).
[1052] N. F. Johnson, P. M. Hui, D. Zheng, and M. Hart, J. Phys. A: Math. Gen. **32**, L427 (1999).
[1053] N. F. Johnson, P. M. Hui, R. Jonson, and T. S. Lo, Phys. Rev. Lett. **82**, 3360 (1999).
[1054] P. Jefferies, M. Hart, N. F. Johnson, and P. M. Hui, J. Phys. A: Math. Gen. **33**, L409 (2000).
[1055] N. F. Johnson, D. J. T. Leonard, P. M. Hui, and T. S. Lo, Physica A **283**, 568 (2000).
[1056] N. F. Johnson, P. M. Hui, D. Zheng, and C. W. Tai, Physica A **269**, 493 (1999).
[1057] N. F. Johnson, S. Jarvis, R. Jonson, P. Cheung, Y. R. Kwong, and P. M. Hui, Physica A **258**, 230 (1998).
[1058] N. F. Johnson, M. Hart, and P. M. Hui, Physica A **269**, 1 (1998).
[1059] M. Hart, P. Jefferies, N. F. Johnson, and P. M. Hui, arXiv:cond-mat/0003486.
[1060] M. Hart, P. Jefferies, N. F. Johnson, and P. M. Hui, Physica A **298**, 537 (2001).
[1061] M. Hart, P. Jefferies, N. F. Johnson, and P. M. Hui, Phys. Rev. E **63**, 017102 (2001).
[1062] M. Hart, P. Jefferies, P. M. Hui, and N. F. Johnson, Eur. Phys. J. B **20**, 547 (2001).
[1063] T. S. Lo, P. M. Hui, and N. F. Johnson, Phys. Rev. E **62**, 4393 (2000).
[1064] M. Hart, P. Jefferies, N. F. Johnson, and P. M. Hui, arXiv:cond-mat/0006141.
[1065] T. S. Lo, S. W. Lim, P. M. Hui, and N. F. Johnson, Physica A **287**, 313 (2000).
[1066] P. M. Hui, T. S. Lo, and N. F. Johnson, Physica A **288**, 451 (2000).
[1067] P. Jefferies, M. L. Hart, and N. F. Johnson, Phys. Rev. E **65**, 016105 (2001).
[1068] K. F. Yip, T. S. Lo, P. M. Hui, and N. F. Johnson, Phys. Rev. E **69**, 046120 (2004).
[1069] N. F. Johnson, S. C. Choe, S. Gourley, T. Jarrett, and P. M. Hui, arXiv:cond-mat/0403158.
[1070] N. F. Johnson, D. M. D. Smith, and P. M. Hui, Europhys. Lett. **74**, 923 (2006).
[1071] D. Lamper, P. Jefferies, M. Hart, and N. F. Johnson, arXiv:cond-mat/0210132.
[1072] P. Jefferies and N. F. Johnson, arXiv:cond-mat/0207523.
[1073] N. F. Johnson, M. Hart, P. M. Hui, and D. Zheng, Int. J. Theor. Appl. Finance **3**, 443 (2000).
[1074] P. Jefferies, M. Hart, P. M. Hui, and N. F. Johnson, Eur. Phys. J. B **20**, 493 (2001).
[1075] M. L. Hart, P. Jefferies, and N. F. Johnson, Physica A **311**, 275 (2002).
[1076] N. F. Johnson, D. Lamper, P. Jefferies, M. L. Hart, and S. Howison, Physica A **299**, 222 (2001).
[1077] K. H. Ho, W. C. Man, F. K. Chow, and H. F. Chau, Phys. Rev. E **71**, 066120 (2005).
[1078] P. Ruch, J. Wakeling, and Y.-C. Zhang, arXiv:cond-mat/0208310.
[1079] P. Laureti, P. Ruch, J. Wakeling, and Y.-C. Zhang, Physica A **331**, 651 (2004).
[1080] M. Marsili, Physica A **299**, 93 (2001).
[1081] V. Alfi, A. De Martino, L. Pietronero, and A. Tedeschi, Physica A **382**, 1 (2007).
[1082] P. Kozłowski and M. Marsili, J. Phys. A: Math. Gen. **36**, 11725 (2003).
[1083] A. De Martino, I. Giardina, and G. Mosetti, J. Phys. A: Math. Gen. **36**, 8935 (2003).
[1084] D. Challet and T. Galla, Quant. Finance **5**, 569 (2005).
[1085] A. De Martino, I. Giardina, A. Tedeschi, and M. Marsili, Phys. Rev. E **70**, 025104 (2004).
[1086] C.-L. Gou, Chinese Phys. **15**, 1239 (2006).
[1087] J. V. Andersen and D. Sornette, Eur. Phys. J. B **31**, 141 (2003).
[1088] F. F. Ferreira and M. Marsili, Physica A **345**, 657 (2005).
[1089] F. Slanina and Y.-C. Zhang, Physica A **289**, 290 (2001).
[1090] R. Manuca, Y. Li, R. Riolo, and R. Savit, arXiv:adap-org/9811005.
[1091] Y. Li, R. Riolo, and R. Savit, Physica A **276**, 234 (2000).
[1092] Y. Li, R. Riolo, and R. Savit, Physica A **276**, 265 (2000).
[1093] G. Bianconi, A. De Martino, F. F. Ferreira, and M. Marsili, Quant. Finance **8**, 225 (2008).
[1094] F. K. Chow and H. F. Chau, Physica A **319**, 601 (2003).
[1095] L. Ein-Dor, R. Metzler, I. Kanter, and W. Kinzel, Phys. Rev. E **63**, 066103 (2001).
[1096] M. Marsili, R. Mulet, F. Ricci-Tersenghi, and R. Zecchina, Phys. Rev. Lett. **87**, 208701 (2001).
[1097] D. Challet, M. Marsili, and A. De Martino, Physica A **338**, 143 (2004).
[1098] D. Challet, A. De Martino, M. Marsili, and I. Pérez Castillo, J. Stat. Mech. P03004 (2006).
[1099] A. De Martino, Eur. Phys. J. B **35**, 143 (2003).
[1100] M. Marsili and M. Piai, Physica A **310**, 234 (2002).
[1101] G. Mosetti, D. Challet, and Y.-C. Zhang, Physica A **365**, 529 (2006).
[1102] M. Pištěk and F. Slanina, Physica A **390**, 2549 (2011).
[1103] S. Moelbert and P. De Los Rios, Physica A **303**, 217 (2002).

[1104] E. Burgos, H. Ceva, and R. P. J. Perazzo, Physica A **337**, 635 (2004).
[1105] E. Burgos, H. Ceva, and R. P. J. Perazzo, Physica A **354**, 518 (2005).
[1106] G. Fagiolo and M. Valente, Computational Economics **25**, 41 (2005).
[1107] T. Kalinowski, H.-J. Schulz, and M. Briese, Physica A **277**, 502 (2000).
[1108] F. Slanina, Physica A **286**, 367 (2000).
[1109] F. Slanina, Physica A **299**, 334 (2001).
[1110] M. Anghel, Z. Toroczkai, K. E. Bassler, and G. Korniss, Phys. Rev. Lett. **92**, 058701 (2004).
[1111] H. Lavička and F. Slanina, Eur. Phys. J. B **56**, 53 (2007).
[1112] S. C. Choe, N. F. Johnson, and P. M. Hui, Phys. Rev. E **70**, 055101 (2004).
[1113] S. Gourley, S. C. Choe, P. M. Hui, and N. F. Johnson, Europhys. Lett. **67**, 867 (2004).
[1114] S. Gourley, S. C. Choe, P. M. Hui, and N. F. Johnson, arXiv:cond-mat/0401527.
[1115] T. S. Lo, H. Y. Chan, P. M. Hui, and N. F. Johnson, Phys. Rev. E **71**, 050101 (2005).
[1116] S. H. Lee and H. Jeong, Phys. Rev. E **74**, 026118 (2006).
[1117] S. H. Lee and H. Jeong, J. Korean Phys. Soc. **48**, S186 (2006).
[1118] L. Shang and X. F. Wang, Physica A **361**, 643 (2006).
[1119] H.-J. Quan, B.-H. Wang, P. M. Hui, and X.-S. Luo, Physica A **321**, 300 (2003).
[1120] H.-J. Quan, B.-H. Wang, W.-S. Yang, W.-N. Wang, and X.-S. Luo, Acta Physica Sinica **51**, 2670 (2002).
[1121] H.-J. Quan, B.-H. Wang, P. M. Hui and X.-S. Luo, Chinese Phys. Lett. **18**, 1156 (2001).
[1122] H. Quan, B.-H. Wang, and P. M. Hui, Physica A **312**, 619 (2002).
[1123] T. S. Lo, H. Y. Chan, P. M. Hui, and N. F. Johnson, Phys. Rev. E **70**, 05610 (2004).
[1124] R. D'Hulst and G. J. Rodgers, Physica A **270**, 514 (1999).
[1125] R. D'Hulst and G. J. Rodgers, arXiv:adap-org/9904003.
[1126] R. D'Hulst and G. J. Rodgers, Physica A **278**, 579 (2000).
[1127] R. D'Hulst and G. J. Rodgers, Physica A **278**, 579 (2000).
[1128] H. F. Chau and F. K. Chow, Physica A **312**, 277 (2002).
[1129] M. Andrecut and M. K. Ali, Phys. Rev. E **64**, 067103 (2001).
[1130] M. Paczuski, K. E. Bassler, and A. Corral, Phys. Rev. Lett. **84**, 3185 (2000).
[1131] G. Reents, R. Metzler, and W. Kinzel, Physica A **299**, 253 (2001).
[1132] R. Mansilla, Phys. Rev. E **62**, 4553 (2000).
[1133] M. A. R. de Cara, O. Pla, and F. Guinea, Eur. Phys. J. B **13**, 413 (2000).
[1134] M. A. R. de Cara, O. Pla, and F. Guinea, Eur. Phys. J. B **10**, 187 (1999).
[1135] D. Zheng and B.-H. Wang, Physica A **301**, 560 (2001).
[1136] A. Vázquez, Phys. Rev. E **62**, R4497 (2000).
[1137] J. Menche and J. R. L. de Almeida, Braz. J. Phys. **33**, 895 (2003).
[1138] J. R. L. de Almeida and J. Menche, arXiv:cond-mat/0312139.
[1139] K. Y. M. Wong, S. W. Lim, and Z. Gao, Phys. Rev. E **70**, 025103 (2004).
[1140] K. Y. M. Wong, S. W. Lim, and Z. Gao, Phys. Rev. E **71**, 066103 (2005).
[1141] B. Zheng, T. Qiu, and F. Ren, Phys. Rev. E **69**, 046115 (2004).
[1142] B. Yuan and K. Chen, Phys. Rev. E **69**, 067106 (2004).
[1143] C. Földy, Z. Somogyvári, and P. Érdi, Physica A **323**, 735 (2003).
[1144] C.-Y. Lee, Phys. Rev. E **64**, 015102 (2001).
[1145] M. Yang, W.-H. Qiu, and X.-Q. Lou, Control and Decision **19**, 420 (2004).
[1146] J. Wakeling and P. Bak, Phys. Rev. E **64**, 051920 (2001).
[1147] S. Hod and E. Nakar, Phys. Rev. E **69**, 066122 (2004).
[1148] S. Hod and E. Nakar, Phys. Rev. E **68**, 026115 (2003).
[1149] S. Hod and E. Nakar, Phys. Rev. Lett. **91**, 189802 (2003).
[1150] S. Hod and E. Nakar, Phys. Rev. Lett. **88**, 238702 (2002).
[1151] E. Nakar and S. Hod, Phys. Rev. E **67**, 016109 (2003).
[1152] E. Burgos, H. Ceva, and R. P. J. Perazzo, Phys. Rev. Lett. **91**, 189801 (2003).
[1153] I. Caridi and H. Ceva, Physica A **317**, 247 (2003).
[1154] E. Burgos, H. Ceva, and R. P. J. Perazzo, Physica A **294**, 539 (2001).
[1155] E. Burgos and H. Ceva, Physica A **284**, 489 (2000).
[1156] H. Ceva, Physica A **277**, 496 (2000).
[1157] E. Burgos, H. Ceva, and R. P. J. Perazzo, Phys. Rev. E **64**, 016130 (2001).
[1158] E. Burgos, H. Ceva, and R. P. J. Perazzo, Phys. Rev. E **65**, 036711 (2002).
[1159] J. Berg, M. Marsili, A. Rustichini, and R. Zecchina, Quant. Finance **1**, 203 (2001).
[1160] M. Marsili and D. Challet, arXiv:cond-mat/0004376.
[1161] D. Challet, Physica A **344**, 24 (2004).
[1162] D. Challet, arXiv:cond-mat/0210319.

[1163] A. De Martino, I. Pérez Castillo, and D. Sherrington, J. Stat. Mech. P01006 (2007).
[1164] D. A. Meyer, Phys. Rev. Lett. **82**, 1052 (1999).
[1165] S. C. Benjamin and P. M. Hayden, Phys. Rev. A **64**, 030301 (2001).
[1166] R. Kay, N. F. Johnson, and S. C. Benjamin, J. Phys. A: Math. Gen. **34**, L547 (2001).
[1167] E. W. Piotrowski and J. Sładkowski, Physica A **308**, 391 (2002).
[1168] N. F. Johnson, Phys. Rev. A **63**, 020302 (2001).
[1169] R. Diestel, *Graph Theory* (Springer, New York, 2000).
[1170] P. Erdős and A. Rényi, Pub. Math. Debrecen **5**, 290 (1959).
[1171] P. Erdős and A. Rényi, Magyar Tud. Akad. Mat. Kutató Int. Közl. **5**, 17 (1960).
[1172] R.Solomonoff and A. Rapoport, Bull. Math. Biophys. **13**, 107 (1951).
[1173] R. J. Riddell Jr. and G. E. Uhlenbeck, J. Chem. Phys. **21**, 2056 (1953).
[1174] E. N. Gilbert, Ann. Math. Stat. **30**, 1141 (1959).
[1175] T. L. Austin, R. E. Fagen, W. F. Penney, and J. Riordan, Ann. Math. Stat. **30**, 747 (1959).
[1176] B. Bollobás and O. M. Riordan, in: *Handbook of Graphs and Networks*, ed. S. Bornholdt and H. G. Schuster, p. 1 (Wiley-VCH, Weinheim, 2003).
[1177] S. Bornholdt and H. G. Schuster (Eds.), *Handbook of Graphs and Networks* (Wiley-VCH, Weinheim, 2003).
[1178] M. Molloy and B. Reed, Random Structures and Algorithms **6**, 161 (1995).
[1179] A. Krzywicki, arXiv:cond-mat/0110574.
[1180] Z. Burda, J. D. Correia, and A. Krzywicki, Phys. Rev. E **64**, 046118 (2001).
[1181] Z. Burda and A. Krzywicki, Phys. Rev. E **67**, 046118 (003).
[1182] S. N. Dorogovtsev, J. F. F. Mendes, and A. N. Samukhin, arXiv:cond-mat/0204111.
[1183] S. N. Dorogovtsev, J. F. F. Mendes, and A. N. Samukhin, Nucl. Phys. B **666**, 396 (2003).
[1184] J. Berg and M. Lässig, Phys. Rev. Lett. **89**, 228701 (2002).
[1185] D. J. Watts and S. H. Strogatz, Nature **393**, 440 (1998).
[1186] M. E. J. Newman, C. Moore, and D. J. Watts, Phys. Rev. Lett. **84**, 3201 (2000).
[1187] M. E. J. Newman and D. J. Watts, Phys. Rev. E **60**, 7332 (1999).
[1188] A.-L. Barabási and R. Albert, Science **286**, 509 (1999).
[1189] A.-L. Barabási, R. Albert, and H. Jeong, Physica A **281**, 69 (2000).
[1190] S. N. Dorogovtsev, J. F. F. Mendes, and A. N. Samukhin, Phys. Rev. Lett. **85**, 4633 (2000).
[1191] P. L. Krapivsky, S. Redner, and F. Leyvraz, Phys. Rev. Lett. **85**, 4629 (2000).
[1192] K. Klemm and V. M. Eguíluz, Phys. Rev. E **65**, 036123 (2002).
[1193] K. Klemm and V. M. Eguíluz, Phys. Rev. E **65**, 057102 (2002).
[1194] G. Bianconi and A.-L. Barabási, Phys. Rev. Lett. **86**, 5632 (2001).
[1195] G. Bianconi and A.-L. Barabási, Europhys. Lett. **54**, 436 (2001).
[1196] H. Jeong, Z. Néda, and A.-L. Barabási, Europhys. Lett. **61**, 567 (2003).
[1197] S.-H. Yook, H. Jeong, A.-L. Barabási, and Y. Tu, Phys. Rev. Lett. **86**, 5835 (2001).
[1198] P. L. Krapivsky and D. Redner, Phys. Rev. E **63**, 066123 (2001).
[1199] P. L. Krapivsky, G. J. Rodgers, and S. Redner, Phys. Rev. Lett. **86**, 5401 (2001).
[1200] S. N. Dorogovtsev and J. F. F. Mendes, Phys. Rev. E **62**, 1842 (2000).
[1201] S. N. Dorogovtsev and J. F. F. Mendes, Phys. Rev. E **63**, 025101 (2001).
[1202] S. N. Dorogovtsev and J. F. F. Mendes, Phys. Rev. E **63**, 056125 (2001).
[1203] S. N. Dorogovtsev, J. F. F. Mendes, and A. N. Samukhin, arXiv:cond-mat/0011077.
[1204] S. N. Dorogovtsev, J. F. F. Mendes, and A. N. Samukhin, arXiv:cond-mat/0106142.
[1205] S. N. Dorogovtsev, J. F. F. Mendes, and A. N. Samukhin, arXiv:cond-mat/0009090.
[1206] S. N. Dorogovtsev, J. F. F. Mendes, and A. N. Samukhin, Phys. Rev. E **63**, 062101 (2001).
[1207] S. Bornholdt and H. Ebel, arXiv:cond-mat/0008465.
[1208] B. A. Huberman and L. A. Adamic, arXiv:cond-mat/9901071.
[1209] B. Tadić, Physica A **293**, 273 (2001).
[1210] A. Vazquez, Europhys. Lett. **54**, 430 (2001).
[1211] G. Ergün and G. J. Rodgers, arXiv:cond-mat/0103423.
[1212] L. Kullmann and J. Kertész, Phys. Rev. E **63**, 051112 (2001).
[1213] S. Mossa, M. Barthélémy, H. E. Stanley, and L. A. N. Amaral, arXiv:cond-mat/0201421.
[1214] M. Catanzaro, G. Caldarelli, and L. Pietronero, Phys. Rev. E **70**, 037101 (2004).
[1215] G. Caldarelli, A. Capocci, P. De Los Rios, and M. A. Muñoz, Phys. Rev. Lett. **89**, 258702 (2002).
[1216] E. Ravasz, A. L. Somera, D. A. Mongru, Z. N. Oltvai, A.-L.Barabási, Science **297**, 1551 (2002).
[1217] S. Maslov and K. Sneppen, Science **296**, 910 (2002).

[1218] J.-D. J. Han, N. Bertin, T. Hao, D. S. Goldberg, G. F. Berriz, L. V. Zhang, D. Dupuy, A. J. M. Walhout, M. E. Cusick, F. P. Roth, and M. Vidal, Nature **430**, 88 (2004).
[1219] K. A. Eriksen, I. Simonsen, S. Maslov, and K. Sneppen, Phys. Rev. Lett. **90**, 148701 (2003).
[1220] A.-L. Barabási, E. Ravasz, and T. Vicsek, Physica A **299**, 559 (2001).
[1221] S. N. Dorogovtsev, A. V. Goltsev, and J. F. F. Mendes, Phys. Rev. E **65**, 066122 (2002).
[1222] E. Ravasz and A.-L. Barabási, Phys. Rev. E **67**, 026112 (2003).
[1223] F. Comellas and M. Sampels, arXiv:cond-mat/0111194.
[1224] S. Jung, S. Kim, and B. Kahng, Phys. Rev. E **65**, 056101 (2002).
[1225] F. Comellas, G. Fertin, and A. Raspaud, Phys. Rev. E **69**, 037104 (2004).
[1226] J. D. Noh and H. Rieger, Phys. Rev. E **69**, 036111 (2004).
[1227] J. C. Nacher, N. Ueda, M. Kanehisa, and T. Akutsu, Phys. Rev. E **71**, 36132 (2005).
[1228] K. Iguchi and H. Yamada, Phys. Rev. E **71**, 036144 (2005).
[1229] D. Nagy, G. Tibély, and J. Kertész, arXiv:cond-mat/0506395.
[1230] J. S. Andrade Jr., H. J. Herrmann, R. F. S. Andrade, L. R. da Silva, arXiv:cond-mat/0406295.
[1231] T. Zhou, G. Yan, P.-L. Zhou, Z.-Q. Fu, and B.-H. Wang, arXiv:cond-mat/0409414.
[1232] J. P. K. Doye and C. P. Massen, Phys. Rev. E **71**, 016128 (2005).
[1233] R. Kumar, P. Raghavan, S. Rajagopalan, and A. Tomkins, Proceedings of the WWW8 conference, http://www8.org/w8-papers/4a-search-mining/trawling/trawling.html (1999).
[1234] R. Albert, H. Jeong, and A.-L. Barabási, Nature **401**, 130 (1999).
[1235] L. A. Adamic and B. A. Huberman, Science **287**, 2115a (2000).
[1236] A.-L. Barabási, R. Albert, H. Jeong, and G. Bianconi, Science **287**, 2115a (2000).
[1237] M. Faloutsos, P. Faloutsos, and C. Faloutsos, ACM SIGCOMM Computer Communication Review **29**, 251 (1999).
[1238] A. Broder, R. Kumar, F. Maghoul, P. Raghavan, S. Rajagopalan, R. Stata, A. Tomkins, and J. Wiener, Computer Networks **33**, 309 (2000).
[1239] S. Milgram, Psychology Today **1**, 61 (1967).
[1240] J. Travers and S. Milgram, Sociometry **32**, 425 (1969).
[1241] M. Granovetter, American Journal of Sociology **78**, 1360 (1973).
[1242] S. Srinivasan, *Advanced Perl Programming* (O' Reilly, Sebastopol, 1997).
[1243] C. Song, S. Havlin, and H. A. Makse, Nature **433**, 392 (2005).
[1244] S. N. Dorogovtsev and J. F. F. Mendes, *Evolution of Networks* (Oxford University Press, Oxford, 2003).
[1245] R. Pastor-Satorras, A. Vázquez, and A. Vespignani, Phys. Rev. Lett. **87**, 258701 (2001).
[1246] S.-H. Yook, H. Jeong, and A.-L. Barabási, Proc. Natl. Acad. Sci. USA **99**, 13382 (2002).
[1247] S. Redner, Eur. Phys. J. B **4**, 131 (1998).
[1248] A. Vazquez, arXiv:cond-mat/0105031.
[1249] H. Ebel, L.-I. Mielsch, and S. Bornholdt, Phys. Rev. E **66**, 035103 (2002).
[1250] G. Csányi and B. Szendrői, Phys. Rev. E **69**, 036131 (2004).
[1251] R. Ferrer i Cancho, C. Janssen, and R. V. Solé, Phys. Rev. E **64**, 046119 (2001).
[1252] D. J. Watts, *Small Worlds* (Priceton University Press, Princeton, 1999).
[1253] E. Ben-Naim, H. Frauenfelder, and Z. Toroczkai (Eds.), *Complex Networks* (Springer, Berlin, 2004).
[1254] S. H. Strogatz, Nature **410**, 268 (2001).
[1255] R. Albert and A.-L. Barabási, Rev. Mod. Phys. **74**, 47 (2002).
[1256] S. N. Dorogovtsev and J. F. F. Mendes, Adv. Phys. **51**, 1079 (2002).
[1257] S. Boccaletti, V. Latora, Y. Moreno, M. Chavez, and D.-U. Hwang, Phys. Rep. **424**, 175 (2006).
[1258] M. A. Serrano, A. Maguitman, M. Boguñá, S. Fortunato, and A. Vespignani, cs.NI/0511035 (2005).
[1259] F. Liljeros, C. R. Edling, L. A. N. Amaral, H. E. Stanley, and Y. Aberg, Nature **411**, 907 (2001).
[1260] L. A. N. Amaral, A. Scala, M. Barthélémy, and H. E. Stanley, Proc. Natl. Acad. Sci. USA **97**, 11149 (2000).
[1261] R. Ferrer i Cancho and R. V. Solé, Proc. Roy. Soc London B **268**, 2261 (2001).
[1262] M. Steyvers and J. B. Tenenbaum, arXiv:cond-mat/0110012.
[1263] M. Sigman and G. A. Cecchi, arXiv:cond-mat/0410609.
[1264] S. N. Dorogovtsev and J. F. F. Mendes, arXiv:cond-mat/0105093.
[1265] P. Uetz et al., Nature **403**, 623 (2000).
[1266] B. Schwikowski, P. Uetz, and S. Fields, Nature Biotechnology **18**, 1257 (2000).
[1267] J.-C. Rain et al., Nature **409**, 211 (2001).

[1268] H. Jeong, S. P. Mason, A.-L. Barabási, Z. N. Oltvai, Nature **411**, 41 (2001).
[1269] R. V. Solé, R. Pastor-Satorras, E. D. Smith, and T. Kepler, Adv. Compl. Syst. **5**, 4 (2002).
[1270] A. Vazquez, A. Flammini, A. Maritan, and A. Vespignani, arXiv:cond-mat/0108043.
[1271] J. Kim, P. L. Krapivsky, B. Kahng, and S. Redner, Phys. Rev. E **66**, 055101 (2002).
[1272] B. L. Chen, D. H. Hall, and D. B. Chklovskii, Proc. Natl. Acad. Sci. USA **103**, 4723 (2006).
[1273] D. M. Post, M. L. Pace, and N. G. Hairston, Nature **405**, 1047 (2000).
[1274] J. M. Montoya and R. V. Solé, arXiv:cond-mat/0011195.
[1275] J. Camacho, R. Guimerà, and L. A. N. Amaral, arXiv:cond-mat/0103114.
[1276] R. J. Williams, N. D. Martinez, E. L. Berlow, J. A. Dunne, and A.-L. Barabási, Proc. Natl. Acad. Sci. USA **99**, 12913 (2002).
[1277] B. Drossel and A. J. McKane, in: *Handbook of Graphs and Networks*, ed. S. Bornholdt and H. G. Schuster, p. 218 (Wiley-VCH, Weinheim, 2003).
[1278] F. Slanina and M. Kotrla, Phys. Rev. Lett. **83**, 5587 (1999).
[1279] R. J. Williams and N. D. Martinez, Nature **404**, 180 (2000).
[1280] J. Camacho, R. Guimerà, and L. A. N. Amaral, arXiv:cond-mat/0102127.
[1281] M. Lässig, U. Bastola, S. C. Manrubia, and A. Valleriani, Phys. Rev. Lett. **86**, 4418 (2001).
[1282] R. Alt and S. Klein, Electronic Markets **21**, 41 (2011).
[1283] P. Laureti, F. Slanina, Y.-K. Yu, and Y.-C. Zhang, Physica A **316**, 413 (2002).
[1284] I. Yang, H. Jeong, B. Kahng, and A.-L. Barabási, Phys. Rev. E **68**, 016102 (2003).
[1285] J. Reichardt and S. Bornholdt, arXiv:physics/0503138.
[1286] I. Yang, E. Oh, and B. Kahng, Phys. Rev. E **74**, 016121 (2006).
[1287] W. Jank and I. Yahav, Annals of Applied Statistics **4**, 151 (2010).
[1288] J. Hou and J. Blodgett, Electronic Markets **20**, 21 (2010).
[1289] R. D'Hulst and G. J. Rodgers, Physica A **294**, 447 (2001).
[1290] I. Yang, B. Kahng, arXiv:physics/0511073.
[1291] F. Slanina, Adv. Compl. Syst. **15**, 1250053 (2012).
[1292] M. E. J. Newman, Phys. Rev. Lett. **89**, 208701 (2002).
[1293] M. E. J. Newman and M. Girvan, in: *Statistical Mechanics of Complex Networks*, ed. R. Pastor-Satorras, J. Rubí and A. Díaz-Guilera, p. 66 (Springer, Berlin, 2003).
[1294] M. E. J. Newman, Phys. Rev. E **67**, 026126 (2003).
[1295] A. Capocci, G. Caldarelli, and P. De Los Rios, Phys. Rev. E **68**, 047101 (2003).
[1296] F. Slanina and Y.-C. Zhang, Acta Phys. Pol. B **36**, 2797 (2005).
[1297] R. Monasson, Eur. Phys. J. B **12**, 555 (1999).
[1298] G. Biroli and R. Monasson, J. Phys. A: Math. Gen. **32**, L255 (1999).
[1299] T. Kottos and U. Smilansky, Phys. Rev. Lett. **85**, 968 (2000).
[1300] I. J. Farkas, I. Derényi, A.-L. Barabási, and T. Vicsek, Phys. Rev. E **64**, 026704 (2001).
[1301] K.-I. Goh, B. Kahng, and D. Kim, Phys. Rev. E **64**, 051903 (2001).
[1302] D. S. Dean, J. Phys. A: Math. Gen. **35**, L153 (2002).
[1303] G. Semerjian and L. F. Cugliandolo, J. Phys. A: Math. Gen. **35**, 4837 (2002).
[1304] S. N. Dorogovtsev, A. V. Goltsev, J. F. F. Mendes, and A. N. Samukhin, Phys. Rev. E **68**, 046109 (2003).
[1305] I. Farkas, I. Derényi, H. Jeong, Z. Néda, Z. N. Oltvai, E. Ravasz, A. Schubert, A.-L. Barabási, and T. Vicsek, Physica A **314**, 25 (2002).
[1306] J. C. Noh and H. Rieger, Phys. Rev. Lett. **92**, 118701 (2004).
[1307] B. J. Kim, H. Hong, and M. Y. Choi, J. Phys. A: Math. Gen. **36**, 6329 (2003).
[1308] I. Simonsen, K. A. Eriksen, S. Maslov, and K. Sneppen, Physica A **336**, 163 (2004).
[1309] C. Kamp and K. Christensen, Phys. Rev. E **71**, 041911 (2005).
[1310] R. Ferrer i Cancho, A. Capocci, and G. Caldarelli, arXiv:cond-mat/0504165.
[1311] I. Simonsen, arXiv:cond-mat/0508632.
[1312] R. Burioni and D. Cassi, J. Phys. A. **38**, R45 (2005).
[1313] F. Zhao, H. Yang, and B. Wang, Phys. Rev. E **72**, 046119 (2005).
[1314] L. Donetti and M. A. Muñoz, arXiv:physics/0504059.
[1315] M. E. J. Newman, Phys. Rev. E **74**, 036104 (2006).
[1316] G. H. Golub and C. F. Van Loan, *Matrix Computations* (The Johns Hopkins University Press, Baltimore, 1989).
[1317] R. Pastor-Satorras, M. Rubi, and A. Díaz-Guilera (Eds.), *Statistical Mechanics of Complex Networks* (Springer, Berlin, 2003).
[1318] R. Pastor-Satorras and A. Vespignani, *Evolution and Structure of the Internet* (Cambridge University Press, Cambridge, 2004).
[1319] G. Caldarelli, *Scale-Free Networks* (Oxford University Press, New York, 2007).

[1320] P. Mika, *Social Networks and the Sematic Web* (Springer, New York, 2007).
[1321] F. Képès (Ed.), *Biological Networks* (World Scientific, Singapore, 2007).
[1322] B. H. Junker and F. Schreiber, *Analysis of Biological Networks* (Wiley, Hoboken, 2008).
[1323] J. Reichardt, *Structure in Complex Networks* (Springer, Berlin, 2009).
[1324] M. E. J. Newman, *Networks* (Oxford University Press, New York, 2010).
[1325] M. E. J. Newman, A.-L. Barabási, and D. J. Watts (Eds.), *The Structure and Dynamics of Networks* (Princeton University Press, Princeton, 2006).
[1326] R. Albert, H. Jeong, and A.-L. Barabási, Nature **406**, 378 (2000).
[1327] R. Cohen, K. Erez, D. ben-Avraham, and S. Havlin, Phys. Rev. Lett. **85**, 4626 (2000).
[1328] R. Cohen, K. Erez, D. ben-Avraham, and S. Havlin, Phys. Rev. Lett. **86**, 3682 (2001).
[1329] D. S. Callaway, M. E. J. Newman, S. H. Strogatz, and D. J. Watts, Phys. Rev. Lett. **85**, 5468 (2000).
[1330] S. N. Dorogovtsev, J. F. F. Mendes, and A. N. Samukhin, Phys. Rev. E **64**, 066110 (2001).
[1331] R. Pastor-Satorras and A. Vespignani, Phys. Rev. Lett. **86**, 3200 (2001).
[1332] R. Pastor-Satorras and A. Vespignani, Phys. Rev. E **63**, 066117 (2001).
[1333] M. Kuperman and G. Abramson, Phys. Rev. Lett. **86**, 2909 (2001).
[1334] Z. Dezső and A.-L. Barabási, Phys. Rev. E **65**, 055103 (2002).
[1335] A. Barrat and M. Weigt, Eur. Phys. J. B **13**, 547 (2000).
[1336] M. Barthélémy and L. A. N. Amaral, Phys. Rev. Lett. **82**, 3180 (1999).
[1337] R. V. Kulkarni, E. Almaas, and D. Stroud, arXiv:cond-mat/9908216.
[1338] R. V. Kulkarni, E. Almaas, and D. Stroud, arXiv:cond-mat/9905066.
[1339] A. Wagner and D. Fell, Santa Fe Working Paper 00-07-041 (2000).
[1340] M. E. J. Newman, arXiv:cond-mat/0001118.
[1341] N. Mathias and V. Gopal, Phys. Rev. E **63**, 021117 (2001).
[1342] M. Marchiori and V. Latora, Physica A **285**, 539 (2000).
[1343] S. Jespersen, I. M. Sokolov, and A. Blumen, Phys. Rev. E **62**, 4405 (2000).
[1344] S. A. Pandit and R. E. Amritkar, Phys. Rev. E **63**, 041104 (2001).
[1345] L. F. Lago-Fernández, R. Huerta, F. Corbacho, and J. A. Sigüenza, Phys. Rev. Lett. **84**, 2758 (2000).
[1346] C. P. Herrero, Phys. Rev. E **65**, 066110 (2002).
[1347] A. Iamnitchi, M. Ripeanu, E. Santos-Neto, and I. Foster, IEEE Transactions on Parallel and Distributed Systems **22**, 1120 (2011).
[1348] J. V. Lopes, Y. G. Pogorelov, J. M. B. L. dos Santos, and R. Toral, Phys. Rev. E **70**, 026112 (2004).
[1349] H. Hong, M. Y. Choi, and B. J. Kim, Phys. Rev. E **65**, 026139 (2002).
[1350] R. Cohen and S. Havlin, Phys. Rev. Lett. **90**, 058701 (2003).
[1351] M. Girvan and M. E. J. Newman, Proc. Natl. Acad. Sci. USA **99**, 7821 (2002).
[1352] M. E. J. Newman and M. Girvan, Phys. Rev. E **69**, 026113 (2004).
[1353] L. Donetti and M. A. Muñoz, J. Stat. Mech. P10012 (2004).
[1354] J. Reichardt and S. Bornholdt, Phys. Rev. Lett. **93**, 218701 (2004).
[1355] F. Radicchi, C. Castellano, F. Cecconi, V. Loreto, and D. Parisi, Proc. Natl. Acad. Sci. USA **101**, 2658 (2004).
[1356] J. P. Bagrow and E. M. Bollt, Phys. Rev. E **72**, 046108 (2005).
[1357] R. Guimerà, M. Sales-Pardo, and L. A. N. Amaral, Phys. Rev. E **70**, 025101 (2004).
[1358] F. Wu and B. A. Huberman, Eur. Phys. J. B **38**, 331 (2004).
[1359] A. Capocci, V. D. P. Servedio, G. Caldarelli, and F. Colaiori, Physica A **352**, 669 (2005).
[1360] J. Duch and A. Arenas, Phys. Rev. E **72**, 027104 (2005).
[1361] L. Danon, J. Duch, A. Arenas, and A. Díaz-Guilera, J. Stat. Mech P09008 (2005).
[1362] L. Danon, J. Duch, A. Arenas, and A. Díaz-Guilera, arXiv:cond-mat/0505245.
[1363] G. Palla, I. Derényi, I. Farkas, and T. Vicsek, Nature **435**, 814 (2005).
[1364] R. Lambiotte and M. Ausloos, Phys. Rev. E **72**, 066107 (2005).
[1365] R. Guimerà and L. A. N. Amaral, Nature **433**, 895 (2005).
[1366] S. Fortunato and M. Barthélémy, arXiv:physics/0607100.
[1367] M. E. J. Newman, arXiv:physics/0602124.
[1368] J. Reichardt and S. Bornholdt, J. Stat. Mech. P06016 (2007).
[1369] M. J. Barber, Phys. Rev. E **76**, 066102 (2007).
[1370] R. Guimerà, M. Sales-Pardo, and L. A. N. Amaral, arXiv:physics/0701151.
[1371] J. Peng and H.-G. Müller, Annals of Applied Statistics **2**, 1056 (2008).
[1372] E. N. Sawardecker, C. A. Amundsen, M. Sales-Pardo, and L. A. N. Amaral, Eur. Phys. J. **72**, 671 (2009).

[1373] A. Srivastava, arXiv:1012.4050.
[1374] A. Lancichinetti, M. Kivelä, J. Saramäki, and S. Fortunato, PLoS ONE **5**, e11976 (2010).
[1375] J. Leskovec, K. J. Lang, and M. W. Mahoney, arXiv:1004.3539.
[1376] A. Capocci, F. Slanina, and Y.-C. Zhang, Physica A **317**, 259 (2003).
[1377] A. Nagurney and K. Ke, Quant. Finance **3**, 71 (2003).
[1378] G. Bianconi, P. Laureti, Y.-K. Yu, and Y.-C. Zhang, Physica A **332**, 519 (2004).
[1379] P. Laureti, J. Mathiesen, and Y.-C. Zhang, Physica A **338**, 596 (2004).
[1380] P. Laureti, L. Moret, Y.-C. Zhang, and Y.-K. Yu, Europhys. Lett. **75**, 1006 (2006).
[1381] M. Blattner, Y.-C. Zhang, and S. Maslov, Physica A **373**, 753 (2007).
[1382] Y.-C. Zhang, M. Blattner, and Y.-K. Yu, Phys. Rev. Lett. **99**, 154301 (2007).
[1383] T. Zhou, J. Ren, M. Medo, and Y.-C. Zhang, Phys. Rev. E **76**, 046115 (2007).
[1384] Y.-C. Zhang, M. Medo, J. Ren, T. Zhou, T. Li, and F. Yang, Europhys. Lett. **80**, 68003 (2007).
[1385] M. Medo, Y.-C. Zhang, and T. Zhou, Europhys. Lett. **88**, 38005 (2009).
[1386] T. Zhou, Z. Kuscsik, J.-G. Liu, M. Medo, J. R. Wakeling, and Y.-C. Zhang, Proc. Natl. Acad. Sci. USA **107**, 4511 (2010).
[1387] G. Cimini, D. Chen, M. Medo, L. Lü, Y.-C. Zhang, and T. Zhou, Phys. Rev. E **85**, 046108 (2012).
[1388] S. Valverde, R. Ferrer i Cancho, and R. V. Solé, Europhys. Lett. **60**, 512 (2002).
[1389] R. Ferrer i Cancho and R. V. Solé, arXiv:cond-mat/0111222.
[1390] G. Palla, I. Derényi, I. Farkas, and T. Vicsek, arXiv:cond-mat/0309556.
[1391] I. Farkas, I. Derényi, G. Palla, and T. Vicsek, arXiv:cond-mat/0401640.
[1392] T. C. Jarrett, D. J. Ashton, M. Fricker, and N. F. Johnson, arXiv:physics/0604183.
[1393] J. Park and M. E. J. Newman, Phys. Rev. E **70**, 066117 (2004).
[1394] M. Rosvall and K. Sneppen, Phys. Rev. Lett. **91**, 178701 (2003).
[1395] M. Marsili, F. Vega-Redondo, and F. Slanina, Proc. Natl. Acad. Sci. U.S.A. **101**, 1439 (2004).
[1396] B. J. Kim, A. Trusina, P. Minnhagen, and K. Sneppen, Eur. Phys. J. B **43**, 369 (2005).
[1397] A. Arenas, A. Díaz-Guilera, and R. Guimerà, Phys. Rev. Lett. **86**, 3196 (2001).
[1398] R. Guimerà, A. Arenas, and A. Díaz-Guilera, arXiv:cond-mat/0103112.
[1399] M. Durand, Phys. Rev. E **73**, 016116 (2006).
[1400] V. Pareto, J. Political Economy **5**, 485 (1897).
[1401] J. C. Stamp, Journal of the Royal Statistical Society **77**, 200 (1914).
[1402] D. H. Macgregor, Economic Journal **46**, 80 (1936).
[1403] G. F. Shirras, Economic Journal **45**, 663 (1935).
[1404] C. Bresciani-Turroni, Journal of the Royal Statistical Society **100**, 421 (1937).
[1405] E. W. Montroll and M. F. Shlesinger, J. Stat. Phys. **32**, 209 (1983).
[1406] M. Levy and S. Solomon, Physica A **242**, 90 (1997).
[1407] K. Okuyama, M. Takayasu, and H. Takayasu, Physica A **269**, 125 (1999).
[1408] H. Aoyama, Y. Nagahara, M. P. Okazaki, W. Souma, H. Takayasu, and M. Takayasu, Fractals **8**, 293 (2000).
[1409] W. Souma, Fractals **9**, 463 (2001).
[1410] A. Drăgulescu and V. M. Yakovenko, Physica A **299**, 213 (2001).
[1411] W. J. Reed and B. D. Hughes, Phys. Rev. E **66**, 067103 (2002).
[1412] H. Aoyama, W. Souma, and Y. Fujiwara, Physica A **324**, 352 (2003).
[1413] E. P. Borges, Physica A **334**, 255 (2004).
[1414] F. Clementi and M. Gallegati, arXiv:cond-mat/0408067.
[1415] O. S. Klass, O. Biham, M. Levy, O. Malcai, and S. Solomon, Economics Letters **90**, 290 (2006).
[1416] A. Y. Abul-Magd, Phys. Rev. E **66**, 057104 (2002).
[1417] A. Drăgulescu and V. M. Yakovenko, Eur. Phys. J. B **17**, 723 (2000).
[1418] A. Drăgulescu and V. M. Yakovenko, Eur. Phys. J. B **20**, 585 (2001).
[1419] A. A. Drăgulescu and V. M. Yakovenko, in: *Modeling of Complex Systems: Seventh Granada Lectures*, AIP Conference Proceedings, **661**, p. 180 (AIP, New York, 2003).
[1420] V. M. Yakovenko, arXiv:cond-mat/0302270.
[1421] A. C. Silva and V. M. Yakovenko, Europhys. Lett. **69**, 304 (2005).
[1422] A. Banerjee, V. M. Yakovenko, and T. Di Matteo, Physica A **370**, 54 (2006).
[1423] W. J. Reed, Physica A **319**, 469 (2003).
[1424] W. Souma, arXiv:cond-mat/0202388.
[1425] T. Mizuno, M. Katori, H. Takayasu, and M. Takayasu, arXiv:cond-mat/0308365.
[1426] T. Mizuno, M. Takayasu, and H. Takayasu, Physica A **332**, 403 (2004).

[1427] W. Souma and M. Nirei, in: *Econophysics of Wealth Distributions. Proceedings of the Econophys-Kolkata I*, ed. A. Chatterjee, S. Yarlagadda, and B. K. Chakrabarti, p. 34 (Springer, Milan, 2005).

[1428] S. Sinha and R. K. Pan, in: *Econophysics of Wealth Distributions. Proceedings of the Econophys-Kolkata I*, ed. A. Chatterjee, S. Yarlagadda, and B. K. Chakrabarti, p. 43 (Springer, Milan, 2005).

[1429] S. Sinha, Physica A **359**, 555 (2006).

[1430] S.-M. Yoon and K. Kim, J. Korean Phys. Soc. **46**, 1037 (2005).

[1431] R. Gibrat, *Les inégalités économiques* (Libr. du Recueil Sirey, Paris, 1931).

[1432] Y. Fujiwara, C. Di Guilmi, H. Aoyama, M. Gallegati, and W. Souma, Physica A **335**, 197 (2004).

[1433] D. G. Champernowne, Economic Journal **63**, 318 (1953).

[1434] Z.-F. Huang and S. Solomon, Physica A **294**, 503 (2001).

[1435] H. Nakao, Phys. Rev. E **58**, 1591 (1998).

[1436] P. Richmond, Eur. Phys. J. B **20**, 523 (2001).

[1437] S. Solomon and M. Levy, arXiv:cond-mat/0005416.

[1438] O. Malcai, O. Biham, P. Richmond, and S. Solomon, Phys. Rev. E **66**, 031102 (2002).

[1439] P. Richmond and L. Sabatelli, Physica A **344**, 344 (2004).

[1440] P. Richmond, P. Repetowicz, S. Hutzler, and R. Coelho, Physica A **370**, 43 (2006).

[1441] R. Zygadło, Phys. Rev. E **77**, 021130 (2008).

[1442] M. Kardar, G. Parisi, and Y.-C. Zhang, Phys. Rev. Lett. **56**, 889 (1986).

[1443] M. Kardar and Y.-C. Zhang, Phys. Rev. Lett. **58**, 2087 (1987).

[1444] B. Derrida and H. Spohn, J. Stat. Phys. **51**, 817 (1988).

[1445] B. Derrida and R.B. Griffiths, Europhys. Lett. **8**, 111 (1989).

[1446] M. Mézard, J. Phys. France **51**, 1831 (1990).

[1447] M. Mézard and G. Parisi, J. Phys. I France **1**, 809 (1991).

[1448] T. Halpin-Healy and Y.-C. Zhang, Phys. Rep. **254**, 215 (1995).

[1449] J.-P. Bouchaud and M. Mézard, Physica A **282**, 536 (2000).

[1450] D.-W. Huang, Phys. Rev. E **69**, 057103 (2004).

[1451] D.-W. Huang, Phys. Rev. E **68**, 048101 (2003).

[1452] W. Souma, Y. Fujiwara, and H. Aoyama, arXiv:cond-mat/0108482.

[1453] T. Di Matteo, T. Aste, and S. T. Hyde, arXiv:cond-mat/0310544.

[1454] D. Garlaschelli and M. I. Loffredo, Physica A **338**, 113 (2004).

[1455] J. C. Maxwell, Philosophical Transactions of the Royal Society of London **157**, 49 (1867).

[1456] S. Ispolatov, P. L. Krapivsky, and S. Redner, Eur. Phys. J. B **2**, 267 (1998).

[1457] A. Chakraborti and B. K. Chakrabarti, Eur. Phys. J. B **17**, 167 (2000).

[1458] A. Chakraborti, Int. J. Mod. Phys. C **13**, 1315 (2002).

[1459] A. Chatterjee, B. K. Chakrabarti, and S. S. Manna, Phys. Scripta **T106**, 36 (2003).

[1460] A. Chatterjee, B. K. Chakrabarti, and S. S. Manna, Physica A **335**, 155 (2004).

[1461] B. K. Chakrabarti and A. Chatterjee, arXiv:cond-mat/0302147.

[1462] A. Das and S. Yarlagadda, Physica Scripta **T106**, 39 (2003).

[1463] M. Patriarca, A. Chakraborti, and K. Kaski, Phys. Rev. E **70**, 016104 (2004).

[1464] M. Patriarca, A. Chakraborti, and K. Kaski, Physica A **340**, 334 (2004).

[1465] A. Chatterjee and B. K. Chakrabarti, in: *Econophysics of Wealth Distributions. Proceedings of the Econophys-Kolkata I*, ed. A. Chatterjee, S. Yarlagadda, and B. K. Chakrabarti, p. 79 (Springer, Milan, 2005).

[1466] A. Chatterjee, B. K. Chakrabarti, and R. B. Stinchcombe, in: *Practical Fruits of Econophysics. Proceedings of The Third Nikkei Econophysics Symposium*, ed. H. Takayasu, p. 333 (Springer, Berlin, 2006).

[1467] A. Chatterjee, B. K. Chakrabarti, and R. B. Stinchcombe, Phys. Rev. E **72**, 026126 (2005).

[1468] A. Das and S. Yarlagadda, Physica A **353**, 529 (2005).

[1469] A. K. Gupta, in: *Econophysics and Sociophysics: Trends and Perspectives*, ed. B. K. Chakrabarti, A. Chakraborti, and A. Chatterjee, p. 161 (Wiley-VCH, Weinheim, 2006).

[1470] A. Chatterjee and B. K. Chakrabarti, Eur. Phys. J. B **54**, 399 (2006).

[1471] A. Chatterjee and B. K. Chakrabarti, Eur. Phys. J. B **60**, 135 (2007).

[1472] A. Chatterjee and B. K. Chakrabarti, Physica A **382**, 36 (2007).

[1473] M. Patriarca, A. Chakraborti, E. Heinsalu, and G. Germano, Eur. Phys. J. B **57**, 219 (2007).

[1474] A. Chatterjee, S. Sinha, and B. K. Chakrabarti, Current Science **92**, 1383 (2007).

[1475] A. Chakraborti and M. Patriarca, Phys. Rev. Lett. **103**, 228701 (2009).

[1476] P. Repetowicz, S. Hutzler, and P. Richmond, Physica A **356**, 641 (2005).

[1477] H. M. Jaeger, S. R. Nagel, and R. P. Behringer, Rev. Mod. Phys. **68**, 1259 (1996).
[1478] A. Puglisi, V. Loreto, U. Marini Bettolo Marconi, and A. Vulpiani, Phys. Rev. E **59**, 5582 (1999).
[1479] T. Antal, M. Droz, and A. Lipowski, Phys. Rev. E **66**, 062301 (2002).
[1480] A. V. Bobylev and C. Cercignani, J. Stat. Phys. **106**, 547 (2002).
[1481] M. H. Ernst and R. Brito, J. Stat. Phys. **109**, 407 (2002).
[1482] P. L. Krapivsky and E. Ben-Naim, J. Phys. A: Math. Gen. **35**, L147 (2002).
[1483] A. Baldassarri, U. Marini Bettolo Marconi, and A. Puglisi, Europhys. Lett. **58**, 14 (2002).
[1484] E. Ben-Naim and P. L. Krapivsky, Phys. Rev. E **61**, R5 (2000).
[1485] M. H. Ernst and R. Brito, arXiv:cond-mat/0111093.
[1486] M. H. Ernst and R. Brito, Phys. Rev. E **65**, 040301 (2002).
[1487] M. H. Ernst and R. Brito, Europhys. Lett. **58**, 182 (2002).
[1488] M. H. Ernst and R. Brito, arXiv:cond-mat/0304608.
[1489] E. Ben-Naim and P. L. Krapivsky, Phys. Rev. E **66**, 011309 (2002).
[1490] E. Ben-Naim and P. L. Krapivsky, in: *Granular Gas Dynamics*, ed. T. Pöschel and N. Brilliantov, p. 65 (Springer, Berlin, 2003).
[1491] E. Barkai, Phys. Rev. E **68**, 055104 (2003).
[1492] D. ben-Avraham, E. Ben-Naim, K. Lindenberg, and A. Rosas, Phys. Rev. E **68**, 050103 (2003).
[1493] A. V. Bobylev, J. A. Carillo, and I. M. Gamba, J. Stat. Phys. **98**, 743 (2000).
[1494] A. V. Bobylev and C. Cercignani, J. Stat. Phys. **110**, 333 (2003).
[1495] M. H. R. Stanley, L. A. N. Amaral, S. V. Buldyrev, S. Havlin, H. Leschhorn, P. Maass, M. A. Salinger, and H. E. Stanley, Nature **379**, 804 (1996).
[1496] R. L. Axtell, Science **293**, 1818 (2001).
[1497] L. A. N. Amaral, S. V. Buldyrev, S. Havlin, P. Maass, M. A. Salinger, H. E. Stanley, and M. H. R. Stanley, Physica A **244**, 1 (1997).
[1498] L. A. N. Amaral, S. V. Buldyrev, S. Havlin, H. Leschhorn, P. Maass, M. A. Salinger, H. E. Stanley, and M. H. R. Stanley, J. Phys. I France **7**, 621 (1997).
[1499] J. Voit, Adv. Compl. Syst. **4**, 149 (2001).
[1500] J. Voit, in: *Modeling Complexity in Economic and Social Systems*, ed. F. Schweitzer, p. 149 (World Scientific, Singapore, 2001).
[1501] E. Gaffeo, M. Gallegati, and A. Palestrini, Physica A **324**, 117 (2003).
[1502] T. Kaizoji, H. Iyetomi, and Y. Ikeda, Evolutionary and Institutional Economic Review **2**, 183 (2006).
[1503] Y. Lee, L. A. N. Amaral, D. Canning, M. Meyer, and H. E. Stanley, Phys. Rev. Lett. **81**, 3275 (1998).
[1504] D. Canning, L. A. N. Amaral, Y. Lee, M. Meyer, and H. E. Stanley, Economics Letters **60**, 335 (1998).
[1505] G. De Fabritiis, F. Pammolli, and M. Riccaboni, Physica A **324**, 38 (2003).
[1506] G. Bottazzi and A. Secchi, Physica A **324**, 213 (2003).
[1507] G. Fagiolo, M. Napoletano, and A. Roventini, Eur. Phys. J. B **57**, 205 (2007).
[1508] Y. Fujiwara, W. Souma, H. Aoyama, T. Kaizoji, and M. Aoki, Physica A **321**, 598 (2003).
[1509] S. V. Buldyrev, L. A. N. Amaral, S. Havlin, H. Leschhorn, P. Maass, M. A. Salinger, H. E. Stanley, and M. H. R. Stanley, J. Phys. I France **7**, 635 (1997).
[1510] L. A. N. Amaral, S. V. Buldyrev, S. Havlin, M. A. Salinger, and H. E. Stanley, Phys. Rev. Lett. **80**, 1385 (1998).
[1511] P. Richmond, S. Hutzler, R. Coelho, and P. Repetowicz, in: *Econophysics and Sociophysics: Trends and Perspectives*, ed. B. K. Chakrabarti, A. Chakraborti, and A. Chatterjee, p. 131 (Wiley-VCH, Weinheim, 2006).
[1512] N. Scafetta, S. Picozzi, and B. J. West, arXiv:cond-mat/0209373.
[1513] N. Scafetta and B. J. West, arXiv:cond-mat/0306579.
[1514] N. Scafetta, S. Picozzi, and B. J. West, Quant. Finance **4**, 353 (2004).
[1515] S. Pianegonda, J. R. Iglesias, G. Abramson, and J. L. Vega, Physica A **322**, 667 (2003).
[1516] J. R. Iglesias, S. Gonçalves, S. Pianegonda, J. L. Vega, and G. Abramson, Physica A **327**, 12 (2003).
[1517] S. Pianegonda and J. R. Iglesias, arXiv:cond-mat/0311113.
[1518] J. R. Iglesias, S. Gonçalves, G. Abramson, and J. L. Vega, arXiv:cond-mat/0311127.
[1519] M. F. Laguna, S. R. Gusman, and J. R. Iglesias, Physica A **356**, 107 (2005).
[1520] M. A. Fuentes, M. Kuperman, and J. R. Iglesias, Physica A **371**, 112 (2006).
[1521] C. Beck, Phys. Rev. Lett. **87**, 180601 (2001).

[1522] A. Ishikawa, T. Suzuki, and M. Tomoyose, arXiv:cond-mat/0203339.

[1523] A. Ishikawa, Physica A **349**, 597 (2005).

[1524] A. Ishikawa, Physica A **371**, 525 (2006).

[1525] A. Ishikawa, Physica A **363**, 367 (2006).

[1526] A. Ishikawa, Physica A **367**, 425 (2006).

[1527] A. Ishikawa, Physica A **383**, 79 (2007).

[1528] M. Gligor and M. Ignat, Eur. Phys. J. B **30**, 125 (2002).

[1529] S. Sinha, arXiv:cond-mat/03043224.

[1530] Y. Ohtaki and H. H. Hasegawa, arXiv:cond-mat/0312568.

[1531] R. Coelho, Z. Néda, J. J. Ramasco, and M. A. Santos, Physica A **353**, 515 (2005).

[1532] A. Fujihara, T. Ohtsuki, and H. Yamamoto, Phys. Rev. E **70**, 031106 (2004).

[1533] Y. Wang and N. Ding, in: *Econophysics of Wealth Distributions. Proceedings of the Econophys-Kolkata I*, ed. A. Chatterjee, S. Yarlagadda, and B. K. Chakrabarti, p. 126 (Springer, Milan, 2005).

[1534] Y. Wang, N. Ding, and N. Xi, in: *Practical Fruits of Econophysics. Proceedings of The Third Nikkei Econophysics Symposium*, ed. H. Takayasu, p. 322 (Springer, Berlin, 2006).

[1535] Y. Wang and H. Qiu, Physica A **353**, 493 (2005).

[1536] N. Xi, N. Ding, and Y. Wang, Physica A **357**, 543 (2005).

[1537] R. Coelho, P. Richmond, J. Barry, and S. Hutzler, Physica A **387**, 3847 (2008).

[1538] A. M. Chebotarev, Physica A **373**, 541 (2007).

[1539] J. Angle, in: *Econophysics of Markets and Business Networks. Proceedings of the Econophys-Kolkata III*, ed. A. Chatterjee and B. K. Chakrabarti, p. 185 (Springer, Milan, 2007).

[1540] J. Mimkes and G. Willis, in: *Econophysics of Wealth Distributions. Proceedings of the Econophys-Kolkata I*, ed. A. Chatterjee, S. Yarlagadda, and B. K. Chakrabarti, p. 61 (Springer, Milan, 2005).

[1541] J. Mimkes and Y. Aruka, in: *Econophysics of Wealth Distributions. Proceedings of the Econophys-Kolkata I*, ed. A. Chatterjee, S. Yarlagadda, and B. K. Chakrabarti, p. 70 (Springer, Milan, 2005).

[1542] J. Mimkes, Physica A **389**, 1665 (2010).

[1543] G. Yaari, D. Stauffer, and S. Solomon, in: *Encyclopedia of Complexity and Systems Science*, 4920 (Springer, New York, 2009).

[1544] G. Yaari, A. Nowak, K. Rakocy, and S. Solomon, Eur. Phys. J. B **62**, 505 (2008).

[1545] F. Pammolli, D. Fu, S. V. Buldyrev, M. Riccaboni, K. Matia, K. Yamasaki, and H. E. Stanley, Eur. Phys. J. B **57**, 127 (2007).

[1546] S. V. Buldyrev, F. Pammolli, M. Riccaboni, K. Yamasaki, D.-F. Fu, K. Matia, and H. E. Stanley, Eur. Phys. J. B **57**, 131 (2007).

[1547] J. Sutton, Physica A **312**, 577 (2002).

[1548] J. Sutton, Physica A **324**, 45 (2003).

[1549] M. Wyart and J.-P. Bouchaud, Physica A **326**, 241 (2003).

[1550] T. Ohtsuki, A. Fujihara, and H. Yamamoto, Phys. Lett. A **324**, 378 (2004).

[1551] R. J. Baxter, *Exactly Solved Models in Statistical Mechanics* (Academic Press, 1982).

[1552] S. Galam, Y. Gefen, and Y. Shapir, J. Math. Sociol. **9**, 1 (1982).

[1553] S. Galam, Physica A **336**, 49 (2004).

[1554] S. Galam, *Sociophysics. A Physicist's Modeling of Psycho-political Phenomena* (Springer, Berlin, 2012).

[1555] J. von Neumann and O. Morgenstern, *Theory of Games and Economic Behavior* (Princeton University Press, Princeton, 1944).

[1556] F. Vega-Redondo, *Evolution, Games, and Economic Behaviour* (Oxford University Press, Oxford, 1996).

[1557] J. F. Nash, Proc. Natl. Acad. Sci. USA **36**, 48 (1950).

[1558] J. F. Nash, Econometrica **18**, 155 (1950).

[1559] J. Berg and M. Weigt, Europhys. Lett. **48**, 129 (1999).

[1560] T. Aste and D. Weaire, *The Pursuit of Perfect Packing* (Taylor and Francis, Boca Raton, 2008).

[1561] R. A. Fisher, *The Genetical Theory of Natural Selection* (Clarendon Press, Oxford, 1930).

[1562] R. C. Lewontin, J. Theor. Biol. **1**, 382 (1961).

[1563] W. D. Hamilton, Science **156**, 477 (1967).

[1564] J. Maynard Smith and G. R. Price, Nature **246**, 15 (1973).

[1565] J. Maynard Smith, *Evolution and the Theory of Games* (Cambridge University Press, Cambridge, 1982).

[1566] M. A. Nowak and K. Sigmund, Science **303**, 793 (2004).
[1567] B. Sinervo and C. M. Lively, Nature **380**, 240 (1996).
[1568] M. Milinski, Nature **325**, 433 (1987).
[1569] D. W. Stephens, C. M. McLinn, and J. R. Stevens, Science **298**, 2216 (2002).
[1570] M. A. Nowak and K. Sigmund, Nature **398**, 367 (1999).
[1571] P. E. Turner and L. Chao, Nature **398**, 441 (1999).
[1572] R. Axelrod, J. Conflict Resolution **24**, 3 (1980).
[1573] R. Axelrod, J. Conflict Resolution **24**, 379 (1980).
[1574] R. Axelrod and W. D. Hamilton, Science **211**, 1390 (1981).
[1575] J. Wu and R. Axelrod, J. Conflict Resolution **39**, 183 (1995).
[1576] K. Lindgren, in: *Artificial Life II*, ed. C. G. Langton, C. Taylor, J. D. Farmer, and S. Rasmussen, p. 295 (Addison-Wesley, 1991).
[1577] K. Lindgren, in: *The Economy as an Evolving Complex System II*, ed. W. B. Arthur, S. N. Durlauf, and D. A. Lane, p. 337 (Perseus, Reading, 1997).
[1578] M.A. Nowak and M. May, Nature **359**, 826 (1992).
[1579] S. Wolfram, Rev. Mod. Phys. **55**, 601 (1983).
[1580] F. Schweitzer, L. Behera, and H. Mühlenbein, Adv. Compl. Syst. **5**, 269 (2002).
[1581] G. Szabó and C. Töke, Phys. Rev. E **58**, 69 (1998).
[1582] J. R. N. Chiappin and M. J. de Oliveira, Phys. Rev. E **59**, 6419 (1999).
[1583] M. H. Vainstein and J. J. Arenzon, Phys. Rev. E **64**, 051905 (2001).
[1584] Y. F. Lim, K. Chen, and C. Jayaprakash, Phys. Rev. E **65**, 026134 (2002).
[1585] G. Szabó, J. Vukov, and A. Szolnoki, Phys. Rev. E **72**, 047107 (2005).
[1586] G. Abramson and M. Kuperman, Phys. Rev. E **63**, 030901 (2001).
[1587] M. G. Zimmermann, V. M. Eguíluz, and M. San Miguel, in: *Economics with Heterogeneous Interacting Agents*, ed. A. Kirman and J.-B. Zimmermann, p. 73 (Springer, Berlin, 2001).
[1588] H. Ebel and S. Bornholdt, arXiv:cond-mat/0211666.
[1589] H. Ebel and S. Bornholdt, Phys. Rev. E **66**, 056118 (2002).
[1590] M. G. Zimmermann and V. M. Eguíluz, Phys. Rev. E **72**, 056118 (2005).
[1591] J. Vukov, G. Szabó, and A. Szolnoki, arXiv:cond-mat/0603419.
[1592] P. Clifford and A. Sudbury, Biometrika **60**, 581 (1973).
[1593] R. A. Holley and T. M. Liggett, Ann. Prob. **3**, 643 (1975).
[1594] T. M. Liggett, *Interacting Particle Systems* (Springer, Berlin, 1985).
[1595] T. M. Liggett, *Stochastic Interacting Systems: Contact, Voter, and Exclusion Processes* (Springer, Berlin, 1999).
[1596] S. Redner, *A Guide to First-passage Processes* (Cambridge University Press, Cambridge, 2001).
[1597] M. Scheucher and H. Spohn, J. Stat. Phys. **53**, 279 (1988).
[1598] P. L. Krapivsky, Phys. Rev. A **45**, 1067 (1992).
[1599] L. Frachebourg and P. L. Krapivsky, Phys. Rev. E **53**, R3009 (1996).
[1600] E. Ben-Naim, L. Frachebourg, and P. L. Krapivsky, Phys. Rev. E **53**, 3078 (1996).
[1601] I. Dornic, H. Chaté, J. Chave, and H. Hinrichsen, Phys. Rev. Lett. **87**, 045701 (2001).
[1602] O. Al Hammal, H. Chaté, I. Dornic, and M. A. Muñoz, Phys. Rev. Lett. **94**, 230601 (2005).
[1603] M. Mobilia and I. T. Georgiev, Phys. Rev. E **71**, 046102 (2005).
[1604] D. ben-Avraham, D. Considine, P. Meakin, S. Redner, and H. Takayasu, J. Phys. A: Math. Gen. **23**, 4297 (1990).
[1605] B. Derrida, V. Hakim, and V. Pasquier, J. Stat. Phys. **85**, 763 (1996).
[1606] V. Spirin, P. L. Krapivsky, and S. Redner, Phys. Rev. E **65**, 016119 (2001).
[1607] C. Castellano and R. Pastor-Satorras, J. Stat. Mech. P05001 (2006).
[1608] C. Itzykson and J.-M. Drouffe, *Statistical Field Theory* (Cambridge University Press, Cambridge, 1989).
[1609] L. D. Landau and E. M. Lifshitz, *The Classical Theory of Fields* (Butterworth-Heinemann, Oxford, 1975).
[1610] J. T. Cox, Ann Prob. **17**, 1333 (1989).
[1611] R. Dickman and R. Vidigal, J. Phys. A: Math. Gen. **35**, 1147 (2002).
[1612] F. Slanina and H. Lavička, Eur. Phys. J. B **35**, 279 (2003).
[1613] S. Wright, Proc. Natl. Acad. Sci. USA **31**, 382 (1945).
[1614] D. Dorninger and H. Langer, Discrete Appl. Math. **6**, 209 (1983).
[1615] C. Muller and R. Tribe, Probab. Theory Rel. **102**, 519 (1995).
[1616] R. Tribe, Probab. Theory Rel. **102**, 289 (1995).
[1617] C. W. Gardiner, *Handbook of Stochastic Methods* (Springer, Berlin, 1985).

[1618] K. Suchecki, V. M. Eguíluz, and M. San Miguel, Europhys. Lett. **69**, 228 (2005).

[1619] V. Sood and S. Redner, Phys. Rev. Lett. **94**, 178701 (2005).

[1620] D. Vilone and C. Castellano, Phys. Rev. E **69**, 016109 (2004).

[1621] C. Castellano, D. Vilone, and A. Vespignani, Europhys. Lett. **63**, 153 (2003).

[1622] K. Suchecki, V. M. Eguíluz, and M. San Miguel, Phys. Rev. E **72**, 036132 (2005).

[1623] C. Castellano, V. Loreto, A. Barrat, F. Cecconi, and D. Parisi, Phys. Rev. E **71**, 066107 (2005).

[1624] C. Castellano, AIP Conference Proceedings **779**, 114 (2005).

[1625] S. Galam, J. Math. Psychol. **30**, 426 (1986).

[1626] S. Galam, J. Stat. Phys. **61**, 943 (1990).

[1627] S. Galam, Physica A **274**, 132 (1999).

[1628] S. Galam, Physica A **285**, 66 (2000).

[1629] S. Galam and S. Wonczak, Eur. Phys. J. B **18**, 183 (2000).

[1630] S. Galam and F. Jacobs, Physica A **381**, 366 (2007).

[1631] B. Schneier, *Applied Cryptography* (John Wiley & Sons, New York, 1996).

[1632] P. L. Krapivsky and S. Redner, Phys. Rev. Lett. **90**, 238701 (2003).

[1633] M. Plischke and B. Bergersen, *Equilibrium Statistical Physics* (World Scientific, Singapore, 1994).

[1634] M. Mobilia, Phys. Rev. Lett. **91**, 028701 (2003).

[1635] M. Mobilia, A. Petersen, and S. Redner, J. Stat. Mech. P08029 (2007).

[1636] S. Galam, Physica A **333**, 453 (2004).

[1637] D. Stauffer, J. S. Sá Martins, Physica A **334**, 558 (2004).

[1638] S. Galam, Quality and Quantity **41**, 579 (2007).

[1639] R. Florian and S. Galam, Eur. Phys. J. B **16**, 189 (2000).

[1640] S. Galam, Eur. Phys. J. B **25**, 403 (2002).

[1641] S. Galam, Eur. Phys. J. B **26**, 269 (2002).

[1642] S. Galam, Physica A **320**, 571 (2003).

[1643] S. Galam, Physica A **330**, 139 (2003).

[1644] S. Galam and A. Mauger, Physica A **323**, 695 (2003).

[1645] S. Galam and A. Vignes, Physica A **351**, 605 (2005).

[1646] S. Galam, Physica A **336**, 56 (2004).

[1647] S. Galam, arXiv:cond-mat/0409484.

[1648] S. Galam, Europhys. Lett. **70**, 705 (2005).

[1649] S. Gekle, L. Peliti, and S. Galam, Eur. Phys. J. B **45**, 569 (2005).

[1650] S. Galam, B. Chopard, A. Masselot, and M. Droz, Eur. Phys. J. B **4**, 529 (1998).

[1651] C. J. Tessone, R. Toral, P. Amengual, H. S. Wio, and M. San Miguel, Eur. Phys. J. B **39**, 535 (2004).

[1652] S. Galam, Phys. Rev. E **71**, 046123 (2005).

[1653] A. O. Sousa, K. Malarz, and S. Galam, Int. J. of Mod. Phys. C **16**, 1507 (2005).

[1654] M. Mobilia, S. Redner, Phys. Rev. E **68**, 046106 (2003).

[1655] P. Chen and S. Redner, Phys. Rev. E **71**, 036101 (2005).

[1656] P. Chen and S. Redner, arXiv:cond-mat/0506068.

[1657] F. W. S. Lima, U. L. Fulco, and R. N. Costa Filho, Phys. Rev. E **71**, 036105 (2005).

[1658] K. Sznajd-Weron and J. Sznajd, Int. J. Mod. Phys. C **11**, 1157 (2000).

[1659] L. Behera and F. Schweitzer, Int. J. Mod. Phys. C **14**, 1331 (2003).

[1660] R. Lambiotte and S. Redner, Europhys. Lett. **82**, 18007 (2008).

[1661] D. Stauffer and P. M. C. de Oliveira, arXiv:cond-mat/0208296.

[1662] D. Stauffer and P. M. C. de Oliveira, Eur. Phys. J. B **30**, 587 (2002).

[1663] D. Stauffer, A. O. Sousa, and S. M. de Oliveira, Int. J. Mod. Phys. C **11**, 1239 (2000).

[1664] A. T. Bernardes, U. M. S. Costa, A. D. Araujo, and D. Stauffer, Int. J. Mod. Phys. C **12**, 159 (2001).

[1665] R. N. Costa Filho, M. P. Almeida, J. S. Andrade Jr., and J. E. Moreira, Phys. Rev. E **60**, 1067 (1999).

[1666] M. L. Lyra, U. M. S. Costa, R. N. Costa Filho, and J. S. Andrade Jr., arXiv:condmat/0211560.

[1667] R. N. Costa Filho, M. P. Almeida, J. E. Moreira, and J. S. Andrade Jr., Physica A **322**, 698 (2003).

[1668] L. E. Araripe, R. N. Costa Filho, H. J. Herrmann, and J. S. Andrade Jr., arXiv:physics/0603271.

[1669] R. Axelrod, J. Conflict Resolution **41**, 203 (1997).

[1670] C. Castellano, M. Marsili, and A. Vespignani, Phys. Rev. Lett. **85**, 3536 (2000).
[1671] J. Hoshen and R. Kopelman, Phys. Rev. B **14**, 3438 (1976).
[1672] D. Vilone, A. Vespignani, and C. Castellano, Eur. Phys. J. B **30**, 399 (2002).
[1673] K. Klemm, V. M. Eguíluz, R. Toral, and M. San Miguel, Journal of Economic Dynamics and Control **29**, 321 (2005).
[1674] K. Klemm, V. M. Eguíluz, R. Toral, and M. San Miguel, arXiv:cond-mat/0210173.
[1675] K. Klemm, V. M. Eguíluz, R. Toral, and M. San Miguel, Physica A **327**, 1 (2003).
[1676] K. Klemm, V. M. Eguíluz, R. Toral, and M. San Miguel, Phys. Rev. E **67**, 026120 (2003).
[1677] K. Klemm, V. M. Eguíluz, R. Toral, and M. San Miguel, Phys. Rev. E **67**, 045101 (2003).
[1678] L. De Sanctis and T. Galla, Phys. Rev. E **79**, 046108 (2009).
[1679] F. Vazquez, P. L. Krapivsky, and S. Redner, J. Phys. A: Math. Gen. **36**, L61 (2003).
[1680] F. Vazquez and S. Redner, J. Phys. A: Math. Gen. **37**, 8479 (2004).
[1681] F. Vazquez and S. Redner, Europhys. Lett. **78**, 18002 (2007).
[1682] M. H. DeGroot, Journal of the American Statistical Association **69**, 118 (1974).
[1683] R. L. Berger, Journal of the American Statistical Association **76**, 415 (1981).
[1684] S. Chatterjee and E. Seneta, J. Appl. Prob. **14**, 89 (1977).
[1685] G. Deffuant, D. Neau, F. Amblard, and G. Weisbuch, Adv. Compl. Syst. **3**, 87 (2000).
[1686] U. Krause, in: *Communications in Difference Equations*, ed. S. Elaydi, G. Ladas, J. Popenda, and J. Rakowski, p. 227 (Gordon and Breach, Amsterdam, 2000).
[1687] R. Hegselmann and U. Krause, J. Artif. Soc. Soc. Simulation **5**, http://jasss.soc.surrey.ac.uk/5/3/2.html (2002).
[1688] F. Slanina, Eur. Phys. J. B **79**, 99 (2011).
[1689] S. Fortunato, Physica A **348**, 683 (2005).
[1690] S. Fortunato, Int. J. Mod. Phys. C **15**, 1021 (2004).
[1691] S. Fortunato, Int. J. Mod. Phys. C **16**, 259 (2005).
[1692] A. Pluchino, V. Latora, and A. Rapisarda, Eur. Phys. J. B **50**, 169 (2006).
[1693] J. Lorenz, Int. J. Mod. Phys. C **18**, 1819 (2007).
[1694] S. Fortunato, arXiv:cond-mat/0501105.
[1695] S. Fortunato and D. Stauffer, in: *Extreme Events in Nature and Society*, ed. S. Albeverio, V. Jentsch, and H. Kantz, p. 233 (Springer, Berlin, 2006).
[1696] S. Fortunato, V. Latora, A. Pluchino, and A. Rapisarda, Int. Jour. Mod. Phys. C **16**, 1535 (2005).
[1697] R. Hegselmann and U. Krause, J. Artif. Soc. Soc. Simulation **9**, http://jasss.soc.surrey.ac.uk/9/3/10.html (2006).
[1698] J. Lorenz, arXiv:0708.3293.
[1699] J. Lorenz, Complexity **15**, No. 4, p. 43 (2010).
[1700] J. Lorenz, arXiv:0806.1587.
[1701] J. Lorenz, European Journal of Economic and Social Systems **19**, 213 (2006).
[1702] J. Lorenz, thesis, Universität Bremen (2007).
[1703] E. Ben-Naim, P. L. Krapivsky, and S. Redner, Physica D **183**, 190 (2003).
[1704] G. Weisbuch, G. Deffuant, F. Amblard, and J.-P. Nadal, arXiv:cond-mat/0111494.
[1705] L. Moreau, IEEE Transactions on Automatic Control **50**, 169 (2005).
[1706] J. Lorenz, Physica A **355**, 217 (2005).
[1707] J. Lorenz, in: *Positive Systems, Lecture Notes in Control and Information Sciences*, Vol. 341, p. 209 (Springer, Berlin, 2006).
[1708] J. Lorenz and D. A. Lorenz, IEEE Transactions on Automatic Control **55**, 1651 (2010).
[1709] E. Bonabeau, G. Theraulaz, and J.-L. Deneubourg, Physica A **217**, 373 (1995).
[1710] A. O. Sousa and D. Stauffer, Int. J. Mod. Phys. C **11**, 1063 (2000).
[1711] D. Stauffer and J. S. Sá Martins, arXiv:cond-mat/0308437.
[1712] C. Schulze and D. Stauffer, arXiv:cond-mat/0405697.
[1713] K. Malarz, D. Stauffer, and K. Kułakowski, arXiv:physics/0502118.
[1714] L. Lacasa and B. Luque, arXiv:physics/0511105.
[1715] G. Weisbuch and D. Stauffer, Physica A **384**, 542 (2007).
[1716] E. Ben-Naim and S. Redner, J. Stat. Mech. L11002 (2005).
[1717] E. Ben-Naim, F. Vazquez, and S. Redner, arXiv:physics/0512144.
[1718] S. Galam and S. Moscovici, Eur. J. Soc. Psychol. **21**, 49 (1991).
[1719] S. Galam, Physica A **238**, 66 (1997).
[1720] S. Galam and J.-D. Zucker, Physica A **287**, 644 (2000).
[1721] F. Schweitzer and J. A. Hołyst, Eur. Phys. J. B **15**, 723 (2000).
[1722] J. A. Hołyst, K. Kacperski, and F. Schweitzer, Physica A **285**, 199 (2000).

[1723] C. Hauert and M. Doebeli, Nature **428**, 643 (2004).
[1724] L.-X. Zhong, D.-F. Zheng, B. Zheng, and P. M. Hui, arXiv:physics/0602039.
[1725] W.-X. Wang, J. Ren, G. Chen, and B.-H. Wang, arXiv:physics/0604103.
[1726] R. L. Riolo, M. D. Cohen, and R. Axelrod, Nature **414**, 441 (2001).
[1727] A. Lipowski and M. Droz, Phys. Rev. E **64**, 031107 (2001).
[1728] A. Lipowski and M. Droz, Phys. Rev. E **65**, 056114 (2002).
[1729] M. Droz, A. L. Ferreira, and A. Lipowski, Phys. Rev. E **67**, 056108 (2003).
[1730] M. Droz and A. Lipowski, Braz. J. Phys. **33**, 526 (2003).
[1731] S. Fortunato and C. Castellano, Phys. Rev. Lett. **99**, 138701 (2007).
[1732] G. Travieso and L. da F. Costa, Phys. Rev. E **74**, 036112 (2006).
[1733] N. A. M. Araújo, J. S. Andrade Jr., and H. J. Herrmann, PLoS ONE **5**, e12446 (2010).
[1734] G. Báez, H. Hernández-Saldaña, and R. A. Méndez-Sánchez, arXiv:physics/0609114.
[1735] A. S. Elgazzar, Int. J. Mod. Phys. C **12**, 1537 (2001).
[1736] A. S. Elgazzar, Physica A **324**, 402 (2003).
[1737] A. T. Bernardes, D. Stauffer, and J. Kertész, Eur. Phys. J. B **25**, 123 (2002).
[1738] J. Bonnekoh, Int. J. Mod. Phys. C **14**, 1231 (2003).
[1739] M. C. González, A. O. Sousa, and H. J. Herrmann, Int. J. Mod. Phys. C **15**, 45 (2004).
[1740] M. C. González, A. O. Sousa, and H. J. Herrmann, arXiv:cond-mat/0510123.
[1741] F. A. Rodrigues and L. da F. Costa, arXiv:physics/0505158.
[1742] Y.-S. Tu, A. O. Sousa, L.-J. Kong, and M.-R. Liu, Int. J. Mod. Phys. C **16**, 1149 (2005).
[1743] Y.-S. Tu, A. O. Sousa, L.-J. Kong, and M.-R. Liu, arXiv:cond-mat/0604055.
[1744] A. O. Sousa and J. R. Sánchez, arXiv:cond-mat/0505318.
[1745] C. Schulze, Physica A **324**, 717 (2003).
[1746] C. Schulze, Int. J. Mod. Phys. C **15**, 867 (2004).
[1747] C. Schulze, Int. J. Mod. Phys. C **15**, 569 (2004).
[1748] T. Antal, P. L. Krapivsky, and S. Redner, Phys. Rev. E **72**, 036121 (2005).
[1749] L. F. C. Pereira and F. G. Brady Moreira, Phys. Rev. E **71**, 16123 (2005).
[1750] P.-P. Li, D.-F. Zheng, and P. M. Hui, Phys. Rev. E **73**, 056128 (2006).
[1751] F. W. S. Lima and K. Malarz, arXiv:cond-mat/0602563.
[1752] R. Lambiotte, M. Ausloos, and J. A. Hołyst, Phys. Rev. E **75**, 030101 (2007).
[1753] F. W. S. Lima, Comm. Comp. Phys. **2**, 358 (2007).
[1754] R. Ochrombel, Int. J. Mod. Phys. C **12**, 1091 (2001).
[1755] K. Sznajd-Weron and R. Weron, Int. J. Mod. Phys. C **13**, 115 (2002).
[1756] L. Sabatelli and P. Richmond, Physica A **344**, 62 (2004).
[1757] K. Sznajd-Weron, and R. Weron, Physica A **324**, 437 (2003).
[1758] C. Schulze, Int. J. Mod. Phys. C **14**, 95 (2003).
[1759] M. Wołoszyn, D. Stauffer, and K. Kułakowski, Physica A **378**, 453 (2007).
[1760] D. Stauffer, arXiv:cond-mat/0207598.
[1761] L. Sabatelli and P. Richmond, Int. J. Mod. Phys. C **14**, 1223 (2003).
[1762] A. A. Moreira, J. S. Andrade Jr., and D. Stauffer, Int. J. Mod. Phys. C **12**, 39 (2001).
[1763] D. Stauffer, J. Artif. Soc. Soc. Simulation **5**,
        http://jasss.soc.surrey.ac.uk/5/1/4.html (2001).
[1764] K. Sznajd-Weron, Phys. Rev. E **66**, 046131 (2002).
[1765] D. Stauffer, Adv. Complex Systems **5**, 97 (2002).
[1766] D. Stauffer, Computing in Science and Engineering **5**, 71 (2003).
[1767] D. Stauffer, Int. J. Mod. Phys. C **13**, 315 (2002).
[1768] D. Stauffer, arXiv:cond-mat/0307133.
[1769] D. Stauffer, arXiv:cond-mat/0307352.
[1770] L. Sabatelli and P. Richmond, Physica A **334**, 274 (2003).
[1771] E. Aydıner and M. Gonulol, arXiv:physics/0510237.
[1772] K. Sznajd-Weron, Acta Physica Polonica B **36**, 2537 (2005).
[1773] N. Klietsch, Int. J. Mod. Phys. C **16**, 577 (2005).
[1774] M. S. de la Lama, J. M. López, and H. S. Wio, Europhys. Lett. **72**, 851 (2005).
[1775] S. Krupa and K. Sznajd-Weron, Int. J. Mod. Phys. C **16**, 177 (2005).
[1776] K. Sznajd-Weron and S. Krupa, arXiv:cond-mat/0603680.
[1777] S. Y. Kim, C. H. Park, and K. Kim, Int. J. Mod. Phys. C **18**, 1429 (2007).
[1778] N. G. F. Medeiros, A. T. C. Silva, and F. G. Brady Moreira, Phys. Rev. E **73**, 046120 (2006).
[1779] M. S. de la Lama, I. G. Szendro, J. R. Iglesias, and H. S. Wio, Eur. Phys. J. B **51**, 435 (2006).
[1780] G. Raffaelli and M. Marsili, Phys. Rev. E **72**, 016114 (2005).
[1781] J. Shao, S. Havlin, and H. E. Stanley, Phys. Rev. Lett. **103**, 018701 (2009).

[1782] P. Curty and M. Marsili, J. Stat. Mech. P03013 (2006).
[1783] C. Schulze, Int. J. Mod. Phys. C **16**, 351 (2005).
[1784] V. Schwämmle, M. C. González, A. A. Moreira, J. S. Andrade Jr., and H. J. Herrmann, Phys. Rev. E **75**, 066108 (2007).
[1785] I. J. Benczik, S. Z. Benczik, B. Schmittmann, and R. K. P. Zia, Europhys. Lett. **82**, 48006 (2008).
[1786] L. da F. Costa, Int. J. Mod. Phys. C **16**, 1001 (2005).
[1787] P. Holme and M. E. J. Newman, Phys. Rev. E **74**, 056108 (2006).
[1788] L. Crochik and T. Tomé, Phys. Rev. E **72**, 057103 (005).
[1789] J. C. González-Avella, M. G. Cosenza, and K. Tucci, Phys. Rev. E **72**, 065102 (2005).
[1790] K. I. Mazzitello, J. Candia, and V. Dossetti, Int. J. Mod. Phys. C **18**, 1475 (2007).
[1791] M. San Miguel, V. M. Eguíluz, R. Toral, and K. Klemm, Computing in Science and Engineering **7**, 67 (2005).
[1792] A. Flache and M. W. Macy, arXiv:physics/0604201.
[1793] M. N. Kuperman, Phys. Rev. E **73**, 046139 (2006).
[1794] J. C. González-Avella, V. M. Eguíluz, M. G. Cosenza, K. Klemm, J. L. Herrera, and M. San Miguel, Phys. Rev. E **73**, 046119 (2006).
[1795] R. Toral and C. J. Tessone, Comm. Comp. Phys. **2**, 177 (2007).
[1796] A. Parravano, H. Rivera-Ramirez, and M. G. Cosenza, Physica A **379**, 241 (2007).
[1797] N. Lanchier, Ann. Appl. Probab. **22**, 860 (2012).
[1798] D. Urbig and J. Lorenz, in: *Proceedings of the Second Conference of the European Social Simulation Association (ISBN 84-688-7964-9)*, (2004).
[1799] M. F. Laguna, G. Abramson, and D. H. Zanette, Complexity **9**, No. 4 p. 31 (2004).
[1800] G. Deffuant, F. Amblard, G. Weisbuch, and T. Faure, J. Artif. Soc. Soc. Simulation **5**, 4 http://jasss.soc.surrey.ac.uk/5/4/1.html (2002).
[1801] M. Porfiri, E. M. Bollt, and D. J. Stilwell, Eur. Phys. J. B **57**, 481 (2007).
[1802] K. Malarz, Int. J. Mod. Phys. C **17**, 1521 (2006).
[1803] M. F. Laguna, G. Abramson, and D. H. Zanette, Physica A **329**, 459 (2003).
[1804] M. Pineda, R. Toral and E. Hernández-García, J. Stat. Mech. P08001 (2009).
[1805] G. Deffuant, F. Amblard, and G. Weisbuch, arXiv:cond-mat/0410199.
[1806] G. Weisbuch, G. Deffuant, and F. Amblard, Physica A **353**, 555 (2005).
[1807] P. Assmann, Int. J. Mod. Phys. C **15**, 1439 (2004).
[1808] S. Fortunato, Int. J. Mod. Phys. C **15**, 1301 (2004).
[1809] S. Fortunato, Int. J. Mod. Phys. C **16**, 17 (2005).
[1810] D. Stauffer, A. O. Sousa, and C. Schulze, arXiv:cond-mat/0310243.
[1811] G. Weisbuch, Eur. Phys. J. B **38**, 339 (2004).
[1812] F. Amblard, and G. Deffuant, Physica A **343**, 725 (2004).
[1813] D. Stauffer and H. Meyer-Ortmanns, J. Mod. Phys. C **15**, 241 (2004).
[1814] D. Jacobmeier, Int. J. Mod. Phys. C **16**, 633 (2005).
[1815] J. C. González-Avella, M. G. Cosenza, K. Klemm, V. M. Eguíluz, and M. San Miguel, J. Artif. Soc. Soc. Simulation **10**, http://jasss.soc.surrey.ac.uk/10/3/9.html (2007).
[1816] A. Soulier and T. Halpin-Healy, Phys. Rev. Lett. **90**, 258103 (2003).
[1817] F. Schweitzer, J. Zimmermann, and H. Mühlenbein, Physica A **303**, 189 (2002).
[1818] M. Pineda, R. Toral, and E. Hernández-García, Eur. Phys. J. D **62**, 109 (2011).
[1819] G. Weisbuch, in: *Econophysics and Sociophysics: Trends and Perspectives*, ed. B. K. Chakrabarti, A. Chakraborti, and A. Chatterjee, p. 339 (Wiley-VCH, Weinheim, 2006).
[1820] D. Jacobmeier, Int. J. Mod. Phys. C **17**, 1801 (2006).
[1821] F. Bagnoli, T. Carletti, D. Fanelli, A. Guarino, and A. Guazzini, Phys. Rev. E **76**, 066105 (2007).
[1822] D. Stauffer, Int. J. Mod. Phys. C **13**, 315 (2002).
[1823] G. Hardin, Science **162**, 1243 (196).
[1824] E. Ben-Naim, Europhys. Lett. **69**, 671 (2005).

Author Index

Subject Index